T0281039

Lecture Notes in Computer Science 11789

Commenced Publication in 1973
Founding and Former Series Editors:
Gerhard Goos, Juris Hartmanis, and Jan van Leeuwen

Advanced Research in Computing and Software Science

Subline of Lecture Notes in Computer Science

More information about this series at http://www.springer.com/series/7407

Ignasi Sau · Dimitrios M. Thilikos (Eds.)

Graph-Theoretic Concepts in Computer Science

45th International Workshop, WG 2019
Vall de Núria, Spain, June 19–21, 2019
Revised Papers

Editors
Ignasi Sau
CNRS, LIRMM, Université de Montpellier
Montpellier, France

Dimitrios M. Thilikos
CNRS, LIRMM, Université de Montpellier
Montpellier, France

ISSN 0302-9743 ISSN 1611-3349 (electronic)
Lecture Notes in Computer Science
ISBN 978-3-030-30785-1 ISBN 978-3-030-30786-8 (eBook)
https://doi.org/10.1007/978-3-030-30786-8

LNCS Sublibrary: SL1 – Theoretical Computer Science and General Issues

This Springer imprint is published by the registered company Springer Nature Switzerland AG
The registered company address is: Gewerbestrasse 11, 6330 Cham, Switzerland

Preface

The 45th International Workshop on Graph-Theoretic Concepts in Computer Science (WG 2019) took place at the Vall de Núria, Catalonia, Spain, a valley very close to the French-Spanish borders, during June 19–21, 2019. About 60 mathematicians and computer scientists from all over the world (Austria, Brazil, Canada, Czech Republic, Denmark, France, Germany, India, Israel, Japan, the Netherlands, Norway, Poland, Romania, Russia, Slovenia, Spain, Switzerland, the United Kingdom, and the United States) attended the conference.

WG has a long standing tradition. Since 1975, WG has taken place 24 times in Germany, 5 times in the Netherlands, 3 times in France, 2 times in Austria, 2 times in Czech Republic, as well as 1 time in Greece, Israel, Italy, Norway, Slovakia, Spain, Switzerland, Turkey, and in the United Kingdom.

WG aims at merging theory and practice by demonstrating how concepts from Graph Theory can be applied to various areas in Computer Science, or by extracting new graph theoretic problems from applications. The goal is to present emerging research results and to identify and explore directions of future research. The conference is well-balanced with respect to established researchers and young scientists.

There were 87 submissions, 3 of which were withdrawn before entering the review process. Each submission was carefully reviewed by at least 3, and on average 3.4, members of the Program Committee (PC). The PC accepted 29 papers – an acceptance ratio of around 35 %. We should stress that, due to the high competition and the limited schedule, there were papers that were not accepted although they deserved to be.

The program also included three excellent invited talks: the first one was given by Marc Noy (Universitat Politècnica de Catalunya and Barcelona Graduate School of Mathematics, Spain) on “Logic of Sparse Random Graphs,” the second one was given by Saket Saurabh (Institute of Mathematical Sciences, Chennai, India, and University of Bergen, Norway) on “Parameterized Algorithms for Geometric Graphs via Decomposition Theorems,” and the third was given by Frédéric Havet (CNRS, I3S, Inria, France) on “Unavoidability and Universality of Digraphs.” The abstracts of the three talks can be found at the beginning of these proceedings.

We wish to thank all those who contributed to the success of WG 2019. While it is impossible to enumerate them all, this certainly includes the authors for submitting high-quality papers, the reviewers and the members of the PC for their detailed work, the speakers for their well-prepared talks, all the participants for their enthusiasm, and the personnel of the Hotel Vall de Núria for the pleasant conference environment and facilities. In particular, we would like to thank Núria Riu for all her help concerning the logistics in Vall de Núria.

We are grateful to all members of the Organizing Committee of WG 2019, namely Raul Wayne Teixera Lopes (Universidade Federal do Ceará, Brazil), Maximilian Wötzel (Universitat Politècnica de Catalunya, Spain), and Vasiliki Velona (Universitat Politècnica de Catalunya and Universitat Pompeu Fabra, Spain).

We are also thankful to Ana Valdés (Creacongresos) for the financial managing of the event. Special thanks to Charalampos Tampakopoulos for his programming, development, and hosting services via IsoftCloud. Finally, our deepest gratitude to Mar Pairó for designing the logo of the conference.

July 2019

Ignasi Sau
Dimitrios M. Thilikos

Organization

Program Committee

Hans L. Bodlaender	Utrecht University and TU/e, The Netherlands
Mohar Bojan	Simon Fraser University, Canada, and University of Ljubljana, Slovenia
L. Chandran Sunil	Indian Institute of Science, India
Maria Chudnovsky	Princeton University, USA
Marek Cygan	University of Warsaw, Poland
Tınaz Ekim Aşici	Bogazici University, Turkey
Jirí Fiala	Charles University, Czech Republic
Loukas Georgiadis	University of Ioannina, Greece
Petr Golovach	Bergen University, Norway
Gregory Gutin	Royal Holloway, University of London, UK
Juraj Hromkovic	ETH Zurich, Switzerland
Michael Kaufmann	University of Tuebingen, Germany
Eun Jung Kim	CNRS, LAMSADE, University of Paris Dauphine, France
Cláudia Linhares Sales	Universidade Federal do Ceara, Brazil
Rolf Niedermeier	Technische Universität Berlin, Germany
Naomi Nishimura	University of Waterloo, Canada
David R. Cheriton	University of Waterloo, Canada
Nicolas Nisse	Inria, I3S, France
Daniël Paulusma	Durham University, UK
Ignaz Rutter	Universität Passau, Germany
Ignasi Sau	CNRS, LIRMM, Universit de Montpellier, France
Maria Serna	Universitat Politecnica de Catalunya and Barcelona Graduate School of Mathematics, Spain
Mordechai Shalom	TelHai College, Israel
Hisao Tamaki	Meiji University, Japan
Dimitrios M. Thilikos	CNRS, LIRMM, Universit de Montpellier, France, and NKUA, Greece

Organizing Committee

Raul Wayne Teixera Lopes
Ignasi Sau
Dimitrios M. Thilikos
Vasiliki Velona
Maximilian Wötzel

Organization Entities

AlGCo project-team, LIRMM, Université de Montpellier, CNRS, France.
Departament de Ciències de la Computació, Universitat Politècnica de Catalunya, Barcelona, Spain.
Departament de Matemàtiques, Universitat Politècnica de Catalunya, Barcelona, Spain.
Creacongresos.
IsoftCloud
Hotel Vall de Núria

Additional Reviewers

Adiga, Abhijin
Agrawal, Akanksha
Aichholzer, Oswin
Almeida, Sheila
Angelini, Patrizio
Araujo, Julio
Babu, Jasine
Bandyapadhyay, Sayan
Barrus, Michael
Basavaraju, Manu
Baste, Julien
Bekos, Michael
Belmonte, Rémy
Benevides, Fabricio S.
Bensmail, Julien
Bentert, Matthias
Bentz, Cédric
Berge, Pierre
Bergougnoux, Benjamin
Bezakova, Ivona
Blažej, Václav
Bläsius, Thomas
Boeckenhauer, Hans-Joachim
Bonnet, Édouard
Bonomo, Flavia
Brause, Christoph
Brettell, Nick
Brewster, Richard
Broersma, Hajo
Bulatov, Andrei
Burjons Pujol, Elisabet
Cabello, Sergio
Campelo, Manoel
Campos, Victor
Carvalho, Marcelo
Chakraborty, Diptarka
Chaplick, Steven
Coudert, David
Crespelle, Christophe
Cseh, Ágnes
Dąbrowski, Konrad Kazimierz
de Lima, Paloma
Dell, Holger
Dibek, Cemil
Doczkal, Christian
Dourado, Mitre
Dross, François
Ducoffe, Guillaume
Dvořák, Zdeněk
Eiben, Eduard
Emek, Yuval
Escoffier, Bruno
Felsner, Stefan
Fluschnik, Till
Francis, Mathew
Frei, Fabian
Froese, Vincent
Förster, Henry
Gajarský, Jakub
Galby, Esther
Giannis, Konstantinos
Gonçalves, Daniel
Gronemann, Martin
Gözüpek, Didem
Harrenstein, Paul
Heeger, Klaus

Hoffmann, Michael
Huang, Shenwci
Igarashi, Ayumi
Issac, Davis
Jaffke, Lars
Jansen, Bart M. P.
Kaczmarczyk, Andrzej
Karanasiou, Aika-terini
Kellerhals, Leon
Khan, Arindam
Kindermann, Philipp
Kita, Nanao
Knauer, Kolja
Knop, Dušan
Kolay, Sudeshna
Kolliopoulos, Stavros
Komm, Dennis
Konstantinidis, Athanasios
Kratsch, Stefan
Kwon, O-Joung
Kynčl, Jan
Lahiri, Abhiruk
Lampis, Michael
Le, Van Bang
Lima, Carlos Vinicius
Lochet, William
Luo, Junjie
Majumdar, Diptapriyo
Masařík, Tomáš
McConnell, Ross
Mchedlidze, Tamara
Mertzios, George
Misra, Neeldhara
Mnich, Matthias
Molter, Hendrik
Montecchiani, Fabrizio
Mouawad, Amer
Mukherjee, Joydeep
Nichterlein, André
Nogueira, Loana
Okamoto, Yoshio
Panolan, Fahad
Papadopoulos, Charis
Paschos, Vangelis
Paul, Christophe
Perarnau, Guillem
Pergel, Martin
Petreschi, Rosella
Pilipczuk, Michał
Pradhan, D.
Radermacher, Marcel
Rajendraprasad, Deepak
Reidl, Felix
Renken, Malte
Richerby, David
Rosenke, Christian
Rossmanith, Peter
Rote, Günter
Roth, Marc
Sajith, P.
Sampaio Rocha, Leonardo
Sampaio, Rudini
Schneck, Thomas
Shachnai, Hadas
Siebertz, Sebastian
Sikora, Florian
Silva, Ana
Simonov, Kirill
Singh, Nitin
Soares, Ronan
Souza, Uéverton
Spirkl, Sophie
Spoerhase, Joachim
Stamoulis, Giannos
Stavropoulos, Konstantinos
Steiner, Raphael
Stumpf, Peter
Suchan, Karol
Suchy, Ondrej
Sun, Kevin
Takaoka, Asahi
Tappini, Alessandra
Thomas, Robin
Ueckerdt, Torsten
Uehara, Ryuhei
Unger, Walter
Verbeek, Kevin
Viennot, Laurent

The Long Tradition of WG

WG 1975 U. Pape – Berlin, Germany
WG 1976 H. Noltemeier – Göttingen, Germany
WG 1977 J. Mühlbacher – Linz, Austria
WG 1978 M. Nagl, H. J. Schneider – Castler Feuerstein, Germany
WG 1979 U. Pape – Berlin, Germany
WG 1980 H. Noltemeier – Bad Honnef, Germany
WG 1981 J. Mühlbacher – Linz, Austria
WG 1982 H. J. Schneider, H. Göttler – Neuenkirchen, Germany
WG 1983 M. Nagl, J. Perl – Haus Ohrbeck near Onasbrück, Germany
WG 1984 U. Pape – Berlin, Germany
WG 1985 H. Noltemeier – Castle Schwanberg near Würzburg, Germany
WG 1986 G. Tinhofer, G. Schmidt – Bernried near Munich, Germany
WG 1987 H. Göttler, H. J. Schneider – Kloster Banz near Bamberg, Germany
WG 1988 J. van Leeuwen – Amsterdam, The Netherlands
WG 1989 M. Nagl – Castle Rolduc, The Netherlands
WG 1990 R. H. Möhring – Berlin, Germany
WG 1991 G. Schmidt, R. Berghammer – Fischbachau near Munich, Germany
WG 1992 E. W Mayr – Wiesbaden-Naurod, Germany
WG 1993 J. van Leeuwen – Utrecht, The Netherlands
WG 1994 G. Tinhofer, E. W. Mayr, G. Schmidt – Herrsching near Munich, Germany
WG 1995 M. Nagl – Aachen, Germany
WG 1996 G. Ausiello, A. Marchetti-Spaccamela – Como, Italy
WG 1997 R. H. Möhring – Berlin, Germany
WG 1998 J. Hromkovič, O. Sýkora – Smolenice Castle, Slovakia Republic
WG 1999 P. Widmayer – Ascona, Switzerland
WG 2000 D. Wagner – Konstanz, Germany
WG 2001 A. Brandstädt, Boltenhagen near Rostock, Germany
WG 2002 L. Kucera – Ceský Krumlov, Czech Republic
WG 2003 H. L. Bodlaender – Elspeet, The Netherlands
WG 2004 J. Hromkovič, M. Nagl – Bad Honnef, Germany
WG 2005 D. Kratsch – Metz, France
WG 2006 F. V. Fomin – Bergen, Norway
WG 2007 A. Brandstädt, D. Kratsch, H. Müller – Dornburg near Jena, Germany
WG 2008 H. Broersma, T. Erlebach – Durham, UK
WG 2009 C. Paul, M. Habib – Montpellier, France
WG 2010 D. M. Thilikos – Zarós, Crete, Greece
WG 2011 J. Kratochvíl – Teplá Monastery, West Bohemia, Czech Republic
WG 2012 M. C. Golumbic, G. Morgenstern, M. Stern, A. Levy – Jerusalem, Israel
WG 2013 A. Brandstädt, K. Jansen, R. Reischuk – Lübeck, Germany
WG 2014 D. Kratsch,, I. Todinca – Zarós, Le Domaine de Chalès, Orléans, France

WG 2015 E. W. Mayr – Munich, Germany
WG 2016 P. Heggernes – Istanbul, Turkey
WG 2017 H. L. Bodlaender, G. J. Woeginger – Eindhoven, The Netherlands
WG 2018 A. Brandstädt, E. Köhler, K. Meer – Zarós, Cottbus, Germany
WG 2019 I. Sau, D. M. Thilikos – Vall de Núria, Catalunya, Spain

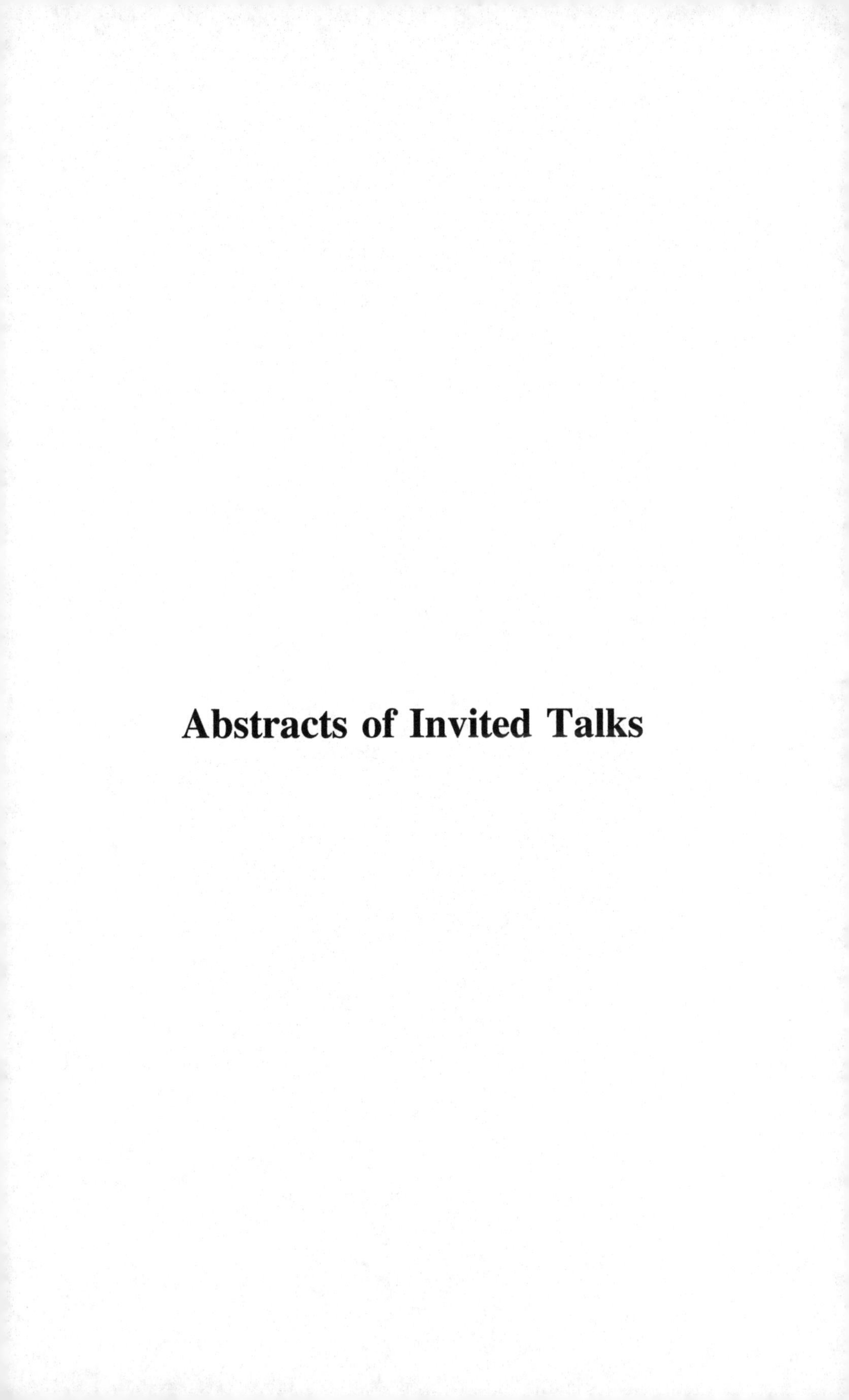

Abstracts of Invited Talks

Logic and Random Graphs

Marc Noy

Department of Mathematics, Universitat Politècnica de Catalunya and Barcelona Graduate School of Mathematics, Edifici Omega, 08034 Barcelona, Spain
marc.noy@upc.edu

Abstract. We survey recent results on limiting probabilities of graph properties expressible in first order logic and monadic second order logic for two models of sparse random graphs: the classical model $G(n,p)$ with $p = c/n$, and random planar graphs and related classes of graphs.

Unavoidability and Universality of Digraphs

Frédéric Havet

CNRS, I3S, and Inria, Université Côte d'Azur, France

Abstract. A digraph F is *n-unadoidable* (resp. *n-universal*) if it is contained in every tournament of order n (resp. n-chromatic digraph). Well-known theorems imply that there is an n_F such that F is n_F-unavoidable (resp. n_F*-universal*) if and only if F is acyclic, (resp. an oriented forest). However, determining the smallest n_F for which it occurs is a challenging question. In this talk, we survey the results on unavoidability and universality with an emphasis on oriented forests. In particular, we shall detail the following new results obtained jointly with F. Dross: every arborescence of order n with k leaves is $(n+k-1)$-unavoidable; every tree of order n with k leaves is $\left(\frac{3}{2}n+\frac{3}{2}k-2\right)$-unavoidable, $\left(\frac{21}{8}n-\frac{47}{16}\right)$-unavoidable. and $(n+144k^2-280k+124)$-unavoidable.

Parameterized Algorithms for Geometric Graphs via Decomposition Theorems

Saket Saurabh

The Institute of Mathematical Sciences, HBNI, Chennai, India
saket@imsc.res.in

Abstract. Parameterized complexity is one of the most established algorithmic paradigms to deal with computationally hard problems. In the first two decades, the field largely focused on problems arising from studies of graphs and networks. However, lately the focus has changed substantially and it has started to permeate into other fields such as computational geometry, and computational social choice theory. In this talk, we will survey some exciting developments in the emerging field of parameterized computational geometry through our contributions. We will focus on designing efficient parameterized algorithms on unit-disk graphs via new graph decomposition theorems.

Parameterized Algorithms for Geometric Graphs via Decomposition Theorems

Contents

Subexponential Algorithms for Variants of Homomorphism Problem in String Graphs

Karolina Okrasa and Paweł Rzążewski(✉)

Faculty of Mathematics and Information Science, Warsaw University of Technology, Warsaw, Poland
{k.okrasa,p.rzazewski}@mini.pw.edu.pl

Abstract. We consider the complexity of finding weighted homomorphisms from intersection graphs of curves (string graphs) with n vertices to a fixed graph H. We provide a complete dichotomy for the problem: if H has no two vertices sharing two common neighbors, then the problem can be solved in time $2^{O(n^{2/3} \log n)}$, otherwise there is no algorithm working in time $2^{o(n)}$, even in intersection graphs of segments, unless the ETH fails. This generalizes several known results concerning the complexity of computational problems in geometric intersection graphs.

Then we consider two variants of graph homomorphism problem, called locally injective homomorphism and locally bijective homomorphism, where we require the homomorphism to be injective or bijective on the neighborhood of each vertex. We show that for each target graph H, both problems can always be solved in time $2^{O(\sqrt{n} \log n)}$ in string graphs. For the locally surjective homomorphism, defined analogously, the situation seems more complicated. We show the dichotomy theorem for simple connected graphs H with maximum degree 2. If H is isomorphic to P_3 or C_4, then the existence of a locally surjective homomorphism from a string graph with n vertices to H can be decided in time $2^{O(n^{2/3} \log^{3/2} n)}$, otherwise, assuming ETH, the problem cannot be solved in time $2^{o(n)}$. As a byproduct, we obtain results concerning the complexity of variants of homomorphism problem in P_t-free graphs – in particular, the weighted homomorphism dichotomy, analogous to the one for string graphs.

Keywords: Graph homomorphism · Subexponential algorithm · String graphs · Segment graphs

1 Introduction

The theory of NP-completeness gives us tools to identify problems which are unlikely to admit polynomial-time algorithms, but it does not give any insight into possible complexities of problems that are considered hard. For example, the

I. Sau and D. M. Thilikos (Eds.): WG 2019, LNCS 11789, pp. 1–13, 2019.
https://doi.org/10.1007/978-3-030-30786-8_1

best algorithms we know for most canonical problems like 3-Coloring, Independent Set, Dominating Set, Vertex Cover, Hamiltonian Cycle, are single-exponential, i.e., with complexity $2^{O(n)}$ (n will always denote the number of vertices in the input graph). On the other hand, in planar graphs these problems are still NP-complete, but they admit *subexponential* algorithms (i.e., working in time $2^{o(n)}$). Indeed, most canonical problems in planar graphs admit a certain "square-root phenomenon", i.e., can be solved in time $2^{\tilde{O}(\sqrt{n})}$[1]. The core building block in construction of subexponential algorithms for planar graphs is the celebrated planar decomposition theorem by Lipton and Tarjan [35], which asserts that every planar graph has a balanced separator of size $O(\sqrt{n})$.

To argue whether those algorithms are asymptotically optimal and, in general, to prove meaningful lower bounds on the complexity of hard problems, we need a stronger assumption than "P $\neq$ NP". Such a stronger assumption, commonly used in complexity theory, is the Exponential Time Hypothesis (ETH) by Impagliazzo and Paturi [27], which implies that 3-Sat with n variables cannot be solved in time $2^{o(n)}$. For example, assuming the ETH, 3-Coloring, Independent Set, Dominating Set, Vertex Cover, Hamiltonian Cycle cannot be solved in time $2^{o(n)}$ in general graphs or in time $2^{o(\sqrt{n})}$ in planar graphs. Thus the algorithms we know are asymptotically tight, unless the ETH fails.

A natural direction of research is to consider restricted graph classes and try to classify problems solvable in subexponential time. Geometric intersection graphs provide a rich family of graph classes, which are potentially interesting from the point of view of fine-grained complexity, as they lie "in between" planar graphs and all graphs. For a family $\mathcal{S}$ of sets, we define its *intersection graph*, whose vertices are in one-to-one correspondence to members of $\mathcal{S}$, and two vertices are adjacent if and only if their corresponding sets intersect. We will be interested in intersection graphs of sets of geometric objects in the plane.

For example, in *unit disk graphs*, i.e., intersection graphs of unit-radius disks in the plane, Independent Set, Hamiltonian Cycle, Vertex Cover can be solved in time $2^{\tilde{O}(\sqrt{n})}$ [1,15,38], and k-Coloring can be solved in time $2^{\tilde{O}(\sqrt{nk})}$ for every k [3,28]. All these bounds are essentially tight under the ETH, up to polylogarithmic factors in the exponent. Many algorithms for (unit) disk graphs use the fact that disk intersection graphs also have small separators. Indeed, Miller *et al.* showed that the intersection graph of a family of n disks, such that at most k of them share a single point, has a balanced separator of size $O(\sqrt{nk})$ [42]. This was later generalized to intersection graphs of families of arbitrary convex shapes that are fat, i.e., with bounded ratio of the radius of the smallest enclosing circle to the radius of the largest enclosed circle [47].

It is perhaps interesting to note that, by the celebrated kissing lemma by Koebe [29], every planar graph is an intersection graph of interior-disjoint disks. Note that in such a representation each point is contained in at most two disks, so the separator theorem for disk graphs implies the planar separator theorem.

In this paper we are interested in intersection graphs of non-fat geometric objects. In particular, we will investigate *string graphs*, i.e., intersection graphs

[1] In the $\tilde{O}(\cdot)$ notation we suppress polylogarithmic factors.

of continuous curves in the plane (see Kratochvíl [30,31]) and *segment graphs*, i.e., intersection graphs of straight-line segments (see Kratochvíl and Matoušek [33]). We can restrict the representation even further and consider k-DIR graphs, which are intersection graphs of segments using at most k distinct slopes [33]. It is known that planar graphs form another subclass of segment graphs [6,18].

General string separator theorems have been proven by Fox and Pach [16] and Matoušek [40]. The following, asymptotically tight version, was shown by Lee [34].

Theorem 1 (Lee [34]). *Every string graph with m edges has a balanced separator of size $O(\sqrt{m})$.* □

Observe that since planar graphs are string graphs and have linear number of edges, Theorem 1 implies the planar separator theorem.

Using the string separator theorem, Fox and Pach [17] showed that Independent Set (and thus Vertex Cover) can be solved in subexponential time in string graphs. Combining Theorem 1 with their approach gives the complexity $2^{\tilde{O}(n^{2/3})}$. The algorithm is a simple win-win strategy: either we have a vertex of large degree and we branch on choosing it to the solution or not, or all degrees are small and thus there exists a small balanced separator, which allows us for one step of divide & conquer. Recently, Marx and Pilipczuk [38] used a different approach to obtain a $2^{O(\sqrt{n})}p^{O(1)}$ algorithm for Independent Set in string graphs, where p is the number of *geometric vertices* in the representation.

While the algorithm of Marx and Pilipczuk seems difficult to generalize to other problems, Bonnet and Rzążewski [5] showed that the win-win strategy of Fox and Pach can be successfully applied to obtain subexponential algorithms for 3-Coloring, Feedback Vertex Set, and Max Induced Matching. Quite surprisingly, they showed that for every $k \geq 4$, k-Coloring cannot be solved in time $2^{o(n)}$, even in 2-DIR graphs, unless the ETH fails. They also showed that assuming the ETH, Dominating Set, Independent Dominating Set, and Connected Dominating Set do not admit subexponential algorithms in segment graphs, and Clique does not admit such an algorithm in string graphs.

This shows that the complexity landscape in string and segment graphs appears to be much more interesting than in planar graphs or intersection graphs of fat objects. In order to understand which problems can be solved in subexponential time, it would be especially desirable to obtain full dichotomy theorems for some natural families of problems, instead of proving ad-hoc results for single problems. A natural language to describe these families in a uniform way is provided by graph homomorphisms. For graphs G and H (with possible loops), a *homomorphism* from G to H, denoted by $h\colon G \to H$, is an edge-preserving mapping from the vertex set of G to the vertex set of H (see the book by Hell and Nešetřil [24]). A homomorphism $h\colon G \to H$ will be often called an *H-coloring* of G and we will think of vertices of H as *colors*. Note that the notion of homomorphisms is flexible and allows us to impose additional restrictions, such as vertex/edge lists [10,25] or vertex/edge weights [21]. This way many well-known problems can be formulated as problems of finding a homomorphism to a certain graph H, possibly with additional constraints. For example, k-Coloring is equivalent to a homomorphism to K_k, and INDEPENDENT SET is equivalent to a weight-maximizing homomorphism to $H =$ ⓪—① (numbers denote weights of vertices of G mapped to particular vertices of H).

Weighted Homomorphisms. Let $H = (V(H), E(H))$ be a fixed graph (with possible loops), and consider the following computational problem called WHOM(H). The instance consists of a graph $G = (V(G), E(G))$, a *weight function* $w\colon (V(G) \times V(H)) \cup (E(G) \times E(H)) \to \mathbb{R}$, and an integer k. For simplicity we also allow $-\infty$ as a weight, but this can be avoided by shifting all weights and using a sufficiently small integer to represent the weight corresponding to a forbidden choice. For a homomorphism $h : G \to H$ and for any $X \subseteq V(G) \cup E(G)$ we define the weight of X by $w_h(X) = \sum_{x \in X} w(x, h(x))$. The *weight of* h is defined as $w_h(V(G) \cup E(G))$. We ask if there exists a homomorphism from G to H whose total weight is at least k. It is straightforward to see that this problem generalizes some well-studied variants of graph homomorphism problem, including LIST HOMOMORPHISM [10] and MIN COST HOMOMORPHISM [21].

We show the following dichotomy theorem for WHOM(H) in string graphs.

Theorem 2. (♠)[2] *Let H be a fixed graph.*

(a) If H has no two vertices with two common neighbors, then the WHOM*(H) problem can be solved in time $2^{O(n^{2/3} \log n)}$ for string graphs.*
(b) Otherwise, the WHOM*(H) problem in NP-complete and cannot be solved in time $2^{o(n)}$ for segment graphs, unless the ETH fails.*

Very recently Groenland *et al.* [20] observed that if H has no two vertices with two common neighbors, then WHOM(H) can be solved in time $2^{O(\sqrt{n})}$ in P_t-free graphs, for every fixed t. Note that there are string graphs with arbitrarily long induced paths, and there are P_t-free graphs that are not string graphs.

The algorithm proving Theorem 2(a) is a slight adaptation of the win-win approach by Fox and Pach [17], later extended by Bonnet and Rzążewski [5], and Groenland *et al.* [20]. The proof of part (b) is divided into a few cases, depending on the structure of H. In our reductions we try not to use the whole expressibility of the WHOM(H) problem, but aim to obtain hardness even for some natural special cases. All hardness proofs follow the same pattern – we start with a grid-like arrangement of segments, inducing a clique or a biclique, and then add constant-size gadgets to encode a specific problem. Note that this requires the objects to be non-fat and gives some intuition why problems in segment graphs tend to be harder than in intersection graphs of fat objects, and how the hardest instances look like. Finally, all graphs we construct are actually P_t-free for some fixed t, which, along with the result of Groenland *et al.* [20], gives the following dichotomy theorem for P_t-free graphs.

Theorem 3 (♠). *Let H be a fixed graph.*

[2] Full proofs of theorems marked with ♠ are omitted, but can be found in the full version of this paper [45].

(a) If H has no two vertices with two common neighbors, then for all fixed t the WHOM*(H) problem can be solved in time $2^{O(\sqrt{n})}$ for P_t-free graphs.*
(b) Otherwise, there is t for which WHOM*(H) is NP-complete and cannot be solved in time $2^{o(n)}$ for P_t-free graphs with n vertices, unless the ETH fails.*

Locally Constrained Homomorphisms. Interesting variants of graph homomorphism problems can be obtained by imposing some additional constrains on the neighborhood of each vertex. A homomorphism h from G to H is called *locally injective* (*locally bijective, locally surjective*) if for every $v \in V(G)$ it induces an injective (bijective, surjective, resp.) mapping between the neighborhood of v and the neighborhood of $h(v)$. Locally bijective homomorphisms have been studied from combinatorial [12,14] and computational point of view [7,11,13,37]. Let LIHOM(H), LBHOM(H), and LSHOM(H) denote, respectively, the computational problems of determining the existence of a locally injective, bijective, and surjective homomorphism from a given graph to H.

Some well-known graph problems can be expressed as locally constrained homomorphism. For example, locally injective homomorphism to the complement of the k-vertex path appears to be equivalent to k-$L(2,1)$-labeling, i.e., a mapping from the vertex set of the input graph to the set $\{1, 2, \ldots, k\}$, in which adjacent vertices get labels differing by at least 2, and vertices with a common neighbor get different labels [19,23]. If H is the complete graph K_k, then LIHOM(H) is exactly the k-coloring of the square of the graph [36,48]. Finally, if H is a complete graph with k vertices, where every vertex has a loop, then LIHOM(H) is equivalent to the injective k-coloring [22,26], in which the only restriction is that no two vertices with a common neighbor get the same color.

We show that, unlike WHOM(H), both LIHOM(H) and LBHOM(H) can always be solved in subexponential time in string graphs.

Theorem 4 (♠). *For every fixed graph H, the* LIHOM*(H) problem and the* LBHOM*(H) problem can be solved in time $2^{O(\sqrt{n}\log n)}$ for string graphs.*

The LSHOM(H) problem appears to be much harder. In particular, we show the following dichotomy for simple graphs H with $\Delta(H) \leq 2$ (observe that if $|H| \leq 2$, the problem can trivially be solved in polynomial time).

Theorem 5. *Let H be a connected simple graph with $\Delta(H) \leq 2$ and $|H| \geq 3$.*

(a) If $H \in \{P_3, C_4\}$, then the LSHOM*(H) problem can be solved in time $2^{O(n^{2/3}\log^{3/2} n)}$ for string graphs, even if geometric representation is not given.*
(b) (♠) Otherwise, the LSHOM*(H) problem cannot be solved in time $2^{o(n)}$ in* 2-DIR *graphs, unless the ETH fails.*

We also show that LSHOM(H) cannot be solved in subexponential time for $H =$  . Note that none of the graphs H, for which we obtain negative results for LSHOM(H) problem, has two vertices with two common neighbors. Thus they are all "easy" cases of WHOM(H).

Representation and Robust Algorithms. When dealing with geometric intersection graphs, we need to be careful, whether the input consist of the graph along with the representation, or just the graph (with a promise that a geometric representation exists). This distinction might be crucial, since finding a representation is often a computationally hard task.

Recognizing string and segment graphs was shown to be NP-hard by Kratochvíl [33], and Kratochvíl and Matoušek [32], respectively. However, for a very long time it was unclear whether these problems are in NP. This is because there are string graphs, whose every representation requires exponential number of crossing points [32] and there are segment graphs, whose every representation requires points with double exponential coordinates [33,41]. Finally, Schaefer, Sedgwick, and Štefankovič showed that recognizing string graph is in NP [44], while recognizing segment graph appears to be complete for the complexity class $\exists\mathbb{R}$ [39,46]. This is a strong indication that the problem might not be in NP.

For these reasons, it is desirable for an algorithm not to require explicit representation. Such algorithms are called *robust* – they either compute a solution, or report that the input graph does not belong to the required class. All algorithms presented in the paper are robust, but can be made slightly faster, if the representation is given. On the other hand, all hardness results hold even if the graph is given along with the geometric representation.

2 Weighted Homomorphism Problem

For every $v \in V(G)$, let $N_G(v)$ denote the set of neighbors of v in G and $N_G(U) = \bigcup_{v \in U} N_G(v)$ for any $U \subseteq V(G)$. Let $d_G(v) = |N_G(v)|$. If the graph is clear from the context, we omit the subscript G and simply write $N(v)$ and $d(v)$. We will say that G (with possible loops) has *property* ($\star$), if $|N(u) \cap N(v)| \leq 1$ for every $u, v \in V(G)$. Note that if G is loopless, then it has property ($\star$) iff it does not have C_4 as a subgraph. We show that property ($\star$) is essential for the existence of subexponential algorithms: for all remaining graphs H, an algorithm solving WHOM(H) for string graphs in subexponential time would contradict the ETH. Observe that we can express the property ($\star$) in terms of forbidden subgraphs.

Observation 6. *A graph has property* ($\star$) *if and only if it does not contain any of the seven graphs shown on the Fig. 1 as an induced subgraph.* □

To prove Theorem 2(b), it is enough to show hardness of WHOM(H) for graphs H shown in Fig. 1. Indeed, let H be an induced subgraph of H' and consider an instance (G, w, k) of WHOM(H). Define $w' \colon V(G) \times V(H') \cup E(G) \times E(H') \to \mathbb{R}$ as follows: for $x \in V(G) \cup E(G)$, if $a \in V(H) \cup E(H)$, then $w'(x, a) = w(x, a)$, otherwise $w'(x, a) = -\infty$. Note that (G, w', k) is an instance of WHOM(H'), equivalent to the instance (G, w, k) of WHOM(H).

Note that the problem of finding a homomorphism to K_4 (the graph (g)) is exactly 4-COLORING. It is known that assuming the ETH, this problem does not admit a subexponential algorithm, even for 2-DIR graphs [5].

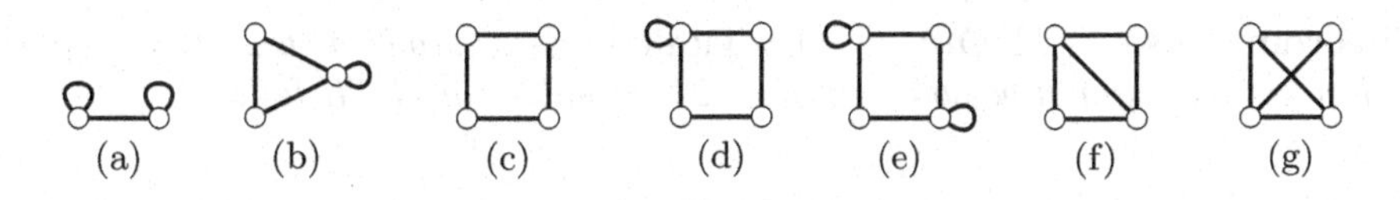

Fig. 1. Characterization of property ($\star$) by forbidden induced subgraphs.

Maximum Cut. In this section, H is the graph (a) from Fig. 1. Note that any function $h\colon V(G) \to V(H)$ is a homomorphism and thus determining the existence of a homomorphism with just vertex lists and weights is trivial. However, it becomes more interesting if we include the edge weights.

We denote the vertices of H by a and b, so we have $E(H) = \{aa, ab, bb\}$ (see Fig. 2). We also define the weight function as follows. Let $w(v,u) = 0$ for every $(v,u) \in V(G) \times V(H)$, and for every $e \in E(G)$ we set $w(e,aa) = w(e,bb) = 0$ and $w(e,ab) = 1$. Note that the value of $w(e,f)$ does not depend on e, so w is in fact an edge-weighting of H (see Fig. 2(a)). Observe that the weight of a homomorphism $h\colon G \to H$ equals the number of edges mapped to ab. Thus finding a homomorphism of maximum weight is equivalent to partitioning the $V(G)$ into two subsets, so that the number of edges crossing this partition is maximized. Such a set of edges is called a *cut* in G and the computational problem of finding the maximum cut is denoted by MAX CUT. It means that for our result, it is enough to show the hardness of MAX CUT on segment graphs.

Fig. 2. Graph H with its corresponding weights defined by w in (a) MAX CUT and (b) ODD CYCLE TRANSVERSAL.

Theorem 7 (♠). *There is no algorithm solving* MAX CUT *in time* $2^{o(n)}$ *for segment graphs, unless the ETH fails.*

Minimum Odd-Cycle Transversal. Now let us consider the case when H is the graph (b) in Fig. 1. This time we will consider a vertex-weighted variant. Denote the vertices of H by a, b, c, where a is the vertex with the loop. All edge-weights are set to 0. For vertex weights, for every $v \in V(G)$ we set $w(v,b) = w(v,c) = 1$ and $w(v,a) = 0$. Note that again the weights do not depend on the choice of v, so we can think of w as a vertex-weighting of H (see Fig. 2(b)).

We observe that finding a homomorphism of maximum weight is equivalent to the problem of finding the maximum number of vertices of G which induce a bipartite subgraph, or, equivalently, the minimum number of vertices, whose removal destroys all odd cycles. This problem is known as ODD CYCLE TRANSVERSAL. We will show the following.

Theorem 8 (♠). *The* Odd Cycle Transversal *problem in* 2-DIR *graphs with n vertices cannot be solved in time $2^{o(n)}$, unless the ETH fails.*

3 Locally Injective and Locally Bijective Homomorphism

Now let us turn our attention to two other variants of the graph homomorphism problem, i.e., locally injective and locally bijective homomorphism. Recall that for a fixed H, the LIHom(H) (LBHom(H), resp.) problem asks if a given graph G admits a homomorphism h to H with a restriction that for every $v \in V(G)$, the mapping h is injective (bijective, resp.) on the set $N_G(v)$. Local injectivity can be equivalently seen as "no two vertices of G with a common neighbor may be mapped to the same vertex of H". Moreover, every locally bijective homomorphism is also locally injective.

We show that unlike the WHom(H), both LIHom(H) and LBHom(H) can be solved in subexponential time on string graphs for every H. The crucial observation is all yes-instances have bounded degree.

Theorem 4 (♠). *For every fixed graph H, the* LIHom*(H) problem and the* LBHom*(H) problem can be solved in time $2^{O(\sqrt{n}\log n)}$ for string graphs.*

As mentioned, locally injective homomorphisms generalize some well-studied graph labeling problems, so Theorem 4 implies the following.

Corollary 9. *For any fixed k, (i) the k-$L(2,1)$-labeling, (ii) the k-coloring of the square of a graph, (iii) the injective k-coloring, can be solved in time $2^{\tilde{O}(\sqrt{n})}$ in string graphs.* □

On the other hand, as every planar graph is a segment graph [6,18], hardness results for planar graphs can be used to derive ETH-lower bounds for LIHom(H) – in particular, the following ones follow from the hardness results for k-$L(2,1)$-labeling [9], 7-coloring of the square of a graph [36], and injective 3-coloring [2].

Theorem 10 (Eggemann *et al.* [9], Ramanathan, Lloyd [36], Bertossi, Bonuccelli [2]). *Unless the ETH fails, there is no algorithm for* LIHom*(H) in segment graphs working in time $2^{o(\sqrt{n})}$, where H is the complement of a path with at least 4 vertices, a complete graph with 7 vertices, a triangle with additional loop on its every vertex.* □

4 Locally Surjective Homomorphism

Now we consider the problem of locally surjective homomorphism, denoted by LSHom(H). For a fixed graph H, the LSHom(H) asks whether a given graph G admits homomorphism to H, which is surjective on $N_G(v)$ for every $v \in V(G)$. In other words, if $h(v) = a$, then every $b \in N_H(a)$ must appear on some $u \in N_G(v)$. We say that vertex $v \in V(G)$ is *happy* if $h(N_G(v)) = N_H(h(v))$. Clearly h is locally surjective if every vertex of G is happy. We write $G \xrightarrow{s} H$ to denote a locally surjective homomorphism from G to H. First, we prove Theorem 5(a), i.e., show a subexponental algorithm for LSHOM(P_3) and LSHOM(C_4).

Let G be an instance of LSHOM(P_3). We can assume that G does not have isolated vertices. Also, G is bipartite with bipartition classes X and Y, as otherwise any homomorphism to P_3 cannot exist. Moreover, in any homomorphism, one of bipartition classes, say Y, will be entirely mapped to the middle vertex of P_3. Note that since no vertex is isolated, vertices of X will always be happy. Thus $G \xrightarrow{s} P_3$ if and only if one can color vertices of X with two colors (1 and 3), so that every vertex from Y has at least one neighbor in each color. We observe that this is exactly the NAE-SAT problem, where G is an incidence graph of the input formula. From this we conclude that LSHOM(P_3) does not have a subexponential algorithm in general graphs, but is solvable in polynomial time in planar graphs, since PLANAR NAE-SAT is in P, see Moret [43]. Moreover, the list variant of LSHOM(P_3) in planar graphs in NP-complete, see Dehghan [8].

The win-win approach of Theorem 2(a) cannot be directly applied for LSHOM(P_3), as there is no good branching on a high-degree vertex. To show that LSHOM(P_3) can be solved in subexponential time in string graphs, we will use the following result.

Theorem 11 (Lee [34]). *There is a constant $c > 0$ such that for every $t \geq 1$, every $K_{t,t}$-free string graph on n vertices has at most $c \cdot n \cdot t \log t$ edges.* □

Theorem 12. LSHOM*(P_3) can be solved in time $2^{O(n^{2/3} \log^{3/2} n)}$ for string graph, even if geometric representation is not given.*

Proof. We assume that an instance graph G has no isolated vertices and is bipartite, with bipartition classes X and Y. Note that G is a yes-instance if and only if there is a homomorphism $h_X \colon G \xrightarrow{s} P_3$, such that $h_X(X) = \{1,3\}$ and $h_X(Y) = \{2\}$, or homomorphism $h_Y \colon G \xrightarrow{s} P_3$, such that $h_Y(Y) = \{1,3\}$ and $h_Y(X) = \{2\}$. Let us assume that X is mapped to $\{1,3\}$, the algorithm will be called twice with roles of X and Y switched. Again, we will solve a more general problem, in which we define an additional function $\sigma : Y \to 2^{\{1,3\}}$ and ask for the existence of a homomorphism $h \colon G \to P_3$, such that $\sigma(y) \subseteq h(N_G(y))$ for every $y \in Y$. Clearly, if $\sigma \equiv \{1,3\}$, then we obtain the LSHOM(P_3) problem.

First, if $|E(G)| \leq c/3 \cdot n^{4/3} \log n$ (where c is a constant from Theorem 11), then we can find a balanced separator S of size $O(n^{2/3} \log^{1/2} n)$ in time $2^{O(n^{2/3} \log^{3/2} n)}$. Denote by V_1, V_2 the sets such that $V(G) = V_1 \uplus V_2 \uplus S$ and there is no V_1-V_2-path in $G - S$. We exhaustively guess $h(x)$ for every $x \in S \cap X$ and the partition $\sigma_1 \uplus \sigma_2$ of $\sigma(y) \setminus h(N(y) \cap S)$ for every $y \in S \cap Y$. Then, for every possibility, we consider graphs $G_1 := G[V_1 \cup S]$ and $G_2 := G[V_2 \cup S]$, in which vertices of S are already colored. For every $y \in S \cap Y$ we set $\sigma(y) = \sigma_1$ if $y \in V(G_1)$, otherwise $\sigma(y) = \sigma_2$. Then, for every $x \in S \cap X$, we remove $h(x)$ from $\sigma(y)$, for $y \in N(x)$, and finally we remove x from the instance. If there exists y for which $\sigma(y) = \phi$, we also remove y. If any isolated vertex $x \in X$ appears, we remove it too, as it means that $\sigma(y)$ of every $y \in N(x)$ was already empty, so the color of x does not matter. We call the algorithm recursively for G_1 and G_2, together with their corresponding functions σ. Note that G_1 or G_2 may contain an isolated vertex $y \in Y$ with $\sigma(y) \neq \phi$, in this case we terminate the current recursive call. The total number of recursive calls is $2^{|X \cap S|}4^{|Y \cap S|} = 2^{O(n^{2/3} \log^{1/2} n)}$ and the overall complexity of this step is $2^{O(n^{2/3} \log^{3/2} n)}$.

If G has more than $c/3 \cdot n^{4/3} \log n$ edges, then, by Theorem 11, it has a bipartite subgraph $K_{n^{1/3},n^{1/3}}$ with bipartition classes $X' \subseteq X$ and $Y' \subseteq Y$. We find it exhaustively in time $n^{O(n^{1/3})} = 2^{O(n^{1/3} \log n)}$. We branch on three possibilities. Either we set $h(x) = 1$ for every $x \in X'$, or $h(x) = 3$ for every $x \in X'$, or we choose $x_1, x_2 \in X'$ and set $h(x_1) = 1$ and $h(x_2) = 3$. In first two cases we proceed with the graph $G - X'$ (and remove $h(X')$ from $\sigma(y)$ of every $y \in N(X')$), and in the last case we remove Y', together with x_1 and x_2 (also adjusting the function σ for their neighbors), as all elements of Y' are happy. Observe that the complexity of this step is $F(n) \leq 2^{O(n^{1/3} \log n)} + 2F(n - n^{1/3}) + n^{2/3}F(n - n^{1/3}) \leq 2^{O(n^{2/3} \log n)}$ and so it the total complexity. □

Using Theorem 12 we can show an analogous result for LSHom(C_4).

Theorem 13. *LSHom(C_4) can be solved in time $2^{O(n^{2/3} \log^{3/2} n)}$ for string graph on n vertices, even if a geometric representation is not given.*

Proof. Again, we assume that an instance graph G is bipartite with bipartition classes X and Y, without isolated vertices. Clearly in any solution h we either have $h(X) = \{1, 3\}$ and $h(Y) = \{2, 4\}$, or $h(X) = \{2, 4\}$ and $h(Y) = \{1, 3\}$.

Claim 1 (♠). There is $h : G \xrightarrow{s} C_4$ such that $h(X) = \{1, 3\}$ iff there are $h_1, h_2 : G \xrightarrow{s} P_3$ for which $h_1(X) = h_2(Y) = \{1, 3\}$ and $h_1(Y) = h_2(X) = \{2\}$.

To solve LSHom(C_4), we run the algorithm from Theorem 12 twice, switching the roles of X and Y. We return true if both calls return true. □

As stated in Theorem 5(b), the existence of subexponential algorithms is unlikely for simple graphs H with $\Delta(H) \leq 2$ and $|H| \geq 3$, different than P_3 and C_4.

5 Consequences for P_t-free Graphs

Let us point out that the graphs constructed in our hardness reductions are P_t-free, for some constant t. This implies the statement (b) of Theorem 3. Recall that the statement (a) follows from the recent result by Groenland *et al.* [20].

Theorem 3 (♠). *Let H be a fixed graph.*

(a) If H has no two vertices with two common neighbors, then the WHom*(H) problem can be solved in time $2^{O(n^{2/3} \log n)}$ for string graphs.*
(b) Otherwise, the WHom*(H) problem in NP-complete and cannot be solved in time $2^{o(n)}$ for segment graphs, unless the ETH fails.*

In particular, we obtain the following result, answering an open problem of Bonamy *et al.* [4].

Corollary 14. Odd Cycle Transversal *problem is NP-complete and cannot be solved in time* $2^{o(n)}$ *in* P_{13}*-free graphs, unless the ETH fails.* □

Bonamy *et al.* [4] considered also a closely related problem Independent Odd Cycle Transversal, where we additionally require that the removed set of vertices is independent. Interestingly, the hardness result of Corollary 14 does not carry over to this problem. Indeed, Independent Odd Cycle Transversal is equivalent to finding a 3-coloring of the input graph, in which the size of one color class is minimized. It is straightforward to see that this problem can be stated as WHom(K_3), where the weight associated with one vertex is 0, the weights associated with two other vertices are 1, and all edge weights are 0. Thus, by Theorem 3, we obtain the following.

Corollary 15. *For every fixed* t*, the* Independent Odd Cycle Transversal *problem can be solved in time* $2^{O(\sqrt{n})}$ *for* P_t*-free graphs on* n *vertices.* □

6 Further Research Directions

Let us point out some directions for further research. First, it would be interesting to obtain a dichotomy for the problems of finding a homomorphism and a list homomorphism from a string graph to a fixed graph H.

Next, we think that obtaining a dichotomy for LSHom(H) in string graphs is an exciting (and probably difficult) task. Let us mention that the NP-hardness proof by Fiala and Paulusma [13] implies that if H is a connected graph with at least two edges, then LSHom(H) cannot be solved in subexponential time in general graphs.

Finally, observe that Theorems 12 and 13 do not directly generalize to P_t-free graphs, as we do not know an analog of Theorem 11 for this class of graphs. It would be very interesting to know whether such a result can be proven, or at least to reprove Theorems 12 and 13 for P_t-free graphs using some other tools.

References

1. Alber, J., Fiala, J.: Geometric separation and exact solutions for the parameterized independent set problem on disk graphs. J. Algorithms **52**(2), 134–151 (2004)
2. Bertossi, A.A., Bonuccelli, M.A.: Code assignment for hidden terminal interference avoidance in multihop packet radio networks. IEEE/ACM Transact. Netw. **3**(4), 441–449 (1995)
3. Biró, C., Bonnet, É., Marx, D., Miltzow, T., Rzążewski, P.: Fine-grained complexity of coloring unit disks and balls. J. of Comp. Geom. **9**, 47–80 (2018)
4. Bonamy, M., Dabrowski, K.K., Feghali, C., Johnson, M., Paulusma, D.: Independent feedback vertex set for P_5-free graphs. Algorithmica **81**(4), 1342–1369 (2019)
5. Bonnet, É., Rzążewski, P.: Optimality program in segment and string graphs. Algorithmica **81**(7), 3047–3073 (2019)

6. Chalopin, J., Gonçalves, D.: Every planar graph is the intersection graph of segments in the plane: extended abstract. In: Proceedings of STOC 2009, pp. 631–638 (2009)
7. Chaplick, S., Fiala, J., van't Hof, P., Paulusma, D., Tesar, M.: Locally constrained homomorphisms on graphs of bounded treewidth and bounded degree. Theor. Comput. Sci. **590**, 86–95 (2015)
8. Dehghan, A.: On strongly planar not-all-equal 3SAT. J. Comb. Optim. **32**(3), 721–724 (2016)
9. Eggemann, N., Havet, F., Noble, S.D.: k-$L(2,1)$-labelling for planar graphs is NP-complete for $k \geq 4$. Disc. Appl. Math. **158**(16), 1777–1788 (2010)
10. Feder, T., Hell, P., Huang, J.: List homomorphisms and circular arc graphs. Combinatorica **19**(4), 487–505 (1999)
11. Fiala, J., Kratochvíl, J.: Locally injective graph homomorphism: lists guarantee dichotomy. In: Fomin, F.V. (ed.) WG 2006. LNCS, vol. 4271, pp. 15–26. Springer, Heidelberg (2006). https://doi.org/10.1007/11917496_2
12. Fiala, J., Maxová, J.: Cantor-Bernstein type theorem for locally constrained graph homomorphisms. Eur. J. Comb. **27**(7), 1111–1116 (2006)
13. Fiala, J., Paulusma, D.: A complete complexity classification of the role assignment problem. Theor. Comput. Sci. **349**(1), 67–81 (2005)
14. Fiala, J., Paulusma, D., Telle, J.A.: Matrix and graph orders derived from locally constrained graph homomorphisms. In: Jędrzejowicz, J., Szepietowski, A. (eds.) MFCS 2005. LNCS, vol. 3618, pp. 340–351. Springer, Heidelberg (2005). https://doi.org/10.1007/11549345_30
15. Fomin, F.V., Lokshtanov, D., Panolan, F., Saurabh, S., Zehavi, M.: Finding, hitting and packing cycles in sub exponential time on unit disk graphs. In: Proceedings of ICALP 2017, LIPIcs, vol. 80, pp. 65:1–65:15 (2017)
16. Fox, J., Pach, J.: A separator theorem for string graphs and its applications. Comb. Probab. Comput. **19**(3), 371–390 (2010)
17. Fox, J., Pach, J.: Computing the independence number of intersection graphs. In: Proceedings of SODA 2011, pp. 1161–1165 (2011)
18. Gonçalves, D., Isenmann, L., Pennarun, C.: Planar Graphs as L-intersection or L-contact graphs. In: Proceedings of SODA 2018, pp. 172–184 (2018)
19. Griggs, J.R., Yeh, R.K.: Labelling graphs with a condition at distance 2. SIAM J. Dis. Math. **5**(4), 586–595 (1992)
20. Groenland, C., Okrasa, K., Rzążewski, P., Scott, A., Seymour, P., Spirkl, S.: H-colouring P_t-free graphs in sub exponential time. Discrete Applied Mathematics (2019, to appear)
21. Gutin, G.Z., Hell, P., Rafiey, A., Yeo, A.: A dichotomy for minimum cost graph homomorphisms. Eur. J. Comb. **29**(4), 900–911 (2008)
22. Hahn, G., Kratochvíl, J., Sirán, J., Sotteau, D.: On the injective chromatic number of graphs. Disc. Math. **256**(1–2), 179–192 (2002)
23. Havet, F., Klazar, M., Kratochvíl, J., Kratsch, D., Liedloff, M.: Exact algorithms for L(2, 1)-labeling of graphs. Algorithmica **59**(2), 169–194 (2011)
24. Hell, P., Nesetril, J.: Graphs and Homomorphisms. Oxford University Press, Oxford (2004)
25. Hell, P., Rafiey, A.: The dichotomy of list homomorphisms for digraphs. In: Proceedings of SODA 2011, pp. 1703–1713 (2011)
26. Hell, P., Raspaud, A., Stacho, J.: On injective colourings of chordal graphs. In: Laber, E.S., Bornstein, C., Nogueira, L.T., Faria, L. (eds.) LATIN 2008. LNCS, vol. 4957, pp. 520–530. Springer, Heidelberg (2008). https://doi.org/10.1007/978-3-540-78773-0_45

27. Impagliazzo, R., Paturi, R.: On the complexity of k-SAT. J. Comput. Syst. Sci. **62**(2), 367–375 (2001)
28. Kisfaludi-Bak, S., van der Zanden, T.C.: On the exact complexity of hamiltonian cycle and q-colouring in disk graphs. In: Fotakis, D., Pagourtzis, A., Paschos, V.T. (eds.) CIAC 2017. LNCS, vol. 10236, pp. 369–380. Springer, Cham (2017). https://doi.org/10.1007/978-3-319-57586-5_31
29. Koebe, P.: Kontaktprobleme der konformen Abbildung. Berichte über die Verhandlungen der Sächsischen Akademie der Wissenschaften zu Leipzig, Mathematisch-Physikalische Klasse **88**, 141–164 (1936)
30. Kratochvíl, J.: String graphs. I. The number of critical nonstring graphs is infinite. J. Comb. Theory Ser. B **52**(1), 53–66 (1991)
31. Kratochvíl, J.: String graphs. II. Recognizing string graphs is NP-hard. J. Comb. Theory Ser. B **52**(1), 67–78 (1991)
32. Kratochvíl, J., Matoušek, J.: String graphs requiring exponential representations. J. Comb. Theory Ser. B **53**(1), 1–4 (1991)
33. Kratochvíl, J., Matoušek, J.: Intersection graphs of segments. J. Comb. Theory Ser. B **62**(2), 289–315 (1994)
34. Lee, J.R.: Separators in region intersection graphs. In: Proceedings of ITCSC 2017, pp. 1:1–1:8 (2017)
35. Lipton, R., Tarjan, R.: A separator theorem for planar graphs. SIAM J. on Appl. Math. **36**(2), 177–189 (1979)
36. Lloyd, E.L., Ramanathan, S.: On the complexity of distance-2 coloring. In: Proceedings of ICCI 1992, pp. 71–74 (1992)
37. MacGillivray, G., Swarts, J.: The complexity of locally injective homomorphisms. Disc. Math. **310**(20), 2685–2696 (2010)
38. Marx, D., Pilipczuk, M.: Optimal parameterized algorithms for planar facility location problems using voronoi diagrams. In: Bansal, N., Finocchi, I. (eds.) ESA 2015. LNCS, vol. 9294, pp. 865–877. Springer, Heidelberg (2015). https://doi.org/10.1007/978-3-662-48350-3_72
39. Matoušek, J.: Intersection graphs of segments and $\exists\mathbb{R}$. CoRR, abs/1406.2636 (2014)
40. Matoušek, J.: Near-optimal separators in string graphs. Comb. Prob. Comput. **23**(1), 135–139 (2014)
41. McDiarmid, C., Müller, T.: Integer realizations of disk and segment graphs. J. Comb. Theory Ser. B **103**(1), 114–143 (2013)
42. Miller, G.L., Teng, S., Thurston, W.P., Vavasis, S.A.: Separators for sphere-packings and nearest neighbor graphs. J. ACM **44**(1), 1–29 (1997)
43. Moret, B.M.: Planar NAE3SAT is in P. SIGACT News **19**(2), 51–54 (1988)
44. Schaefer, M., Sedgwick, E., Štefankovič, D.: Recognizing string graphs in NP. J. Comput. Syst. Sci. **67**(2), 365–380 (2003)
45. Okrasa, K., Rzążewski, P.: Subexponential algorithms for variants of homomorphism problem in string graphs. CoRR, abs/1809.09345 (2018)
46. Schaefer, M., Štefankovič, D.: Fixed points, nash equilibria, and the existential theory of the reals. Theory Comput. Syst. **60**(2), 172–193 (2017)
47. Smith, W.D., Wormald, N.C.: Geometric separator theorems and applications. In: Proceedings of FOCS 1998, pp. 232–243 (1998)
48. van den Heuvel, J., McGuinness, S.: Coloring the square of a planar graph. J. Graph Theory **42**(2), 110–124 (2003)

The 4-Steiner Root Problem

Guillaume Ducoffe[1,2,3](✉)

[1] Faculty of Mathematics and Computer Science, University of Bucharest, Bucharest, Romania
[2] The Research Institute of the University of Bucharest ICUB, Bucharest, Romania
[3] National Institute for Research and Development in Informatics, Bucharest, Romania
guillaume.ducoffe@ici.ro

Abstract. The kth-power of a graph G is obtained by adding an edge between every two distinct vertices at a distance $\leq k$ in G. We call G a *k-Steiner power* if it is an induced subgraph of the kth-power of some tree T. In particular, G is a *k-leaf power* if all vertices in $V(G)$ are leaf-nodes of T. Our main contribution is a polynomial-time recognition algorithm of 4-Steiner powers, thereby extending the decade-year-old results of (Lin, Kearney and Jiang, *ISAAC'00*) for $k = 1, 2$ and (Chang and Ko, *WG'07*) for $k = 3$. As a byproduct, we give the first known polynomial-time recognition algorithm for 6-leaf powers. Our work combines several new algorithmic ideas that help us overcome the previous limitations on the usual dynamic programming approach for these problems.

Keywords: k-Leaf powers · k-Steiner powers · Clique-tree · Clique-arrangement · Dynamic programming · Maximum matching

1 Introduction

A basic problem in computational biology is, given some set of species and a dissimilarity measure in order to compare them, find a *phylogenetic tree* that explains their respective evolution. Namely, such a rooted tree starts from a common ancestor and branches every time there is a separation between at least two of the species we consider. In the end, the leaves of the phylogenetic tree should exactly represent our given set of species. We study a related problem that has attracted some attention in Graph theory:

Problem 1 (k-LEAF POWER).

Input: a graph $G = (V, E)$.
Question: Is there a tree T whose leaf-nodes are the vertices in V and such that $uv \in E \iff dist_T(u, v) \leq k$?

This work was supported by an ICUB Fellowship for Young Researchers and a grant of Romanian Ministry of Research and Innovation CCCDI-UEFISCDI. project no. 17PCCDI/2018.

I. Sau and D. M. Thilikos (Eds.): WG 2019, LNCS 11789, pp. 14–26, 2019.
https://doi.org/10.1007/978-3-030-30786-8_2

The yes-instances of Problem 1 are called *k-leaf powers*. Their structural and algorithmic properties have been intensively studied (*e.g.*, see [1,8,12] and the papers cited therein). However, the complexity of k-LEAF POWER is a longstanding open problem. Very recently, parameterized (FPT) algorithms were proposed for every fixed k on the graphs with degeneracy at most d, where the parameter is $k + d$ [11]. For general graphs, polynomial-time recognition algorithms are known only for $k \leq 5$ [4,6,9]. Characterizations are known only for $k \leq 4$. Several variations of k-leaf powers were introduced in the literature [5,7,9,13]. In this work, we consider *k-Steiner* powers: a natural relaxation of k-leaf powers where the vertices in the graph may also be internal nodes in the tree T. Interestingly, there is a linear-time reduction from k-LEAF POWER to $(k-2)$-STEINER POWER [6]. However, there only exist polynomial-time recognition algorithms for k-Steiner powers for $k \leq 3$ [9,13].

Our Results. We obtain the first improvement on the recognition of k-Steiner powers in a decade, by solving the case $k = 4$. Combining our main result with the aforementioned reduction from k-LEAF POWER to $(k-2)$-STEINER POWER [6], we also improve the state of the art for k-leaf powers.

Theorem 1. *There is a polynomial-time algorithm for the problems* 4-STEINER POWER *and* 6-LEAF POWER. *For the yes-instances, this algorithm also outputs a corresponding tree T.*

Proving this above Theorem 1, while it may look like a modest improvement in our understanding of the k-LEAF POWER and k-STEINER POWER problems, was technically challenging. In the full version of this paper, we further discuss why the dominant approach for k-STEINER POWER, based on dynamic programming, was already showing its limitations with 4-STEINER POWER. We so believe that one of the main merits of our work is to bring several new ideas in order to tackle with these aforementioned limitations. As such, we expect further uses of these ideas in the study of k-leaf powers and their relatives.

Organization of the Paper. We refer to Sect. 2 for any missing definition in what follows. As our starting point we restrict our study to *chordal graphs* and *strongly chordal graphs*, that are two well-known classes in algorithmic graph theory of which k-Steiner powers form a particular subclass [1]. Doing so, we can use various properties of these classes of graphs, such as: the existence of a tree-like representation of chordal graphs, that is called a *clique-tree* [2] and is commonly used in the design of dynamic programming algorithms on this class of graphs; and an auxiliary data structure which is called "*clique arrangement*" and is polynomial-time computable on strongly chordal graphs [16]. Roughly, this clique arrangement encodes all possible intersections of a subset of maximal cliques in a graph. It is worth noticing that clique arrangements were introduced in the same paper as leaf powers, under the different name of "clique graph" [17].

Given a k-Steiner power G, let us call *k-Steiner root* a corresponding tree T. In Sect. 3 we present new results on the structure of k-Steiner roots that we use in the analysis of our algorithm. Specifically, we show in Sect. 3.1 that in any k-Steiner root T of a graph G, any intersection of maximal cliques in G must be contained in a particular subtree where no other vertex of G can be present. Furthermore, the inclusion relationships between these "clique-intersections" in G are somewhat reflected by the diameter of their corresponding subtrees in T. This extends prior results from [9,17]. Then, we focus in Sect. 3.2 on the case $k = 4$. For every clique-intersection X in a chordal graph G, we classify the vertices in X into two main categories: "free" and "constrained", that depend on the other clique-intersections these vertices are contained in. Our study shows that free vertices cause a combinatorial explosion of the number of partial solutions we should store in a naive dynamic programming algorithm. However, we overcome this issue by proving that there always exists a "well-structured" 4-Steiner root where such free vertices are leaves with very special properties.

Sections 4, 5, 6 and 7 are devoted to the main steps of the algorithm. We start by presenting a constructive proof of a rooted clique-tree with quite constrained properties in Sect. 4. Roughly we carefully control the ancestor/descendant relationships between the edges that are labelled by different minimal separators of the graph. These technicalities are the cornerstone of our approach in Sect. 6 in order to bound the number of partial solutions that we should store in our dynamic programming algorithm.

Then given our special rooted clique-tree T_G, we recall that the maximal cliques and the minimal separators of G can be mapped to the nodes and edges of T_G, respectively. For every node and edge in T_G, we consider the corresponding clique-intersection in G and we precompute by dynamic programming all possible subtrees to which it could be mapped in some well-structured 4-Steiner root of G. Of particular importance is Sect. 5.1 where for any minimal separator S, we give a polynomial-time algorithm in order to generate all the candidate smallest subtrees that could contain S in a well-structured 4-Steiner root of G. The result is then easily extended to the maximal cliques that appear as leaves in our clique-tree (Sect. 5.2). Correctness of these two first parts follows from Sect. 3.2. Finally, in Sect. 5.3 we give a more complicated representation of a family of candidate subtrees $T\langle K_i \rangle$ for all the other maximal cliques K_i. This part is based on a careful analysis of clique-intersections in K_i and several additional tricks. Roughly, our representation in Sect. 5.3 is composed of partially constructed subtrees and of "problematic" subsets that need to be inserted to these subtrees in order to complete the construction. The exact way these insertions must be done is postponed until the very end of the algorithm (Sect. 7).

Section 6 is devoted to the encoding of partial solutions in our dynamic programming. Specifically, instead of computing partial solutions at each node of the clique-tree and storing their encodings, we rather pre-compute a polynomial-size subset of *allowed* encodings for each node. Then, the problem becomes to decide whether given such an encoding, there exists a corresponding partial solution. We formalize our approach by introducing an intermediate problem where the goal is to compute a 4-Steiner root with additional constraints on its structure and the distances between some sets of nodes. Finally, we detail in Sect. 7 the resolution of our intermediate problem, thereby completing the presentation of our algorithm. An all new contribution in this part is a greedy procedure, based on MAXIMUM-WEIGHT MATCHING, in order to ensure some distances' constraints are satisfied by the solutions we generate during the algorithm.

This is only an extended abstract. Full proofs can be found in our technical report [10]. Due to their intricacy we gave up optimizing the running-time of our algorithm. A very rough upper bound would be $\mathcal{O}(n^{16}m^5)$-time.

2 Preliminaries

For standard graph terminology, see [3]. All graphs in this study are finite, simple, unweighted and connected. Given a graph $G = (V, E)$, let $n := |V|$ and $m := |E|$. The neighbourhood of a vertex $v \in V$ is defined as $N_G(v) := \{u \in V \mid uv \in E\}$. By extension, we define the neighbourhood of a set $S \subseteq V$ as $N_G(S) := \left(\bigcup_{v \in S} N_G(v)\right) \setminus S$. The subgraph induced by any subset $U \subseteq V$ is denoted by $G[U]$. For every $u, v \in V$, we denote by $dist_G(u, v)$ the minimum length (number of edges) of a uv-path. The eccentricity of vertex v is defined as $ecc_G(v) := \max_{u \in V} dist_G(u, v)$. The radius and the diameter of G are defined, respectively, as $rad(G) := \min_{v \in V} ecc_G(v)$ and $diam(G) := \max_{v \in V} ecc_G(v)$. We denote by $\mathcal{C}(G)$ the center of G, *a.k.a.* the vertices with minimum eccentricity.

Steiner Roots. The kth-*power* of G, denoted G^k has same vertex-set V as G and edge-set $\{uv \mid 0 < dist_G(u, v) \leq k\}$. If there is some tree T such that G is an induced subgraph of T^k then, we call G a *k-Steiner power* and T a *k-Steiner root* of G. Nodes in $V(G)$ are called *real*, whereas nodes in $V(T) \setminus V(G)$ are called *Steiner*. We so define, for any $S \subseteq V(T)$: $Real(S) := S \cap V(G)$ and $Steiner(S) := S \setminus V(G)$. Two (sub)trees T, T' are *Steiner-equivalent*, denoted $T \equiv_G T'$, if and only if $Real(T) = Real(T') = S$ and there exists an isomorphism $\iota : V(T) \to V(T')$ such that $\iota(v) = v$ for any $v \in S$ (the trees are equal up to an appropriate identification of their Steiner nodes). Finally, given a node-subset $X \subseteq V(T)$, $T\langle X\rangle$ is the smallest subtree of T such that $X \subseteq V(T\langle X\rangle)$.

(Strongly) Chordal Graphs. A *clique-tree* is a tree T_G whose nodes are the maximal cliques of G and such that for every $v \in V$, the maximal cliques containing v induce a subtree of T_G. A graph $G = (V, E)$ is called *chordal* if and only if it has a clique-tree. Moreover if G is chordal then, we can construct a clique-tree for G in $\mathcal{O}(m)$-time [2]. An uv-separator is a subset $S \subseteq V \setminus \{u, v\}$ such that u and v are disconnected in $G \setminus S$. If in addition, no strict subset of S is an uv-separator then, S is a *minimal* uv-separator. A *minimal separator* of G is a minimal uv-separator for some $u, v \in V$. For a chordal graph G and *any* clique-tree T_G of G, S is a minimal separator of G if and only if there exist two maximal cliques K_i, K_j such that: $K_iK_j \in E(T_G)$ and $K_i \cap K_j = S$ [2]. We define $E_S(T_G) := \{K_iK_j \in E(T_G) \mid K_i \cap K_j = S\}$ (edges labeled by S).

A *clique-intersection* of G is the intersection of a subset of maximal cliques in G. The families of all clique-intersections, maximal cliques and minimal separators of G are denoted by $\mathcal{X}(G)$, $\mathcal{K}(G)$ and $\mathcal{S}(G)$, respectively. For a superclass of k-Steiner powers known as *strongly chordal graphs*, the family $\mathcal{X}(G)$ has polynomial size and can be computed in polynomial time [15].

Step 0 (Initialization). Given $G = (V, E)$, we check whether G is strongly chordal. If this is not the case then, G cannot be a 4-Steiner power, and we stop. Otherwise we compute $\mathcal{X}(G)$.

3 Structure Theorems

Some relationships between k-Steiner roots and clique-intersections are proved in Sect. 3.1. These structural results are the cornerstone of our algorithm and its analysis. Then, we refine our results for the special case $k = 4$ in Sect. 3.2.

3.1 Playing with the Root

The following result is a generalization of [9, Lemma 1] to any k. We prove it by using some intricate properties of the eccentricity function on trees [14].

Theorem 2. *Given $G = (V, E)$ and T any k-Steiner root of G, the following properties hold for any clique-intersection $X \in \mathcal{X}(G)$:*

- *We have $Real(T\langle X\rangle) = X$ and $diam(T\langle X\rangle) \leq k$;*
- *If $T' \supset T\langle X\rangle$ then, either $X = Real(T')$ or $diam(T') > diam(T\langle X\rangle)$;*
- *If $k = 2k'$ is even then, for any two different maximal cliques $K_i, K_j \in \mathcal{K}(G)$ we have $\mathcal{C}(T\langle K_i\rangle) \cap \mathcal{C}(T\langle K_j\rangle) = \emptyset$.*

3.2 Well-Structured 4-Steiner Roots

For $k = 4$, we introduce new notions which only depend on the clique-intersections of G. Roughly, given $X \in \mathcal{X}(G)$ and an arbitrary root T, we introduce some operations in order to modify $T\langle X\rangle$. Doing so, we wish to force this subtree to have some more structure, thereby avoiding a combinatorial explosion of the number of possibilities to consider. Therefore, we carefully study the situations when a vertex $v \in X$ may not be arbitrarily movable inside $T\langle X\rangle$ (in which case we call v *X-constrained*). The most natural case is when there is a $X' \in \mathcal{X}(G)$ s.t. $X' \subset X$, $v \in X'$ and $|X'| \geq 2$ (v is *internally X-constrained*)[1]. However, more subtle cases occur when X is a minimal separator, or more generally X is contained in some larger clique-intersection. We say that v is *(X, X_1, X_2)-sandwiched* if $X_1, X_2 \in \mathcal{X}(G)$ are s.t. $X \subset X_1$ and $X \cap X_2 = \{v\} \subset X_1 \cap X_2$. Our study reveals that X-constrained vertices have a very rigid structure. Finally, a vertex that does not fall in one of these two above cases is called *X-free*. We prove that we can always force the X-free vertices to be *leaves* of the subtree $T\langle X\rangle$, thereby considerably reducing the number of possibilities for the latter.

[1] When X is a *maximal clique*, the internally X-constrained vertices can be characterized in terms of simplicial vertices and a subset of the cut-vertices.

Theorem 3. *Let $G = (V, E)$ be a 4-Steiner power. There always exists a well-structured 4-Steiner root T of G where, for any clique-intersection $X \in \mathcal{X}(G)$:*

- *all the X-free vertices are leaves of $T\langle X\rangle$ with maximum eccentricity $diam(T\langle X\rangle)$;*
- *there is a node $c \in \mathcal{C}(T\langle X\rangle)$ such that for every X-free vertex v, except maybe one, $dist_T(v, c) = dist_T(v, \mathcal{C}(T\langle X\rangle))$;*
- *all the internal nodes on a path between $\mathcal{C}(T\langle X\rangle)$ and a X-free vertex are Steiner nodes of degree two;*
- *and if $X \in \mathcal{K}(G)$ and it has a X-free vertex then, $diam(T\langle X\rangle) = 4$.*

4 A Special Rooted Clique-Tree

We now present Step 1 of our algorithm so as to show all the steps in chronological order. However, please note that in the next Sect. 5, *any* clique-tree could be used. Indeed, we will only start using the peculiar properties of our rooted clique-tree in Sect. 6.

Step 1 (Construction of the rooted clique-tree). We construct a clique-tree T_G of G that we root in some $K_0 \in \mathcal{K}(G)$. In order to give the main intuition behind its construction, let us consider an arbitrary maximal clique K_i that is not the root, and its parent node $K_{p(i)}$. Let G_i be induced by the maximal cliques in the subtree of T_G rooted at K_i. If G has a 4-Steiner root then, by heredity, so does G_i. Roughly, we would like to bound the number of partial solutions for G_i that we will need to store for our dynamic programming algorithm. By Theorem 3, one first step for doing so would be to force most vertices in $S_i := K_i \cap K_{p(i)}$ to be K_i-*free* in the subgraph G_i. More specifically, for every descendant K_j of K_i in T_G we would like to impose $S_j \not\subseteq S_i$ and $S_i \not\subseteq S_j$[2]. However, both objectives are conflicting and so, we need to find a trade-off. Admittedly, our proposed solution is quite technical.

Given a clique-tree T_G of $G = (V, E)$, we say that a minimal separator S is *weakly T_G-convergent* if there exists some maximal clique K_S that is incident to all edges in $\bigcup_{S', S \subset S'} E_{S'}(T_G)$. S is termed *T_G-convergent* if it is weakly T_G-convergent and the maximal clique K_S is also incident to all edges in $E_S(T_G)$. The relationship between these notions and 4-Steiner roots is as follows:

Lemma 1. *Let T be any 4-Steiner root of G, and let $S \in \mathcal{S}(G)$. If $T\langle S\rangle$ is a non-edge star then, S is weakly T_G-convergent for* any *clique-tree T_G of G.*

Sketch Proof. We may assume that S is strictly contained in some minimal separator S'. By Theorem 2, $T\langle S'\rangle$ has diameter three. This implies $\mathcal{C}(T\langle S\rangle) \subset \mathcal{C}(T\langle S'\rangle)$. Furthermore, we can prove that S' is contained in exactly two maximal cliques K_i, K_j and $\mathcal{C}(T\langle K_i\rangle) \cup \mathcal{C}(T\langle K_j\rangle) = \mathcal{C}(T\langle S'\rangle)$. Let us assume w.l.o.g.

[2] Observe that if $S_j \subset S_i$, and $|S_j| \geq 2$, then the vertices of S_j are S_i-constrained.

that $\mathcal{C}(T\langle S\rangle) = \mathcal{C}(T\langle K_i\rangle)$. Then, any minimal separator S'' that strictly contains S is contained in K_i and one other maximal clique $K_{S''}$. Let T_G be a clique-tree of G. We have $E_{S''}(T_G) \neq \emptyset$, and so $K_iK_{S''} \in E(T_G)$. By setting $K_S := K_i$, we get that S is weakly T_G-convergent. □

Therefore, weak convergence is a *necessary* condition for a $S \in \mathcal{S}(G)$ to be contained in a star in some 4-Steiner root (that is the hardest case to deal with in our algorithm). If furthermore there is convergence then, we needn't store any costly information about the separators that strictly contain S in the encoding of partial solutions. Indeed, any "inclusion issue" between S and these separators can be handled with when we process the maximal clique K_S. So, we want to force weak convergence to imply convergence. Our construction in what follows applies to any S of size at least three. – For smaller separators, we can use much simpler counting arguments in order to bound the number of possible partial solutions that we will need to consider by a *constant*. See Sect. 6 for details. –

Theorem 4. *For any chordal graph G, we can compute in polynomial time a rooted clique-tree T_G where, for any $S_i := K_i \cap K_{p(i)}$:*

- *If S_i is weakly T_G-convergent and $|S_i| \geq 3$ then, S_i is T_G-convergent;*
- *Any minimal separator of G_i that is contained in S_i is T_G-convergent, has at least three vertices and is* strictly *contained in a minimal separator of G_i.*

5 A Family of Subtrees for the Clique-Intersections

Step 2 (Candidate set generation). We exploit a result of Sect. 3.1 which states that, for any 4-Steiner root T of G and for any clique-intersection X, the smallest subtree containing X does not contain any other real nodes. Then, our goal is, for every $X \in \mathcal{X}(G)$, to compute a polynomial-size family $\mathcal{T}_X$ of "candidate subtrees" whose real nodes are exactly X. Intuitively, $\mathcal{T}_X$ should contain all possibilities for $T\langle X\rangle$ in a *well-structured* 4-Steiner root T (such a root must satisfy additional properties given in Sect. 3.2). Note that we only need to compute this above family for *minimal separators* and *maximal cliques.*

5.1 Case of Minimal Separators

The following result serves as a brick-basis construction for computing all the other families of candidate subtrees.

Theorem 5. *In $\mathcal{O}(n^5m)$-time we can construct a collection $(\mathcal{T}_S)_{S\in\mathcal{S}(G)}$ such that, for any well-structured 4-Steiner root T of G, and for any $S \in \mathcal{S}(G)$, $T\langle S\rangle$ is Steiner-equivalent to some subtree in $\mathcal{T}_S$.*

Sketch Proof. Let us describe the main difficulty we had to face on in order to prove this above result. Given $S \in \mathcal{S}(G)$ the difficulty in generating $\mathcal{T}_S$ comes from the bistars (diameter-three subtrees), as a brute-force generation of all possibilities would take time exponential in $|S|$. Let $\mathcal{X}(S) = \{X \in \mathcal{X}(G) \mid$

$X \subset S$, $|X| \geq 2\}$. Based on a careful analysis of the intersection graph $I_S = (\mathcal{X}(S), \{XX' \mid X \cap X' \neq \emptyset\})$, we can bound the number of possible mappings of the internally S-constrained vertices to the nodes of a bistar by an $\mathcal{O}(|S|^2)$. We can also bipartition the sandwiched vertices in such a way that each group should be mapped to a different side of the bistar; each group should in fact correspond to one of the two maximal cliques containing S. Then, we use the fact that in a well-structured 4-Steiner root of G, S-free vertices are leaves of such a bistar with all of them, except maybe one, adjacent to the same central node. For a fixed mapping of the S-constrained vertices, this only gives us $\mathcal{O}(|S|)$ possibilities in order to map the S-free vertices. Overall, we reduce the number of possible bistars to an $\mathcal{O}(|S|^5)$. □

5.2 Case of a Leaf Node

Theorem 6. *Given $G = (V, E)$ and a rooted clique-tree T_G of G, let $K_i \in \mathcal{K}(G)$ be a leaf. We can construct, in time polynomial in $|K_i|$, a set $\mathcal{T}_i$ of 4-Steiner roots for $G_i := G[K_i]$ with the following additional property: In any* well-structured *4-Steiner root T of G, there exists a $T_i' \in \mathcal{T}_i$ Steiner-equivalent to $T\langle K_i \rangle$.*

Sketch Proof. We use a well-known decomposition of K_i into a unique minimal separator $S_i := K_i \cap K_{p(i)}$ and a set of simplicial vertices. Given any fixed possibility for $T\langle S_i \rangle$, there are $\mathcal{O}(|S_i|)$ possibilities for $T\langle S_i \cup \mathcal{C}(T\langle K_i \rangle) \rangle$. Then, we use the fact that all simplicial vertices are K_i-free. Since we already fixed $\mathcal{C}(T\langle K_i \rangle)$, by Theorem 3, there is essentially one way to add the K_i-free vertices in order to complete the construction (up to Steiner equivalence). □

5.3 Case of an Internal Node

Finally, we consider the maximal cliques K_i that are internal nodes of T_G. Unsurprisingly, several new difficulties arise in the construction of $\mathcal{T}_{K_i}$. Our bottleneck is solving the following subproblem: compute (up to Steiner equivalence) all possible central nodes and their neighbourhood in any subtree $T\langle K_i \rangle$ of diameter four. We solved this subproblem in most situations, *e.g.*, when there is a minimal separator $S \subseteq K_i$ such that $T\langle S \rangle$ must be a bistar (diameter-three subtree). For that, we combine some key arguments in the proof of Theorem 5 with the transformation techniques that we used in the proof of Theorem 3.

Lemma 2. *For any graph G, let $S \in \mathcal{S}(G)$, let K be a maximal clique containing S and let R, c be such that $R \subset S$ and either $c \in R$ or c is Steiner. We can compute in $\mathcal{O}(nm \log n)$-time a node c' with the following properties: For any well-structured 4-Steiner root T of G s.t. $T\langle S \rangle$ is a bistar, $c \in \mathcal{C}(T\langle S \rangle) \backslash \mathcal{C}(T\langle K \rangle)$, and $Real(N_T[c]) = R$, there exists a well-structured root T' with the same properties s.t. $\mathcal{C}(T'\langle K \rangle) = \{c'\}$, and $dist_{T'}(u, v) \geq dist_T(u, v)$ for every $u, v \in V$; moreover, either $T \equiv_G T'$, or $\sum_{u,v \in V} dist_{T'}(u, v) > \sum_{u,v \in V} dist_T(u, v)$.*

In order to better understand the significance of Lemma 2, assume that $T\langle S \rangle$ should be a bistar in the final solution we want to compute, and that we already identified one of its center node c and the set of real nodes R to which c must be adjacent. What this above property says is that there is essentially one canonical way to compute the bistar given R and c. The more technical condition $dist_{T'}(u,v) \geq dist_T(u,v)$ is simply there in order to ensure that by doing so, we cannot miss a solution of an intermediate problem we call DISTANCE-CONSTRAINED ROOT (*i.e.*, see Sect. 6). Finally, our condition on the potential function $\sum_{u,v \in V} dist_{T'}(u,v)$ increasing ensures that we can repeatedly apply our "canonical completion" method for arbitrarily many minimal separators S. By using this above method, we obtain the following intermediate construction:

Lemma 3. *For any chordal graph G, let T_G be a rooted clique-tree and let K_i be a maximal clique of $G = (V,E)$ with no K_i-free vertex. In $\mathcal{O}(|K_i|^6 \cdot n^3 m \log n)$-time, we can compute a family $\mathcal{B}_i$ with the following special property: For any well-structured 4-Steiner root T of G where for at least one minimal separator $S \subset K_i$, $T\langle S \rangle$ is a bistar, there is a T' such that $T'\langle S_i \rangle \equiv_G T\langle S_i \rangle$, $T'\langle K_i \rangle \in \mathcal{B}_i$ and $dist_{T'}(r, V(G_i) \setminus S_i) \geq dist_T(r, V(G_i) \setminus S_i)$ for every $r \in V(T\langle S_i \rangle)$*[3].

Note that we do not capture *all* well-structured roots with this above lemma, but only those maximizing certain distances' conditions. In the remaining cases when there are no minimal separators S that are mapped to a bistar, our construction is less satisfying. Specifically, we are left with some "problematic subsets" called *thin branches*: with exponentially many possible ways to include them in candidate subtrees. As a way to circumvent this combinatorial explosion, we also include in $\mathcal{T}_{K_i}$ some partially constructed subtrees where the thin branches are omitted. We will greedily decide how to include the thin branches in these subtrees at Step 4 (Sect. 7).

6 Deciding the Partial Solutions to Store

Step 3 (Selection of the encodings). For the remainder of the algorithm, let $(K_q, K_{q-1}, \ldots, K_0)$ be a post-ordering of the maximal cliques (*i.e.*, obtained by depth-first-search traversal of our rooted clique-tree T_G). We consider the maximal cliques $K_i \in \mathcal{K}(G)$ sequentially, from $i = q$ downto $i = 0$. The next two Sections are devoted to the computation of a subset $\mathcal{T}_i$ of 4-Steiner roots for G_i. Specifically, for any 4-Steiner root T_i of G_i we define the following encoding:

$$\texttt{encode}(T_i) := \Big[\; T_i\langle S_i \rangle \;\mid\; (dist_{T_i}(r, V(G_i) \setminus S_i))_{r \in V(T_i\langle S_i \rangle)} \; \Big].$$

In what follows, we compute a polynomial-size subset of *allowed* encodings for the partial solutions in $\mathcal{T}_i$. That is, we only want to add in $\mathcal{T}_i$ some partial solutions for which the encoding is in the list. Formally, we define an auxiliary problem called DISTANCE-CONSTRAINED ROOT, where given an encoding as input, we ask whether there exists a corresponding 4-Steiner root of G_i.

[3] Recall that S_i and G_i were defined in Sect. 4. By convention, $S_i = \emptyset$ if K_i is the root.

Problem 2 (DISTANCE-CONSTRAINED ROOT).

Input: a graph $G = (V, E)$ with a rooted clique-tree T_G, a maximal clique K_{i_j}, a tree $T_{S_{i_j}}$ s.t. $Real(T_{S_{i_j}}) = S_{i_j}$, and a sequence $(d_r)_{r \in V(T_{S_{i_j}})}$ of positive integers.

Output: Either a 4-Steiner root T_{i_j} of G_{i_j} s.t. $T_{S_{i_j}} \equiv_G T_{i_j}\langle S_{i_j}\rangle$ and, $\forall r \in V(T_{S_{i_j}})$: $dist_{T_{i_j}}(r, V(G_{i_j}) \setminus S_{i_j}) \geq d_r$; Or $\perp$ if there is no such a T_{i_j} which can be extended to some *well-structured* 4-Steiner root T of G.

Theorem 7. *Given $G = (V, E)$ chordal and a rooted clique-tree T_G as in Theorem 4, let K_i be an internal node with children $K_{i_1}, K_{i_2}, \ldots, K_{i_p}$. If we can solve* DISTANCE-CONSTRAINED ROOT *in time $P(n, |S_{i_j}|)$ for some polynomial P then, we can compute in time $\mathcal{O}(n|K_i|^5 P(n, |K_i|))$ a family $\mathcal{T}_{i_1}, \mathcal{T}_{i_2}, \ldots, \mathcal{T}_{i_p}$ of 4-Steiner roots for $G_{i_1}, G_{i_2}, \ldots, G_{i_p}$, respectively, such that:*

1. *For any $j \in \{1, 2, \ldots, p\}$, $|\mathcal{T}_{i_j}| = \mathcal{O}(|S_{i_j}|^5)$;*
2. *For any well-structured 4-Steiner root T of G, there exists a T' such that: $T\langle K_i\rangle \equiv_G T'\langle K_i\rangle$, $T'\langle V(G_{i_j})\rangle \in \mathcal{T}_{i_j}$ for any $j \in \{1, 2, \ldots, p\}$, and $dist_{T'}(r, V(G_i) \setminus S_i) \geq dist_T(r, V(G_i) \setminus S_i)$ for any node $r \in V(T\langle S_i\rangle)$.*

Sketch Proof. We process the children nodes K_{i_j} sequentially by non-decreasing size of the minimal separators S_{i_j}. For that, we start constructing the family $\mathcal{T}_{S_{i_j}}$ of Theorem 5, and we consider the subtrees $T_{S_{i_j}} \in \mathcal{T}_{S_{i_j}}$ sequentially. We divide the proof into several cases depending on $|S_{i_j}|$ and on $diam(T_{S_{i_j}})$.

Case $|S_{i_j}| \leq 2$. There can only be $\mathcal{O}(1)$ different possibilities for the distances $(d_r)_{r \in V(T_{S_{i_j}})}$. We could solve DISTANCE-CONSTRAINED ROOT for all these possibilities, thereby obtaining the family $\mathcal{T}_{i_j}$. But in fact, this seemingly simple case hides a time bomb that will detonate during the second part of the proof (*i.e.*, when we consider larger minimal separators). To understand why through an example, let us assume the existence of a large separator S_{i_k} of which every vertex is also a cut-vertex. Then, one possibility for $T_{S_{i_k}}$ is a star with a Steiner central node. For every leaf-node v of that star, let us consider a maximal clique K_{i_j} s.t. $S_{i_j} = \{v\}$. The star $T_{S_{i_k}}$ can only be compatible with solutions T_{i_j} s.t. $dist_{T_{i_j}}(v, V(G_{i_j}) \setminus \{v\}) \geq 5 - dist_{T_{S_{i_k}}}(v, S_{i_k} \setminus \{v\}) = 3$. In particular, we may have up to *two* compatible solutions T_{i_j}, and that gives us in turn two different possibilities for the constraint $dist_{T_{i_k}}(v, V(G_{i_k}) \setminus S_{i_k})$. But then, since this is true for any $v \in S_{i_k}$, we are left with $2^{|S_{i_k}|}$ possibilities for the distance constraints $(d_r)_{r \in V(T_{S_{i_k}})}$! We can resolve this issue by always choosing any compatible solution which *maximizes* $dist_{T_{i_j}}(v, V(G_{i_j}) \setminus \{v\})$. Specifically, if $S_{i_j} = \{v\}$ is a cut-vertex then, we only keep in the family $\mathcal{T}_{i_j}$ the partial solution maximizing $dist_{T_{i_j}}(v, V(G_{i_j}) \setminus S_{i_j})$. In the same way, if $S_{i_j} = \{u, v\}$ and $T_{S_{i_j}}$ is an edge then, we only need to keep two solutions, namely: among all those maximizing $dist_{T_{i_j}}(v, V(G_{i_j}) \setminus S_{i_j})$ (resp., $dist_{T_{i_j}}(u, V(G_{i_j}) \setminus S_{i_j})$) the one maximizing $dist_{T_{i_j}}(u, V(G_{i_j}) \setminus S_{i_j})$ (resp., $dist_{T_{i_j}}(v, V(G_{i_j}) \setminus S_{i_j})$).

Case $|S_{i_j}| \geq 3$**.** The processing of large minimal separators S_{i_j} is more intricate. For a fixed $T_{S_{i_j}}$ we define a family of shorter encodings with only $|S_{i_j}|^{\mathcal{O}(1)}$ possibilities, that essentially summarizes at "guessing" the central nodes of $T\langle K_i\rangle$ and $T\langle K_{i_j}\rangle$. Assuming a correct guess of these above central nodes, for any partial solution T_{i_j} that is compatible with $T_{S_{i_j}}$, we show how to extract a *constant-number* of distance constraints from $\texttt{encode}(T_{i_j})$, in such a way that all other constraints can be retrieved from those $\mathcal{O}(1)$ that we keep in the short encoding or proved to be irrelevant. Overall, we show that it is sufficient to store only one solution per possible short encoding. For the purpose of illustration, let us focus on the case when $T_{S_{i_j}}$ is a star (the case of bistars is similar, but simpler). We first assume that no minimal separator of G_{i_j} contains S_{i_j}. We may further assume that no minimal separator of G_{i_j} can be contained in S_{i_j} (otherwise, by the second property of Theorem 4 such separators should have size at least 3, whereas since $T_{S_{i_j}}$ is a star they should have size at most 2). In our first subcase, we assume that the center c of the star will *not* end in $\mathcal{C}\left(T_{i_j}\langle K_{i_j}\rangle\right)$. Then, we prove that for every leaf v of the star except maybe one, $dist_{T_{i_j}}(v, V(G_{i_j}) \setminus S_{i_j}) = dist_{T_{i_j}}(c, V(G_{i_j}) \setminus S_{i_j}) + 1$ (two distances to store in the short encoding). Otherwise, $c \in \mathcal{C}\left(T_{i_j}\langle K_{i_j}\rangle\right)$. Our previous formula for $dist_{T_{i_j}}(v, V(G_{i_j}) \setminus S_{i_j})$ stays true unless v is also contained in a minimal separator of G_{i_j}. In this latter case, such a minimal separator must *overlap* S_{i_j}, and so we can prove that we always have $dist_{T_{i_j}}(v, V(G_{i_j}) \setminus S_{i_j}) = 1$. Finally, we assume that a minimal separator of G_{i_j} contains S_{i_j}. We derive from both properties of Theorem 4 that $c \in \mathcal{C}\left(T_{i_j}\langle K_{i_j}\rangle\right)$ and S_{i_j} is T_G-convergent, with $K_{S_{i_j}} = K_{i_j}$. In particular, $S_{i_j} \not\subseteq S_i$, and so, $|S_{i_j} \cap S_i| \leq 2$. Recall that we started by guessing $\mathcal{C}\left(T\langle K_i\rangle\right)$ and $\mathcal{C}\left(T\langle K_{i_j}\rangle\right)$. We include in our short encoding from the previous case the distances $dist_{T_{i_j}}(v, V(G_{i_j}) \setminus S_{i_j})$, $v \in S_{i_j} \cap (S_i \cup \mathcal{C}\left(T\langle K_i\rangle\right))$, plus some *fixed* additional constraints that are derived from the smaller separators contained in S_{i_j}. We stress that this approach could not work with an *arbitrary* T_G. □

7 The Dynamic Programming

Step 4 (Greedy strategy). While we execute Step 3 for its father node $K_{p(i)}$, we compute for K_i a polynomial-size subset of allowed encodings for the 4-Steiner roots of G_i which we want to compute. For all the constraints in such encodings, we are left to decide whether there exists a 4-Steiner root of G_i which satisfies all of them (*i.e.*, we must solve DISTANCE-CONSTRAINED ROOT).

Theorem 8. *For every strongly chordal graph G, let $||G|| := \sum_{K_i \in \mathcal{K}(G)} |K_i|$. Let T_G be a rooted clique-tree as in Theorem 4 and let $K_i \in \mathcal{K}(G)$. There is some polynomial P such that, after a pre-processing in time $\mathcal{O}(n||G_i||^5 P(n))$, we can solve* DISTANCE-CONSTRAINED ROOT *for any input $T_{S_i}, (d_r)_{r \in V(T_{S_i})}$ in time $\mathcal{O}(P(n))$.*

Sketch Proof. If K_i is a leaf of T_G then, we construct the family given by Theorem 6. We keep the trees $T_i \in \mathcal{T}_i$ that satisfy the constraints we have. From now on, let us assume K_i is internal with children $K_{i_1}, K_{i_2}, \ldots, K_{i_p}$. We start by computing $\mathcal{T}_{i_1}, \mathcal{T}_{i_2}, \ldots, \mathcal{T}_{i_p}$ as in Theorem 7. We also need to construct a representation of the family $\mathcal{T}_{K_i}$, as sketched in Sect. 5.3. Roughly, the elements in this representation are of the form $(T_{Y_i}, \mathcal{C}_i)$ where $Y_i \subseteq K_i$ and $\mathcal{C}_i$ must represent the center of $T\langle K_i \rangle$ (missing vertices of $K_i \setminus Y_i$ are supposed to be located in thin branches). This ends the pre-processing step for K_i. In what follows let T_{S_i} and $(d_r)_{r \in T_{S_i}}$ be fixed. In order to solve DISTANCE-CONSTRAINED ROOT, we start by enumerating all pairs $(T_{Y_i}, \mathcal{C}_i)$. Our construction ensures that $S_i \subseteq Y_i$ and so, we can check whether $T_{S_i} \equiv_G T_{Y_i}\langle S_i \rangle$. If this not the case then, we can withdraw this pair and continue. Due to lack of space, we now only sketch the case $Y_i = K_i$ (no thin branch). For every $r \in T_{S_i}$ we check whether we have: $dist_{T_{Y_i}}(r, K_i \setminus S_i) \geq d_r$ (otherwise, we violate our distances' constraints). In the same way, for every $j \in \{1, \ldots, p\}$, we remove from $\mathcal{T}_{i_j}$ any partial solution T_{i_j} s.t. either $T_{i_j}\langle S_{i_j} \rangle \not\equiv T_{Y_i}\langle S_{i_j} \rangle$ or the distances' constraints are violated. We finally explain how to greedily construct a solution (if any), starting from $T_i := T_{Y_i}$. The procedure is divided into a constant number of phases. Every time we complete one of these phases, we select a $T_{i_j} \in \mathcal{T}_{i_j}$, for some j, then we remove from all other $\mathcal{T}_{i_k}$'s the uncompatible partial solutions.

Phase 1: Processing the Cut-Vertices. We consider all the indices j s.t. $S_{i_j} = \{v\}$ is a cut-vertex. There is *one* solution left in $\mathcal{T}_{i_j}$, so we need to add it.

Phase 2: Processing the Edges. We consider all the indices j s.t. $S_{i_j} = \{u, v\}$ and $T_{Y_i}\langle S_{i_j} \rangle$ is an edge. We show that we can almost proceed similarly as for Phase 1 provided we know which among u or v will be closest to $V(G_i) \setminus V(G_{i_j})$ in a final solution. Therefore, computing this information is the main objective of this phase. In general, we will pick a vertex of S_{i_j} which is the closest to $\mathcal{C}_i$, but several cases need to be considered before we can validate such a choice.

Phase 3: Processing the Bistars. We consider all the indices j s.t. $T_{Y_i}\langle S_{i_j} \rangle$ is a bistar. A careful analysis shows that in all cases but one degenerate, we can select any $T_{i_j} \in \mathcal{T}_{i_j}$ s.t. $dist_{T_{i_j}}(\mathcal{C}_i, V(G_{i_j}) \setminus S_{i_j})$ is maximized.

Phase 4: Processing the Stars. We finally consider all the indices j s.t. $T_{Y_i}\langle S_{i_j} \rangle$ is a star. Let $\mathcal{C}\left(T_{Y_i}\langle S_{i_j} \rangle\right) = \{c\}$. Due to lack of space, we only describe the subcase $c \in \mathcal{C}_i$, which is simpler[4]. We first prove that for every unprocessed $S_{i_k} \neq S_{i_j}$, a best possible choice would be to pick a $T_{i_j} \in \mathcal{T}_{i_j}$ s.t. $dist_{T_{i_j}}(\mathcal{C}\left(T_{i_j}\langle K_{i_j} \rangle\right), V(G_{i_j}) \setminus S_{i_j})$ is maximized. However, we also need to account for the other indices k such that $S_{i_k} = S_{i_j}$. For that, let $J = \{j' \mid S_{i_{j'}} = S_{i_j}\}$. The solutions $T_{i_{j'}}, j' \in J$ that we will choose must have diameter four, and the center nodes $v_{j'}$ in $T_{i_{j'}}\langle K_{i_{j'}} \rangle$ must be pairwise different. We do a reduction to MAXIMUM-WEIGHT MATCHING where we create a bipartite graph with respective partite sets J and all possible central nodes. For every $j' \in J$ and $T_{i_{j'}} \in \mathcal{T}_{i_{j'}}$ of diameter four, we add an edge $\{j', \mathcal{C}\left(T_{i_{j'}}\right)\}$ of weight $dist_{T_{i_{j'}}}(\mathcal{C}\left(T_{i_j}\langle K_{i_{j'}} \rangle\right), V(G_{i_{j'}}) \setminus S_{i_{j'}})$. □

[4] All the missing cases, that includes the addition of thin branches to T_{Y_i}, are solved by using the same matching-based approach as in this subcase.

References

1. Arumugam, S., Brandstädt, A., Nishizeki, T., Thulasiraman, K.: Handbook of Graph Theory, Combinatorial Optimization, and Algorithms (2016)
2. Blair, J., Peyton, B.: An introduction to chordal graphs and clique trees. In: George, A., Gilbert, J.R., Liu, J.W.H. (eds.) Graph Theory and Sparse Matrix Computation, pp. 1–29. Springer, New York (1993). https://doi.org/10.1007/978-1-4613-8369-7_1
3. Bondy, J.A., Murty, U.S.R.: Graph Theory (2008)
4. Brandstädt, A., Le, V.: Structure and linear time recognition of 3-leaf powers. Inf. Process. Letters **98**(4), 133–138 (2006)
5. Brandstädt, A., Le, V., Rautenbach, D.: Exact leaf powers. Theor. Comput. Sci. **411**(31–33), 2968–2977 (2010)
6. Brandstädt, A., Le, V., Sritharan, R.: Structure and linear-time recognition of 4-leaf powers. ACM Transact. Algorithms (TALG) **5**(1), 11 (2008)
7. Brandstädt, A., Wagner, P.: Characterising (k, ℓ)-leaf powers. Discrete Appl. Math. **158**(2), 110–122 (2010)
8. Calamoneri, T., Sinaimeri, B.: Pairwise compatibility graphs: a survey. SIAM Rev. **58**(3), 445–460 (2016)
9. Chang, M., Ko, M.: The 3-Steiner root problem. In: WG, pp. 109–120 (2007)
10. Ducoffe, G.: Polynomial-time Recognition of 4-Steiner Powers. Technical Report arXiv:1810.02304, ArXiv (2018)
11. Eppstein, D., Havvaei, E.: Parameterized leaf power recognition via embedding into graph products. In: IPEC 2018, pp. 16:1–16:14 (2019)
12. Fellows, M., Meister, D., Rosamond, F., Sritharan, R., Telle, J.: Leaf powers and their properties: using the trees. In: ISAAC, pp. 402–413 (2008)
13. Jiang, T., Kearney, P., Lin, G.: Phylogenetic k-root and steiner k-root. In: ISAAC, pp. 539–551 (2000)
14. Jordan, C.: Sur les assemblages de lignes. J. Reine Angew. Math **70**(185), 81 (1869)
15. Nevries, R., Rosenke, C.: Characterizing and computing the structure of clique intersections in strongly chordal graphs. Discr. Appl. Math. **181**, 221–234 (2015)
16. Nevries, R., Rosenke, C.: Towards a characterization of leaf powers by clique arrangements. Graphs and Combinatorics **32**(5), 2053–2077 (2016)
17. Nishimura, N., Ragde, P., Thilikos, D.: On graph powers for leaf-labeled trees. J. Algorithms **42**(1), 69–108 (2002)

Hamiltonicity Below Dirac's Condition

Bart M. P. Jansen[1], László Kozma[2(✉)], and Jesper Nederlof[1]

[1] Eindhoven University of Technology, Eindhoven, Netherlands
{b.m.p.jansen,j.nederlof}@tue.nl
[2] Freie Universität Berlin, Berlin, Germany
laszlo.kozma@fu-berlin.de

Abstract. Dirac's theorem (1952) is a classical result of graph theory, stating that an n-vertex graph ($n \geq 3$) is Hamiltonian if every vertex has degree at least $n/2$. Both the value $n/2$ and the requirement for *every vertex* to have high degree are necessary for the theorem to hold.

In this work we give efficient algorithms for determining Hamiltonicity when either of the two conditions are relaxed. More precisely, we show that the Hamiltonian Cycle problem can be solved in time $c^k \cdot n^{O(1)}$, for a fixed constant c, if at least $n - k$ vertices have degree at least $n/2$, or if all vertices have degree at least $n/2 - k$. The running time is, in both cases, asymptotically optimal, under the exponential-time hypothesis (ETH).

The results extend the range of tractability of the Hamiltonian Cycle problem, showing that it is fixed-parameter tractable when parameterized below a natural bound. In addition, for the first parameterization we show that a kernel with $O(k)$ vertices can be found in polynomial time.

Keywords: Hamiltonicity · Fixed-parameter tractability · Kernelization

1 Introduction

The Hamiltonian Cycle problem asks whether a given undirected graph has a cycle that visits each vertex exactly once. It is a central problem of graph theory, operations research, and computer science, with an early history that well predates these fields (see e.g. [29]).

Several conditions that guarantee the existence of a Hamiltonian cycle in a graph are known. Perhaps best known among these is Dirac's theorem from 1952 [15]. It states that a graph with n vertices ($n \geq 3$) is Hamiltonian if every

A full version of the paper is available on arXiv [26].

B.M.P. Jansen—Supported by NWO Gravitation grant "Networks".
L. Kozma—Supported by ERC Consolidator Grant No 617951.
J. Nederlof—Supported by NWO Gravitation grant "Networks" and NWO Grant No 639.021.438.

I. Sau and D. M. Thilikos (Eds.): WG 2019, LNCS 11789, pp. 27–39, 2019.
https://doi.org/10.1007/978-3-030-30786-8_3

vertex has degree at least $n/2$. Various extensions and refinements of Dirac's theorem have been obtained, often involving further graph parameters besides minimum degree (see e.g. the book chapters [14, § 10], [31, § 11] and survey articles [18,30,32] for an overview). We remark that a polynomial-time verifiable condition for Hamiltonicity cannot be both necessary and sufficient, unless $\mathsf{P} = \mathsf{NP}$ [27]. In its stated form, Dirac's theorem is as strong as possible. In particular, if we replace $n/2$ by $\lfloor n/2 \rfloor$, the graph may fail to be two-connected—a precondition for Hamiltonicity. (Consider two $\lceil n/2 \rceil$-cliques with a common vertex).

In this paper we relax the conditions of Dirac's theorem and consider input graphs in which (1) at least $n - k$ vertices have degree at least $n/2$ (the degrees of the remaining vertices can be arbitrarily small), or (2) all vertices have degree at least $n/2 - k$.

For both relaxations we show that Hamiltonian Cycle can be solved deterministically, in time $c^k \cdot n^{O(1)}$, for some fixed constant c. This establishes the fixed-parameter tractability of Hamiltonian Cycle when parameterized by the distance from Dirac's bound, for two natural ways of measuring this distance.

The known exact algorithms for Hamiltonian Cycle in general graphs have exponential running time (the problem is one of the original 21 NP-hard problems [27]). The best deterministic running time of $O(2^n \cdot n^2)$ is achieved by the dynamic programming algorithm of Bellman [4], and Held and Karp [24], and has not been improved since the 1960s. Among randomized algorithms, the current-best running time of $O(1.657^n)$ is achieved by the more recent algorithm of Björklund [6] based on determinants. Improving these bounds remains a central open question of the field.

Assuming the exponential-time hypothesis (ETH) [25], there is no algorithm for Hamiltonian Cycle with running time $2^{o(n)}$. In both parameterizations considered in this paper, $k \leq n$ holds. Thus, under ETH, a running time of the form $2^{o(k)} \cdot n^{O(1)}$ is ruled out, and our algorithms are optimal, up to the base of the exponential. Furthermore, there exists a fixed constant $\alpha > 0$, such that our parameterized bounds asymptotically improve the current-best bounds for Hamiltonian Cycle, if the value of k is at most $\alpha \cdot n$.

For the first parameterization, we show that Hamiltonian Cycle admits a kernel with $O(k)$ vertices, computable in polynomial time. In other words, the input graph can be compressed (roughly) to the order of its sparse part, while preserving Hamiltonicity.

Our results show that checking Hamiltonicity becomes tractable as we approach the degree-bound of Dirac's theorem. The crude intuition behind Dirac's theorem (and many of its generalizations) is that *having many edges* makes a graph Hamiltonian. It is a priori far less obvious why approaching the Dirac bound would make the *algorithmic problem* easier; one may even expect that the more edges there are, the harder it becomes to certify *non-Hamiltonicity*. To provide some intuition why this is not the case, we give a brief informal summary of the arguments.

When $n - k$ vertices have degree at least $n/2$, i.e. in the first case, our algorithm takes advantage of the fact that, by a result of Bondy and Chvátal, the subgraph induced by the high-degree vertices can be completed to a clique without changing the Hamiltonicity of the graph; all relevant structure is thus in the sparse part and its interconnection with the dense part. Then, we find a subset of the vertices in the clique that are well-connected to the sparse part (by solving a matching problem in an auxiliary graph), and we ignore the remainder of the clique. Finally, we show how a Hamiltonian cycle on this smaller, well-connected subgraph, can be extended to a Hamiltonian cycle of the entire graph, guided by the alternating paths of the matching. For this parameterization we are not aware of a comparable result in the literature.

When all vertices have degree at least $n/2-k$, i.e. in the second case, a result of Nash-Williams implies that either a Hamiltonian cycle, or a sufficiently large independent set can be found in polynomial time. In the latter case, we certify non-Hamiltonicity by showing (roughly) that the complement of the independent set is not coverable by a certain number of disjoint paths. This argument is essentially the same as the one given by Häggkvist [23] towards his algorithm with running time $O(n^{5k})$ for the same parameterization. (Häggkvist states this algorithmic result as a corollary of structural theorems. He does not describe the details of the algorithm or its analysis, but these are not hard to reconstruct.) Here we improve the running time of Häggkvist's algorithm to the stated (asymptotically optimal) $c^k \cdot n^{O(1)}$ by more efficiently solving the arising path cover subproblem. The case $k = 1$ of this parameterization was also considered by Büyükçolak et al. [9].

Statement of Results. Our first result shows that if a graph has a "relaxed" Dirac property, it can be compressed while preserving its Hamiltonicity.

Theorem 1. *Let G be an n-vertex graph such that at least $n - k$ vertices of G have degree at least $n/2$. There is a deterministic algorithm that, given G, constructs in time $O(n^3)$ a $3k$-vertex graph G', such that G is Hamiltonian if and only if G' is Hamiltonian.*

Equivalently stated in the language of parameterized complexity, the Hamiltonian cycle problem parameterized by k has a *kernel* with a linear number of vertices. To determine the Hamiltonicity of a graph G, we simply apply the algorithm of Theorem 1 to compress G, and use an exponential-time algorithm (for instance, the Held-Karp algorithm) to solve Hamiltonian Cycle directly on the compressed graph. We thus obtain the following result.

Corollary 1. *If at least $n - k$ vertices of an n-vertex graph G have degree at least $n/2$, then* Hamiltonian Cycle *with input G can be solved in deterministic time $O(8^k \cdot k^2 + n^3)$.*

As an alternative, we may also use an approach based on inclusion-exclusion [28] to solve the reduced Hamiltonian Cycle instance, achieving the overall running time $O(8^k \cdot k^3 + n^3)$, with *polynomial space.*

Our result for the second relaxation of Dirac's theorem is as follows.

Theorem 2. *If every vertex of an n-vertex graph G has degree at least $n/2 - k$, then* Hamiltonian Cycle *with input G can be solved in deterministic time $O(2^{34k} \cdot n^3)$.*

The running time of the Bellman-Held-Karp algorithm for Hamiltonian Cycle is $O(2^n \cdot n^2)$. Denoting $\alpha = k/n$, our results represent an asymptotic improvement if $\alpha < 1/3$ in the first parameterization, and if $\alpha < 0.0294$ in the second parameterization.

As a counterpoint to our results, we observe that Hamiltonian Cycle remains hard (in both parameterizations) for arbitrarily small values of α.

Theorem 3. *Assuming ETH,* Hamiltonian Cycle *cannot be solved in time $2^{o(n)}$ in n-vertex graphs with at least $(1 - \alpha) \cdot n$ vertices of degree at least $n/2$, and in n-vertex graphs with minimum degree $(1 - \alpha) \cdot n/2$, for arbitrary fixed $0 < \alpha < 1/2$.*

Proof. In both cases we construct a graph with the given degree-requirements that embeds a hard instance of Hamiltonian Path with $\alpha \cdot n$ vertices. For the second statement we can use the construction from the NP-hardness proof of Dahlhaus, Hajnal, and Karpinski [13, Theorem 3.1]. For the first statement, consider an $\alpha \cdot n$-vertex instance of Hamiltonian Path, connected by two disjoint edges to an $(1 - \alpha) \cdot n$-vertex clique. □

Related Work. In general, parameterized complexity [12,17] allows a finer-grained understanding of algorithmic problems than classical, univariate complexity. No new insight is gained, however, if the chosen parameter k is large in all interesting cases. For example, in planar graphs, the Four Color Theorem guarantees the existence of an independent set of size $n/4$. As a consequence, any exponential-time algorithm for maximum independent set trivially achieves fixed-parameter tractability in terms of the solution size.

To deal with this issue, Mahajan and Raman [33] introduced the method of parameterizing problems *above* or *below* a guaranteed bound. (Similar considerations motivate the "distance from triviality" framework of Guo, Hüffner, and Niedermeier [19].) In the example of planar independent set, an interesting parameter is the amount by which the solution size exceeds $n/4$. Similar ideas have successfully been applied to several problems (see e.g. [2,5,11,20,21,34]). Our results also fall in the framework of "above/below" parameterization, with the remark that our parameter of interest is not the value to be optimized but a structural property of the input, which we parameterize near its "critical value".

Perhaps closest to our work is the recent result of Gutin and Patel [22] on the Traveling Salesman problem, parameterized below the cost of the *average* tour. Although it concerns Hamiltonian cycles (in an edge-weighted complete graph), the result of Gutin and Patel is not directly comparable with our results. In particular, averaging arguments do not seem to help when studying the *existence* of Hamiltonian cycles, which is often determined by local structure in the graph. For instance, Hamiltonian Cycle remains NP-hard even in graphs with average degree αn for any constant $\alpha < 1$. (Consider a clique of $\sqrt{\alpha}n$ vertices, connected by two disjoint edges to the remaining graph that encodes a hard instance of HAMILTONIAN PATH on $(1 - \sqrt{\alpha})n$ vertices).

2 Preliminaries

We use standard graph-theoretic notation (see e.g. [14]). An edge between vertices u and v is written simply as uv or vu. The *neighborhood* of a vertex v in graph G is denoted by $N_G(v)$. The *degree* of v in G is $d_G(v) = |N_G(v)|$, and the minimum degree of G is $\delta_G = \min_{v \in V(G)} d_G(v)$. We conveniently omit the subscript G whenever possible. For a set $S \subseteq V(G)$ of vertices, $G[S]$ denotes the subgraph induced by S on G.

We state Dirac's theorem and a strengthened statement due to Ore. Let G be an n-vertex undirected graph, with $n \geq 3$.

Lemma 1 (Dirac [15]). *If $\delta \geq n/2$, then G is Hamiltonian.*

Lemma 2 (Ore [36]). *If $d_G(u) + d_G(v) \geq n$ for every non-adjacent pair of vertices u, v of G, then G is Hamiltonian.*

We state a theorem of Bondy and Chvátal that we use in the proofs of both Theorems 1 and 2.

Lemma 3 (Bondy-Chvátal [7]). *Let G be an n-vertex graph, and let G' be obtained from G by adding an edge uv to G for some pair of non-adjacent vertices u, v such that $d_G(u) + d_G(v) \geq n$. Then G' is Hamiltonian if and only if G is Hamiltonian. Moreover, given a Hamiltonian cycle of G', a Hamiltonian cycle of G can be obtained in linear time.*

Lemma 3 implies both Lemmas 1 and 2, as in both cases we can iterate the edge-augmentation step until obtaining a complete graph.

Finally, we state yet another strengthening of Dirac's theorem, due to Nash-Williams [35]. We write this result in a slightly non-standard, explicitly algorithmic form. Our use of this result in proving Theorem 2 is the same as in the argument of Häggkvist [23].

Lemma 4 (Nash-Williams [35]). *Let G be a 2-connected graph with n vertices, with $\delta \geq (n+2)/3$. Then, we can find in G, in time $O(n^3)$, either a Hamiltonian cycle, or an independent set of size $\delta + 1$.*

A simpler proof of Lemma 4 was given by Bondy [8], sketched in [31, § 11]. To make our discussion self-contained, we spell out an explicitly algorithmic form of this proof in the full version [26] of the paper (this requires minor changes with respect to [8]).

3 Relaxing the Cardinality-Constraint (Theorem 1)

Let $C \subseteq V(G)$ denote the set of high-degree vertices of G (those with degree at least $n/2$), and let $S = V(G) \setminus C$ denote the remaining (i.e. low-degree) vertices.

Observe that $|S| \leq k$. By Lemma 3, we may add all edges between vertices in C, without changing the Hamiltonicity of G. Assume therefore that C is a clique.

The proof of the following theorem is inspired by the *crown reductions* [1,10, 16] used to obtain kernels for VERTEX COVER and SAVING k COLORS.

Theorem 4. *There is a polynomial-time algorithm that, given a graph G and a nonempty set $S \subseteq V(G)$ such that $G - S$ is a clique, outputs an induced subgraph G' of G on at most $3|S|$ vertices such that G is Hamiltonian if and only if G' is Hamiltonian.*

Proof. Given a graph G let $S \subseteq V(G)$, such that $C := V(G) \setminus S$ is the vertex set of a clique in G. If $|C| \leq 2|S|$ then $G' := G$ suffices, so we assume $|C| > 2|S|$ in the remainder. Let $S' := \{v_1, v_2 \mid v \in S\}$ be a set containing two representatives for each vertex of S. Construct a bipartite graph H on vertex set $C \cup S'$. For each edge $cv \in E(G)$ with $c \in C$ and $v \in S$, add the edges cv_1, cv_2 to H. Compute a maximum matching $M \subseteq E(H)$ in graph H, for example using the Edmonds-Karp algorithm. Let C^* be the vertices of C saturated (matched) by M. If $|C^*| \geq |S| + 1$ then let $C' := C^*$, and otherwise let $C' \subseteq C$ be a superset of C^* of size $|S| + 1$. Output the graph $G' := G[C' \cup S]$ as the result of the reduction.

Claim 1. *Graph G' has at most $3|S|$ vertices.*

Proof. Since each vertex of C^* is matched to a distinct vertex in S', with $|S'| = 2|S|$, it follows that $|C^*| \leq 2|S|$ which implies $|C'| \leq 2|S|$. As $V(G') = C' \cup S$, the claim follows. ⌟

The output graph G' therefore satisfies the size bound. It remains to prove that it is equivalent to G with respect to Hamiltonicity. We first prove the simpler implication.

Claim 2. *If G' is Hamiltonian, then G is Hamiltonian.*

Proof. Suppose that G' is Hamiltonian, and let $F \subseteq E(G)$ be a Hamiltonian cycle in G'. Fix an arbitrary orientation of F. As each vertex from C' has a unique successor on F, while $|C'| > |S|$ by definition, it follows that some vertex $x \in C'$ has a successor from C' along the cycle; let this be $y \in C'$. Then we can transform F into a Hamiltonian cycle in G by removing the edge xy and replacing it by a path through all the clique-vertices of $C \setminus C'$. ⌟

The remainder of the proof is aimed at proving the reverse implication. For this, we introduce some terminology. For a vertex set S^* in a graph G^*, we define a *path cover of S^* in G^** as a set of pairwise vertex-disjoint simple paths $P_1, \ldots, P_\ell$ in G^*, such that each vertex of S^* belongs to exactly one path P_i. For a vertex set C^* in G^*, we say the path cover *has C^*-endpoints* if the endpoints of each path P_i belong to C^*. We will sometimes interpret a subgraph in which each connected component is a path as a path cover, in the natural way.

Claim 3. *If there is a path cover of S in G' having C'-endpoints, then G' is Hamiltonian.*

Proof. Any path cover of S consists of at least one path (since S is nonempty by assumption) and the endpoints of the paths are all distinct. Hence a path cover consisting of $\ell \geq 1$ paths has exactly 2ℓ distinct endpoints $\{s_1, t_1, \ldots, s_\ell, t_\ell\}$, which are vertices in the clique C'. Let $P_{\ell+1}$ be a simple path in G' visiting all vertices that are not touched by the path cover; such a path exists because the only vertices not touched by the path cover belong to the clique C'. Then one can obtain a Hamiltonian cycle in G' by taking the edges of $P_1, \ldots, P_\ell, P_{\ell+1}$, together with edges connecting the end of path P_i to the beginning of path P_{i+1} for all relevant values of i. ⌟

To prove that Hamiltonicity of G implies Hamiltonicity of G', we will construct a path cover of S in G' having C'-endpoints, using a hypothetical Hamiltonian cycle in G. To do so we need several properties enforced by the matching M in H, which we now explore.

Let U_C be the vertices of C that are not saturated by M. Let R denote the vertices of H that are reachable from U_C by an M-alternating path in the bipartite graph H (which necessarily starts with a non-matching edge), and define $R_C := R \cap C$ and $R_{S'} := R \cap S'$.

Claim 4. *The sets $R, R_C, R_{S'}$ satisfy the following.*

1. *Each M-alternating path in H from U_C to a vertex in $R_{S'}$ (resp. R_C) ends with a non-matching (resp. matching) edge.*
2. *Each vertex of $R_{S'}$ is matched by M to a vertex in R_C.*
3. *For each vertex $x \in R_C$ we have $N_H(x) \subseteq R_{S'}$.*
4. *For each vertex $v \in S$ we have $v_1 \in R_{S'} \Leftrightarrow v_2 \in R_{S'}$.*
5. *For each vertex $v \in S' \setminus R_{S'}$, we have $N_H(v) \cap R_C = \emptyset$ and each vertex of $N_H(v)$ is saturated by M.*

Proof. (1) An M-alternating path starting in U_C must start with a non-matching edge, since U_C consists of unsaturated vertices, and it starts from the C-partite set of H. Hence such a path moves to the S'-partite set over non-matching edges, and moves back to the C-partite set over matching edges.

(2) If a vertex $x \in R_{S'} \subseteq R$ is not saturated, then the M-alternating path from U_C witnessing $x \in R$ starts and ends with a non-matching edge (by (1)) and is in fact an M-augmenting path. This contradicts that M is a maximum matching. Hence each $x \in R_{S'}$ is matched by M to some vertex y. By (1) the M-alternating path from U_C to x that witnesses $x \in R_{S'}$ ends with a non-matching edge, so together with the matching edge xy this forms an M-alternating path witnessing $y \in R_C$.

(3) Consider a vertex $x \in R_C$ and an M-alternating path P from U_C witnessing $x \in R$. By (1) the last edge on P (if any) is a matching edge. Hence if x is saturated by M, then its matching partner y is the predecessor of x on P and a prefix of P witnesses $y \in R$ and hence $y \in R_{S'}$. For any vertex $z \in N_H(x)$ that is not the matching partner of x, we can augment P by the edge xz to obtain an M-alternating path from U_C to z witnessing $z \in R_{S'}$. Together, these two arguments show $N_H(x) \subseteq R_{S'}$.

(4) Suppose $v_1 \in R_{S'}$ and let P be an M-alternating path from U_C to v_1. By (1) path P ends with a non-matching edge xv_1. If $xv_2 \in M$, then v_2 is the predecessor of x on P, and therefore $v_2 \in R_{S'}$. Otherwise, since v_1 and v_2 have identical neighborhoods in H, we can replace the last edge of P by xv_2 to obtain an M-alternating path witnessing $v_2 \in R_{S'}$. The case that $v_2 \in R_{S'}$ is symmetric.

(5) Consider $v \in S' \setminus R_{S'}$. If there is a vertex $x \in N_H(v) \cap R_C$, then (3) implies $v \in R_{S'}$, a contradiction. Hence $N_H(v) \cap R_C = \emptyset$. An unsaturated H-neighbor x of v would imply $x \in N_H(v) \cap U_C \subseteq N_H(v) \cap R_C$, so each vertex of $N_H(v)$ is saturated by M. ⌟

Using these structural insights we can now prove the desired converse to Claim 2. Before we give the formal proof, we present the main idea. To prove that G' is Hamiltonian if G is, we take a Hamiltonian cycle F in G and turn it into a path cover of S in G' with C'-endpoints. Any Hamiltonian cycle F in G yields a path cover of S with S-endpoints, by simply taking the restriction of F onto the vertices of S. The challenge is to extend this path cover with edges into C' to give it the desired C'-endpoints: if the Hamiltonian cycle F used an edge to jump from S to C, we have to provide a similar jump in G'. If F jumps from a vertex $v \in S$ whose corresponding copies $v_1, v_2 \in S'$ do not belong to $R_{S'}$, then by (5) the C-endpoint of the jumping edge is saturated by M, belongs to C' and therefore to G', and can be used to provide the analogous jump in G'. On the other hand, for all vertices $v \in S$ whose copies v_1, v_2 belong to $R_{S'}$, we will globally assign new jumping edges based on the matching H. The properties of a matching will ensure that these jumping edges lead to distinct targets and give a valid path cover of S in G' having C'-endpoints. We now formalize these ideas.

Claim 5. *If G is Hamiltonian, then G' is Hamiltonian.*

Proof. Let F be a Hamiltonian cycle in G. By Claim 3 it suffices to build a path cover of S in G' with C'-endpoints. View F as a 2-regular subgraph of G, and let $F_1 := F[S]$ be the subgraph of F induced by S. Since F spans G and all vertices of S are present in G', it follows that F_1 is a path cover of S in G'. However, the paths in F_1 have their endpoints in S rather than in C'. We resolve this issue by inserting edges into F_1 to turn it into an acyclic subgraph F_2 of G' in which each vertex of S has degree exactly two. This structure F_2 must be a path cover of S in G' with C'-endpoints, since the degree-two vertices S cannot be endpoints of the paths. To do the augmentation, initialize F_2 as a copy of F_1. Define $R_S := \{v \in S \mid v_1 \in R_{S'} \vee v_2 \in R_{S'}\}$ and proceed as follows.

- For each vertex $v \in R_S$, we have $v_1, v_2 \in R_{S'}$ by Claim 4(4), which implies by Claim 4(2) that both v_1 and v_2 are matched to distinct vertices x_1, x_2 in R_C. If v has degree zero in subgraph F_1, then add the edges vx_1, vx_2 to F_2. If v has degree one in F_2 then only add the edge vx_1. Do not add any edges if v already has degree two in F_1.
- For each vertex $v \in S \setminus R_S$, we claim that $N_G(v) \cap R_C = \emptyset$. This follows from the fact that $N_G(v) = N_H(v_1) = N_H(v_2)$ and Claim 4(5), using that $v \notin R_S$ implies $v_1, v_2 \notin R_{S'}$. Hence the (up to two) neighbors that $v \in S \setminus R_S$ has in C on the Hamiltonian cycle F do not belong to R_C, while Claim 4(5) ensures that all vertices of $N_G(v)$ are saturated by H and hence belong to C'. For each vertex $v \in S \setminus R_S$, for each edge from v to $C \cap C'$ incident on v in F, we insert the corresponding edge into F_2.

It is clear that the above procedure produces a subgraph F_2 in which all vertices of S have degree exactly two. To see that F_2 is indeed a path cover, having no vertex of degree larger than two, it suffices to notice that the edges inserted for $v \in R_S$ connect to *distinct* vertices in $C' \cap R_C$, while the edges inserted for $v \in S \setminus R_S$ connect to $C' \setminus R_C$ in the same way as in the Hamiltonian cycle F. Hence F_2 forms a path cover of S in G' having C'-endpoints, which implies that G' is Hamiltonian and proves Claim 5. ⌟

Claims 2 and 5 prove the correctness of the reduction and Claim 1 gives the desired size bound. Since the reduction can easily be performed in polynomial time, this completes the proof of Theorem 4. □

Observe that the proof of Claim 2 explicitly constructs the Hamiltonian cycle in case of a "yes"-answer. The running time of the reduction is dominated by the bipartite matching step, and the process of undoing the Bondy-Chvátal augmentations (Lemma 3), if a cycle of the original graph is to be constructed. Both tasks can be performed in time $O(n^3)$.

4 Relaxing the Degree-Constraint (Theorem 2)

The outline of the proof largely follows an earlier argument of Häggkvist [23]. We improve the $O(n^{5k})$ running time of Häggkvist's algorithm to $c^k \cdot n^{O(1)}$.

The algorithm either finds a Hamiltonian cycle or constructs a certificate of non-Hamiltonicity, in the form of a cut (S, T) of the graph, such that the vertices of T can not be covered by $|S|$ vertex-disjoint paths, and this certificate can be verified within the required running time. (Observe that a Hamiltonian cycle induces such a path cover for an arbitrary cut; paths consisting of single vertices are allowed).

Assume that $k < n/34$, and thus $\delta > 8n/17$. (Otherwise we revert to a standard exponential-time algorithm, and the running time in this case gives the bound stated in Theorem 2.) Furthermore, $\delta < n/2$ may be assumed, as otherwise G is Hamiltonian by Dirac's theorem. Also assume that G is 2-connected (otherwise it is not Hamiltonian).

Start by running the procedure from the proof of Lemma 4, either obtaining a Hamiltonian cycle, or an independent set of size $\delta+1$. Assume that the latter is the case, and label the obtained independent set as A_1.

Partition $V(G)$ into sets A_1, A_2, and A_3, where A_2 denotes the set of vertices in $v \in V(G) \setminus A_1$ such that $|N_G(v) \cap A_1| \geq \delta/2$, and $A_3 = V(G) \setminus (A_1 \cup A_2)$. In words, A_2 contains vertices that are sufficiently highly connected to the obtained independent set, and A_3 contains the remaining vertices.

Lemma 5 ([23], Theorem 2). *Given sets A_1, A_2, A_3 as defined, we can find a set of vertices $S \subseteq V(G)$ such that $|S| \geq 3\delta - n + 2$, and $G[V(G) \setminus S]$ can be covered by $|S|$ vertex-disjoint paths if and only if G is Hamiltonian.*

We sketch the argument in the full version [26] of the paper, referring to Häggkvist [23, p. 32–33] for the full details. Setting $T := V(G) \setminus S$, it remains to verify whether $G[T]$ can be covered by $|S|$ vertex-disjoint paths.

Lemma 6. *Given an n-vertex graph G, we can find in time $O(c^t \cdot n^3)$ a cover of G with $n-t$ vertex-disjoint paths, or report that no such cover exists, for arbitrary $c > (2e)^2$.*

Proof. We apply color-coding [3], [12, § 5.2]. Call a path *nontrivial* if it has more than one vertex. Say that a coloring is *good* for a cover by vertex-disjoint paths, if all vertices that appear in a nontrivial path receive a different color. Clearly, if there is a cover by at most $n-t$ paths then there is a cover by exactly $n-t$ paths, and in such a cover there are at most $2t$ vertices that appear in a nontrivial path. So if there is a path cover with $n-t$ paths, a random coloring with $2t$ colors is *good* for this cover with probability e^{-2t}. (See e.g. [12, Lemma 5.4]).

On a vertex-colored graph with color set $C = \{1, \ldots 2t\}$, we solve the following problem by dynamic programming: for a set $X \subseteq C$ and $v \in V(G)$, let $T[X, v]$ be the smallest number q for which there exists a collection $P_1, \ldots, P_q$ of vertex-disjoint paths in G, such that P_q ends in vertex v and the multiset of colors used in $P_1, \ldots, P_q$ is exactly equal to X. (In particular, this implies that no two vertices in a path may have the same color, for it would appear twice in the multiset and only once in the set X).

Let the $2t$-coloring of G be given by $f\colon V(G) \to [2t]$. Then $T[X, v]$ satisfies the following recurrence:

- $T[\{c\}, v] = 1$ if $f(v) = c$,
- $T[X, v] = +\infty$ if $f(v) \notin X$,
- $T[X,v] = \min\left\{1 + \min_{u \in V(G) \setminus \{v\}} T\big[X \setminus \{f(v)\}, u\big], \min_{u \in N_G(v)} T\big[X \setminus \{f(v)\}, u\big]\right\}$, otherwise.

Intuitively, the interesting part of the recurrence has two cases: either we can let v be a trivial path (so we pay 1 for having a path with v, and then need a collection of paths that can end at any other vertex u that covers the remaining colors), or we take a system of paths covering the remaining colors that ends in a neighbor u of v, and add the edge uv to the end of that path.

Now, observe that for any color-subset X and vertex v, there is a cover of G with $T[X, v] + (n - |X|)$ paths: we cover $|X|$ vertices, one of each color in X, by $T[X, v]$ paths and cover the remaining $(n - |X|)$ vertices by trivial paths. So if we encounter a set X and vertex v for which $T[X, v] + (n - |X|) \leq n - t$, or equivalently, $T[X, v] \leq |X| - t$, then the answer is "yes". On the other hand, if G has a path cover by $n - t$ paths and a coloring is good for this cover, then letting X be the set of colors of vertices that appear in a nontrivial path and v an endpoint of such a path, we obtain $T[X, v] + (n - |X|) \leq n - t$.

So by trying e^{2t} random colorings and solving the dynamic program for each one, we solve the "cover by $n - t$ disjoint paths" problem with constant success probability. With c^{2t} independent runs for $c > e$, we can boost the success probability arbitrarily close to 1. The dynamic program can be solved in time $O(2^{2t}n^3)$. The claimed running time follows.

We may de-randomize the algorithm by replacing the randomized coloring by a deterministic construction, e.g. via *splitters*. We omit the details of this, by now standard, technique [12, § 5.6]. □

In our application of Lemma 6, we need to cover $G[T]$ by $|S|$ vertex-disjoint paths. Observe that $|S| \geq n/2 - 3k$, and consequently $|T| \leq n/2 + 3k$. The difference between the order of the graph $G[T]$ and the number of paths t with which we want to cover it, is therefore at most $6k$.

Applying Lemma 6, the running time of this step is thus $O(c^{6k} \cdot n^3)$, for arbitrary $c > (2e)^2$.

To construct a Hamiltonian cycle, find the set S using Lemmas 4 and 5, find an appropriate path cover using Lemma 6, and recover the Hamiltonian cycle of G by undoing the Bondy-Chvátal steps in Lemma 5. A running time of $O(30^{6k} \cdot n^3)$ follows by adding up the corresponding terms and by using straightforward data structuring. As $30^6 < 2^{34}$, the case $k \geq n/34$ dominates the overall running time.

5 Conclusion

A natural question left open by our work is whether the two parameterizations can be combined, to obtain a generalization of both.

Conjecture 1. If at least $n - k$ vertices of G have degree at least $n/2 - k$, then Hamiltonian Cycle with input G can be solved in time $c^k \cdot n^{O(1)}$ for some constant c.

The results of this paper can be extended with minimal changes to similar parameterizations of Ore's theorem (Lemma 2). Extending the results to generalizations of Dirac's and Ore's theorems to *digraphs* would be interesting. More generally, finding new algorithms by parameterizing structural results of graph theory (whether related to Hamiltonicity or not) is a promising direction.

Acknowledgement. We thank Naomi Nishimura, Ian Goulden, and Wendy Rush for obtaining a copy of Bondy's 1980 research report [8].

References

1. Abu-Khzam, F.N., Fellows, M.R., Langston, M.A., Suters, W.H.: Crown structures for vertex cover kernelization. Theory Comput. Syst. **41**(3), 411–430 (2007). https://doi.org/10.1007/s00224-007-1328-0
2. Alon, N., Gutin, G., Kim, E.J., Szeider, S., Yeo, A.: Solving MAX-r-SAT above a tight lower bound. Algorithmica **61**(3), 638–655 (2011). https://doi.org/10.1007/s00453-010-9428-7
3. Alon, N., Yuster, R., Zwick, U.: Color-coding. J. ACM **42**(4), 844–856 (1995). https://doi.org/10.1145/210332.210337
4. Bellman, R.: Dynamic programming treatment of the travelling salesman problem. J. Assoc. Comput. Mach. **9**, 61–63 (1962)
5. Bezáková, I., Curticapean, R., Dell, H., Fomin, F.V.: Finding detours is fixed-parameter tractable. In: Proceedings of 44th ICALP. pp. 54:1–54:14 (2017). https://doi.org/10.4230/LIPIcs.ICALP.2017.54
6. Björklund, A.: Determinant sums for undirected Hamiltonicity. SIAM J. Comput. **43**(1), 280–299 (2014). https://doi.org/10.1137/110839229
7. Bondy, J.A., Chvátal, V.: A method in graph theory. Discrete Math. **15**(2), 111–135 (1976)
8. Bondy, J.: Longest Paths and Cycles in Graphs of High Degree. Research report, Department of Combinatorics and Optimization, University of Waterloo (1980)
9. Büyükçolak, Y., Gözüpek, D., Özkan, S., Shalom, M.: On one extension of Dirac's theorem on Hamiltonicity. Discrete Appl. Math. **252**, 10–16 (2019). https://doi.org/10.1016/j.dam.2017.01.011
10. Chor, B., Fellows, M., Juedes, D.W.: Linear kernels in linear time, or how to save k colors in $O(n^2)$ steps. In: Proceedings of 30th WG, pp. 257–269 (2004). https://doi.org/10.1007/978-3-540-30559-0_22
11. Crowston, R., Jones, M., Mnich, M.: Max-cut parameterized above the Edwards-Erdős bound. Algorithmica **72**(3), 734–757 (2015). https://doi.org/10.1007/s00453-014-9870-z
12. Cygan, M., et al.: Parameterized Algorithms. Springer, Cham (2015). https://doi.org/10.1007/978-3-319-21275-3
13. Dahlhaus, E., Hajnal, P., Karpinski, M.: On the parallel complexity of Hamiltonian cycle and matching problem on dense graphs. J. Algorithms **15**(3), 367–384 (1993). https://doi.org/10.1006/jagm.1993.1046
14. Diestel, R.: Graph Theory. Graduate Texts in Mathematics, vol. 173, 4th edn. Springer, Heidelberg (2012)
15. Dirac, G.A.: Some theorems on abstract graphs. Proc. London Math. Soc. s3 **2**(1), 69–81 (1952). https://doi.org/10.1112/plms/s3-2.1.69
16. Fellows, M.R.: Blow-Ups, Win/Win's, and crown rules: some new directions in *FPT*. In: Bodlaender, H.L. (ed.) WG 2003. LNCS, vol. 2880, pp. 1–12. Springer, Heidelberg (2003). https://doi.org/10.1007/978-3-540-39890-5_1
17. Flum, J., Grohe, M.: Parameterized Complexity Theory. TTCSAES. Springer, Heidelberg (2006). https://doi.org/10.1007/3-540-29953-X
18. Gould, R.J.: Recent advances on the Hamiltonian problem: survey III. Graphs Comb. **30**(1), 1–46 (2014). https://doi.org/10.1007/s00373-013-1377-x
19. Guo, J., Hüffner, F., Niedermeier, R.: A structural view on parameterizing problems: distance from triviality. In: Proceedings of 1st IWPEC, pp. 162–173 (2004). https://doi.org/10.1007/978-3-540-28639-4_15

20. Gutin, G.Z., Kim, E.J., Lampis, M., Mitsou, V.: Vertex cover problem parameterized above and below tight bounds. Theory Comput. Syst. **48**(2), 402–410 (2011). https://doi.org/10.1007/s00224-010-9262-y
21. Gutin, G.Z., Kim, E.J., Szeider, S., Yeo, A.: A probabilistic approach to problems parameterized above or below tight bounds. J. Comput. Syst. Sci. **77**(2), 422–429 (2011). https://doi.org/10.1016/j.jcss.2010.06.001
22. Gutin, G.Z., Patel, V.: Parameterized traveling salesman problem: beating the average. SIAM J. Discrete Math. **30**(1), 220–238 (2016). https://doi.org/10.1137/140980946
23. Häggkvist, R.: On the structure of non-Hamiltonian graphs I. Comb. Probab. Comput. **1**(1), 27–34 (1992). https://doi.org/10.1017/S0963548300000055
24. Held, M., Karp, R.M.: A dynamic programming approach to sequencing problems. J. Soc. Indust. Appl. Math. **10**, 196–210 (1962). https://doi.org/10.1137/0110015
25. Impagliazzo, R., Paturi, R., Zane, F.: Which problems have strongly exponential complexity? J. Comput. Syst. Sci. **63**(4), 512–530 (2001). https://doi.org/10.1006/jcss.2001.1774
26. Jansen, B.M.P., Kozma, L., Nederlof, J.: Hamiltonicity below Dirac's condition. CoRR abs/1902.01745 (2019). http://arxiv.org/abs/1902.01745
27. Karp, R.M.: Reducibility among combinatorial problems. In: Miller, R.E., Thatcher, J.W., Bohlinger, J.D. (eds.) IRSS, pp. 85–103. Springer, Heidelberg (1972). https://doi.org/10.1007/978-1-4684-2001-2_9
28. Karp, R.M.: Dynamic programming meets the principle of inclusion and exclusion. Oper. Res. Lett. **1**(2), 49–51 (1982). https://doi.org/10.1016/0167-6377(82)90044-X
29. Knuth, D.: The art of computer programming: updates; pre-fascicle 8A, A draft of section 7.2.2.4: Hamiltonian paths and cycles. In: Addison-Wesley Series in Computer Science and Information Proceedings, vol. 4. Addison-Wesley (2018). https://www-cs-faculty.stanford.edu/~knuth/fasc8a.ps.gz
30. Kühn, D., Osthus, D.: Hamilton cycles in graphs and hypergraphs: an extremal perspective. CoRR abs/1402.4268 (2014). http://arxiv.org/abs/1402.4268
31. Lawler, E., Shmoys, D., Kan, A., Lenstra, J.: The Traveling Salesman Problem. Wiley, Hoboken (1985)
32. Li, H.: Generalizations of Dirac's theorem in Hamiltonian graph theory-a survey. Discrete Math. **313**(19), 2034–2053 (2013). https://doi.org/10.1016/j.disc.2012.11.025
33. Mahajan, M., Raman, V.: Parameterizing above guaranteed values: MaxSat and MaxCut. J. Algorithms **31**(2), 335–354 (1999). https://doi.org/10.1006/jagm.1998.0996
34. Mahajan, M., Raman, V., Sikdar, S.: Parameterizing above or below guaranteed values. J. Comput. Syst. Sci. **75**(2), 137–153 (2009). https://doi.org/10.1016/j.jcss.2008.08.004
35. Nash-Williams, C.: Edge-disjoint Hamiltonian circuits in graphs with large valency. In: Mirksy, L. (ed.) Studies in Pure Mathematics, pp. 157–183. Academic Press, London (1971)
36. Ore, O.: Note on Hamilton circuits. Am. Math. Monthly **67**(1), 55 (1960). http://www.jstor.org/stable/2308928

Maximum Independent Sets in Subcubic Graphs: New Results

Ararat Harutyunyan[1], Michael Lampis[1(✉)], Vadim Lozin[2,3], and Jérôme Monnot[1]

[1] Université Paris-Dauphine, Université PSL, CNRS, LAMSADE, Paris, France
michail.lampis@dauphine.fr
[2] University of Warwick, Coventry, UK
[3] University of Nizhny Novgorod, Nizhny Novgorod, Russia

Abstract. We consider the complexity of the classical INDEPENDENT SET problem on classes of subcubic graphs characterized by a finite set of forbidden induced subgraphs. It is well-known that a necessary condition for INDEPENDENT SET to be tractable in such a class (unless P = NP) is that the set of forbidden induced subgraphs includes a subdivided star $S_{k,k,k}$, for some k. Here, $S_{k,k,k}$ is the graph obtained by taking three paths of length k and identifying one of their endpoints.

It is an interesting open question whether this condition is also sufficient: is INDEPENDENT SET tractable on all hereditary classes of subcubic graphs that exclude some $S_{k,k,k}$? A positive answer to this question would provide a complete classification of the complexity of INDEPENDENT SET on all classes of subcubic graphs characterized by a finite set of forbidden induced subgraphs. The best currently known result of this type is tractability for $S_{2,2,2}$-free graphs. In this paper we generalize this result by showing that the problem remains tractable on $S_{2,k,k}$-free graphs, for any fixed k. Along the way, we show that subcubic INDEPENDENT SET is tractable for graphs excluding a type of graph we call an "apple with a long stem", generalizing known results for apple-free graphs.

Keywords: Independent set · Sub-Cubic graphs · Apple-Free graphs

1 Introduction

In a graph, an *independent set* is a subset of vertices no two of which are adjacent. The maximum independent set problem asks to find in a graph G an independent set of maximum size. The size of a maximum independent set in G is called the *independence number* of G and is denoted $\alpha(G)$.

The maximum independent set problem is one of the first problems that were shown to be NP-hard. Moreover, the problem remains NP-hard under substantial restrictions. In particular, it is NP-hard for graphs of vertex degree at most 3, also known as *subcubic* graphs. In terms of vertex degree, this is the strongest

I. Sau and D. M. Thilikos (Eds.): WG 2019, LNCS 11789, pp. 40–52, 2019.
https://doi.org/10.1007/978-3-030-30786-8_4

possible restriction under which the problem remains NP-hard, since for graphs of vertex degree at most 2 the problem is solvable in polynomial time. However, with respect to other parameters the restriction to subcubic graphs is not best possible, as the problem remains NP-hard for subcubic graphs of girth at least k for any fixed value of k [10], where the girth of a graph is the size of a smallest cycle. In other words, the problem is NP-hard for $(C_3, \ldots, C_k)$-free subcubic graphs for each value of k, where C_k is a chordless cycle of length k. The idea behind this conclusion is quite simple: it is not difficult to see that a double subdivision of an edge increases the independence number of the graph by exactly one, and hence, by repeatedly subdividing the edges of a subcubic graph G we destroy all small cycles in G, i.e. we transform G into a graph of large girth.

Let us observe that by means of edge subdivisions we can also destroy small copies of some other graphs, in particular, graphs of the form H_k represented in Fig. 1 (left). Therefore, the maximum independent set problem remains NP-hard for $(C_3, \ldots, C_k, H_1, \ldots, H_k)$-free subcubic graphs for each value of k.

Let us denote by S_k the class of $(C_3, \ldots, C_k, H_1, \ldots, H_k)$-free subcubic graphs and by $\kappa(G)$ the maximum k such that $G \in S_k$. If G belongs to no class S_k, then $\kappa(G)$ is defined to be 0, and if G belongs to all classes S_k, then $\kappa(G)$ is defined to be ∞. Also, for a set of graphs M, $\kappa(M)$ is defined as $\kappa(M) = \sup\{\kappa(G) : G \in M\}$. With this notation, we can derive the following conclusion from the above discussion (see e.g. [7]).

Theorem 1. *Let M be a set of graphs. If $\kappa(M) < \infty$, then the maximum independent set problem is NP-hard in the class of M-free subcubic graphs.*

This theorem suggests that, unless $P = NP$, the maximum independent set problem is solvable in polynomial time in the class of M-free graphs only if the parameter κ is unbounded in the set M. There are three basic ways to unbind this parameter in M:

1. include in M a graph G with $\kappa(G) = \infty$;
2. include in M graphs with arbitrarily large induced cycles;
3. include in M graphs with arbitrarily large induced subgraphs of the form H_k.

To give an example of a polynomial-time result of the first type, let us observe that $\kappa(G) = \infty$ if and only if every connected component of G has the form $S_{i,j,k}$ represented in Fig. 1 (right). We call any graph of the form $S_{i,j,k}$ a *tripod*.

In other words, if the set M of forbidden induced subgraphs is finite, then M must contain a graph for which every component is a tripod for the maximum independent set problem in the class of M-free subcubic graphs to be polynomial-time solvable (assuming P $\neq$ NP). In [5], it was conjectured that this condition is also sufficient. Moreover, for graphs of bounded vertex degree the problem can be easily reduced to connected forbidden induced graphs, in which case the conjecture can be restated as follows.

Conjecture 1. The maximum independent set problem is polynomial-time solvable for G-free subcubic graphs if and only if G is a tripod.

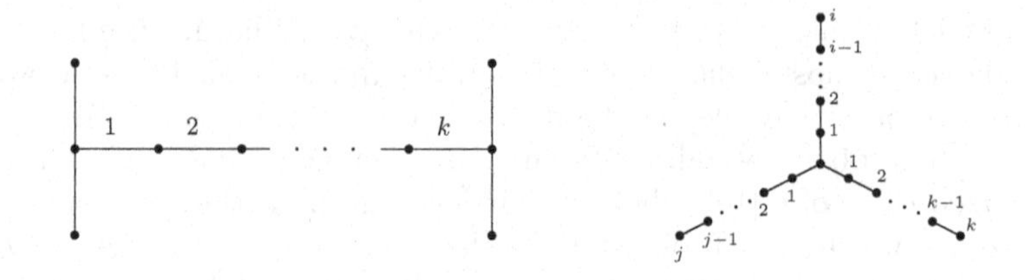

Fig. 1. The graphs H_k (left) and $S_{i,j,k}$ (right)

One of the minimal non-trivial tripods is the claw $S_{1,1,1}$. The problem can be solved for the claw-free graphs in polynomial time even without the restriction to bounded degree graphs [9]. In [6], the result for claw-free graphs was extended to $S_{1,1,2}$-free graphs, also known as fork-free graphs, and again without the restriction to bounded degree graphs. However, any further extension becomes much harder even for bounded degree graphs, and only recently a solution was found for $S_{2,2,2}$-free subcubic graphs [8]. Currently, this is one of the few maximal subclasses of subcubic graphs with polynomial-time solvable independent set problem.

Now we turn to polynomial-time solutions of the second type, i.e. classes of graphs where forbidden induced subgraphs contain arbitrarily large chordless cycles. Clearly, in this case the set of forbidden induced subgraphs must be infinite. A typical example of this type deals with classes of bounded chordality, i.e. classes excluding *all* chordless cycle of length at least k for a constant k. Without a restriction to bounded degree graphs a solution of this type is known only for $k = 4$, i.e. for chordal graphs [4], and is unknown for larger values of k. Together with the restriction to bounded degree graphs bounded chordality implies bounded tree-width [2] and hence polynomial-time solvability of the maximum independent set problem. In other words, the problem can be solved for $(C_k, C_{k+1}, \ldots)$-free graphs of bounded vertex degree for each value of $k \geq 3$.

An *apple* A_k, $k \geq 4$, is a graph formed of a chordless cycle C_k and an additional vertex, called the *stem*, which has exactly one neighbour on the cycle C_k. The class of $(A_4, A_5, \ldots)$-free graphs generalizes both chordal graphs and claw-free graphs, and a solution for the maximum independent set problem in this class was presented in [3]. In case of bounded degree graphs this solution can be extended to graphs without *large* apples, i.e. to $(A_k, A_{k+1}, \ldots)$-free graphs of bounded vertex degree for any fixed value of k [7].

Generalizing both the subcubic graphs without large apples and $S_{2,2,2}$-free subcubic graphs, in the present paper we prove polynomial-time solvability of the maximum independent set problem for subcubic graphs excluding large apples with a long stem. An *apple with a long stem* A_k^* is obtained from an apple A_k by adding one more vertex which is adjacent to the stem of A_k only. We show that for any fixed value of k, the maximum independent set problem in the class of $(A_k^*, A_{k+1}^*, \ldots)$-free subcubic graphs can be solved in polynomial time. Observe that this class contains all $S_{2,p,p}$-free subcubic graphs for any fixed $p < k$ (as long as $k \geq 6$) and hence our result brings us much closer to the proof of Conjecture 1.

2 Preliminaries

All graphs in this paper are simple, i.e. undirected, without loops and multiple edges. The vertex set and the edge set of a graph G is denoted by $V(G)$ and $E(G)$, respectively. The *neighbourhood* $N(v)$ of a vertex $v \in V(G)$ is the set of vertices of G adjacent to v. The *degree* of $v \in V(G)$ is the number of its neighbours, i.e. $|N(v)|$. As usual, P_n and C_n denote a chordless path and a chordless cycle with n vertices, respectively,

A subgraph of G induced by a subset $U \subseteq V(G)$ is denoted $G[U]$. If G contains no induced subgraph isomorphic to a graph H, we say that G is H-free.

Outline of the Proof. To prove polynomial-time solvability of the maximum independent set problem in the class of $(A_k^*, A_{k+1}^*, \ldots)$-free subcubic graphs,

1. We start by checking if the input graph G has an induced copy of $S_{2,2,2}$. If G is $S_{2,2,2}$-free, then the problem can be solved for G in polynomial time [8]. Otherwise, we proceed to checking whether G has an induced cycle of length at least $p = 300k$. This can be done in polynomial time, as shown in Lemma 1 below. If G does not contain induced cycles of length at least p, then the tree-width of G is bounded by a function of k [2] and hence the problem can be solved in polynomial time for G.
2. If G contains an induced copy of $S_{2,2,2}$ and a large induced cycle C, then in the absence of large induced apples with long stems we prove that it must contain a large extended cycle C^*, which is a graph obtained from C by adding two vertices that create a C_6 together with four consecutive vertices of C (see Fig. 7 in Sect. 4). This is shown in Sect. 3. An important ingredient of this proof is the assumption that the input graph G is connected and has no separating cliques, i.e. cliques whose removal disconnects the graph. A polynomial-time reduction of the maximum independent set problem to graphs without separating cliques can be found in [11,12].
3. After the previous two steps we can assume that our graph contains a large extended cycle. In Sect. 4 we show how to destroy such a large extended cycle by means of various local reductions. Each of them transforms G into a smaller graph G' in the same class with a fixed difference $\alpha(G) - \alpha(G')$. The set of reductions is described in Sect. 4.1 and their application to a graph G containing a large extended cycle is described in Sect. 4.2. By destroying the large extended cycle C^*, we destroy either the cycle C or the induced copy of $S_{2,2,2}$ (or both) and return to Step 1 to check if there are other copies of a large induced cycle or an induced $S_{2,2,2}$.

The first step of the proof outline above is rather straight-forward and relies on Lemma 1, stated below. The main difficulties lie in the second step (showing that if the graph has an $S_{2,2,2}$ and a large induced cycle, then it has a long extended cycle), which is handled in Sect. 3; and in the third step (showing how to deal with a large extended cycle), which is handled in Sect. 4. Due to space constraints, some proofs have been moved to the appendix.

Lemma 1. *For each p there is an algorithm running in time $n^{O(p)}$ which decides if a given n-vertex graph contains an induced cycle of length at least p.*

3 From Large Cycles to Extended Large Cycles

We recall that C^* denotes an *extended cycle*, i.e. the graph obtained from a cycle C by adding two vertices that create a C_6 together with four consecutive vertices of C (see Fig. 7 in Sect. 4). Also, A_p^* denotes an *apple with a long stem*, where p stands for the size of the cycle in the apple. An apple with a long stem consisting of a cycle C and two stem vertices x, y will be denoted $C_{x,y}$.

The main goal of this section is to show that if G contains a large induced cycle and an induced copy of $S_{2,2,2}$, then it contains either a large induced extended cycle or a large induced apple with a long stem. This will be shown in two steps in Lemmas 2 and 3. Since we are dealing with graphs which do not contain large induced apples with long stems, the result of this section is that we may assume that our graph contains a large induced extended cycle. We note that throughout this section we will assume that our graph does not contain any separating cliques; in case it does, it is known how to reduce solving INDEPENDENT SET to smaller graphs that do not contain such cliques [11,12].

Lemma 2. *Let G be a subcubic graph without separating cliques. If G has an induced cycle C of length p and an induced copy of $S_{2,2,2}$, then G has an induced cycle of length at least $p/12$ containing the center of an induced $S_{2,2,2}$.*

Lemma 3. *Let G be a subcubic graph without separating cliques. If G has an induced cycle C of length p containing the center of an induced $S_{2,2,2}$, then G has an induced extended cycle C_t^* or an induced apple with a long stem A_t^* with $t \geq p/8$.*

4 Destroying Large Extended Cycles

According to the previous section, if an $(A_k^*, A_{k+1}^*, \ldots)$-free subcubic graph G contains a large induced cycle and an induced copy of $S_{2,2,2}$, then it must contain a large extended cycle C^*. The goal of the present section is to show how to destroy large extended cycles by means of various local graph reductions. We describe these reductions in Sect. 4.1 and apply them to large extended cycles in Sect. 4.2.

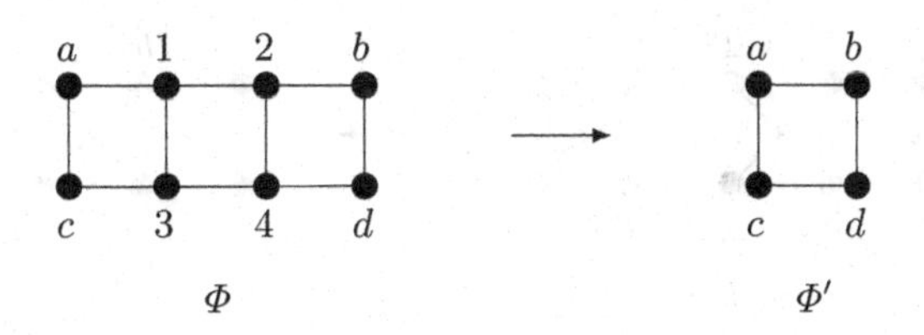

Fig. 2. Φ-reduction

4.1 Graph Reductions

Φ-Reduction and *House*-Reduction. We start with the Φ-reduction introduced in [8]. It applies to a graph G containing an induced copy of the graph Φ represented on the left of Fig. 2 and consists in replacing Φ by the graph on the right of Fig. 2.

Lemma 4. *By applying the Φ-reduction to an $(A^*_k, A^*_{k+1}, \ldots)$-free subcubic graph G, we obtain an $(A^*_k, A^*_{k+1}, \ldots)$-free subcubic graph G' with $\alpha(G') = \alpha(G) - 2$.*

A *house* is the complement of a P_5. If a graph G contains an induced *house*, the *house*-reduction consists in removing from G the vertices that form a triangle in the *house*. It was shown in [8] that if G is a subcubic graph, then the *house*-reduction reduces $\alpha(G)$ by exactly 1.

Π-Reduction. Now we introduce the Π-reduction illustrated in Fig. 3. In a graph G, an induced Π is the graph represented on the left of Fig. 3. We observe that vertex f can be missing, in which case vertices a and c have no other neighbours in G. However, if f exists, that is, if one of a, c has a neighbour outside of $\{1, 3, e\}$, then f is a common neighbour of a, c. Similarly, vertex h can be missing, in which case vertices b and d have no other neighbours in G.

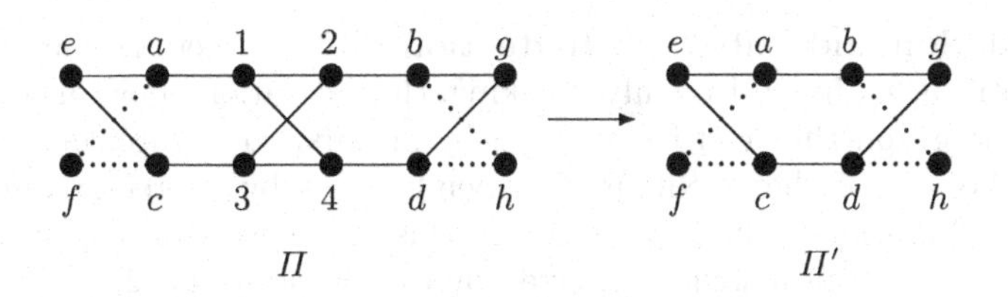

Fig. 3. Π-reduction

Lemma 5. *By applying the Π-reduction to an $(A^*_k, A^*_{k+1}, \ldots)$-free subcubic graph G, we obtain an $(A^*_k, A^*_{k+1}, \ldots)$-free subcubic graph G' with $\alpha(G') = \alpha(G) - 2$.*

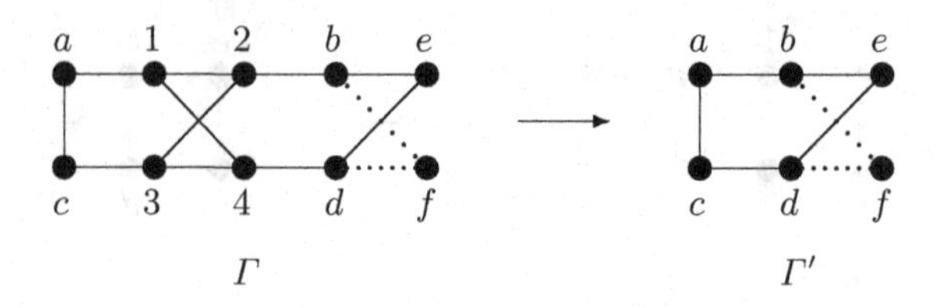

Fig. 4. Γ-reduction

Γ-Reduction. One more reduction is illustrated in Fig. 4. We will refer to it as Γ-reduction. Again, vertex f can be missing, in which case vertices b and d have degree 2 in the graph, but if f exists it is a common neighbour of b, d.

Lemma 6. *By applying the Γ-reduction to an $(A_k^*, A_{k+1}^*, \ldots)$-free subcubic graph G, we obtain an $(A_k^*, A_{k+1}^*, \ldots)$-free subcubic graph G' with $\alpha(G') = \alpha(G) - 2$.*

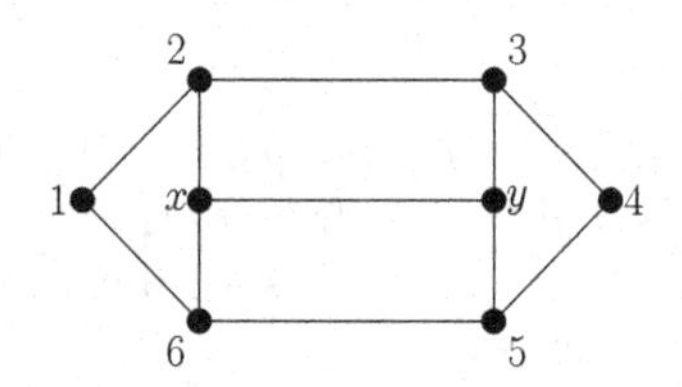

Fig. 5. Θ graph

Θ-Reduction

Lemma 7. *If a subcubic graph G contains an induced Θ (see Fig. 5), then the deletion of vertices x, y reduces the independence number of G by exactly 1.*

Total Struction and Subgraph Reduction. Total struction is an operation that was introduced in [1]. Roughly speaking, this operation allows us to identify a part of the graph that can be replaced by an auxiliary graph in a way that decreases the size of the maximum independent set by a precise value. Even though this operation is quite powerful, in this paper we will only need to use two special cases of total struction, given by Corollaries 1 and 2.

Corollary 1. *For any graph $G = (V, E)$ and $H \subseteq V$ let $N[H]$ denote the set of vertices at distance at most 1 from H. Then, we have the following: if $\alpha(G[H]) = \alpha(G[N[H]])$, then $\alpha(G[V \setminus N[H]]) = \alpha(G) - \alpha(G[H])$.*

Informally, Corollary 1 gives rise to the following transformation: if we can find a set of vertices H such that $G[H]$ and $G[N[H]]$ have the same maximum independent set, then we simply select an independent set of H in our solution

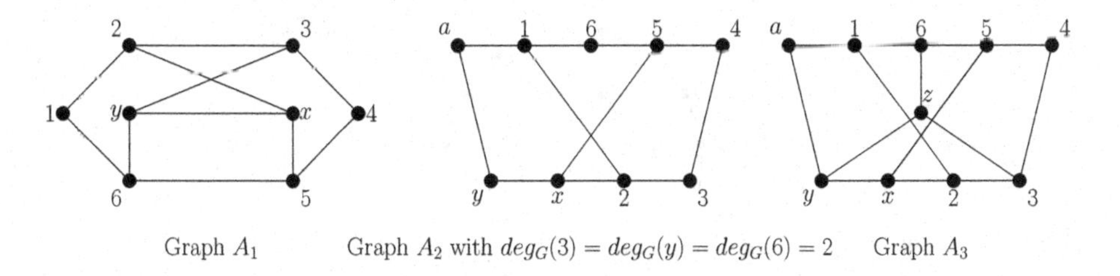

Fig. 6. Graphs A_1, A_2 and A_3

and delete all vertices of $N[H]$. The deletion of $N[H]$ in the case when $\alpha(G[H]) = \alpha(G[N[H]])$ was called in [8] the H-subgraph reduction.

It is not difficult to check that if A_1, A_2, or A_3 (see Fig. 6) is an induced subgraph of a subcubic graph, then we can use Corollary 1 as we have:

- $\alpha(A_1[\{2, 3, 5, 6, x, y\}]) = \alpha(A_1) = 3$,
- $\alpha(A_2[\{1, 2, 3, 5, 6, x, y\}]) = \alpha(A_2) = 4$,
- $\alpha(A_3[\{1, 2, 3, 5, 6, x, y\}]) = \alpha(A_3) = 4$.

Lemma 8. *If A_1, A_2, or A_3 is an induced subgraph of a subcubic graph G, then $\alpha(G - A_1) = \alpha(G) - 3$, $\alpha(G - A_2) = \alpha(G) - 4$, $\alpha(G - A_3) = \alpha(G) - 4$.*

Corollary 2. *Let $G = (V, E)$ be a subcubic graph and $K \subseteq V$ such that $G[K]$ induces a $K_{2,3}$. Then, if G' is the graph obtained from G by deleting the vertices of K and introducing a new vertex z connected to $N(K)$, we have (i) $\alpha(G') = \alpha(G) - 2$ and (ii) if G' contains an apple with a long stem A^*_p, then G also contains an apple with a long stem $A^*_{p'}$, with $p' \geq p$.*

4.2 Applying Graph Reductions to Large Extended Cycles

Let G be an $(A^*_k, A^*_{k+1}, \ldots)$-free subcubic graph. For ease of terminology and notation we will refer to any A^*_t with $t \geq k$ simply as a large apple with a long stem. According to Sect. 3, we may assume that G contains a large extended cycle C^*_p, i.e. a graph that consists of an induced cycle of length p, plus two extra vertices which form a C_6 together with four consecutive vertices of the cycle and have no other neighbours in C^*_p. We denote the vertices of an extended cycle as shown in Fig. 7, where we have given labels to the vertices of the C_6, plus some other interesting vertices. In the remainder we use simply C^* to denote the extended cycle and C_6 to denote the set of vertices $\{1, 2, 3, 4, 5, 6\}$. Without loss of generality, we assume that $p \geq 3k$.

We will now go through a sequence of cases that covers all possible ways in which C^* may be connected to the rest of the graph.

Case 0: Vertices 2 and 3 both have degree 2 in G. In this case we delete 2, 3 from the graph and add the edge connecting 1 to 4. This decreases $\alpha(G)$ by exactly 1. Also, it is not difficult to check that this transformation does not create any new forbidden induced subgraphs.

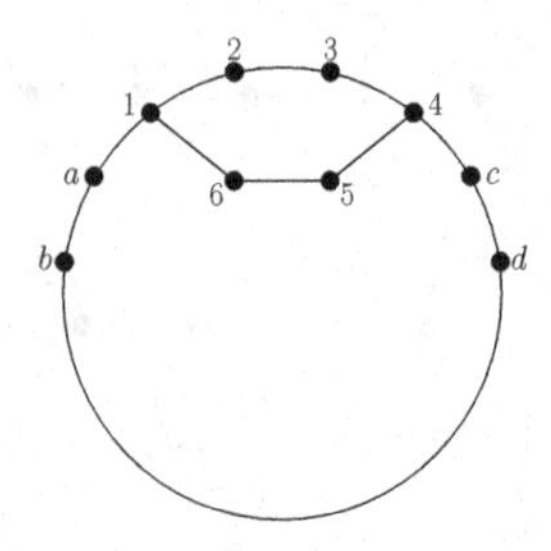

Fig. 7. An extended cycle

Because of the above we can assume that the set $\{2, 3\}$ has a neighbour outside of C^*. We call this vertex x. Without loss of generality we assume that x is connected to 2. Let us consider how x is connected to the rest of C^*. The rest of the cases are defined as follows.

- *Case 1.1*: $N(x) \cap C^* = \{2\}$
- *Case 1.2*: $N(x) \cap C_6 = \{2\}$ and x has exactly one neighbour in $C^* \setminus C_6$
- *Case 1.3*: $N(x) \cap C_6 = \{2\}$ and x has two neighbours in $C^* \setminus C_6$

If we rule out the above cases we conclude that x has at least two neighbours in C_6. Since the degrees of $1, 4$ are already three in C^*, we conclude that x has at least two neighbours in $\{2, 3, 5, 6\}$. Let us also rule out two further cases.

- *Case 1.4*: $N(x) \cap C_6 = \{2, 3\}$;
- *Case 1.5*: $|N(x) \cap C_6| = 3$

Lemma 9. *If one of Cases 1.1–1.5 applies, then the instance can be simplified in polynomial time. If none of Cases 1.1–1.5 applies, then either $N(x) \cap C_6 = \{2, 5\}$ or $N(x) \cap C_6 = \{2, 6\}$.*

Thus, we may suppose: $N(x) \cap C_6 = \{2, 5\}$ or $N(x) \cap C_6 = \{2, 6\}$. We handle these two cases separately in the following subsections.

x is Adjacent to 2 and 6.

Lemma 10. *Let x be a vertex adjacent to 2 and 6 and assume x has a neighbour y not in C^*. Then G contains an induced Φ or an induced Π or an induced Γ or an induced Θ.*

Proof. If y is adjacent to 3, then by Lemma 9 (and symmetry) y is also adjacent to 5 and hence vertices $1, 2, 3, 4, 5, 6, x, y$ induce a Θ.

If y is adjacent to c, then vertices $2, 3, 4, x, y, c$ create a cycle of length 6 which, together with the path $1ab \ldots d$ gives a second large extended cycle. Therefore, by Lemma 9 applied to this extended cycle, vertex 5 must be adjacent to y and hence vertices $1, 2, x, 6, y, 5, c, 4$ induce a Φ.

If y is adjacent to a, then vertices $a, y, 1, 2, x, 6, 3, 4, 5$ induce a Γ with a possible missing common neighbour of 3 and 5 (any neighbour of these vertices must be common by Lemma 9).

If y is adjacent to b and not adjacent to a, then vertices $a, b, y, 1, 2, x, 6, 3, 4, 5$ induce a Π with a possible missing common neighbour of 3 and 5 (any neighbour of these vertices must be common by Lemma 9).

From now on, we assume y has no neighbours in $\{3, 5, a, b, c\}$. If y has neighbours on $C^* \setminus C_6$, then we can distinguish at most 3 cycles containing y as shown in Fig. 8 (if y has only 1 neighbour on $C^* \setminus C_6$, the cycle C^2 is missing).

We observe that at least one of the cycles C^1, C^2, C^3 is large, i.e. has length at least $p/3$. Then G contains a large apple with a long stem

- $C^* \cup \{x, y\} \setminus \{5, 6\}$ if y has no neighbours on $C^* \setminus C_6$,
- $C^1 \cup \{3, 4\}$ if C^1 is large,
- $C^2 \cup \{x, 2\}$ if C^2 is large,
- $C^3 \cup \{1, a\}$ if C^3 is large.

A contradiction in all cases shows that y has a neighbour in $\{3, 5, a, b, c\}$ and hence G contains an induced Φ or an induced Π or an induced Γ or an induced Θ. □

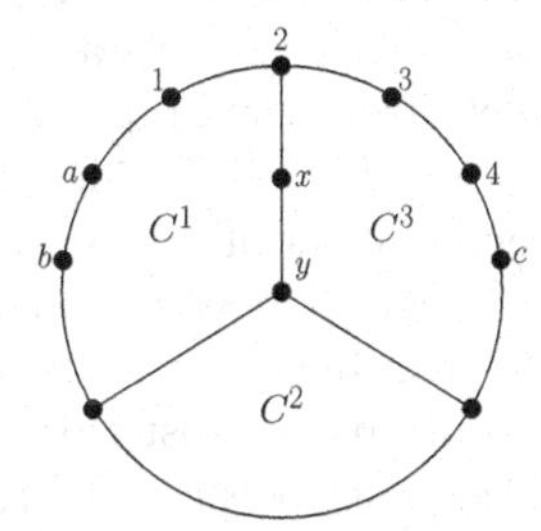

Fig. 8. Vertex y has neighbours on C

We therefore find ourselves in the following context: $N(x) \cap C_6 = \{2, 6\}$ and $N(x) \setminus C^* = \emptyset$. Before we proceed, let us identify another relevant vertex. If 3 has a neighbour outside C^* we call that vertex y. By Lemma 9 (and appropriate symmetry) y is also connected to 5. We have also argued that x and y are not adjacent. We will in the remainder assume that the degree of x is at least as large as the degree of y. This is without loss of generality, as the two vertices can be exchanged by an appropriate automorphism of C^*. In what follows, we analyze all possible adjacencies of x and y to the vertices of C^*.

Case 2.1: If x has degree 2 and y does not exist (therefore 3, 5 have degree 2), then we apply the H-subgraph reduction (Corollary 1) with $H = \{x, 3, 5\}$, in which case $\alpha(G[H]) = \alpha(G[N[H]]) = 3$ and hence the removal of $N[H]$ decreases $\alpha(G)$ by 3.

Case 2.2: Assume x has degree 2 and y exists (therefore, y is connected to 3, 5). We have assumed without loss of generality that x has at least as high degree as y, therefore y has no other neighbour. We delete from the graph vertices

$2, 3, x, y$. If G' is the new graph, we claim that $\alpha(G') = \alpha(G) - 2$. The inequality $\alpha(G') \geq \alpha(G) - 2$ is clear, since no independent set can take more than two of the deleted vertices. To see that $\alpha(G) \geq \alpha(G') + 2$, take a maximum independent set in G'. If it contains vertex 5, then it does not contain 4 or 6. Therefore, we can augment it with $x, 3$. If it contains 6, we can augment it similarly by adding $y, 2$. Finally, if it contains neither 5 nor 6, we augment it with x, y.

Case 2.3: If x is connected to a, $\{x, 1, a, 2, 6\}$ induces a $K_{2,3}$, we can therefore invoke Corollary 2 to simplify the graph.

Case 2.4: If x is connected to c, then $x61ab \ldots cx$ together with $3, 4$ form a large apple with a long stem.

Case 2.5: If x is connected to d, then $x21ab \ldots dx$ together with $3, 4$ form a large apple with a long stem.

Case 2.6: If x is connected to a vertex f of C^* in the path from b to d (but not b or d), then: if f is closer to a than to c, we take the path $xf \ldots dc432x$ plus $1, a$; otherwise we take $xf \ldots ba12x$ plus $3, 4$. In both cases these form a large apple with a long stem.

Case 2.7: If x is connected to b and y does not exist, then we apply the H-subgraph reduction with $H = \{x, 1, 3, 5\}$. It is not hard to check that $\alpha(G[H]) = \alpha(G[N[H]]) = 4$ and hence the removal of $N[H]$ decreases $\alpha(G)$ by 4.

Case 2.8: Assume x is connected to b, y exists and it has degree 2 (that is, y is connected only to $3, 5$). We delete from the graph the vertices $\{x, y, 1, 2, 3, 5, 6\}$ and add a new vertex z adjacent to $a, b, 4$. We claim $\alpha(G') = \alpha(G) - 3$. To see that $\alpha(G) \geq \alpha(G') + 3$ take an independent set of the new graph. If it does not include z then we augment it with $\{2, 6, y\}$; if it does include z, it does not contain any of $a, b, 4$, so we replace z with $\{1, x, 3, 5\}$. To see that $\alpha(G') \geq \alpha(G) - 3$ take an independent set of G. If it contains at most three of the deleted vertices we are done. If it contains four, these must be $\{1, x, 3, 5\}$, therefore the set does not contain any of $a, b, 4$; in this case we replace the deleted vertices by z.

The new graph does not have a large apple with a long stem that uses z and both a, b, since that would induce a triangle. If, on the other hand, it has an apple with a long stem that uses z and at most two of its neighbours, then G also has a subdivided copy of the same subgraph if we replace z with $1, 2, 3$.

Case 2.9: Finally, suppose x is connected to b, y exists and y has degree 3. Since x and y have the same degree, we may exchange their roles, and by symmetry and the same case analysis that we did for x we conclude that y must be connected to d (otherwise one of the previous cases applies). We transform the graph as follows: we delete the vertices $1, 2, 3, 4, 5, 6, x, y$ and add two new vertices z, w such that z, w are connected to each other, z is connected to a, b, and w is connected to c, d. We claim that $\alpha(G') = \alpha(G) - 3$. First, to obtain $\alpha(G') \geq \alpha(G) - 3$, take a maximum independent set of G. If it contains a vertex from a, b and a vertex from c, d, then it contains at most three of the deleted vertices, since the six deleted vertices which are not adjacent to a vertex of the independent set induce a cycle of length 6. In all other cases, the independent set in G contains at most four of the deleted vertices. However, if the set does not contain any of a, b, we can augment it with z in G', while if it does not contain any of c, d we can add to it w. To see that $\alpha(G) \geq \alpha(G') + 3$, take a maximum independent set in G'. If it is using z, then it does not contain a or b. In G we replace z with $1, x, 3, 5$. The situation is symmetric if the set contains w. Finally, if it does not contain either z or w, we observe that deleting the neighbours of the set among the removed vertices gives a cycle of length 6, of which we can select three vertices. The transformation does not introduce a new large apple with a long stem, since the closed neighbourhoods of z, w include a triangle, therefore if one or two of these vertices is used in the apple we can replace them with an appropriate induced path through the deleted vertices in G.

$x$is Adjacent to 2 and 5

Lemma 11. *Let x be a vertex adjacent to 2 and 5 and assume x has a neighbour y not in C^*. Then G contains an induced A_1 or an induced A_2 or an induced A_3 (Fig. 6).*

Proof. If y is adjacent to 3 or 6, then y is adjacent to both 3 and 6 (Lemma 9) and hence G contains an induced A_1. Assume y is adjacent to a. Then, if all three vertices $3, 6, y$ have degree 2 in G, then G contains an induced A_2. If vertex 3 has degree three, it has a common neighbour with 6 (by Lemma 9), call this neighbour z. We claim that z must also be connected to y, which will give an induced A_3. To see this, consider the set of vertices $(C^* \setminus \{2,3\}) \cup \{x, y\}$. This set induces an extended cycle, where the C_6 is now formed by $a, 1, 6, 5, x, y$. Since z is connected to 6, it must be connected to one of $\{x, y\}$ (Lemma 9). However, x already has three neighbours $(2, 5, y)$, therefore, z is connected to y.

If y is adjacent to c this is symmetric to y being adjacent to a. So, we suppose that y is adjacent to none of $3, 6, a, c$. The rest of the proof is similar to that of Lemma 10 with the only difference that if y is adjacent only to b this time we can find a large apple with a long stem, where the stem is $\{1, 6\}$ and the cycle goes through $byx234cd \dots b$. □

Lemma 12. *Let x be a vertex adjacent to 2 and 5 and assume x has a neighbour in $C^* \setminus C_6$. Then this neighbour is one of a and c.*

To complete the case analysis, we prove the following lemma.

Lemma 13. *Let x be a vertex adjacent to 2 and 5 and suppose that if x has a neighbour in $C^* \setminus C_6$, then this neighbour is a. Then we can in polynomial time reduce our instance to a smaller instance.*

5 Conclusion

Summarizing the discussion in the previous sections, we make the following conclusion, which extends several previously known results.

Theorem 2. *Maximum independent set can be solved in polynomial time in the class of $(A_k^*, A_{k+1}^*, \dots)$-free subcubic graphs for any fixed value of k.*

Since A_t^* contains $S_{2,k,k}$ for any $t > 2k+1$, we derive the following corollary

Corollary 3. *Maximum independent set can be solved in polynomial time in the class of $S_{2,k,k}$-free subcubic graphs for any fixed value of k.*

This result brings us closer to the dichotomy of Conjecture 1. However, proving this conjecture in its whole generality remains a challenging open problem.

Acknowledgment. Vadim Lozin acknowledges support from the Russian Science Foundation Grant No. 17-11-01336.

References

1. Alexe, G., Hammer, P.L., Lozin, V.V., de Werra, D.: Struction revisited. Discrete Appl. Math. **132**(1–3), 27–46 (2003)
2. Bodlaender, H.L., Thilikos, D.M.: Treewidth for graphs with small chordality. Discrete Appl. Math. **79**(1–3), 45–61 (1997)
3. Brandstädt, A., Lozin, V.V., Mosca, R.: Independent sets of maximum weight in apple-free graphs. SIAM J. Discrete Math. **24**(1), 239–254 (2010)
4. Gavril, F.: Algorithms for minimum coloring, maximum clique, minimum covering by cliques, and maximum independent set of a chordal graph. SIAM J. Comput. **1**(2), 180–187 (1972)
5. Lozin, V.V.: From matchings to independent sets. Discrete Appl. Math. **231**, 4–14 (2017)
6. Lozin, V.V., Milanic, M.: A polynomial algorithm to find an independent set of maximum weight in a fork-free graph. J. Discrete Algorithms **6**(4), 595–604 (2008)
7. Lozin, V.V., Milanic, M., Purcell, C.: Graphs without large apples and the maximum weight independent set problem. Graphs Combinatorics **30**(2), 395–410 (2014)
8. Lozin, V.V., Monnot, J., Ries, B.: On the maximum independent set problem in subclasses of subcubic graphs. J. Discrete Algorithms **31**, 104–112 (2015)
9. Minty, G.J.: Minty on maximal independent sets of vertices in claw-free graphs. J. Comb. Theory Ser. B **28**(3), 284–304 (1980)
10. Murphy, O.J.: Computing independent sets in graphs with large girth. Discrete Appl. Math. **35**(2), 167–170 (1992)
11. Tarjan, R.E.: Decomposition by clique separators. Discrete Math. **55**(2), 221–232 (1985)
12. Whitesides, S.: An algorithm for finding clique cut-sets. Inf. Process. Lett. **12**(1), 31–32 (1981)

Cyclewidth and the Grid Theorem for Perfect Matching Width of Bipartite Graphs

Meike Hatzel(✉), Roman Rabinovich, and Sebastian Wiederrecht(✉)

TU Berlin, Berlin, Germany
{meike.hatzel,roman.rabinovich,sebastian.wiederrecht}@tu-berlin.de

Abstract. A connected graph G is called *matching covered* if every edge of G is contained in a perfect matching. *Perfect matching width* is a width parameter for matching covered graphs based on a branch decomposition. It was introduced by Norine and intended as a tool for the structural study of matching covered graphs, especially in the context of Pfaffian orientations. Norine conjectured that graphs of high perfect matching width contain a large grid as a matching minor, similar to the result on treewidth by Robertson and Seymour.

In this paper we obtain the first results on perfect matching width since its introduction. For the restricted case of bipartite graphs, we show that perfect matching width is equivalent to directed treewidth and thus, the Directed Grid Theorem by Kawarabayashi and Kreutzer for directed treewidth implies Norine's conjecture.

Keywords: Branch decomposition · Perfect matching · Directed treewidth · Matching minor

1 Introduction

The concept of width parameters, or decompositions of graphs into tree-like structures has proven to be a powerful tool in both structural graph theory and for coping with computational intractability. The shining star among these concepts is the *treewidth* of undirected graphs introduced in its popular form in the Graph Minor series by Robertson and Seymour (see [RS10]).

Tree decompositions are a way to decompose a given graph into loosely connected small subgraphs of bounded size that, in many algorithmic applications, can be dealt with individually instead of considering the graph as a whole. This concept allows the use of dynamic programming and other techniques to solve many hard computational problems (see for example [Bod96, Bod97, Bod05, DF16]). A major milestone in the graph minor project was the

This work has been supported by the European Research Council (ERC) under the European Union's Horizon 2020 research and innovation programme (ERC consolidator grant DISTRUCT, agreement No. 648527).

I. Sau and D. M. Thilikos (Eds.): WG 2019, LNCS 11789, pp. 53–65, 2019.
https://doi.org/10.1007/978-3-030-30786-8_5

Grid Theorem [RS86], which states that if the treewidth of a graph is sufficiently high, the graph has a large grid as a minor.

A natural generalisation of treewidth to directed graphs, *directed treewidth*, was introduced by Reed in [Ree99] and by Johnson et al. in [JRST01] along with the conjecture of a directed version of the Grid Theorem. After being open for several years, the conjecture was proven by Kawarabayashi and Kreutzer [KK15]).

It is possible to go further and to consider even more general structures than directed graphs. One of the ways to do this is to characterise (strongly connected) directed graphs by pairs of undirected bipartite graphs and *perfect matchings*. The generalisation (up to the strong connectivity) is then to drop the condition on the graphs to be bipartite. The theory of *matching minors* in matching covered graphs is deeply connected to the theory of *butterfly minors* in strongly connected directed graphs. This connection can be used to show structural results on directed graphs by using matching theory (see [McC00, GT11]).

The corresponding branch of graph theory was developed from the theory of *tight cuts* and *tight cut decompositions* of *matching covered* graphs introduced by Kotzig, Lovász and Plummer [Lov87, LP09, Kot60]. A graph is matching covered if it is connected and each of its edges is contained in a perfect matching. One of the main incentives of the field is the question of Pfaffian orientations; see [McC04, Tho06] for an overview on the subject. Matching minors can be used to characterise the bipartite Pfaffian graphs [McC04, RST99]. The characterisation implies a polynomial time algorithm for the problem to decide whether a matching covered bipartite graph is Pfaffian or not.

While no polynomial time algorithm for recognising general Pfaffian graphs is known, Norine defines a branch decomposition for matching covered graphs and gives an algorithm that decides whether a graph from a class of bounded *perfect matching width* is Pfaffian in XP-time. Norine and Thomas also conjecture a grid theorem for this new width parameter (see [Nor05, Tho06]). Based on the above mentioned ties between bipartite matching covered graphs and directed graphs, Norine conjectures in his thesis that the Grid Theorem for digraphs would at least imply the conjecture in the bipartite case. Whether perfect matching width and directed treewidth could be seen as equivalent was unknown at that time.

Contribution. We prove that high perfect matching width of a bipartite graph implies that it has a large cylindrical grid as a matching minor, which settles the *Matching Grid Conjecture* for the bipartite case. We also show that the reverse direction holds. To do so, in Sect. 2, we introduce a branch decomposition and a corresponding new width parameter for directed graphs, *cyclewidth*, and prove its equivalence to directed treewidth. This new width measure is very interesting in itself and not only because, while equivalent to treewidth, it is much better behaved since it is closed under minors.

As an advantage of cyclewidth we consider that its introduction leads to a straightforward proof of the Matching Grid Theorem for bipartite graphs: in Sect. 3 we show that cyclewidth and perfect matching width are within a constant factor of each other; this immediately implies the Matching Grid Theorem for bipartite graphs. Our proofs are algorithmic and thus also imply an approximation algorithm for perfect matching width on bipartite graphs, which is the first known result on this matter.

We also show that the perfect matching width of a matching minor of a matching covered bipartite graph is at most two times higher than the perfect matching width of the original graph.

A width parameter like cyclewidth never considers an edge that is not contained in a directed cycle. So, when studying cyclewidth and related topics, one might restrict themselves to strongly connected digraphs. Similarly, an edge that is not contained in any perfect matching is, in most cases, irrelevant for the matching theoretic properties of the graph. For this reason it is common to only consider matching covered graphs as this does not pose any loss of generality.

We take the freedom to rename Norine's *matching-width* [Nor05] to *perfect matching width* to better distinguish it from related parameters such as *maximum matching width* (see [JST17]).

1.1 Preliminaries

We consider finite graphs and digraphs without multiple edges and use standard notation (see [Die17]). For a graph G, its vertex set is denoted by $V(G)$ and its edge set by $E(G)$, and similarly for digraphs where we call arcs edges.

Let $X \subseteq V(G)$ be a non-empty set of vertices in a graph G. The *cut around* X is the set $\partial(X) \subseteq E(G)$ of all edges joining vertices of X to vertices of $V(G) \setminus X$. We call X and $V(G) \setminus X$ the *shores* of $\partial(X)$. A set $E \subseteq E(G)$ is a *cut* if $E = \partial(X)$ for some X. Note that in connected graphs the shores are uniquely defined.

A *matching* of a graph G is a set $M \subseteq E(G)$ such that no two edges in M share a common endpoint. If $e = xy \in M$, e is said to *cover* the two vertices x and y. A matching M is called *perfect* if every vertex of G is covered by an edge of M. We denote by $\mathcal{M}(G)$ the set of all perfect matchings of a graph G. A graph G is called *matching covered* if G is connected and for every edge $e \in E(G)$ there is an $M \in \mathcal{M}(G)$ with $e \in M$.

Definition 1.1. *Let $G = (A \cup B, E)$ be a bipartite graph and let $M \in \mathcal{M}(G)$ be a perfect matching of G. The M-*direction $\mathcal{D}(G, M)$ *of G is defined as follows (see also Fig. 1). Let $M = \{a_1b_1, \ldots, a_{|M|}b_{|M|}\}$ with $a_i \in A, b_i \in B$ for $1 \leq i \leq |M|$. Then,*

(i) $V(\mathcal{D}(G, M)) := \{v_1, \ldots, v_{|M|}\}$ *and*
(ii) $E(\mathcal{D}(G, M)) := \{(v_i, v_j) \mid a_ib_j \in E(G)\}$.

Thus, the *M-direction* $\mathcal{D}(G, M)$ of G is defined by contracting the edges of M, and orienting the remaining edges of G from A to B. The following is a well known observation about M-directions.

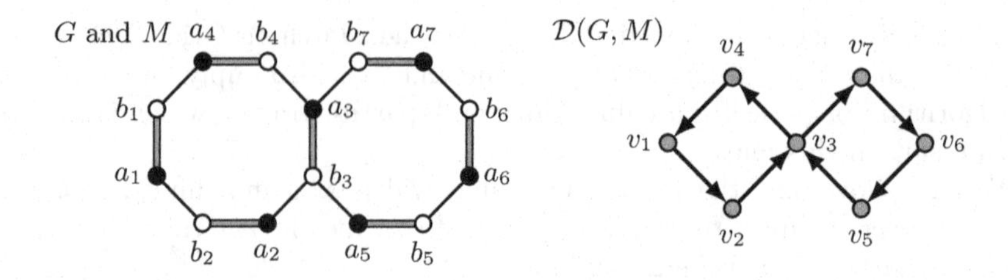

Fig. 1. A bipartite graph $G = (A \cup B, E)$ with perfect matching M and its M-direction.

Lemma 1.2. *A digraph D is strongly connected if and only if there is a bipartite matching covered graph G and a perfect matching $M \in \mathcal{M}(G)$ such that D is isomorphic to $\mathcal{D}(G, M)$. Furthermore, the pair (G, M) is uniquely defined by D.*

This shows why matching covered graphs are such a meaningful graph class, because they correspond to the strongly connected directed graphs.

2 Directed Treewidth and Cyclewidth

Since perfect matching width is defined via a branch decomposition, our first step towards showing the asymptotic equivalence of directed treewidth and perfect matching width of bipartite graphs is to relate directed treewidth to *cyclewidth*, a directed branchwidth parameter. In Sect. 2.1, we introduce cyclewidth and show that it provides a linear lower bound on the directed treewidth. Then, in Sect. 2.2, we show that taking butterfly minors does not increase the cyclewidth and that large cylindrical grids have large cyclewidth. The Directed Grid Theorem then implies that there exists a function that bounds the cyclewidth of a digraph from below by its directed treewidth.

2.1 Cyclewidth: A Branch Decomposition for Digraphs

An *arborescence* is a directed tree T with a root r_0 and all edges directed away from r_0. Let D be a digraph and let $Z \subseteq V(D)$. A set $S \subseteq V(D) - Z$ is *Z-normal* if there is no directed walk in $D - Z$ with the first and last vertex in S that uses a vertex of $D - (Z \cup S)$.

Definition 2.1. *A* directed tree decomposition *of a digraph D is given by a triple (T, β, γ), where T is an arborescence, $\beta\colon V(T) \to 2^{V(D)}$ and $\gamma\colon E(T) \to 2^{V(D)}$ are functions such that*

(i) $\{\beta(t) \mid t \in V(T)\}$ is a partition of $V(D)$ into possibly empty sets, and
(ii) if $e \in E(T)$, then $\bigcup\{\beta(t) \mid t \in V(T), t > e\}$ is $\gamma(e)$-normal.

For any $t \in V(T)$ we define $\Gamma(t) := \beta(t) \cup \bigcup\{\gamma(e) \mid e \in E(T), e \sim t\}$, where $e \sim t$ means that e is incident with t. The width *of (T, δ, γ) is the least integer w such* *that* $|\Gamma(t)| \leq w+1$ *for all* $t \in V(T)$. *The* directed treewidth $\mathsf{dtw}(D)$ *of* D *is the least integer* w *such that* D *has a directed tree decomposition of width* w. *The sets* $\beta(t)$ *are called* bags *and the sets* $\gamma(e)$ *are called the* guards *of the directed tree decomposition. For a subtree* T' *of* T *we write* $\beta(T')$ *for* $\bigcup_{v \in V(T)} \beta(v)$.

Definition 2.2 (Butterfly Minor). *Let* D *be a digraph. An edge* $e = (u, v) \in E(D)$ *is* butterfly-contractible *if* e *is the only outgoing edge of* u *or the only incoming edge of* v. *A* butterfly contraction *is the operation of identifying the endpoints of a butterfly-contractible edge and deleting any resulting multiple edges and loops. A digraph* D' *is a* butterfly-minor *of* D, *if it can be obtained from a subgraph of* D *by butterfly contractions.*

Definition 2.3 (Cylindrical Grid). *A* cylindrical grid $\mathrm{D}_k^{\circlearrowright}$ *of order* k *consists of* k *concentric directed cycles and* $2k$ *paths connecting the cycles in alternating directions, see Fig. 2.*

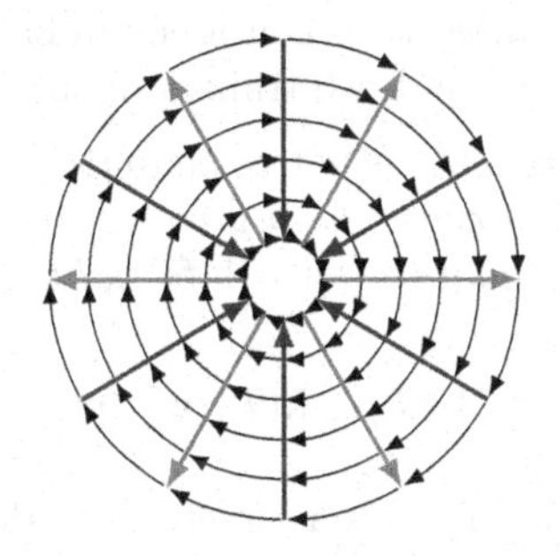

Fig. 2. A cylindrical grid of order 6.

Theorem 2.4 (Kawarabayashi and Kreutzer [KK15]). *There is a function* $f\colon \mathbb{N} \to \mathbb{N}$ *such that every digraph* D *either satisfies* $\mathsf{dtw}(D) \leq k$, *or contains the cylindrical grid of order* $f(k)$ *as a butterfly minor.*

The problem of relating directed treewidth and perfect matching width is that the latter is defined by a branch decomposition where the value of a cut is determined locally with respect to the cut while, in a directed tree decomposition, the guards of a tree edge may appear almost everywhere in the graph.

In order to approach this problem, we introduce a new width parameter, which is defined over a branch decomposition.

Let T be a tree and $e = tt' \in E(T)$. Then $T \ltimes e := (T_1, T_2)$ where T_1 is the subtree containing t and T_2 the subtrees containing t' in $T - e$. Let $L(T)$ denote the set of leaves of T.

Definition 2.5 (Cyclewidth). *Let* D *be a digraph. A* cycle decomposition *of* D *is a tuple* (T, φ), *where* T *is a cubic tree (i.e. all inner vertices have degree three)* *and* $\varphi\colon L(T) \to V(D)$ *is a bijection. For a subtree* T' *of* T *we use* $\varphi(T') := \{\varphi(t) \mid t \in V(T') \cap L(T)\}$. *Let* $t_1t_2 \in E(T)$ *and let* $(T_1, T_2) := T \ltimes t_1t_2$. *Let* $\partial(t_1t_2) := \partial(\varphi(T_1))$. *The* cyclic porosity *of the edge* t_1t_2 *is defined as*

$$\mathrm{cp}(\partial(t_1t_2)) := \max_{\substack{\mathcal{C}\ \text{family of pairwise} \\ \text{disjoint directed cycles} \\ \text{in } D}} \left| \partial(t_1t_2) \cap \bigcup_{C \in \mathcal{C}} E(C) \right|.$$

The width *of a cycle decomposition* (T, φ) *is given by* $\max_{t_1t_2 \in E(T)} \mathrm{cp}(\partial(t_1t_2))/2$ *and the* cyclewidth *of* D *is defined as*

$$\mathsf{cw}(D) := \min_{\substack{(T,\varphi)\ \text{cycle decomposition} \\ \text{of } D}} \max_{t_1t_2 \in E(T)} \mathrm{cp}(\partial(t_1t_2))/2.$$

The factor of $1/2$ might seem arbitrary and it kind of is. We added it because otherwise cyclewidth would only take even numbers which is a quite strange property for a width measure.

In order to show that directed treewidth bounds cyclewidth from above by a function, we construct a cycle decomposition from a directed tree decomposition in two steps. First, we push all vertices contained in bags of inner vertices of the arborescence into leaf bags. Second, we transform the result into a cubic tree.

Lemma 2.6. *Let* (T, β, γ) *be a directed tree decomposition of a digraph* D. *There is a linear time algorithm that computes a directed tree decomposition* (T', β', γ') *of* D *of the same width such that* $|\beta'(\ell)| = 1$ *for all* $\ell \in L(T')$ *and* $\beta'(t) = \emptyset$ *for all inner vertices of* T.

Proof (sketch). For every $t \in V(T) \setminus L(T)$ we introduce a new child t' with $\beta'(t') := \beta(t)$ and $\gamma'(tt') := \gamma(st)$ where $st \in E(T)$. Finally, $\beta'(t) := \emptyset$. It is easy to confirm that the construction yields a directed tree decomposition of the same width as (T, β, γ). □

We call a directed tree decomposition like the one we obtain from Lemma 2.6 a *leaf directed tree decomposition.* For the second step we need to further manipulate this decomposition. A cycle decomposition requires a cubic tree, therefore we have to transform the arborescence of our decomposition into one of total degree 3 at every vertex. To achieve this we replace every high degree vertex by a long path and attach the children one at a time while maintaining the guards. Then, for every vertex of degree two we contract one of its incident edges.

Lemma 2.7. *If the directed treewidth of a digraph* D *is at most* k, *then there is a cubic leaf directed tree decomposition of width* k *for* D.

Both constructions in the proofs of Lemmata 2.6 and 2.7 can be computed in linear time. It remains to observe that if we forget the orientation of the edges as well as the guard function of the resulting decomposition, we obtain a cycle decomposition of bounded width.

Proposition 2.8. *For every digraph* D *holds* $\mathsf{cw}(D) \leq \mathsf{dtw}(D)$. *Moreover, a cycle decomposition of* D *of width at most* k *can be computed from a directed tree decomposition of* D *of width* k.

This can be done in polynomial time.

2.2 Cyclewidth and Cylindrical Grids

The goal of this subsection is to establish a lower bound on the cyclewidth of a digraph in terms of its directed treewidth. Here we face a special challenge. Most width parameters, including directed treewidth, imply separations of bounded size, namely in the width of the decomposition. For cyclewidth it is not immediately clear whether there exists a function $f\colon \mathbb{N} \to \mathbb{N}$ such that, given a digraph D and a cut $\partial(X)$ in D, there always is a set $S \subseteq V(D)$ that hits all directed cycles crossing $\partial(X)$ and satisfying $|S| \leq f(\mathrm{cp}(\partial(X)))$.

We take a different approach. We show that the cyclewidth of cylindrical grids is unbounded and that cyclewidth is closed under taking butterfly minors (a property that does not hold for directed treewidth as shown by Adler [Adl07]). The Directed Grid Theorem then implies the desired result.

Theorem 2.9. *If D is a digraph and D' is a butterfly minor of D, then* $\mathsf{cw}(D') \leq \mathsf{cw}(D)$.

For some intuition behind Theorem 2.9 note that, if an edge e is butterfly contractible, every directed cycle containing one of the two endpoints of e must contain e itself. Therefore, by contracting the edge no new cycles are generated and so if there is a family of directed cycles in D' that witnesses the cycle porosity of some cut in D', it corresponds to a family of cycles in D witnessing the cycle porosity of the corresponding cut. This property is especially interesting, because directed treewidth itself is not closed under butterfly minors.

Lemma 2.10. *The cylindrical grid of order k has cyclewidth at least $k/3$.*

Proof (sketch). Every cycle decomposition of every digraph D contains an edge e that induces a bipartition of $V(D)$ where both partitions contain at least a third of $V(D)$. For a cycle decomposition of a large cylindrical grid, consider such an edge. In case that both shores contain a concentric cycle C_1 and C_2 of the grid, the cycle starting at C_1, going to C_2, returning to C_1 and so on intersects the cut with every change between C_1 and C_2. In the other case one of the shores contains no concentric cycle completely. The other shore can contain at most two third of the concentric cycles completely, i.e. the remaining (at least $k/3$) cycles cross the cut. □

We conclude this section by stating and proving its main result: the equivalence between directed treewidth and cyclewidth.

Theorem 2.11. *A class $\mathcal{D}$ of digraphs is a class of bounded directed treewidth if and only if it is a class of bounded cyclewidth.*

Proof. Let $\mathcal{D}$ be a class of digraphs. Suppose $\mathcal{D}$ has unbounded directed treewidth, then for each $n \in \mathbb{N}$ there is a digraph $D'_n \in \mathcal{D}$ such that $\mathsf{dtw}(D') \geq n$. By Theorem 2.4, we can conclude that for every $n \in \mathbb{N}$ there is a digraph $D_n \in \mathcal{D}$ that contains the cylindrical grid of order n as a butterfly minor. Therefore, $\mathsf{cw}(D_n) \geq \frac{n}{3}$ by Lemma 2.10 and Theorem 2.9. Thus, $\mathcal{D}$ has also unbounded cyclewidth. Vice versa, assume $\mathcal{D}$ is of bounded directed treewidth. Then it also is of bounded cyclewidth due to Proposition 2.8. □

3 Perfect Matching Width

We now leave the world of directed graphs and start to consider undirected graphs with perfect matchings. As seen in Lemma 1.2, strongly connected directed graphs correspond to matching covered bipartite graphs with a fixed perfect matching. We discuss this correspondence in more detail in this section. The goal of this section is to establish a connection between the perfect matching width of bipartite matching covered graphs and the directed treewidth of their M-directions.

A set $S \subseteq V(G)$ of vertices is called *conformal* if $G - S$ has a perfect matching. Given a perfect matching $M \in \mathcal{M}(G)$, a set $S \subseteq V(G)$ is called *M-conformal* if M contains a perfect matching of both $G - S$ and $G[S]$. A subgraph $H \subseteq G$ is *conformal* if $V(H)$ is a conformal set and it is called *M-conformal* if $V(H)$ is M-conformal. If a cycle C is M-conformal, there is another perfect matching $M' \neq M$ with $E(C) \setminus M \subseteq M'$. Hence, if needed, we say C is *M-M'-conformal* to indicate that M and M' form a partition of the edges of C.

3.1 Perfect Matching Width and Directed Cycles

Definition 3.1 (Perfect Matching Width). *Let G be a matching covered graph. We define the* matching-porosity *of $\partial(X)$ as follows:*

$$\mathrm{mp}(\partial(X)) := \max_{M \in \mathcal{M}(G)} |M \cap \partial(X)|.$$

A perfect matching decomposition of G is a tuple (T, δ) where T is a cubic tree and $\delta \colon L(T) \to V(G)$ a bijection. Let e be an edge in T. Removing the edge e splits T in two subtrees T_1 and T_2. Let

$$X_i := \{\delta(t) \mid t \in L(T) \cap V(T_i)\}$$

be the two classes of that partition. Note that $\partial(X_1) = \partial(X_2)$ defines an edge cut in G, we refer to it by $\partial(e)$. The width of (T, δ) is given by $\max_{e \in E(T)} \mathrm{mp}(e)$ and the perfect matching width *of G is then defined as*

$$\mathsf{pmw}(G) := \min_{\substack{(T,\delta) \text{ perfect matching} \\ \text{decomposition of } G}} \max_{e \in E(T)} \mathrm{mp}(\partial(e)).$$

If we consider the M-direction of a matching covered bipartite graph G with $M \in \mathcal{M}(G)$, then any cycle decomposition (T, φ) of $\mathcal{D}(G, M)$ can be interpreted as a decomposition of G where φ is a bijection between $L(T)$ and M. Then, every edge in T induces a bipartition of $V(G)$ into M-conformal sets. The next definition relates this observation to perfect matching decompositions.

Definition 3.2 (M-Perfect Matching Width). *Let G be a matching covered graph and $M \in \mathcal{M}(G)$. Define S as the set of all perfect matching decompositions (T, δ) of G such that for every inner edge e holds if $(T_1, T_2) = T \ltimes e$,* *then* $\delta(L(T_1))$ *and* $\delta(L(T_2))$ *are* M*-conformal. The* M-perfect matching width, M-pmw, *is defined as*

$$M\text{-pmw}(G) := \min_{(T,\delta)\in S} \max_{e\in E(T)} \text{mp}(\partial(e))/2.$$

Here again the factor $1/2$ avoids the measure to take only even numbers.

Proposition 3.3. *Let* G *be a matching covered graph and* $M \in \mathcal{M}(G)$*. Then,* $\text{pmw}(G)/2 \leq M\text{-pmw}(G) \leq \text{pmw}(G)$*.*

Proof (sketch). The first inequality is trivial. For the second one consider a cut $\partial(X)$ of matching porosity k. Then, our desired perfect matching M has at most k edges in $\partial(X)$. Therefore, we have to shift at most k vertices from X to $\overline{X}$ in order to obtain a cut between two M-conformal sets. Since we shift at most k vertices for every cut, the matching porosity increases by at most k, so it ends up to be $2k$. Thus the M-perfect matching width is k again. □

We now need the following observation. Let $G = (A \cup B, E)$ be a bipartite matching covered graph and $M, M' \in \mathcal{M}(G)$ two distinct perfect matchings. Then the graph induced by $M \cup M'$ consists only of isolated edges and M-M'-conformal cycles. Moreover, the isolated edges are exactly the set $M \cap M'$. Let C be such an M-M'-conformal cycle. Then in both, $\mathcal{D}(G, M)$ and $\mathcal{D}(G, M')$, C corresponds to a directed cycle. On the other hand let $N \in \mathcal{M}(G)$ and let C be a directed cycle in $\mathcal{D}(G, N)$. Then C corresponds to an N-conformal cycle C_N in G of exactly double the length, where $E(C)$ and $E(C_N) \setminus N$ coincide (if we forget the direction of edges in C). Thus $(N \setminus E(C_N)) \cup (E(C_N) \setminus N)$ is a perfect matching of G.

So there is a one-to-one correspondence between the directed cycles in $\mathcal{D}(G, M)$ and the M-conformal cycles in G. Using this insight we can translate an M-perfect matching decomposition of G to a cycle decomposition of $\mathcal{D}(G, M)$ and back, which yields the next lemma.

Lemma 3.4. *Let* G *be a bipartite and matching covered graph and* $M \in \mathcal{M}(G)$*. Then* $M\text{-pmw}(G) = \text{cw}(\mathcal{D}(G, M))$*.*

The following theorem is an immediate corollary of Proposition 3.3 and Lemma 3.4.

Theorem 3.5. *Let* G *be a bipartite and matching covered graph and* $M \in \mathcal{M}(G)$*. Then* $\text{pmw}(G)/2 \leq \text{cw}(\mathcal{D}(G, M)) \leq \text{pmw}(G)$*.*

3.2 The Bipartite Matching Grid

We now almost have all pieces in place to deduce the grid theorem for matching covered bipartite graphs. The only missing piece is a minor concept for matching covered graphs. The standard concept of contractions in graphs reduces the number of vertices by exactly one. Thus, it does not preserve the property whether the graph contains a perfect matching. However, if we always consider conformal subgraphs and contract two edges at a time, parity is not an issue.

The idea of matching minors appears in the work of McGuaig [McC01], but the formal framework and the actual name were introduced by Norine and Thomas in [NT07].

Definition 3.6 (Bicontraction). *Let G be a graph and let v_0 be a vertex of G of degree two incident to the edges $e_1 = v_0v_1$ and $e_2 = v_0v_2$. Let H be obtained from G by contracting both e_1 and e_2 and deleting all resulting parallel edges. We say H is obtained from G by* bicontraction *or* bicontracting the vertex v_0.

Definition 3.7 (Matching Minor). *Let G and H be graphs. We say that H is a* matching minor *of G if H can be obtained from a conformal subgraph of G by repeatedly bicontracting vertices of degree two.*

There is a strong relation between matching minors of bipartite matching covered graphs and butterfly minors of strongly connected digraphs.

Lemma 3.8 (McGuaig [McC00]). *Let G and H be bipartite matching covered graphs. Then H is a matching minor of G if and only if there exist perfect matchings $M \in \mathcal{M}(G)$ and $M' \in \mathcal{M}(H)$ such that $\mathcal{D}(H, M')$ is a butterfly minor of $\mathcal{D}(G, M)$.*

We want to establish a relation between the perfect matching width of a matching covered graph and the perfect matching width of its matching minors. By using our result on the relation of the cyclewidth of digraphs and the cyclewidth of their butterfly minors, we are able to derive the next result.

Proposition 3.9. *Let G and H be matching covered bipartite graphs. If H is a matching minor of G, then $\mathsf{pmw}(H) \leq 2\mathsf{pmw}(G)$.*

Proof. Let H be a matching minor of G. Then, Lemma 3.8 provides the existence of perfect matchings $M \in \mathcal{M}(G)$ and $M' \in \mathcal{M}(H)$ such that $\mathcal{D}(H, M')$ is a butterfly minor of $\mathcal{D}(G, M)$. The M-perfect matching width of G is at most $\mathsf{pmw}(G)$ by Proposition 3.3 and, by Lemma 3.4, $M\text{-}\mathsf{pmw}(G) = \mathsf{cw}(\mathcal{D}(G, M))$. Since $\mathcal{D}(H, M')$ is a butterfly minor of $\mathcal{D}(G, M)$, Theorem 2.9 gives us $\mathsf{cw}(\mathcal{D}(H, M')) \leq \mathsf{cw}(\mathcal{D}(G, M))$. At last, using Lemma 3.4 again and combining the above inequalities, we obtain $\mathsf{pmw}(H) \leq 2\mathsf{pmw}(G)$. □

As we are going for a cylindrical grid and derive the grid theorem for bipartite matching covered graphs from the Directed Grid Theorem anyway, it makes sense to derive our grid from the directed case as well. The following definition defines the bipartite matching grid by providing a procedure that allows us to obtain it from the directed cylindrical grid. Let E_o be the set of edges of the outermost cycle containing exactly those edges which are the sole outgoing edges of their tails. Let E_i be the set of edges of the innermost cycle containing exactly those edges which are the sole incoming edges of their heads.

Definition 3.10 (Bipartite Matching Grid). *Let $k \in \mathbb{N}$ be a positive integer. Let $\widehat{\mathrm{D}}_k^{\circlearrowright}$ be the digraph obtained from $\mathrm{D}_k^{\circlearrowright}$ by butterfly contracting every edge from E_o and every edge from E_i. The* bipartite matching grid *of order k is the unique bipartite matching covered graph $\mathrm{G}_k^{\mathcal{M}}$ that has a perfect matching $M \in \mathcal{M}\left(\mathrm{G}_k^{\mathcal{M}}\right)$ such that $\mathcal{D}\left(\mathrm{G}_k^{\mathcal{M}}, M\right) = \widehat{\mathrm{D}}_k^{\circlearrowright}$.*

The uniqueness of $\mathrm{G}_k^{\mathcal{M}}$ and M follows from Lemma 1.2. Figure 3 provides an example. Note that all contractible edges are contained in the innermost and in the outermost cycle of $\mathrm{D}_k^{\circlearrowright}$ and provide a perfect matching of those two cycles. Here, we construct $\mathrm{G}_3^{\mathcal{M}}$ from $\mathrm{D}_3^{\circlearrowright}$. The contractible edges are highlighted.

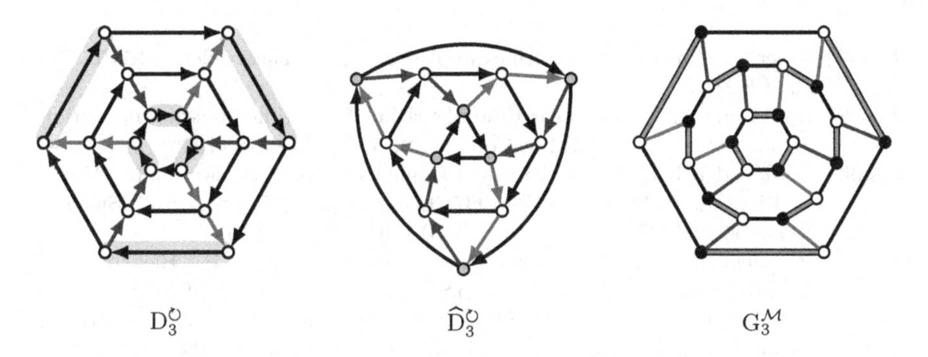

Fig. 3. The construction of the bipartite matching grid of order 3 from the (directed) cylindrical grid of order 3

Assume that a bipartite matching covered graph G has high perfect matching width. By Theorem 3.5, this implies high cyclewidth for all M-directions of G. This, in turn, implies large cylindrical grids as butterfly minors on those M-directions. Now Lemma 3.8 allows us to translate these cylindrical grids into matching minors of G and as the perfect matching width of G is bounded from below by the width of its matching minors (Proposition 3.9), we obtain the grid theorem for bipartite matching covered graphs.

Proposition 3.11. *There is a function $f\colon \mathbb{N} \to \mathbb{N}$ such that every matching covered bipartite graph G either satisfies $\mathsf{pmw}(G) \leq k$, or contains the bipartite matching grid of order $f(k)$ as a matching minor.*

4 Conclusion

We provide a first proof for the bipartite version of the Matching Grid Conjecture. However, this proof relies heavily on machinery found and used in directed graph structure theory and it would be nice to have a completely matching theoretical proof. Moreover, there is probably no hope to extend the methods used for proving the Directed Grid Theorem to the non-bipartite matching covered case which remains open. Most likely, we need a completely novel approach in order to attempt solving this second and much more difficult part of the conjecture.

A smaller and probably more accessible problem occurring in this paper is the question for a nice lower bound on the cyclewidth in terms of directed treewidth. The current proof uses the Directed Grid Theorem and therefore the lower bound is equally exponential as the function provided by the theorem. We conjecture that cyclewidth and directed treewidth are actually within a constant factor of each other, a result that would, most likely, enable us to prove Erdős-Pósa-type results for matching covered bipartite graphs.

References

[Adl07] Adler, I.: Directed tree-width examples. J. Comb. Theory Ser. B **97**(5), 718–725 (2007)

[Bod96] Bodlaender, H.L.: A linear-time algorithm for finding tree-decompositions of small treewidth. SIAM J. Comput. **25**(6), 1305–1317 (1996)

[Bod97] Bodlaender, H.L.: Treewidth: algorithmic techniques and results. In: Prívara, I., Ružička, P. (eds.) MFCS 1997. LNCS, vol. 1295, pp. 19–36. Springer, Heidelberg (1997). https://doi.org/10.1007/BFb0029946

[Bod05] Bodlaender, H.L.: Discovering treewidth. In: Vojtáš, P., Bieliková, M., Charron-Bost, B., Sýkora, O. (eds.) SOFSEM 2005. LNCS, vol. 3381, pp. 1–16. Springer, Heidelberg (2005). https://doi.org/10.1007/978-3-540-30577-4_1

[DF16] Downey, R.G., Fellows, M.R.: Fundamentals of Parameterized Complexity, 1st edn. Springer, London (2013). https://doi.org/10.1007/978-1-4471-5559-1

[Die17] Diestel, R.: Graph Theory, vol. 173. Springer, Heidelberg (2017)

[GT11] Guenin, B., Thomas, R.: Packing directed circuits exactly. Combinatorica **31**(4), 397–421 (2011)

[JRST01] Johnson, T., Robertson, N., Seymour, P.D., Thomas, R.: Directed tree-width. J. Comb. Theory Ser. B **82**(1), 138–154 (2001)

[JST17] Jeong, J., Sæther, S.H., Telle, J.A.: Maximum matching width: new characterizations and a fast algorithm for dominating set. Discrete Appl. Math. **248**, 114–124 (2017)

[KK15] Kawarabayashi, K., Kreutzer, S.: The directed grid theorem. In: Proceedings of the Forty-Seventh Annual ACM on Symposium on Theory of Computing, STOC 2015, pp. 655–664 (2015)

[Kot60] Kotzig, A.: On the theory of finite graphs with a linear factor I–III [Slovak with German summary]. Fyz. Casopis Slovensk. Akad. Vied, 9–10 (1959/1960)

[Lov87] Lovász, L.: Matching structure and the matching lattice. J. Comb. Theory Ser. B **43**(2), 187–222 (1987)

[LP09] Lovász, L., Plummer, M.D.: Matching Theory, vol. 367. American Mathematical Society, Providence (2009)

[McC00] McCuaig, W.: Even dicycles. J. Graph Theory **35**(1), 46–68 (2000)

[McC01] McCuaig, W.: Brace generation. J. Graph Theory **38**(3), 124–169 (2001)

[McC04] McCuaig, W.: Pólya's permanent problem. Electron. J. Comb. **11**(1), 79 (2004)

[Nor05] Norine, S.: Matching structure and Pfaffian orientations of graphs. Ph.D. thesis. Georgia Institute of Technology (2005)

[NT07] Norine, S., Thomas, R.: Generating bricks. J. Comb. Theory Ser. B **97**(5), 769–817 (2007)

[Ree99] Reed, B.: Introducing directed tree width. Electron. Notes Discrete Math. **3**, 222–229 (1999)

[RS86] Robertson, N., Seymour, P.D.: Graph minors. V. Excluding a planar graph. J. Comb. Theory Ser. B **41**(1), 92–114 (1986)

[RS10] Robertson, N., Seymour, P.D.: Graph minors I–XXIII. J. Comb. Theory Ser. B (1982–2010)

[RST99] Robertson, N., Seymour, P.D., Thomas, R.: Permanents, Pfaffian orientations, and even directed circuits. Ann. Math. **150**(3), 929–975 (1999)

[Tho06] Thomas, R.: A survey of Pfaffian orientations of graphs. In: Proceedings of the International Congress of Mathematicians, vol. 3, pp. 963–984. Citeseer (2006)

Local Approximation of the Maximum Cut in Regular Graphs

Étienne Bamas[1] and Louis Esperet[2(✉)]

[1] School of Computer and Communication Sciences, École Polytechnique Fédérale de Lausanne, Lausanne, Switzerland
etienne.bamas@epfl.ch
[2] Laboratoire G-SCOP (CNRS, Univ. Grenoble Alpes), Grenoble, France
louis.esperet@grenoble-inp.fr

Abstract. This paper is devoted to the distributed complexity of finding an approximation of the maximum cut in graphs. A classical algorithm consists in letting each vertex choose its side of the cut uniformly at random. This does not require any communication and achieves an approximation ratio of at least $\frac{1}{2}$ in average. When the graph is d-regular and triangle-free, a slightly better approximation ratio can be achieved with a randomized algorithm running in a single round. Here, we investigate the round complexity of *deterministic* distributed algorithms for MaxCut in regular graphs. We first prove that if G is d-regular, with d even and fixed, no deterministic algorithm running in a constant number of rounds can achieve a constant approximation ratio. We then give a simple one-round deterministic algorithm achieving an approximation ratio of $\frac{1}{d}$ for d-regular graphs with d odd. We show that this is best possible in several ways, and in particular no deterministic algorithm with approximation ratio $\frac{1}{d} + \epsilon$ (with $\epsilon > 0$) can run in a constant number of rounds. We also prove results of a similar flavour for the MaxDiCut problem in regular oriented graphs, where we want to maximize the number of arcs oriented from the left part to the right part of the cut.

Keywords: Maximum cut · Distributed approximation · Local algorithm

1 Introduction

Although the maximum cut problem (MaxCut) is fundamental in combinatorial optimization, it has not been intensively studied from the perspective of distributed algorithms. The folklore algorithm consisting in choosing uniformly at random one side of the cut for each vertex of a graph G can however be seen as a distributed randomized algorithm with no rounds of communication.

Partially supported by ANR Projects GATO (ANR-16-CE40-0009-01) and GrR (ANR-18-CE40-0032), and LabEx PERSYVAL-Lab (ANR-11-LABX-0025). The full version of the paper is available at arXiv:1902.04899.

I. Sau and D. M. Thilikos (Eds.): WG 2019, LNCS 11789, pp. 66–78, 2019.
https://doi.org/10.1007/978-3-030-30786-8_6

By the linearity of expectation, this algorithm gives a cut (a bipartition of the vertex set) of size at least $m/2$ in average, where m is the number of edges of G. Here, by the *size* of the cut, we mean the number of edges connecting the two parts of the bipartition. Since every cut in G contains at most m edges, this algorithm has *approximation ratio* at least $\frac{1}{2}$ in average, which means that the size of the cut given by the algorithm is at least $\frac{1}{2}$ of the size of the maximum cut in average.

A natural question is whether a better approximation ratio can be obtained if more rounds of communications are allowed. This question was answered positively by Shearer [26] in the case of triangle-free d-regular graphs. A *d-regular* graph is a graph in which every vertex has degree d. In the case of triangle-free d-regular graphs, Shearer gave a simple randomized algorithm finding a cut of size at least $m\cdot(\frac{1}{2}+\frac{0.177}{\sqrt{d}})$ in average, and thus achieving an approximation ratio of $\frac{1}{2}+\frac{0.177}{\sqrt{d}}$ in average. Shearer's algorithm uses a single round of communication, messages consisting of a single bit, and at most 3 random bits per vertex. This was recently improved by Hirvonen, Rybicki, Schmid and Suomela [15], who obtained a simpler algorithm finding a cut of size at least $m\cdot\left(\frac{1}{2}+\frac{0.28125}{\sqrt{d}}\right)$ in average. Their algorithm uses a single round of communication, messages consisting of a single bit, and a single random bit per vertex.

The case where d is small and the *girth* (length of a shortest cycle) is large has also been considered: for 3-regular graphs, Kardoš, Král' and Volec [16] showed that when the girth is at least 637789, there exists a randomized distributed algorithm that outputs a cut of average size at least 0.88672 m in at most 318894 rounds (the important value here is the size of the cut). This was improved by Lyons [19], who proved a lower bound of 0.89 m for cubic graphs of girth at least 655. The best known lower bound for cubic graphs of large girth, 0.90 m, was proved by Gamarnik and Li [9], using a result of Csóka, Gerencsér, Harangi, and Virág [3]. The bound of Lyons [19] holds for any d-regular graphs of large enough (but constant) girth: such graphs have a cut of size at least $m\cdot(\frac{1}{2}+\frac{2}{\pi\sqrt{d}})\approx m\cdot(\frac{1}{2}+\frac{0.637}{\sqrt{d}})$. On the other hand, Dembo, Montanari and Sen [6] showed that in random d-regular graphs, the maximum cut has size $m\cdot(\frac{1}{2}+\frac{0.763+o(1)}{\sqrt{d}})+o(m)$ with high probability, proving a conjecture of [9]. The existence of this constant $\approx$0.763 is also connected to a conjecture of Hatami, Lovász and Szegedy [14] on limits of sparse graphs (see also the conclusion of [24] where the conjecture is strongly disproved for maximum independent sets, improving on an earlier result of [10]).

All the results mentioned above (except the result of Gamarnik and Li [9]) can be translated into efficient algorithms working in the CONGEST model. In this model, each node of the graph corresponds to a processor with infinite computational power and has a unique ID (each ID is an integer between 1 and poly(n), where n denotes the number of vertices in the graph). Nodes can communicate with their neighbors in the graph in synchronous rounds until each node outputs 0 or 1, corresponding to its side in the cut. In the CONGEST model, each message sent by a node to a neighbor has size $O(\log n)$, while in some of the algorithms above, the messages have size at most 1. Let us call CONGEST(B) the variant of the CONGEST model in which messages are restricted to have size at most B (instead of $O(\log n)$), and let us say that an algorithm is *local* in a model if it runs in a constant number of rounds in this model. In particular the results of [15,16,26] mentioned above can be translated into local algorithms in the CONGEST($O(1)$) model, while the results of [3,19] can be translated into local algorithms in the CONGEST model.

Note that some of our lower bounds are also valid in the less restricted LOCAL model where the size of each message is not limited. In the following, we will make it clear if this applies. On the other hand, all our algorithms can be implemented in the PO model (anonymous network with port numbering and orientations), which is significantly stronger than the CONGEST model (see [13] for some results on local algorithms in PO and CONGEST).

We now review recent results on distributed approximation of MaxCut. On the deterministic side, Censor-Hillel, Levy, and Shachnai [2] designed a deterministic $\frac{1}{2}$-approximation that runs in $\tilde{O}(\Delta + \log^* n)$ rounds in the CONGEST model on any graph of maximum degree at most Δ. More recently, Kawarabayashi and Schwartzman [17] improved the complexity for constant factor approximation by providing a deterministic $(\frac{1}{2} - \epsilon)$-approximation that runs in $O(\log^* n)$ rounds (for any $\epsilon > 0$), in the CONGEST model. However, no deterministic *local* approximation for MaxCut (i.e. running in a constant number of rounds) in the CONGEST model is known.

There is a similar gap between randomized and deterministic approximations for the maximum directed cut problem. Censor-Hillel, Levy, and Shachnai [2] provided a deterministic algorithm running in $O(\Delta + \log^* n)$ rounds that guarantees a $\frac{1}{3}$-approximation as well as a randomized $\frac{1}{2}$-approximation with the same round complexity. The round complexities were improved by Kawarabayashi and Schwartzman [17] who provided a deterministic $(\frac{1}{3} - \epsilon)$-approximation running in $O(\log^* n)$ rounds as well as a randomized $(\frac{1}{2} - \epsilon)$-approximation in $O(\epsilon^{-1})$ rounds. All these results are stated in the CONGEST model. Similarly, no deterministic local algorithm is known to achieve a constant factor approximation for this problem.

1.1 Our Results

Our work focuses on bridging the gap between extremely efficient randomized local algorithms and slower deterministic algorithms for MaxCut. It should be noted that there are generic tools to derandomize distributed algorithms (see [4,11] for recent results in this direction) but existing techniques mainly apply to *locally checkable* problems (problem for which a solution can be checked locally), which is not the case of (approximations of) MaxCut.

In Sect. 2 we show that any deterministic algorithm that guarantees a constant factor approximation for MaxCut on the class of bipartite *d-regular* graphs when d is a (fixed) even integer requires $\Omega(\log^* n)$ rounds, which matches the complexity of the algorithm of Kawarabayashi and Schwartzman [17] mentioned above. When d is odd, we show that one cannot achieve a approximation ratio better than $\frac{1}{d}$ in a constant number of rounds. Our proofs use an elementary graph construction and then apply Ramsey's theorem [25]. Both these arguments are not new in distributed algorithms: our construction is inspired from Linial's seminal paper [18] that provides a lower bound on the round complexity of coloring cycles and from a more recent paper by Åstrand, Polishchuk, Rybicki, Suomela, and Uitto [1] which applies Ramsey's theorem in a similar setting to prove that there is no deterministic and local constant factor approximation for the maximum matching problem. It was pointed out to us that similar arguments were also used by Czygrinow, Hanckowiak, and Wawrzyniak [5] to prove lower bounds for the approximation of maximum independent sets in cycles. Here, our results hold for any d-regular graph (d is not necessarily equal to 2), so some additional work needs to be done compared to the simple case of cycles.

In Sect. 3, we show that this barrier of $\frac{1}{d}$ when d is odd is sharp: we first remark that a result of Naor and Stockmeyer [22] on weak 2-coloring of graphs directly gives a deterministic local algorithm that guarantees a $\frac{1}{d}$-approximation. We then provide a much simpler and faster deterministic local algorithm achieving the same approximation ratio. It runs in a single round with messages of size $O(\log n)$ and we also argue that this cannot be improved.

For the Maximum Directed Cut problem in d-regular graphs, we prove that a similar situation occurs. If d is even, a constant factor approximation cannot be achieved in $o(\log^* n)$ rounds, and if d is odd, no $(\frac{2}{d} + \epsilon)$-approximation can be achieved in $o(\log^* n)$ rounds (for any $\epsilon > 0$). On the other hand, if d is odd, a $\frac{2}{d+1/d}$ factor approximation can be achieved in 0 round, and a $\frac{2}{d+1/d-\Omega(1/d^2)}$ factor approximation can be achieved in 2 rounds. Note that there is a small gap between the lower bounds and the upper bound of $\frac{2}{d}$, and we explain some obstacles towards closing the gap.

Our results imply that while finding a constant factor approximation for the (directed) maximum cut problem in regular graphs of even degree does not require any communication for randomized distributed algorithm (i.e. it can be solved in 0 round), for deterministic algorithms an unbounded number of rounds are needed in this case. Note that this separation is not possible for locally checkable problems (see Theorem 3 in [4]). The (perhaps) surprising aspect is that in the case of regular graphs of *odd* degree, the problem can be solved by a *deterministic* algorithm without communication (if some orientation is given).

Note that another example of non locally checkable problem with such a separation between the randomized and deterministic complexities was given in [11]. Their problem consists in marking $(1 + o(1))\sqrt{n}$ vertices of an n-cycle; the randomized version can also be solved in 0 round, while the deterministic version needs $\Omega(\sqrt{n})$ rounds.

1.2 Definitions

A *cut* in a graph G is a bipartition (A, B) of its vertex set $V(G)$. We usually refer to A and B as the left side and the right side of the cut, respectively. The *size* of a cut (A, B) is the number of edges with one end in A and the other in

B. The MAXCUT problem in a graph G consists in finding a cut in G whose size is maximum.

Given an oriented graph G, a *directed cut* is again a bipartition (A, B) of the vertex set of G, and the size of the directed cut (A, B) is the number of arcs with their tail in A and their head in B. The MAXDICUT problem in an oriented graph G consists in finding a directed cut in G whose size is maximum.

Our results in this paper mainly concern d-regular graph, i.e. graphs in which each vertex has degree d. When we refer to an oriented d-regular graph G, we mean that the underlying unoriented graph is d-regular (the out-degrees can be arbitrary).

For an integer $k \geqslant 1$, the *tower function* twr_k is the function defined as $\mathrm{twr}_1(x) = x$ and $\mathrm{twr}_k(x) = 2^{\mathrm{twr}_{k-1}(x)}$ for $k \geqslant 2$. The *iterated logarithm* of an integer n, denoted by $\log^* n$ is defined as 0 if $n \leqslant 1$, and as $1 + \log^*(\log n)$ otherwise (here and everywhere else in the paper, log denotes the logarithm base 2). The following can be easily derived by induction on k.

Claim 1. *For any $k, n \geqslant 1$:*

$$\log^*(\mathrm{twr}_k(n)) = k - 1 + \log^*(n)$$

2 Many Rounds for Deterministic Constant Factor Approximation in Regular Graphs

As mentioned in the introduction of this paper, Kawarabayashi and Schwartzman [17] provided a deterministic approximation running in $O(\log^* n)$ rounds for both problems studied here. In this section, we show with simple arguments based on bounds on Ramsey numbers that their bound is best possible.

In this section, we set $[n] = \{1, \ldots, n\}$. The *q-color Ramsey number* $r_k(n; q)$ is the minimum N such that in any q-coloring of the k-element subsets of $[N]$, there is an n-element subset S of $[N]$ such that all k-element subsets of S have the same color (see [21] for a recent survey on Ramsey numbers).

Theorem 2 ([7,8]). *There exists $c > 0$ such that for any positive integers q, k, and n, we have $r_k(n; q) \leqslant \mathrm{twr}_k(c \cdot n \cdot q \log q)$.*

We will also need two simple constructions of d-regular bipartite graphs.

We first assume that d is even. We consider a cycle C of size n, with n even, and then add an edge between each pair of vertices that are at distance exactly i in C for every $i \in \{3, 5, 7, \ldots, d-1\}$. This graph, which we denote by C_n^d, is certainly bipartite (the bipartition corresponds to the vertices at even distance from some arbitrary vertex in C, and the vertices at odd distance from this vertex). See Fig. 1 for an example of this graph. By a slight abuse of language, we say that two (or more) vertices of C_n^d are *consecutive* if they are consecutive in C. Similarly, when we refer to the *clockwise order* around C_n^d, we indeed refer to the clockwise order around C.

Assume now d is odd. We take two disjoint copies of C_n^{d-1} and assume that the vertices of the cycle C in the first copy are $u_1, u_2, \ldots, u_n$, in clockwise order, and the vertices of the cycle C in the second copy are $v_1, v_2, \ldots, v_n$ in clockwise order. We then connect u_i and v_i by an edge, for any $1 \leqslant i \leqslant n$. This graph, which we denote by D_{2n}^d, is clearly bipartite and d-regular, see Fig. 2 for an example.

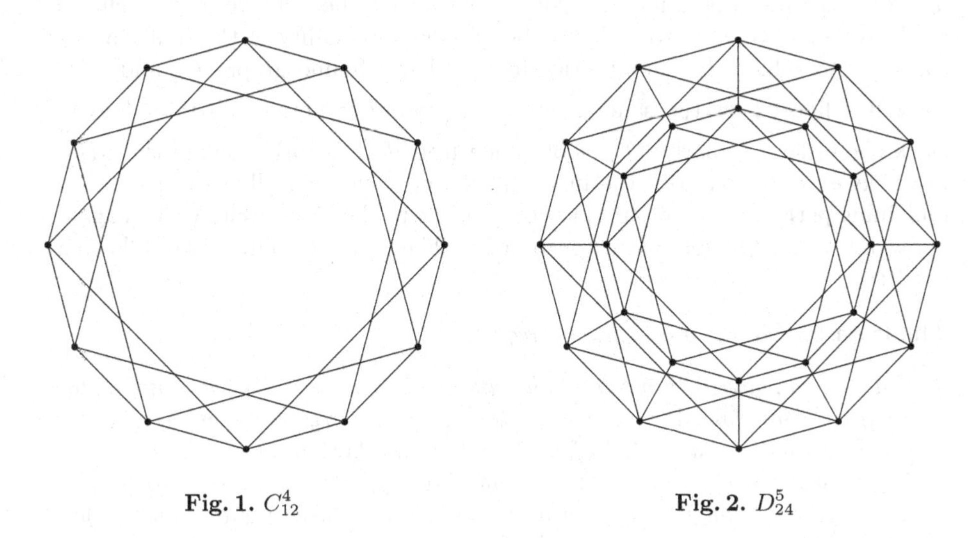

Fig. 1. C_{12}^4 **Fig. 2.** D_{24}^5

We are now ready to state the main result of this section.

Theorem 3. *Let $d \geqslant 2$ be a fixed integer.*

- *If d is even, then any deterministic algorithm in the* LOCAL *model that guarantees a constant factor approximation for* MaxCut *on the class of bipartite d-regular n-vertex graphs runs in $\Omega(\log^* n)$ rounds.*
- *If d is odd, then for any $\epsilon > 0$, any deterministic $\left(\frac{1}{d} + \epsilon\right)$-approximation algorithm in the* LOCAL *model for* MaxCut *on the class of bipartite d-regular n-vertex graphs runs in $\Omega(\log^* n)$ rounds.*

Note that since the LOCAL model is less restrictive than the CONGEST model, this theorem is also valid in the CONGEST model.

Due to space limitation, the full proof of Theorem 3 is given in the full version of the paper. The idea is to use the 2 color version of Theorem 2 to find a large number of blocks of ID's that behave similarly with respect to a given algorithm, and to place these blocks of ID's consecutively around the graphs C_n^d and D_{2n}^d. In both cases a significant portion of the edges of a maximum cut will be missed. In the second case the barrier of $\frac{1}{d}$ comes from the perfect matching between the outer cycle and the inner cycle, on which little can be said.

A direct consequence of our theorem is the following corollary that matches the round complexity obtained by Kawarabayashi and Schwartzman [17]:

Corollary 4. *Deterministic constant factor approximation on general graphs for* MAXCUT *in the LOCAL model requires $\Omega(\log^* n)$ rounds.*

2.1 Directed Cut

In this section, we consider the similar problem MAXDICUT where edges are oriented and we only count the edges going from the left side of the cut to the right side. We can prove similar bounds on the quality of the solution one can hope to achieve by simply orienting our lower bound graphs C_n^d and D_{2n}^d: we will define $\overrightarrow{C_n^d}$ as the same graph as C_n^d where we orient all the edges in clockwise order. Similarly, $\overrightarrow{D_{2n}^d}$ is obtained from D_{2n}^d by orienting all the edges in clockwise order on both the inner and outer cycle, and all the edges in the remaining perfect matching from the outer cycle to the inner cycle. We can again apply Ramsey's theorem as in the proof of Theorem 3 to obtain the following result:

Theorem 5. *Let $d > 0$ be a fixed integer.*

- *If d is even, any deterministic algorithm that guarantees a constant factor approximation for* MAXDICUT *on the class of d-regular bipartite n-vertex oriented graphs requires $\Omega(\log^* n)$ rounds in the LOCAL model.*
- *If d is odd, then for any $\epsilon > 0$, any deterministic $\left(\frac{2}{d} + \epsilon\right)$-approximation of* MAXDICUT *on the class of d-regular bipartite n-vertex oriented graphs requires $\Omega(\log^* n)$ rounds in the LOCAL model.*

We note a slight difference with Theorem 3 in the case where d is odd. In Theorem 5 the approximation ratio is only bounded by $\frac{2}{d}$, instead $\frac{1}{d}$. This happens because with our definition of $\overrightarrow{D_{2n}^d}$, one can check that the optimal directed cut is of size $\frac{nd}{4} = \frac{m}{2}$ instead of m in the undirected case.

3 Matching the Approximation Ratio When d Is Odd

3.1 Weak-Coloring

In a landmark paper, Naor and Stockmeyer [22] addressed the issue of what can or cannot be computed locally. In particular, they proved one result that turns out to be relevant in our case.

A *weak coloring* of a graph is a coloring of its vertices such that each vertex has at least one neighbor with a different color. Observe that a weak coloring using only 2 colors is a $\frac{1}{d}$-approximation of the MAXCUT problem when the graph is d-regular. Let O_d be the class of graphs of maximum degree d where the degree of every vertex is odd. Naor and Stockmeyer proved the following theorem.

Theorem 6 ([22]). *There is a constant b such that, for every d, there is a deterministic algorithm with round complexity $\log^* d + b$ in the CONGEST model that solves the weak 2-coloring problem in the class O_d.*

As discussed above, this result directly implies that one can produce a local deterministic $\frac{1}{d}$-approximation of the MaxCut problem on d-regular graphs. However, the result given here is much stronger than what we are looking for as in this case *every* vertex has at least one incident edge in the cut. A natural question is whether a faster algorithm (of round complexity that does not depend on d) exists for the MaxCut problem on d-regular graphs with d odd. In the next section, we prove that such an algorithm exists.

3.2 A Simpler and Faster Algorithm

Consider the following algorithm: every vertex v collects the list of IDs of its neighbors, then v chooses its side of the cut depending on whether the median value of this list is higher or lower than its own ID. We call this algorithm the *median algorithm*. It runs in a single round and we prove the following theorem:

Theorem 7. *When the input is a d-regular graph on n vertices, with d odd, the median algorithm finds in 1 round a $\frac{1}{d}$-approximation for the* MaxCut *problem in the CONGEST model.*

We will actually give two different proofs of this result, the first proof shows a slightly better result in term of size of the cut while the second proof holds even for graphs in O_d. In the first one, we prove the following slightly stronger statement.

Theorem 8. *When the input is a d-regular graph on n vertices, with d odd, the median algorithm outputs in 1 round (in the CONGEST model) a cut of size at least $\frac{n}{2} + \frac{(d-1)(d+1)}{4}$.*

The proof of Theorem 8 is given in the full version of the paper. The idea is to orient each edge from the lower ID to the higher ID. The obtained orientation is acyclic, and the median algorithm places the vertices with out-degree higher than in-degree on one side of the cut and the other vertices on the other side of the cut. In the analysis the fact that the orientation is acyclic is important to obtain the additive term $\frac{(d-1)(d+1)}{4}$.

An interesting aspect of Theorem 8 is that it shows that in Theorem 3, it is crucial that d is a fixed constant (independent of n). Indeed, if $d = \Omega(\sqrt{n})$, then $\frac{n}{2} + \frac{(d-1)(d+1)}{4} \geqslant (1 + \Omega(1))\frac{n}{2}$ and thus the median algorithm achieves a $\frac{1+\epsilon}{d}$-approximation, for some $\epsilon > 0$. This is impossible when d is a constant, as shown by Theorem 3.

The median algorithm is based on finding an (acyclic) orientation of the input graph. Here, we do it by simply orienting the edges from the end with lower ID to the end with higher ID. This costs a single round of communication, with messages of size $\log n$ (since vertices have to send their ID to their neighbors).

It follows that in the more restricted $\mathsf{CONGEST}(b)$ model, where messages have size at most b, our algorithm takes $\frac{\log n}{b}$ rounds. In particular, if only messages of size 1 are allowed, our algorithm takes $\log n$ rounds. It turns out that this is close to best possible.

Theorem 9. *Let $D^d = \{D^d_{2n}, n > 0\}$ for d odd. Any deterministic constant factor approximation of* MaxCut *on the class D^d requires at least $(1-o(1))\log n$ rounds in the $\mathsf{CONGEST}(1)$ model.*

The proof of Theorem 9 is given in the full version of the paper. It is based on a simple symmetry argument.

3.3 Directed Cuts

Given a bipartition (V_1, V_2) of an oriented graph G, the set of arcs oriented from V_1 to V_2 (the *directed cut* from V_1 to V_2) is denoted by $\overrightarrow{E}(V_1, V_2)$. The maximum cardinality of a directed cut in G is denoted by $\text{maxdicut}(G)$.

Let G be an oriented graph. For each vertex v, we define the *deficit* of v as $\delta(v) = d^+(v) - d^-(v)$, where $d^+(v)$ and $d^-(v)$ denote the out-degree and in-degree of v, respectively. We define the *sign* of a vertex v as the sign of $\delta(v)$, and we say that that a vertex is *positive* or *negative* accordingly. The set of positive vertices is denoted by V^+ and the set of negative vertices is denoted by V^-. Note that if all the vertices of G have odd degree (in particular if G is d-regular with d odd), then every vertex is positive or negative and this case V^+, V^- form a bipartition of the vertex set V of G.

Note that the median algorithm described in the previous subsection can be rephrased as: find an acyclic orientation of G and then choose the cut (V^-, V^+) with respect to this orientation. Our second proof of Theorem 7 will be a direct consequence of the following general result (which proves that not only the cut, but also the *directed* cut between V^+ and V^- has size at least $n/2$, and that the original orientation does not need to be acyclic).

Theorem 10. *Let G be an n-vertex oriented d-regular graph with d odd, and let V^+ and V^- be defined as above. Then the directed cut $\overrightarrow{E}(V^+, V^-)$ contains at least $\max\{\frac{n}{2}, \frac{2}{d+1/d} \cdot \text{maxdicut}(G)\}$ arcs.*

The proof of Theorem 10 is given in the full version of the paper. From now on, we call the 0-round algorithm resulting from Theorem 10 the *oriented median algorithm*. The factor $\frac{2}{d+1/d}$ might seem a little surprising, but it turns out to be sharp, in the following sense: there are infinite families of d-regular oriented graphs G for which the oriented median algorithm outputs a cut of size precisely $\frac{2}{d+1/d}\text{maxdicut}(G)$ (an example is given in the full version of the paper). So the problem does not come from the analysis of the algorithm, but rather from the algorithm itself.

To overcome this issue and close the gap with the $\frac{2}{d}$ bound, one might be tempted to consider local improvements. In the following, a vertex will be *stable* if it has at least one neighbor on the other side of the cut. Otherwise it will be *unstable*. We now consider the following simple algorithm: at every round, every unstable vertex changes side. When we perform one round of this algorithm, we say we perform a *flip* (as this algorithm can be seen as a variant of the well known FLIP algorithm that is further discussed in the conclusion).

Theorem 11. *Assume that $d \geqslant 3$ is odd. Then the 2-round algorithm consisting of the oriented median algorithm followed by two flips provides a $\frac{2}{d+1/d-c/d^2}$-approximation (for some $c > 0$) for the* MaxDiCut *problem in d-regular graphs.*

The proof of Theorem 11 is given in the full version of the paper. Theorem 11 proves that after 2 flips, we can slightly improve on the approximation ratio of Theorem 10. A natural question is whether the same can be achieved after a single flip (in the full version of the paper, we give an example showing that the answer is negative).

4 Conclusion

4.1 FLIP

In Sect. 3, we have designed a very simple one-round algorithm approximating MaxCut in regular graphs (with odd degrees). Once a solution has been obtained, it might be tempting to run a few more rounds of computation to see if the solution can be improved locally.

We have already seen a simple way to improve the quality of a solution (by moving the so-called unstable vertices to the other side of the cut), but the notion of stability we used was specifically designed to improve the approximation ratio in a small number of rounds. Another simple way to locally improve a cut (in the sequential setting this time) is to take a vertex with more neighbors in its own part than in the other part, and change its side. If this is done until no such vertex exists, the resulting cut is *maximal*, and in this case is a $\frac{1}{2}$-approximation of the maximum cut. This operation, called *FLIP*, has been studied for a long time. When the edges are weighted, it was proved by Poljak [23] that any sequence of FLIPs takes only polynomially many steps before reaching a maximal cut in cubic graphs, while Monien and Tscheuschner [20] proved that there are graphs of maximum degree 4 for which a sequence of FLIPs can take exponentially many steps to reach a maximal cut. In the unweighted case however, since each flip improves the cut by at least one, the maximum number of flips before reaching a maximal cut is bounded by the number edges (which is linear in n in bounded degree graphs). In the distributed framework, it might be tempting to consider running some rounds of the *distributed FLIP* dynamics: at each round, each vertex with more neighbors in its own part than in the other part changes side. The graph D^d_{2n} constructed in the previous sections shows that it might not be helpful at all: if all the vertices of the outer cycle are in one side of the cut, and all the vertices of the inner cycle are on the other side of the cut, then at each round, all the vertices of the graph would change side, not improving the solution.

It might be worth noting that in our application of the median algorithm, not all vertices of the outer cycle of D_{2n}^d are on the same side of the cut (given some bad labelling): due to some side-effects, roughly d vertices in the outer cycle are not on the same side of the cut as the others, and similarly for the inner cycle. It can then be checked that if we run the distributed FLIP dynamics in this instance, the solution does improve over time, but improving the approximation ratio from $\frac{1}{d}$ to $\frac{1}{d} + \epsilon$ requires $\Omega(\epsilon n)$ rounds, which is extremely unpractical. This has to be compared with the lower bound of Theorem 3, which says that in order to achieve an approximation ratio of $\frac{1}{d} + \epsilon$ in general, one needs a number of rounds of the order of $\Omega(\log^* n)$.

4.2 SLOCAL vs LOCAL Model in the Deterministic Setting

Recently introduced by Ghaffari, Kuhn and Maus in [12], the SLOCAL model is designed to study the influence of the two major issues in LOCAL algorithms separately: in this model nodes are processed sequentially in any order. When a node v is processed, it can access its t-neighborhood and eventually additional information stored by vertices in this neighborhood that have been processed before v. In this model, symmetry breaking becomes free (the given order already breaks the symmetry) and only locality remains challenging.

Although finding an approximate maximum cut is not locally checkable, one can make an interesting parallel in this case. A very simple deterministic approximation algorithm for the maximum cut can be run in the SLOCAL model: simply process vertices in any order and when a vertex is processed put it on the side maximizing its local cut according to already processed neighbors. Note that each vertex loses at most half of its edges at each step, and each edge is counted once, so this is indeed a $\frac{1}{2}$-approximation with locality 1 and although we are studying a problem that is not locally checkable, this suggests that symmetry breaking (and not locality) is the bottleneck in our case.

Acknowledgement. We would like to thank Jérémie Chalopin and Keren Censor-Hillel for their remarks on the complexity of finding an orientation using very small messages in the CONGEST model. We also thank Michal Dory for calling reference [5] to our attention, and David Gamarnik for pointing out references [3,6,9,19] to us.

References

1. Åstrand, M., Polishchuk, V., Rybicki, J., Suomela, J., Uitto, J.: Local algorithms in (weakly) coloured graphs. CoRR abs/1002.0125 (2010)
2. Censor-Hillel, K., Levy, R., Shachnai, H.: Fast distributed approximation for max-cut. In: Fernández Anta, A., Jurdzinski, T., Mosteiro, M.A., Zhang, Y. (eds.) ALGOSENSORS 2017. LNCS, vol. 10718, pp. 41–56. Springer, Cham (2017). https://doi.org/10.1007/978-3-319-72751-6_4

3. Csóka, E., Gerencsér, B., Harangi, V., Virág, B.: Invariant gaussian processes and independent sets on regular graphs of large girth. Random Struct. Algorithms **47**, 284–303 (2015)
4. Chang, Y.-J., Kopelowitz, T., Pettie, S.: An exponential separation between randomized and deterministic complexity in the LOCAL model. In: IEEE 57th Annual Symposium on Foundations of Computer Science (2016)
5. Czygrinow, A., Hańćkowiak, M., Wawrzyniak, W.: Fast distributed approximations in planar graphs. In: Taubenfeld, G. (ed.) DISC 2008. LNCS, vol. 5218, pp. 78–92. Springer, Heidelberg (2008). https://doi.org/10.1007/978-3-540-87779-0_6
6. Dembo, A., Montanari, A., Sen, S.: Extremal cuts of sparse random graphs. Ann. Probab. **45**(2), 1190–1217 (2017)
7. Erdős, P., Rado, R.: Combinatorial theorems on classifications of subsets of a given set. Proc. Lond. Math. Soc. **3**, 417–439 (1952)
8. Erdős, P., Szekeres, G.: A combinatorial problem in geometry. Compos. Math. **2**, 463–470 (1935)
9. Gamarnik, D., Li, Q.: On the max-cut of sparse random graphs. Random Struct. Algorithms **52**(2), 219–262 (2018)
10. Gamarnik, D., Sudan, M.: Limits of local algorithms over sparse random graphs. In: Proceedings of Innovations in Theoretical Computer Science (ITCS), pp. 369–376 (2014)
11. Ghaffari, M., Harris, D.G., Kuhn, F.: On derandomizing local distributed algorithms. In: IEEE Symposium on Foundations of Computer Science (2018)
12. Ghaffari, M., Kuhn, F., Maus, Y.: On the complexity of local distributed graph problems. In: 49th Annual ACM Symposium on Theory of Computing, pp. 784–797 (2017)
13. Göös, M., Hirvonen, J., Suomela, J.: Lower bounds for local approximation. J. ACM **60**, #39 (2013)
14. Hatami, H., Lovász, L., Szegedy, B.: Limits of local-global convergent graph sequences. Geom. Funct. Anal. **24**(1), 269–296 (2014)
15. Hirvonen, J., Rybicki, J., Schmid, S., Suomela, J.: Large cuts with local algorithms on triangle-free graphs. Electron. J. Combin. **24**(4), #P4.21 (2017)
16. Kardoš, F., Král', D., Volec, J.: Maximum edge-cuts in cubic graphs with large girth and in random cubic graphs. Random Struct. Algorithms **41**(4), 506–520 (2012)
17. Kawarabayashi, K.-I., Schwartzman, G.: Adapting local sequential algorithms to the distributed setting. In: 32nd International Symposium on Distributed Computing, pp. 35:1–35:17 (2018)
18. Linial, N.: Locality in distributed graph algorithms. SIAM J. Comput. **21**(1), 193–201 (1992)
19. Lyons, R.: Factors of IID on trees. Combin. Probab. Comput. **26**(2), 285–300 (2017)
20. Monien, B., Tscheuschner, T.: On the power of nodes of degree four in the local max-cut problem. In: Calamoneri, T., Diaz, J. (eds.) CIAC 2010. LNCS, vol. 6078, pp. 264–275. Springer, Heidelberg (2010). https://doi.org/10.1007/978-3-642-13073-1_24
21. Mubayi, D., Suk, A.: A survey of hypergraph Ramsey problems. ArXiv e-prints (2017)
22. Naor, M., Stockmeyer, L.: What can be computed locally?. In: 25th Annual ACM Symposium on Theory of Computing, pp. 184–193 (1993)
23. Poljak, S.: Integer linear programs and local search for max-cut. SIAM J. Comput. **21**(3), 450–465 (1995)

24. Rahman, M., Virág, B.: Local algorithms for independent sets are half-optimal. Ann. Probab. **45**(3), 1543–1577 (2017)
25. Ramsey, F.P.: On a problem of formal logic. Proc. Lond. Math. Soc. **30**, 264–286 (1930)
26. Shearer, J.B.: A note on bipartite subgraphs of triangle-free graphs. Random Struct. Algorithms **3**(2), 223–226 (1992)

Fixed-Parameter Tractability of Counting Small Minimum (S, T)-Cuts

Pierre Bergé[1(✉)], Benjamin Mouscadet[2], Arpad Rimmel[2], and Joanna Tomasik[2]

[1] LRI, Université Paris-Sud, Université Paris-Saclay, Orsay, France
Pierre.Berge@lri.fr
[2] LRI, CentraleSupélec, Université Paris-Saclay, Orsay, France
Benjamin.Mouscadet@supelec.fr, {Arpad.Rimmel,Joanna.Tomasik}@lri.fr

Abstract. The parameterized complexity of counting minimum cuts stands as a natural question because Ball and Provan showed its #P-completeness. For any undirected graph $G = (V, E)$ and two disjoint sets of its vertices S, T, we design a fixed-parameter tractable algorithm which counts minimum edge (S, T)-cuts parameterized by their size p. Our algorithm operates on a transformed graph instance. This transformation, called drainage, reveals a collection of at most $n = |V|$ successive minimum (S, T)-cuts Z_i. We prove that any minimum (S, T)-cut X contains edges of at least one cut Z_i. This observation, together with Menger's theorem, allows us to build the algorithm counting all minimum (S, T)-cuts with running time $2^{O(p^2)}n^{O(1)}$. Initially dedicated to counting minimum cuts, it can be modified to obtain an FPT sampling of minimum edge (S, T)-cuts.

Keywords: Fixed-parameter tractability · Counting problems · Minimum cuts

1 Introduction

The issue of counting minimum cuts in graphs has been drawing attention over the years due to its practical applications. Indeed, the number of minimum cuts is an important factor for the network reliability analysis [2–4, 24]. Thereby, the probability that a stochastic graph is connected may be computed [3]. Furthermore, cuts on planar graphs are used for image segmentation [8]. An image is seen as a planar graph where vertices represent pixels and edges connect two neighboring pixels if they are similar. Counting minimum cuts provides an estimation of the number of segmentations.

We focus on the problem of counting minimum edge (S, T)-cuts in undirected graphs $G = (V, E)$, $S, T \subseteq V$. We call it COUNTING MINCUTS (Definition 1) as it is the counting variant of the classical problem MINCUT, which asks for a minimum (S, T)-cut in graph G. Ball and Provan showed in [26] that COUNTING MINCUTS is unlikely solvable in polynomial time as it is #P-complete. They

I. Sau and D. M. Thilikos (Eds.): WG 2019, LNCS 11789, pp. 79–92, 2019.
https://doi.org/10.1007/978-3-030-30786-8_7

also devised a polynomial-time algorithm for COUNTING MINCUTS on planar graphs [3]. Bezáková and Friedlander [6] generalized it with an $O(n\mu + n \log n)$-time algorithm on weighted planar graphs, where μ is the length of the shortest (s,t)-paths. For general graphs, some upper bounds on the number of minimum cuts have been given [10] in function of parameters such as the radius, the maximum degree, etc. Two fixed-parameter tractable (FPT) algorithms have been proposed for COUNTING MINCUTS. Bezáková *et al.* [5] built an algorithm for both directed and undirected graphs with small treewidth λ; its time complexity is $O(2^{3\lambda}\lambda n)$. Moreover, Chambers *et al.* [9] designed an algorithm for directed graphs embedded on orientable surfaces of genus g: its execution time is $O(2^g n^2)$. We study the fixed-parameter tractability of COUNTING MINCUTS, parameterized by the size p of the minimum (S,T)-cuts.

Definition 1 (Counting mincuts).

Input: *Undirected graph* $G = (V, E)$*, sets of vertices* $S, T \subsetneq V$*,* $S \cap T = \emptyset$*.*

Output: *The number of minimum edge* (S,T)*-cuts.*

The minimum (S,T)-cut size for a COUNTING MINCUTS instance $\mathcal{I} = (G, S, T)$ is obtained in polynomial time [17]. A brute force XP algorithm computes the number $C(\mathcal{I})$ of minimum (S,T)-cuts in time $n^{O(p)}$ by enumerating all edge sets of size p and picking up those which are (S,T)-cuts. More efficient exponential algorithms exist, as the one of Nagamochi *et al.*, in time $O\left(pn^2 + pnC(\mathcal{I})\right)$, in [24]. Our contribution, summarized in the theorem below, is an algorithm efficient for small values of p.

Theorem 1. *The counting of minimum edge* (S,T)*-cuts can be solved in time* $O(2^{p(p+2)}pmn^3)$ *on undirected graphs* $G = (V, E)$*.*[1]

An FPT$\langle p \rangle$ algorithm can be deduced from the results in two articles [5,21] and its execution time is $O^*\left(2^{H(p)}\right)$ where $H(p) = \Omega\left(\frac{2^p}{\sqrt{p}}\right)$. The treewidth reduction theorem established by Marx *et al.* in [21] says that there is a linear-time reduction transforming graph G into another graph G' which conserves the (S,T)-cuts of size p and such that the treewidth of G', $\tau(G')$, verifies $\tau(G') = 2^{O(p)}$. After this transformation, the number of minimum (S,T)-cuts of G' is obtained thanks to the algorithm given in [5]. The overall time taken with this method is $O^*\left(2^{2^p}\right)$. Our result, Theorem 1, improves this exponential factor.

This result highlights a complexity gap between the counting and the enumeration, as the latter cannot be FPT parameterized by p. Indeed, certain instances contain a number $C(\mathcal{I}) = (\frac{n-1}{p})^p$ of minimum cuts, as in case of graph G made of p vertex-disjoint (S,T)-paths with $S = \{s\}$ and $T = \{t\}$.

Our algorithm is based on a cut-decomposition $\mathcal{Z}(\mathcal{I}) = (Z_1, \ldots, Z_k)$ of instance $\mathcal{I}$, $1 \leq k < n$, called the *drainage* where for every $1 \leq i \leq k$, edge

[1] The proofs in Sects. 3 and 4 are withdrawn due to the restriction on the number of pages. The full version can be found here: https://arxiv.org/abs/1907.02353.

set Z_i is a minimum (S,T)-cut. Set $R(Z_i,S)$ denotes the vertices which are reachable from S after the removal of edges in Z_i. The reachable sets of Z_i are included one into another: $R(Z_1,S) \subsetneq R(Z_2,S) \subsetneq \ldots \subsetneq R(Z_k,S)$. The drainage fulfils the following property: if X is a minimum (S,T)-cut, some edges B_i of a certain Z_i belong to X, $B_i = X \cap Z_i \neq \emptyset$, and no other edge of X has one endpoint in $R(Z_i,S)$. The set B_i is called the *front dam* of cut X. The key idea of the recursive counting we propose is that any minimum cut X is the union of its front dam with a minimum cut of a sub-instance, called *dry instance*, of $\mathcal{I}$. These techniques work as well on multigraphs, *i.e.* on graphs with multiple edges. After modifications, our algorithm also samples minimum edge (S,T)-cuts.

To design the drainage $\mathcal{Z}(\mathcal{I})$, we use the concept of *important cuts* [20] which is the key ingredient of many FPT algorithms to solve cuts problems [7,12,14, 20,22]. An (S,T)-cut Y is *important* if there is no other (S,T)-cut Y' such that $|Y'| \leq |Y|$ and $R(Y,S) \subsetneq R(Y',S)$. There is a unique minimum important (S,T)-cut and it can be identified in polynomial time [20].

The second concept used in our algorithm is Menger's theorem [23]. It states that the size of minimum edge (S,T)-cuts in an undirected graph is equal to the maximum number of edge-disjoint (S,T)-paths. As the max-flow min-cut theorem [17] generalizes Menger's theorem, one of the largest sets of edge-disjoint (S,T)-paths is found in polynomial time.

To close this introductory chapter, we give a "table of contents" of our article. Section 2 introduces the notations used. Section 3 explains the construction of the drainage $\mathcal{Z}(\mathcal{I}) = (Z_1,\ldots,Z_k)$. In Sect. 4, we propose our algorithm and compute its time complexity. Finally, we conclude and give ideas about future research.

2 Definitions and Notation

We summarize basic concepts of parameterized and counting complexity but also introduce the notation we will use.

Fixed-Parameter Tractability. NP-hard problems are unlikely to be solvable with polynomial time algorithms. However, solving them efficiently may become possible when parameters are associated to problem instances and the values of these parameters are small.

Referring to Downey and Fellows [15] and Niedermeier [25], a parameterized problem is said *fixed-parameter tractable* (FPT) if there is an algorithm solving it in time $O(f(p)P(n)) = O^*(f(p))$, where p is a parameter, n is the instance size, P is a polynomial function, and f is an arbitrary computable function. As a problem may be studied for different parameters $p_1, p_2, \ldots$, the notation "FPT" becomes ambiguous. If there is an algorithm solving a problem in time $O(f(p_1)P(n))$, then it is FPT$\langle p_1 \rangle$. In this study, the parameter p of COUNTING MINCUTS is the size of the minimum (S,T)-cut.

Counting Problems. The study of #P complexity class and the counting problems it contains, started with Valiant [27]. Class #P is the set of counting problems such that their decision version is in class NP. The subclass #P-complete contains counting problems such that all problems in #P can be reduced to them with a polynomial-time counting reduction. No #P-complete problem can be solved in polynomial time unless P = NP. Moreover, there are decision problems such as CNF-2SAT [19] which are solvable in polynomial time but their associated counting problem is #P-complete [27]. The complexity of counting problems has been extended via the parameterized complexity framework [13,16]. A relevant question to ask about a #P-complete problem is whether there is an FPT algorithm counting all its solutions. For example, with G and H as input, FPT algorithms counting the number of occurrences of H as a subgraph of G have been intensively studied [1,18,28].

Cuts in Undirected Graphs. We study undirected graphs $G = (V, E)$, where $n = |V|$ and $m = |E|$. For any set of vertices $U \subseteq V$, we denote by $E[U]$ the set of edges of G with two endpoints in U and $G[U]$ the subgraph of G induced by U: $G[U] = (U, E[U])$. Notation $G \backslash U$ refers to the graph deprived of vertices in U. For any set of edges $E' \subseteq E$, the graph G deprived of edges E' is denoted by $G \backslash E'$:

$$G \backslash U = G[V \backslash U] \quad \text{and} \quad G \backslash E' = (V, E \backslash E').$$

A *path* is a sequence of pairwise different vertices $v_1 \cdot v_2 \cdot v_3 \cdots v_i \cdot v_{i+1} \cdots$, where two successive vertices (v_i, v_{i+1}) are adjacent in G. To improve readability, we abuse notations: $v_1 \in Q$ and $(v_1, v_2) \in Q$ mean that vertex v_1 and edge (v_1, v_2) are on path Q, respectively.

Cut problems usually consist in finding the smallest set of edges $X \subseteq E$ which splits the graph $G \backslash X$ into connected components meeting certain criteria. Given two sets of vertices S (sources) and T (targets), set $X \subseteq E$ is an (S, T)-cut if there is no path connecting a vertex from S with a vertex from T in $G \backslash X$. An (S, T)-cut X is said to be *minimum* if there is no (S, T)-cut X' such that $|X'| < |X|$. For any (S, T)-cut X, its *source side* $R(X, S)$ is the set of vertices that are reachable from S in $G \backslash X$. Its *target side* $R(X, T)$ contains the vertices reachable from T in $G \backslash X$. We define two sets $V^S(X)$ and $V^T(X)$:

- set $V^S(X) = \{u \in R(X, S); (u, v) \in X\}$, *i.e.* the vertices of $R(X, S)$ incident to cut X,
- set $V^T(X) = \{u \in R(X, T); (u, v) \in X\}$, *i.e.* the vertices of $R(X, T)$ incident to cut X.

Important and Closest Cuts. As defined in [20], an (S, T)-cut X is *important* if there is no other (S, T)-cut X' such that $|X'| \leq |X|$ and $R(X, S) \subsetneq R(X', S)$. Intuitively, an important (S, T)-cut is such that there is no other cut smaller in size which is closer to T. The number of important (S, T)-cuts of size at most p depends only on p [11] and there is no more than one minimum important (S, T)-cut [20]. Although the proofs in [20] handle vertex cuts, an edge-to-vertex reduction preserves these properties on edge cuts [7,20].

Lemma 1 (Unicity of minimum important cuts [20]). *For disjoint sets of vertices S and T, there is a unique minimum important (S, T)-cut and it may be found in polynomial time.*

On undirected graphs, we say that an important (T, S)-cut is a *closest* (S, T)-*cut*. Figure 1 gives an example of graph G with two (S, T)-cuts X_1 and X_2, where $S = \{s_1, s_2\}$ and $T = \{t\}$. Cut X_1 is not closest as the edges incident to S form a cut Z_1 smaller than X_1 and $R(Z_1, S) \subsetneq R(X_1, S)$. Cut X_2 is closest because there is no cut with at most three edges whose reachable set of vertices is contained in $R(X_2, S)$.

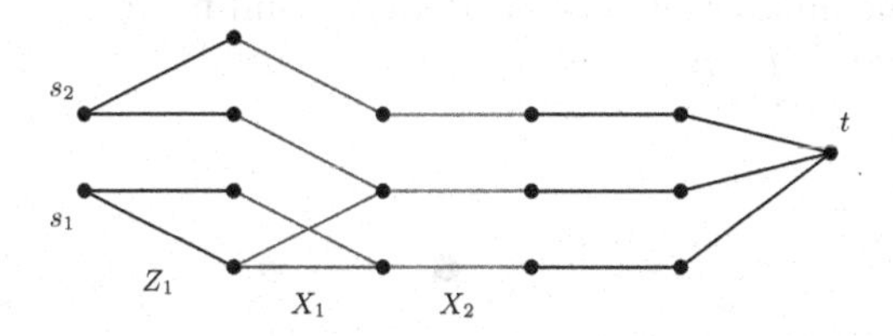

Fig. 1. Illustration of Definition 2 for closest (S, T)-cuts: X_2 is closest whereas X_1 is not.

Definition 2. *An (S, T)-cut X is closest if there is no other (S, T)-cut X' such that $|X'| \leq |X|$ and $R(X', S) \subsetneq R(X, S)$.*

As a minimum closest (S, T)-cut is also a minimum important (T, S)-cut on undirected graphs, there is a unique minimum closest (S, T)-cut according to Lemma 1. Since the graph is uncapacitated, computing the minimum closest (S, T)-cut is made in time $O(mp)$, using p iterations of Ford-Fulkerson's algorithm [17].

3 Framework: Drainage and Menger's Paths

We introduce tools needed to design an algorithm solving COUNTING MINCUTS in FPT$\langle p \rangle$ time, where p is the size of any minimum (S, T)-cut. We build the *drainage*, a collection of minimum cuts Z_i, $i \in \{1, \ldots, k\}$, where $k < n$, such that at least one edge of any minimum (S, T)-cut X belongs to $\bigcup_{i=1}^{k} Z_i$. Then, we highlight properties coming from Menger's theorem.

3.1 Construction of the Drainage

The drainage $\mathcal{Z}(\mathcal{I}) = (Z_1, \ldots, Z_k)$ of an instance $\mathcal{I} = (G, S, T)$ is a collection of minimum (S, T)-cuts Z_i, $|Z_i| = p$, satisfying the following properties:

- there are less than n cuts Z_i, *i.e.* $1 \leq k < n$,
- the reachable sets of cuts Z_i fulfil $R(Z_i, S) \subsetneq R(Z_{i+1}, S)$ for $i \in \{1, \ldots, k-1\}$,
- for any minimum (S, T)-cut X, there is at least one cut Z_i which has edges with X in common: $X \cap Z_i \neq \emptyset$.

We construct the drainage iteratively. Let $S_1 = S$ and Z_1 be the minimum closest (S_1, T)-cut. We fix $R_1 = R(Z_1, S)$. Let S_2 be the set of vertices incident to edges of Z_1 inside $R(Z_1, T)$: $S_2 = V^T(Z_1) = \{v \notin R_1, (u, v) \in Z_1\}$.

Next, we construct Z_2 which is the minimum closest (S_2, T)-cut in $G \backslash R(Z_1, S)$. If $|Z_2| > p$, the drainage construction stops. Otherwise, if $|Z_2| = p$, set R_2 follows the same scheme as R_1, $R_2 = R(Z_2, S_2)$ in graph $G \backslash R(Z_1, S)$. We repeat the process until no more minimum (S_i, T)-cut Z_i of size p can be found. We denote by k the number of cuts Z_i produced and fix $R_{k+1} = R(Z_k, T)$. Cuts Z_i form the *minimum drainage cuts* of $\mathcal{I}$.

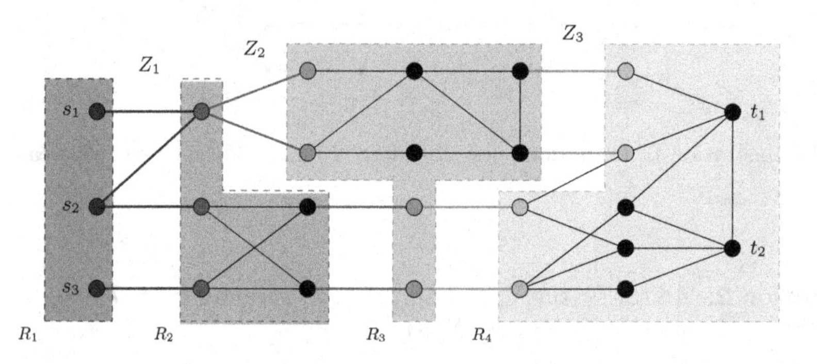

Fig. 2. The drainage (cuts Z_i, sets R_i and S_i) for an instance containing graph G, sources $S = \{s_1, s_2, s_3\}$ and targets $T = \{t_1, t_2\}$. Here, $R_1 = S_1$ (in general, $R_1 \supseteq S_1$). (Color figure online)

Figure 2 provides us with an example of graph G with $S = \{s_1, s_2, s_3\}$ and $T = \{t_1, t_2\}$ and indicates its drainage. The size of minimum (S, T)-cuts is $p = 4$. Blue, red, and green edges represent minimum drainage cuts Z_1, Z_2, and Z_3, respectively. Similarly, blue, red, green, and yellow vertices represent sets $S_1 = S$, S_2, S_3, and S_4. Reachable sets R_1, R_2, R_3, and R_4 are also appropriately colored. As the size of the minimum cut between S_4 (yellow vertices) and T in graph $G \backslash R(Z_3, S)$ is greater than p, we have $k = 3$.

We emphasize that set R_i, which is $R(Z_i, S_i)$ taken in $G \backslash R(Z_{i-1}, S)$, and set $R(Z_i, S)$ are different for $i \neq 1$. On the one hand, set $R(Z_i, S) = \bigcup_{\ell=1}^{i} R_\ell$ contains the vertices reachable from S in graph G deprived of Z_i. On the other hand, set R_i can be written $R_i = R(Z_i, S) \backslash R(Z_{i-1}, S)$. Sets R_i and R_{i+1} are disjoint and nonempty, as $S_i \subseteq R_i$ and $S_{i+1} \subseteq R_{i+1}$. Moreover, the minimum drainage cuts are disjoint: $Z_i \cap Z_j = \emptyset$. The number k of minimum drainage cuts is less than n and the running time needed to construct the drainage is in $O(mnp)$. The reachable vertex sets of cuts Z_i are included one into another: $R(Z_i, S) \subsetneq R(Z_{i+1}, S)$. The following theorem shows that, for any minimum (S, T)-cut X, there is a cut Z_i containing edges of X. Among cuts Z_i sharing edges with X, we are interested in the one with the smallest index.

Definition 3 (Front of X). *Front of X, $i(X)$, $1 \le i(X) \le k$ is the smallest index i such that $Z_i \cap X \neq \emptyset$.*

The next theorem states the properties of $i(X)$ for any minimum (S,T)-cut.

Theorem 2. *Any minimum (S,T)-cut X admits a front $i(X)$ and $X \cap E\left[R(Z_{i(X)}, S)\right] = \emptyset$.*

The reader can verify that any minimum (S,T)-cut of G contains some edges of at least one cut Z_1, Z_2, or Z_3 in Fig. 2.

3.2 Menger's Paths

Menger's theorem states that the size of the minimum edge (S,T)-cuts is equal to the maximum number of edge-disjoint (S,T)-paths [23]. One of these largest sets of edge-disjoint (S,T)-paths can be found in polynomial time using flow techniques [17]. We denote by $\mathcal{Q} = \{Q_1, \ldots, Q_p\}$ such a set of p edge-disjoint (S,T)-paths, taken arbitrarily. We call paths from $\mathcal{Q}$ *Menger's paths*, to distinguish them from other paths in graph G.

Set $\mathcal{Q}$ is used to identify minimum (S,T)-cuts. It is fixed throughout the course of the proofs in this article. The observation that edges of all minimum (S,T)-cuts belong to the paths from $\mathcal{Q}$ is formulated in:

Lemma 2. *For any minimum (S,T)-cut X, each Menger's path Q_j contains one edge of X. If $Q_j : v_1^{(j)} \cdot v_2^{(j)} \cdots v_\ell^{(j)} \cdot v_{\ell+1}^{(j)} \cdots$ and $(v_\ell^{(j)}, v_{\ell+1}^{(j)}) \in X$, then $v_\ell^{(j)} \in R(X,S)$ and $v_{\ell+1}^{(j)} \in R(X,T)$.*

Its consequence is that each edge of a cut Z_i belongs to a Menger's path. Figure 3 illustrates the Menger's paths on the instance (G,S,T) already introduced in Fig. 2. As the minimum (S,T)-cut size is four, there are four edge-disjoint (S,T)-paths distinguished with colors. Menger's paths are naturally oriented from sources to targets.

4 Counting Minimum Edge (S,T)-Cuts in Undirected Graphs

We describe our algorithm which counts all minimum (S,T)-cuts in an undirected graph G. Based on the concepts introduced in Sect. 3, we prove not only that it achieves its objective but also that its time complexity is FPT.

4.1 Dams and Dry Areas

We begin by the definition of *dams* which are subsets of cuts Z_i of the drainage of G.

Definition 4 (Dam). *A dam B_i is a nonempty subset of a minimum drainage cut Z_i,* i.e. $B_i \subseteq Z_i$, $B_i \neq \emptyset$.

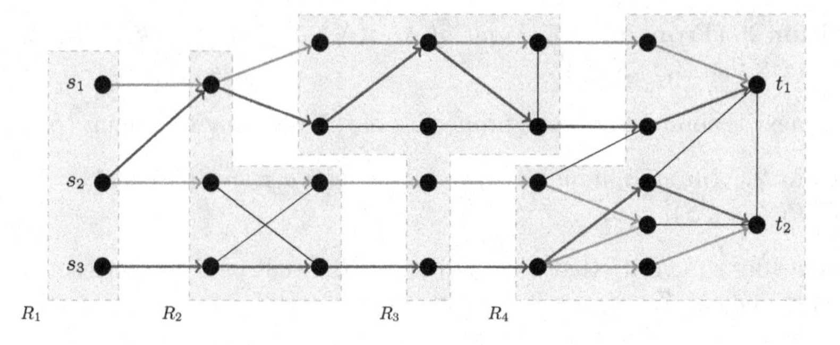

Fig. 3. Menger's paths in graph G with sources $S = \{s_1, s_2, s_3\}$, targets $T = \{t_1, t_2\}$. (Color figure online)

Thanks to this definition, Theorem 2 together with the concept of the front makes us observe that any minimum (S, T)-cut X contains a *front dam*:

Definition 5 (Front dam). *The front dam of a minimum (S, T)-cut X is $B_{i(X)} = X \cap Z_{i(X)}$.*

We know that all edges in $X \backslash B_{i(X)}$ belong to the target side of $Z_{i(X)}$, $E\left[R(Z_{i(X)}, T)\right]$, and the source side of $Z_{i(X)}$ is empty, $X \cap E\left[R(Z_{i(X)}, S)\right] = \emptyset$. If $X \backslash B_{i(X)} = \emptyset$, then $X = Z_{i(X)}$. A dam B_i is characterized by:

- its *level*, *i.e.* the index i of the cut Z_i it belongs to,
- its *signature* $\sigma(B_i) = \{Q_j : B_i \cap Q_j \neq \emptyset\}$, *i.e.* the set of Menger's paths passing through it.

Choking graph G with dam $B_{i(X)}$ puts in evidence a subgraph which still connects S and T through $X \backslash B_{i(X)}$. Our idea is to dam a graph gradually in order to dry it completely.

The description of the method we devised to reach this goal requires a transformation of G into G_D which is actually G with certain edges directed (G_D is a mixed graph). If edge e does not belong to a Menger's path, it stays undirected. For path $Q_j : v_1^{(j)} \cdot v_2^{(j)} \cdot v_3^{(j)} \cdots$, edges $(v_i^{(j)}, v_{i+1}^{(j)})$ become arcs $(v_i^{(j)}, v_{i+1}^{(j)})$, respecting the natural flow from sources to targets.

Figure 3 illustrates graph G_D. Arrows indicate the arcs while bare segments represent its edges. According to Lemma 2, any minimum (S, T)-cut of G is made up of arcs in G_D. Minimum drainage cuts Z_i are thus composed of arcs, directed from R_i to R_{i+1}. We insist on the fact that graph G_D is only used to define the notion of dry area, we do not count minimum cuts in it.

Definition 6 (Dry area). *The dry area of B_i is the set $A^*(B_i)$ which contains the vertices of G which are not reachable from S in graph G_D deprived of B_i,* i.e. $G_D \backslash B_i$.

In a less formal way, set $A^*(B_i)$ keeps vertices which are dried as B_i is the only means to irrigate them. The definition of the dry instance follows.

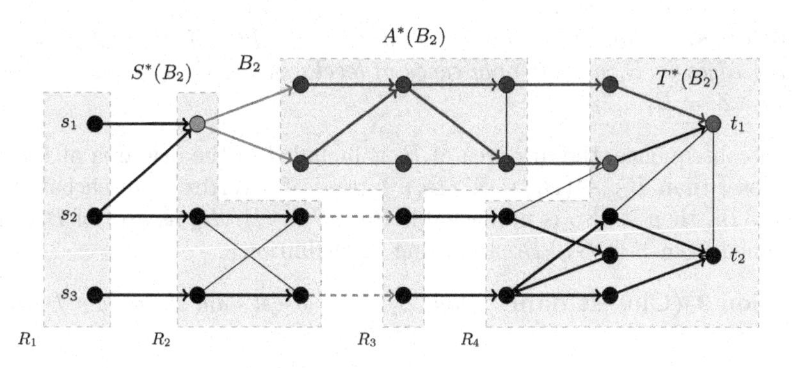

Fig. 4. An example of dam B_2 and its dry instance $\mathcal{D}(\mathcal{I}, B_2) = (G^*(B_2), S^*(B_2), T^*(B_2))$ (Color figure online)

Definition 7 (Dry instance). *The dry instance induced by a dam B_i is an instance $\mathcal{D}(\mathcal{I}, B_i) = (G^*(B_i), S^*(B_i), T^*(B_i))$ with graph $G^*(B_i) = (V^*(B_i), E^*(B_i))$. In particular,*

- *set $S^*(B_i)$ keeps vertices reachable from S "just before" dam B_i. Formally, it contains the tails of arcs in B_i: $S^*(B_i) = \{u : (u,v) \in B_i\}$,*
- *set $T^*(B_i)$ keeps vertices placed "after" dam B_i which become irrigated in $G_D \backslash B_i$. Formally, it contains the heads of arcs which have their tail either inside $S^*(B_i)$ or inside $A^*(B_i)$ and their head outside: $T^*(B_i) = \{v \notin A^*(B_i) : (u,v) \in E, u \in S^*(B_i) \cup A^*(B_i)\}$,*
- *set $V^*(B_i)$ is the union: $V^*(B_i) = S^*(B_i) \cup A^*(B_i) \cup T^*(B_i)$,*
- *set $E^*(B_i)$ stores edges of G which lie inside the dry area of B_i or on its border (one endpoint is outside) in G_D. Formally, it is composed of edges with two endpoints in $V^*(B_i)$ and at least one of them in $A^*(B_i)$: $E^*(B_i) = \{(u,v) \in E : u \in A^*(B_i), v \in V^*(B_i)\}$.*

Figure 4 gives an example of dam $B_2 \subseteq Z_2$ and the dry instance it induces in G. Its arcs are drawn in red, arcs of $Z_2 \backslash B_2$ are red and dashed. Blue vertices represent the vertices unreachable from S in $G_D \backslash B_2$, *i.e.* set $A^*(B_2)$. Sets $S^*(B_2)$ and $T^*(B_2)$ are drawn in green and purple, respectively. Set $E^*(B_2)$ is composed of dam B_2 (red arcs) and blue edges/arcs.

An important property of dry areas is that there is no arc (u,v) of G_D "entering" in the dry area $A^*(B_i)$, except for arcs in B_i.

Lemma 3. *For any dam B_i, there is no arc (u,v) in G_D such that $u \notin A^*(B_i)$ and $v \in A^*(B_i)$, except for arcs in B_i. Moreover, there is no undirected edge with exactly one endpoint in $A^*(B_i)$.*

In Theorem 3, we provide a characterization of any minimum (S,T)-cut which is based on dry instances and on *closest dams*. We start by:

Definition 8. *A dam B_h is closer than dam B_i if (i) $h < i$, (ii) $\sigma(B_h) = \sigma(B_i)$, and (iii) edges in B_i are the only edges of level i inside the dry instance of B_h: $E^*(B_h) \cap Z_i = B_i$.*

As a consequence, the dry area of B_i is included in the dry area of B_h when B_h is closer than B_i: $A^*(B_i) \subsetneq A^*(B_h)$. Indeed, if a vertex is unreachable from S in $G_D \backslash B_i$, then it also is unreachable from S in $G_D \backslash B_h$ as arcs of B_i cannot be attained from S in $G_D \backslash B_h$ according to Definition 8.

Definition 9 (Closest dam). *Dam B_i is a closest dam if no dam B_h, $h < i$ is closer than B_i.*

For any dam B_i, either B_i is a closest dam or there is a closest dam $B_h \neq B_i$, closer than B_i. Each dam B_i admits a closest dam (itself or B_h) which is unique.

Lemma 4. *Any dam B_i has a unique closest dam.*

Moreover, if B_h is a closest dam then its complement $\overline{B}_h = Z_h \backslash B_h$ is also a closest dam. This property will be used to prove the fixed-parameter tractability of our algorithm.

Lemma 5. *If B_h is a closest dam, then $\overline{B}_h = Z_h \backslash B_h$ is also a closest dam.*

Observe that the dry areas of a dam B_i and its complement, $A^*(B_i)$ and $A^*(\overline{B_i})$ respectively, are disjoint because any vertex is reachable from S either in $G \backslash B_i$ or in $G \backslash \overline{B}_i$ or in both of them.

4.2 A Characterization of Minimum Cuts with Dry Instances

Theorem 3 provides us with the keystone to build our FPT$\langle p \rangle$ algorithm. It combines the concepts of dry instance and closest dam: given a minimum (S, T)-cut X and its front dam $B_{i(X)}$, either $X \backslash B_{i(X)} = \emptyset$ and $X = Z_{i(X)}$ or edges in $X \backslash B_{i(X)} \neq \emptyset$ belong to the dry instance of the dam $\overline{B}_{h(X)} = Z_{h(X)} \backslash B_{h(X)}$, where $B_{h(X)}$ is the closest dam of $B_{i(X)}$.

Theorem 3. *If $X \neq Z_{i(X)}$ is a minimum cut for $\mathcal{I}$, $B_{i(X)}$ its front dam, and $B_{h(X)}$ the closest dam of $B_{i(X)}$, then set $X \backslash B_{i(X)}$ is a minimum cut for the dry instance of $\overline{B}_{h(X)} = Z_{h(X)} \backslash B_{h(X)}$,* i.e. $\mathcal{D}\left(\mathcal{I}, \overline{B}_{h(X)}\right)$.

Therefore, any minimum (S, T)-cut X, which is not a minimum drainage cut $Z_{i(X)}$ itself, can be partitioned into two sets, $B_{i(X)}$ and $X \backslash B_{i(X)}$, such that:

- $B_{i(X)}$ is a minimum cut of instance $\mathcal{D}\left(\mathcal{I}, B_{h(X)}\right)$ and a dam of $\mathcal{I}$,
- $X \backslash B_{i(X)}$ is a minimum cut of instance $\mathcal{D}\left(\mathcal{I}, \overline{B}_{h(X)}\right)$ and all its edges belong to the target side of $Z_{i(X)}$, $E\left[R(Z_{i(X)}, T)\right]$.

Conversely, given a closest dam B_h and its complement $\overline{B}_h = Z_h \backslash B_h$, the union $B_i \cup X_{\overline{B}_h}$, where the closest dam of B_i is B_h and $X_{\overline{B}_h}$ is a minimum cut of $\mathcal{D}\left(\mathcal{I}, \overline{B}_h\right)$, separates S from T.

Theorem 4. *Let B_h be a closest dam of $\mathcal{Z}(\mathcal{I})$ and $\overline{B}_h = Z_h \backslash B_h$. Let B_i be a dam such that B_h is closer than B_i and $X_{\overline{B}_h}$ a minimum cut of $\mathcal{D}(\mathcal{I}, \overline{B}_h)$. Then, $B_i \cup X_{\overline{B}_h}$ is a minimum (S,T)-cut for instance $\mathcal{I}$.*

We now prove a stronger result for set $X \backslash B_{i(X)}$. In fact, edges of set $X \backslash B_{i(X)}$ lie in the target side of level $i(X) - h(X) + 1$ in the drainage of instance $\mathcal{D}(\mathcal{I}, \overline{B}_{h(X)})$. This statement is formulated in the theorem below.

Theorem 5. *Let X be a minimum (S,T)-cut of G and let $(Z'_1, \ldots, Z'_{k'})$ be the drainage of instance $\mathcal{D}(\mathcal{I}, \overline{B}_{h(X)})$. Then, set $Z'_{i(X)-h(X)+1}$ is equal to $\overline{B}_{i(X)} = Z_{i(X)} \backslash B_{i(X)}$ and edges $X \backslash B_{i(X)}$ belong to the target side of $Z'_{i(X)-h(X)+1}$ inside instance $\mathcal{D}(\mathcal{I}, \overline{B}_{h(X)})$.*

4.3 Description of the Algorithm

Our algorithm starts by computing the drainage $\mathcal{Z}(\mathcal{I})$ and the Menger's paths of input instance $\mathcal{I}$. For all dams B_i, it counts the minimum cuts of size p in $\mathcal{I}$ which admit the front dam B_i. If $B_i \neq Z_i$, it does this recursively by counting the minimum cuts in instance $\mathcal{D}(\mathcal{I}, \overline{B}_h)$ which only contains edges from the target side of the internal level $i - h + 1$ of $\mathcal{D}(\mathcal{I}, \overline{B}_h)$, where B_h is the closest dam of B_i. The minimum cut size in $\mathcal{D}(\mathcal{I}, \overline{B}_h)$ is at most $p - 1$.

We denote by $C_0(\mathcal{I}) = C(\mathcal{I})$ the total number of minimum (S,T)-cuts of instance $\mathcal{I}$. We define $C_\ell(\mathcal{I})$ as the number of minimum cuts of instance $\mathcal{I}$ which are composed of edges from $E[R(Z_\ell, T)]$ only. For example, $C_2(\mathcal{I})$ gives the number of minimum (S,T)-cuts in instance $\mathcal{I}$ without edges of $Z_1 \cup Z_2$. Value $C_\ell(\mathcal{I})$, $0 \leq \ell \leq k-1$, can be written:

$$C_\ell(\mathcal{I}) = k - \ell + \sum_{\substack{\text{Closest} \\ \text{dam } B_h \subsetneq Z_h}} \; \sum_{\substack{i\,:\,i>\ell, \\ \exists B_i : B_h \\ \text{closer than } B_i}} C_{i-h+1}\left(\mathcal{D}\left(\mathcal{I}, \overline{B}_h\right)\right). \tag{1}$$

The first $k - \ell$ cuts are the minimum drainage cuts of $\mathcal{I}$ with level greater than ℓ, *i.e.* cuts $Z_{\ell+1}, \ldots, Z_k$. The second term counts cuts taking edges only from $E[R(Z_\ell, T)]$ and admitting a front dam $B_{i(X)} \neq Z_{i(X)}$. Theorems 3 and 5 guarantee that any of these minimum (S,T)-cuts is counted at least once. Indeed, for any front dam B_i and its closest dam B_h, we compute the number of cuts in instance $\mathcal{D}(\mathcal{I}, \overline{B}_h)$ such that all their edges belong to the target side of $\overline{B}_i$, which is the internal level $i - h + 1$ in $\mathcal{D}(\mathcal{I}, \overline{B}_h)$. In the event that the drainage of $\mathcal{D}(\mathcal{I}, \overline{B}_h)$ has less than $i-h+1$ levels, then $C_{i-h+1}(\mathcal{D}(\mathcal{I}, \overline{B}_h)) = 0$, as it means no minimum cut of $\mathcal{I}$ has the front dam B_i.

Conversely, the unicity of a closest dam ensures us that each minimum cut is counted exactly once. A minimum (S,T)-cut $X \neq Z_{i(X)}$ has a unique front dam $B_{i(X)}$ and the closest dam $B_{h(X)}$ of $B_{i(X)}$ is unique (Lemma 4). Finally, Theorem 4 secures that all cuts counted with Eq. (1) are minimum (S,T)-cuts.

Value $C_0(\mathcal{I})$ is computed thanks to recursive calls on multiple instances $\mathcal{D}(\mathcal{I}, \overline{B}_h)$. From now on, we distinguish the input instance $\mathcal{I}$ (for which we want to compute $C_0(\mathcal{I})$) with other instances (denoted by $\mathcal{J}$ later on) of the recursive tree. The base cases of the recursion, *i.e.* the leaves of the recursive tree, are the computation of values $C_\ell(\mathcal{J})$ either in instances $\mathcal{J}$ where the minimum cut size is one or in instances where no minimum cut admits a front dam $B_i \neq Z_i$, $i > \ell$. In both cases, the only minimum cuts of $\mathcal{J}$ are its minimum drainage cuts. Each recursive call of the algorithm makes the minimum cut size decrease: for example, if the minimum cut size of $\mathcal{J}$ is q, then it is $\left|\overline{B}_h\right| < q$ for an instance $\mathcal{D}\left(\mathcal{J}, \overline{B}_h\right)$. Therefore, the recursive tree is not deeper than $p-1$.

To terminate this algorithm analysis, we prove that its execution time is FPT$\langle p \rangle$. To count minimum cuts for an instance of depth d in the recursive tree, our algorithm counts not only its minimum drainage cuts but also minimum cuts of multiple instances of depth $d+1$. Consequently, the time complexity of our algorithm depends on the number of instances in the recursive tree. The following theorem states it is upper-bounded by $2^{p^2}m$.

Theorem 6. *There are at most $2^{p^2}m$ instances in the recursive tree.*

For any instance $\mathcal{J}$ of the recursive tree, the algorithm computes its drainage $\mathcal{Z}(\mathcal{J})$, its Menger's paths and all instances $\mathcal{D}(\mathcal{J}, B_h)$ where B_h is a closest dam of $\mathcal{Z}(\mathcal{J})$. This third operation is done by enumerating all dams B_i of $\mathcal{Z}(\mathcal{J})$, verifying whether there is another dam B_h which is closer than B_i, and (if B_i is a closest dam) identifying the vertices/edges of $\mathcal{D}(\mathcal{J}, B_i)$ thanks to a depth-first search in $G_D \backslash B_i$. As there are at most $2^p n$ dams in $\mathcal{Z}(\mathcal{J})$, its execution time is $O(2^{2p}n^3)$. The overall complexity is $O\left(2^{p^2}m(mnp + 2^{2p}n^3)\right) = O\left(2^{p(p+2)}pmn^3\right)$.

5 Conclusion

In this study, we were interested in the parameterized complexity of counting the minimum (S,T)-cuts in undirected graphs. The conclusion is that this problem is FPT$\langle p \rangle$ as we devised an algorithm running in $O(2^{p(p+2)}pmn^3)$. Our algorithm starts by "draining" the graph: the drainage is made of $k < n$ minimum cuts Z_i. For any minimum cut of the instance, at least one of the minimum drainage cuts Z_i contains edges of X. For this reason, we believe that the drainage could be used on other cut problems.

In the future, we will focus on the fixed-parameter tractability of counting minimum vertex (S,T)-cuts. Although the concepts used for the edge version can be generalized to vertex cuts (the drainage, Menger's paths, dry areas, and closest dams), the generalization of our algorithm to vertex cuts is compromised for one major reason: Theorem 3 is not true anymore.

References

1. Arvind, V., Raman, V.: Approximation algorithms for some parameterized counting problems. In: Bose, P., Morin, P. (eds.) ISAAC 2002. LNCS, vol. 2518, pp. 453–464. Springer, Heidelberg (2002). https://doi.org/10.1007/3-540-36136-7_40

2. Ball, M.O., Colbourn, C.J., Provan, J.S.: Network reliability. In: Handbooks in Operations Research and Management Science, vol. 7, pp. 673–762. Elsevier (1995)
3. Ball, M.O., Provan, J.S.: Calculating bounds on reachability and connectedness in stochastic networks. Networks **13**(2), 253–278 (1983)
4. Ball, M.O., Provan, J.S.: Computing network reliability in time polynomial in the number of cuts. Oper. Res. **32**(3), 516–526 (1984)
5. Bezáková, I., Chambers, E.W., Fox, K.: Integrating and sampling cuts in bounded treewidth graphs. In: Letzter, G., et al. (eds.) Advances in the Mathematical Sciences. AWMS, vol. 6, pp. 401–415. Springer, Cham (2016). https://doi.org/10.1007/978-3-319-34139-2_20
6. Bezáková, I., Friedlander, A.J.: Counting and sampling minimum (S, T)-cuts in weighted planar graphs in polynomial time. Theoret. Comput. Sci. **417**, 2–11 (2012)
7. Bousquet, N., Daligault, J., Thomassé, S.: Multicut is FPT. In: Proceedings of STOC, pp. 459–468 (2011)
8. Boykov, Y., Veksler, O.: Graph cuts in vision and graphics: theories and applications. In: Paragios, N., Chen, Y., Faugeras, O. (eds.) Handbook of Mathematical Models in Computer Vision, pp. 79–96. Springer, Boston (2006). https://doi.org/10.1007/0-387-28831-7_5
9. Chambers, E.W., Fox, K., Nayyeri, A.: Counting and sampling minimum cuts in genus g graphs. In: Proceedings of SoCG, pp. 249–258 (2013)
10. Chandran, L.S., Ram, L.S.: On the number of minimum cuts in a graph. In: Proceedings of COCOON, pp. 220–229 (2002)
11. Chen, J., Liu, Y., Lu, S.: An improved parameterized algorithm for the minimum node multiway cut problem. Algorithmica **55**(1), 1–13 (2009)
12. Chitnis, R.H., Hajiaghayi, M., Marx, D.: Fixed-parameter tractability of directed multiway cut parameterized by the size of the cutset. SIAM J. Comput. **42**(4), 1674–1696 (2013)
13. Curticapean, R.: Counting problems in parameterized complexity. In: Proceedings of IPEC, pp. 1–18 (2018)
14. Cygan, M., Lokshtanov, D., Pilipczuk, M., Pilipczuk, M., Saurabh, S.: Minimum bisection is fixed parameter tractable. In: Proceedings of STOC, pp. 323–332 (2014)
15. Downey, R.G., Fellows, M.R.: Parameterized Complexity. Monographs in Computer Science. Springer, New York (1999). https://doi.org/10.1007/978-1-4612-0515-9
16. Flum, J., Grohe, M.: The parameterized complexity of counting problems. SIAM J. Comput. **33**(4), 892–922 (2004)
17. Ford, L.R., Fulkerson, D.R.: Maximal flow through a network. Can. J. Math. **8**, 399–404 (1956)
18. Guillemot, S., Sikora, F.: Finding and counting vertex-colored subtrees. Algorithmica **65**(4), 828–844 (2013)
19. Krom, M.R.: The decision problem for a class of firstorder formulas in which all disjunctions are binary. Math. Logic Q. **13**(12), 15–20 (1967)
20. Marx, D.: Parameterized graph separation problems. Theoret. Comput. Sci. **351**(3), 394–406 (2006)
21. Marx, D., O'Sullivan, B., Razgon, I.: Finding small separators in linear time via treewidth reduction. ACM Trans. Algorithms **9**(4), 30:1–30:35 (2013)
22. Marx, D., Razgon, I.: Fixed-parameter tractability of multicut parameterized by the size of the cutset. In: Proceedings of STOC, pp. 469–478 (2011)
23. Menger, K.: Zur allgemeinen Kurventheorie. Fundamenta Mathematicæ **10**(1), 96–115 (1927)

24. Nagamochi, H., Sun, Z., Ibaraki, T.: Counting the number of minimum cuts in undirected multigraphs. IEEE Trans. Reliab. **40**, 610–614 (1991)
25. Niedermeier, R.: Invitation to Fixed-Parameter Algorithms. Oxford Lecture Series in Mathematics and Its Applications. OUP, Oxford (2006)
26. Provan, J.S., Ball, M.O.: The complexity of counting cuts and of computing the probability that a graph is connected. SIAM J. Comput. **12**(4), 777–788 (1983)
27. Valiant, L.G.: The complexity of counting the permanent. Theoret. Comput. Sci. **8**, 189–201 (1979)
28. Williams, V.V., Williams, R.: Finding, minimizing, and counting weighted subgraphs. SIAM J. Comput. **42**(3), 831–854 (2013)

Fast Breadth-First Search in Still Less Space

Torben Hagerup(✉)

Institut für Informatik, Universität Augsburg, 86135 Augsburg, Germany
hagerup@informatik.uni-augsburg.de

Abstract. It is shown that a breadth-first search in a directed or undirected graph with n vertices and m edges can be carried out in $O(n+m)$ time with $n \log_2 3 + O((\log n)^2)$ bits of working memory.

Keywords: Graph algorithms · Space efficiency · BFS · Succinct data structures · Choice dictionaries · In-place chain technique

1 Introduction

1.1 Space-Bounded Computation

The study of the amount of memory necessary to solve specific computational problems has a long tradition. A fundamental early result in the area is the discovery by Savitch [14] that the s-t connectivity problem (given a graph G and two vertices s and t in G, decide whether G contains a path from s to t) can be solved with $O((\log n)^2)$ bits of memory on n-vertex graphs. In order for this and related results to make sense, one must distinguish between the memory used to hold the input and the working memory, which is the only memory accounted for. The working memory is usable without restrictions, but the memory that holds the input is read-only and any output is stored in write-only memory. Informally, these conventions serve to forbid "cheating" by using input or output memory for temporary storage. They are all the more natural when, as in the original setting of Savitch, the input graph is present only in the form of a computational procedure that can test the existence of an edge between two given vertices.

Savitch's algorithm is admirably frugal as concerns memory, but its (worst-case) running time is superpolynomial. It was later generalized by Barnes, Buss, Ruzzo and Schieber [4], who proved, in particular, that the s-t connectivity problem can be solved on n-vertex graphs in $n^{O(1)}$ time using $O(n/2^{b\sqrt{\log n}})$ bits for arbitrary fixed $b > 0$. In the special case of undirected graphs, a celebrated result of Reingold [13] even achieves polynomial time with just $O(\log n)$ bits. The running times of the algorithms behind the latter results, although polynomial, are "barely so" in the sense that the polynomials are of high degree. A more recent research direction searches for algorithms that still use memory as sparingly as

I. Sau and D. M. Thilikos (Eds.): WG 2019, LNCS 11789, pp. 93–105, 2019.
https://doi.org/10.1007/978-3-030-30786-8_8

possible but are nonetheless fast, ideally as fast as the best algorithms that are not subject to space restrictions. The quest to reduce space requirements and running time simultaneously is motivated in practical terms by the existence of small mobile or embedded devices with little memory, by memory hierarchies that allow smaller data sets to be processed faster, and by situations in which the input is too big to be stored locally and must be accessed through query procedures running on a remote server. The Turing machine models running time on real computers rather crudely, so the model of computation underlying the newer research is the random-access machine and, more specifically, the word RAM.

1.2 The Breadth-First-Search Problem

This paper continues an ongoing search for the best time and space bounds for carrying out a breadth-first search or BFS in a directed or undirected graph. Formally, we consider the BFS problem to be that of computing a shortest-path spanning forest of an input graph $G = (V, E)$ consistent with a given permutation of V in top-down order, a somewhat tedious definition of which can be found in [8]. Suffice it here to say that if all vertices of the input graph G are reachable from a designated start vertex $s \in V$, the task at hand essentially is to output the vertices in V in an order of nondecreasing distance from s in G. The BFS problem is important in itself, but has also served as a yardstick with which to gauge the strength of new algorithmic and data-structuring ideas in the realm of space-efficient computing.

In the following consider an input graph $G = (V, E)$ and take $n = |V|$ and $m = |E|$. The algorithms of Savitch [14] and of Barnes et al. [4] are easily adapted, within the time and space bounds cited above, to compute the actual distance from s to t (∞ if t is not reachable from s). As a consequence, the BFS problem can be solved on n-vertex graphs with $O((\log n)^2)$ bits or in $n^{O(1)}$ time with $n/2^{\Omega(\sqrt{\log n})}$ bits. Every reasonably fast BFS algorithm known to the author, however, can be characterized by an integer constant $c \geq 2$, dynamically assigns to each vertex in V one of c states or *colors*, and maintains the color of each vertex explicitly or implicitly. Let us call such an algorithm a *c-color* BFS algorithm. E.g., the classic BFS algorithm marks each vertex as visited or unvisited and stores a subset of the visited vertices in a FIFO queue, which makes it a 3-color algorithm: The unvisited vertices are *white*, the visited vertices in the FIFO queue are *gray*, and the remaining visited vertices are *black*. Because the distribution of colors over the vertices can be nearly arbitrary, a c-color BFS algorithm with an n-vertex input graph must spend at least $n \log_2 c$ bits on storing the vertex colors. The classic BFS algorithm uses much more space since the FIFO queue may hold nearly n vertices and occupy $\Theta(n \log n)$ bits.

Similarly as Dijkstra's algorithm can be viewed as an abstract algorithm turned into a concrete algorithm by the choice of a particular priority-queue data structure, Elmasry, Hagerup and Kammer [6] described a simple abstract 4-color BFS algorithm that uses $O(n + m)$ time plus $O(n + m)$ calls of operations in an appropriate data structure that stores the vertex colors. This allowed them to derive a first BFS algorithm that works in $O(n+m)$ time with $O(n)$ bits. Using the same abstract algorithm with a different data structure, Banerjee, Chakraborty and Raman [2] lowered the space bound to $2n+O(n\log\log n/\log n)$ bits. Concurrently, Hagerup and Kammer [10] obtained a space bound of $n\log_2 3+O(n/(\log n)^t)$ bits, for arbitrary fixed $t\ge 1$, by stepping to a better so-called *choice-dictionary* data structure but, more significantly, by developing an abstract 3-color BFS algorithm to work with it. The algorithm uses the three colors white, gray and black and, for an undirected graph in which all vertices are reachable from the start vertex s, can be described via the code below. No output is mentioned, but a vertex can be output when it is colored gray.

```
Color all vertices white;
Color s gray;
while some vertex is gray do
  for each gray vertex u do    (* exploration round *)
    if u = s or u has a black neighbor then
      Color gray all white neighbors of u;
  for each gray vertex u do    (* consolidation round *)
    if u has no white neighbor then
      Color u black;
```

Roughly speaking, the white vertices have not yet been encountered by the search, the black vertices are completely done with, and the gray vertices form the layer of currently active vertices at a common distance from s. The two inner loops of the algorithm iterate over the gray vertices in order to replace them by their white neighbors, which form the next gray layer. Both iterations are *dynamic* in the sense that the set of gray vertices changes while it is being iterated over. The first iteration (the *exploration round*) colors additional vertices gray, and we would prefer for these newly gray vertices not to be enumerated by the iteration. Satisfying this requirement is not easy for a space-efficient algorithm, however, and therefore the iteration instead tests each enumerated vertex for being "old"—exactly then does it equal s or have a black neighbor—and ignores the other gray vertices. Similarly, the second iteration (the *consolidation round*) colors black only those gray vertices that are no longer needed as neighbors of white vertices—these include all "old" gray vertices. Even so, the choice dictionary must support dynamic iteration suitably. This represents the biggest challenge for a space-efficient implementation of the abstract algorithm.

For a directed graph, the changes are slight: "black neighbor" should be replaced by "black inneighbor", and each of the two occurrences of "white neighbor" should be replaced by "white outneighbor". If not all vertices are reachable from s, the code above, except for its first line, must be wrapped in a standard way in an outer loop that steps s through all vertices in a suitable order and restarts the BFS at every vertex found to still be white when it is chosen as s. This leads to no additional complications and will be ignored in the following.

1.3 Recent Work and Our Contribution

Starting with the algorithm of Hagerup and Kammer [10], all new BFS algorithms have space bounds of the form $n \log_2 3 + s(n)$ bits for some function s with $s(n) = o(n)$. In a practical setting the leading factor of $\log_2 3$ is likely to matter more than the exact form of s, so that the progress since the algorithm of Hagerup and Kammer could be viewed as insignificant. However, at least from a theoretical point of view it is interesting to explore how much space is needed beyond the seemingly unavoidable $n \log_2 3$ bits required to store the vertex colors. If a 3-color BFS algorithm uses $n \log_2 3 + s(n)$ bits, we will therefore say that it works with $s(n)$ *extra* bits. If its running time is $t(n, m)$, we may summarize its resource requirements in the pair $(t(n, m), s(n))$. Adapting the notion of pareto dominance, we say that an algorithm with the resource pair $(t(n, m), s(n))$ *dominates* an algorithm with the resource pair $(t'(n, m), s'(n))$ if $t(n, m) = O(t'(n, m))$ and $s(n) = o(s'(n))$ or $t(n, m) = o(t'(n, m))$ and $s(n) = O(s'(n))$.

Banerjee, Chakraborty, Raman and Satti [3] indicated a slew of 3-color BFS algorithms with the following resource pairs: $(n^{O(1)}, o(n))$, $(O(nm), O((\log n)^2))$, $(O(m(\log n)^2), o(n))$ and $(O(m \log n f(n)), O(n/f(n)))$ for certain slowly growing functions f. The first of these algorithms is dominated by that of Barnes et al. [4], which also uses polynomial time but $o(n)$ bits altogether, not just $o(n)$ extra bits. The third algorithm of [3] is dominated by the algorithm of Hagerup and Kammer [10], whose resource pair is $(O(n + m), O(n/(\log n)^t))$ for arbitrary fixed $t \geq 1$. Instantiating the 3-color abstract algorithm of [10] with a new choice dictionary, Hagerup [8] obtained an algorithm that has the resource pair $(O(n \log n + m \log \log n), O((\log n)^2))$ and dominates the two remaining algorithms of [3]. Another algorithm of [8] is faster but less space-efficient and has the resource pair $(O(n \log n + m), n^\epsilon)$ for arbitrary fixed $\epsilon > 0$.

Here we present a new data structure, designed specifically to be used with the abstract 3-color BFS algorithm of [10], that leads to a concrete BFS algorithm operating in $O(n + m)$ time using $n \log_2 3 + O((\log n)^2)$ bits of working memory. The new algorithm combines the best time and space bounds of all previous algorithms with running-time bounds of $O(nm)$ or less, and therefore dominates all of them. It is also simpler than several previous algorithms. We obtain a slightly more general result by introducing a tradeoff parameter $t \geq 1$: The running time is $O((n + m)t)$, and the space bound is $n \log_2 3 + O((\log n)^2/t + \log n)$ bits. If the degrees of the vertices $1, \ldots, n$ of the input graph G form a nondecreasing sequence or if G is approximately regular, we achieve a running time of $O((n + m) \log \log n)$ with just $n \log_2 3 + O(\log n)$ bits.

The technical contributions of the present paper include:

- A new representation of vertex colors
- A new approach to dynamic iteration
- A refined analysis of the abstract 3-color BFS algorithm of [10]
- An amortized analysis of the new data structure.

Conversely, we draw on [10] not only for its abstract 3-color BFS algorithm, but also for setting many of the basic concepts straight and for a technical lemma. Another crucial component is the in-place chain technique of Katoh and Goto [12], as developed further in [8,9,11]. A fundamental representation of n colors drawn from $\{0, 1, 2\}$ in close to $n \log_2 3$ bits so as to support efficient access to individual colors is due to Dodis, Pătraşcu and Thorup [5].

2 Preliminaries

We do not need to be very specific about the way in which the input graph $G = (V, E)$ is presented to the algorithm. With $n = |V|$ and $m = |E|$, we assume that n can be determined in $O(n + m)$ time and that $V = \{1, \ldots, n\}$. If G is undirected, we also assume that for each vertex $u \in V$, it is possible to iterate over the neighbors of u in at most constant time plus time proportional to their number. If G is directed, the assumption is the same, but now the neighbors of u include both the inneighbors and the outneighbors of u, and inneighbors should (of course) be distinguishable from outneighbors.

Our model of computation is a word RAM [1,7] with a word length of w bits, where we assume that w is large enough to allow all memory words in use to be addressed. As part of ensuring this, we assume that $w \geq \log_2 n$. The word RAM has constant-time operations for addition, subtraction and multiplication modulo 2^w, division with truncation ($(x, y) \mapsto \lfloor x/y \rfloor$ for $y > 0$), left shift modulo 2^w ($(x, y) \mapsto (x \ll y) \bmod 2^w$, where $x \ll y = x \cdot 2^y$), right shift ($(x, y) \mapsto x \gg y = \lfloor x/2^y \rfloor$), and bitwise Boolean operations (AND, OR and XOR (exclusive or)).

3 The Representation of the Vertex Colors

This section develops a data structure for storing a color drawn from the set {white, gray, black} for each of the n vertices in $V = \{1, \ldots, n\}$. As we will see, the data structure enables linear-time execution of the abstract 3-color BFS algorithm of [10] and occupies $n \log_2 3 + O((\log n)^2)$ bits. An inspection of the algorithm reveals that the operations that must be supported by the data structure are reading and updating the colors of given vertices—this by itself is easy—and dynamic iteration over the set of gray vertices. A main constraint for the latter operation is that the iteration must happen in time proportional to the number of gray vertices, i.e., we must be able to find the gray vertices efficiently.

Let us encode the color white as $1 = \mathtt{01}_2$, gray as $0 = \mathtt{00}_2$ and black as $2 = \mathtt{10}_2$. In the following we will not distinguish between a color and its corresponding integer or 2-bit string. Take $\Lambda = \lfloor \log_2 n \rfloor$, $q = 10\Lambda$ and $\lambda = \lceil \log_2 q \rceil = \Theta(\log \log n)$. In the interest of simplicity let us assume that n is large enough to make $\lambda^2 \leq \Lambda$. In order to keep track of the colors of the vertices in V we divide the sequence of n colors into $N = \lfloor n/q \rfloor$ *segments* of exactly q colors each, with at most $q - 1$ colors left over. Each segment is represented via a *big integer* drawn from $\{0, \ldots, 3^q - 1\}$. Because $q = O(\log n)$, big integers can be manipulated in constant time. The N big integers are in turn maintained in an instance of the data structure of Lemma 1 below, which occupies $N \log_2 3^q + O((\log N)^2) = n \log_2 3 + O((\log n)^2)$ bits.

Lemma 1 ([5], Theorem 1). *There is a data structure that, given arbitrary positive integers N and C with $C = N^{O(1)}$, can be initialized in $O(\log N)$ time and subsequently maintains an array of N elements drawn from $\{0, \ldots, C-1\}$ in $N \log_2 C + O((\log N)^2)$ bits such that individual array elements can be read and updated in constant time.*

3.1 Containers and Their Structure and Operations

We view the N big integers as objects with a nontrivial internal structure and therefore use the more suggestive term *container* to denote the big integer in a given position in the sequence of N big integers. We shall say that each of the q vertices whose colors are stored in a container is *located* in the container. A container may represent q colors $a_0, \ldots, a_{q-1}$ in several different ways illustrated in Fig. 1. The most natural representation is as the integer $x = \sum_{j=0}^{q-1} a_j 3^j$. We call this the *regular representation*, and a container is *regular* if it uses the regular representation (Fig. 1(b)). When a vertex u is located in a regular container, we can read and update the color of u in constant time, provided that we store a table of the powers $3^0, 3^1, \ldots, 3^{q-1}$. E.g., with notation as above, $a_j = \lfloor x/3^j \rfloor \bmod 3$ for $j = 0, \ldots, q-1$. The table occupies $O((\log n)^2)$ bits and can be computed in $O(\log n)$ time.

We allow a variant in which a regular container D is a *master* (Fig. 1(a)). The difference is that the 3Λ most significant bits of the big integer corresponding to D are relocated to a different container D', said to be the *slave* corresponding to D, and stored there. This frees $3\Lambda - 1$ bits in D for other uses (the most significant bit is fixed at 0 to ensure that the value of the big digit does not exceed $3^q - 1$). Since $\lceil \log_2(N+1) \rceil \le \log_2(n/8) + 1 = \log_2 n - 2 \le \lfloor \log_2 n \rfloor - 1 = \Lambda - 1$, we can store a pointer (possibly *null*) to a container in $\Lambda - 1$ bits, so a master has room for three such pointers. One of these designates the slave D', while the use of the two other pointers, called *iteration pointers*, is explained later. Even though it may be necessary to access the data relocated to the slave, a master still allows vertex colors to be read and updated in constant time.

When it is desired to iterate over the gray vertices in a regular container, the container is first converted to the *loose representation*, in which the 2-bit strings corresponding to the q color values are simply concatenated to form a string of $2q$ bits. Since $\log_2 q \le \lambda$, this can be done in $O(\lambda)$ time with the algorithm of Lemma 2 below, used with $c = 3$, $d = 4$ and $s = q$. The algorithm is a word-parallel version (i.e., essentially independent computations take place simultaneously in different regions of a word) of a simple divide-and-conquer procedure.

Lemma 2 ([8], Lemma 3.3 with $f = 2$ and $p = 1$). *Given integers c, d and s with $2 \le c, d \le 4$, $s \ge 1$ and $s = O(w)$ and an integer of the form $\sum_{j=0}^{s-1} a_j c^j$, where $0 \le a_j < \min\{c, d\}$ for $j = 0, \ldots, s-1$, the integer $\sum_{j=0}^{s-1} a_j d^j$ can be computed in $O(\log(s+1))$ time.*

Conversely, using the lemma instead with $c = 4$ and $d = 3$, we can convert from the loose to the regular representation, again in $O(\lambda)$ time. Once a container

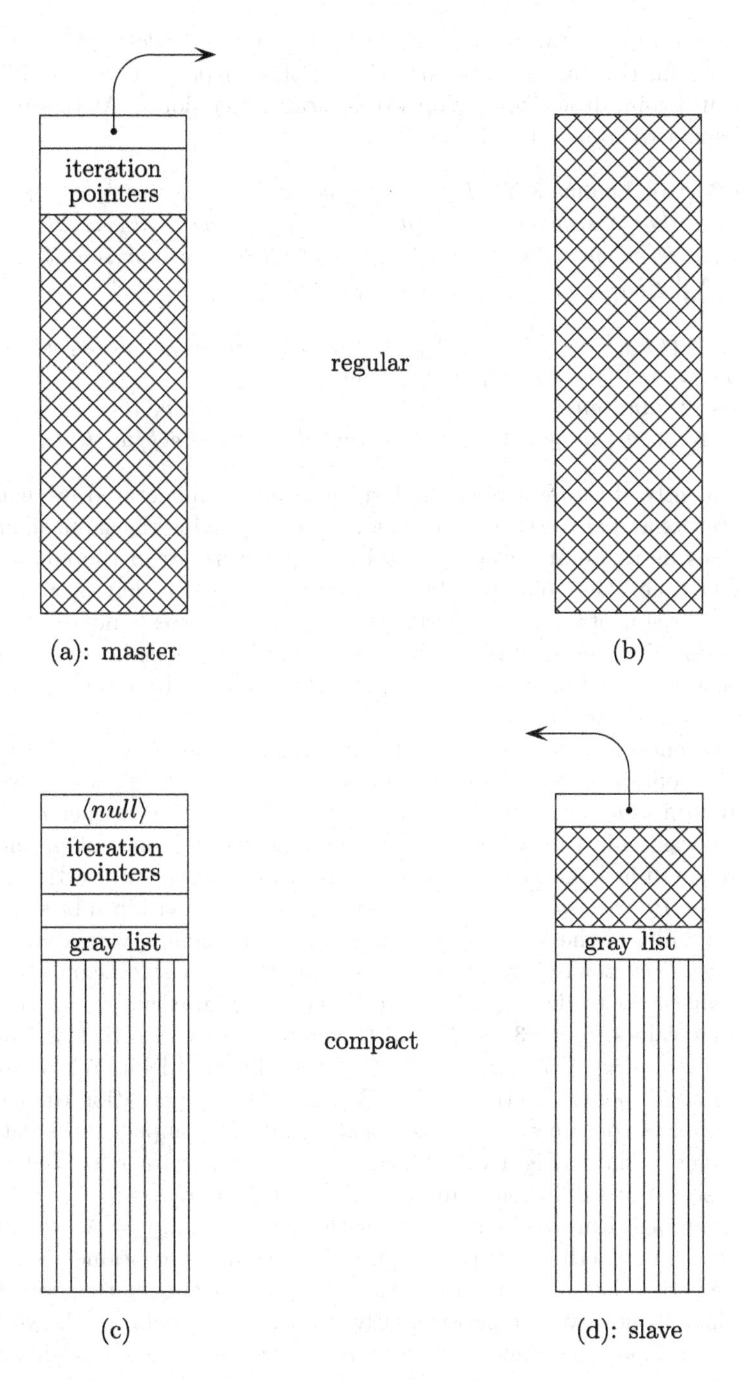

Fig. 1. Four different representations in containers. Crosshatched areas symbolize colors white, gray and black stored to base 3, while vertically striped areas symbolize black-and-white vectors.

is in the loose representation, we can locate the first (smallest) gray vertex in the container in constant time with the algorithm of part (a) of the following lemma that, again, draws heavily on word-parallel techniques. At this point we use the lemma with $m = q$ and $f = 2$.

Lemma 3 ([10], Lemma 3.2). *Let m and f be given integers with $1 \le m, f < 2^w$ and suppose that a sequence $A = (a_1, \ldots, a_m)$ with $a_i \in \{0, \ldots, 2^f - 1\}$ for $i = 1, \ldots, m$ is given in the form of the (mf)-bit binary representation of the integer $\sum_{i=0}^{m-1} 2^{if} a_{i+1}$. Then the following holds:*

(a) Let $I_0 = \{i \in \{1, \ldots m\} : a_i = 0\}$. Then, in $O(1 + mf/w)$ time, we can test whether $I_0 = \emptyset$ and, if not, compute $\min I_0$.
(b) If $m < 2^f$ and an integer $k \in \{0, \ldots, 2^f - 1\}$ is given, then $rank(k, A) = |\{i \in \{1, \ldots, m\} : k \ge a_i\}|$ can be computed in $O(1 + mf/w)$ time.

Subsequently, if we remember the last grey vertex enumerated, we can shift out that vertex and all vertices preceding it before applying the same algorithm. This enables us to iterate over the set V_g of gray vertices in the container in $O(|V_g| + 1)$ time. The colors of the at most $q - 1$ vertices left over from the division into segments are kept permanently in what corresponds to the loose representation. This uses $O(\log n)$ bits, and it will be obvious how to adapt the various operations to take these vertices and their colors into account, for which reason we shall ignore them in the following.

If a container is a slave (Fig. 1(d)), we require the number n_g of gray vertices in the container to be bounded by $\lambda - 1$, and we store its gray vertices separately in a *gray list*. The gray list takes the form of the integer n_g, stored (somewhat wastefully) in λ bits, followed by a sorted sequence of n_g integers, each represented in λ bits, that indicate the positions of the gray vertices within the container. By the assumption $\lambda^2 \le \Lambda$, the gray list fits within Λ bits. Because of the availability of the gray list, we can store the remaining vertex colors in a *black-and-white vector* of just q bits by dropping the most significant bit, which normally allows us to distinguish between the colors gray and black, from all q 2-bit color values. Since $3^2 \ge 2^3$ and therefore $q \log_2 3 \ge 15\Lambda$, this leaves at least $15\Lambda - \Lambda - q = 4\Lambda$ bits, which are used to hold the 3Λ bits relocated from the master and a pointer to the master. We call this representation the *compact representation*. A container may be *compact*, i.e., in the compact representation, without being a slave (Fig. 1(c)). Then, instead of the data relocated from a master, it stores two iteration pointers and a null pointer.

Using the algorithm of Lemma 3(b) with $m = n_g$ and $f = \lambda$, we can test in constant time whether a vertex located in a compact container is gray by checking whether its number within the container occurs in the gray list of the container. If not, we can subsequently determine the color of the vertex in constant time from the black-and-white vector. Similarly, we can change the color of a given vertex in constant time. This may involve creating a gap for the new vertex in the gray list or, conversely, closing such a gap, which is easily accomplished with a constant number of bitwise Boolean and shift operations.

It is also easy to see that we can iterate over the n_g gray vertices in $O(n_g + 1)$ time.

If a color change increases the number n_g of gray vertices in a compact container to λ, the container must be converted to the regular representation. For this it will be convenient if the black-and-white vector stores the q least significant bits of the vertex colors not in their natural order, but in the shuffled order obtained by placing the first half of the bits, in the natural order, in the odd-numbered positions of the black-and-white vector and the last half in the even-numbered positions. With this convention, we can still read and update vertex colors in constant time. We can also unshuffle the black-and-white vector in constant time, creating 1-bit gaps for the most significant bits, by separating the bits in the odd-numbered positions from those in the even-numbered positions and concatenating the two sequences. Subsequently each most significant bit is set to be the complement of its corresponding least significant bit to represent the colors white and black according to the loose representation. Going through the gray list, we can then introduce the gray colors one by one. Thus we can convert from the compact to the loose and from there to the regular representation in $O(n_g + \lambda) = O(\lambda)$ time. Conversely, if a container in the loose representation has fewer than λ gray vertices, it can be converted to the compact representation in $O(\lambda)$ time.

3.2 The In-place Chain Technique for Containers

The overall organization of the containers follows the in-place chain technique [8, 9,11,12]. By means of an integer $\mu \in \{0, \ldots, N\}$ equal to the number of compact containers, the sequence $D_1, \ldots, D_N$ of containers is dynamically divided into a *left part*, consisting of $D_1, \ldots, D_\mu$, and a *right part*, $D_{\mu+1}, \ldots, D_N$. A regular container is a master if and only if it belongs to the left part, and a compact container is a slave if and only if it belongs to the right part. Thus the two representations shown on the left in Fig. 1((a) and (c)) can occur only in the left part, while the two representations shown on the right can occur only in the right part. In particular, every container in the left part has iteration pointers.

Call a container *gray-free* if no vertex located in the container is gray. The iteration pointers are used to join all containers in the left part, with the exception of the gray-free compact containers, into a doubly-linked *iteration list* whose first and last elements are stored in $O(\log n)$ bits outside of the containers.

When an update of a vertex color causes a container D_i to switch from the compact to the regular representation, μ decreases by 1, say from μ_0 to $\mu_0 - 1$. If $i = \mu_0$, D_i belongs to the left part before the switch and to the right part after the switch, i.e., in terms of Fig. 1, the switch is from (c) to (b). If $i \neq \mu_0$, the switch is more complicated, in that it involves other containers. If $i < \mu_0$ (Fig. 1, (c) to (a)), D_i becomes a master, whereas if $i > \mu_0$ (Fig. 1, (d) to (b)), D_i stops being a slave. In both cases there now is a master D_m without a slave, a situation that must be remedied. However, D_{μ_0} also switches, namely either from (a) to (b) or from (c) to (d). In the case "(a) to (b)" D_{μ_0} stops being a master, and its former slave can become the slave of D_m. In the case "(c) to (d)" D_{μ_0} becomes a slave and can serve as the slave of D_{m}. Thus in all cases masters and slaves can again be matched up appropriately. Altogether, the operation involves changing some pointers and moving some relocated data in at most four containers. After the conversion of D_i, this takes constant time.

In some circumstances that still need to be specified, a container D_i may switch from the regular to the compact representation, which causes μ to increase by 1, say from μ_0 to $\mu_0 + 1$. We can handle this situation similarly as above. If $i = \mu_0 + 1$, the switch is from (b) to (c) in Fig. 1, and nothing more must be done. Otherwise, whether the switch is from (a) to (c) or from (b) to (d), there will be a slave without a master. Simultaneously D_{μ_0+1} switches either from (b) to (a) (it becomes the needed master) or from (d) to (c) (it stops being a slave, and its former master takes on the new slave). Again, after the conversion of D_i, the matching between masters and slaves can be updated in constant time.

4 BFS Algorithms

4.1 The Basic Algorithm

To execute the first line of the abstract 3-color BFS algorithm with the vertex-color data structure developed in the previous section, we initialize the data structure as follows: All vertices are white, all containers are compact, but not slaves, all have empty gray lists, the iteration list is empty, and $\mu = N$.

It was already described how to read and update vertex colors. If a color change causes a compact container D in the left part to become gray-free, D is shunted out of the iteration list. Conversely, if a compact container in the left part stops being gray-free, it is inserted at the end of the iteration list. The case in which a container enters or leaves the left part because of a change in μ is handled analogously. All of this can happen in constant time. The only exception is if a color change forces a container to switch from the compact to the regular representation, which takes $O(\lambda)$ time.

Recall that the abstract 3-color BFS algorithm alternates between *exploration rounds*, in which it iterates over the gray vertices and colors some of their white neighbors gray, and *consolidation rounds*, in which it iterates over the gray vertices and colors some of them black. Each iteration is realized by iterating over two lists of containers: the explicitly maintained iteration list and the implicit *right list*, which consists of the containers $D_N, D_{N-1}, \ldots, D_{\mu+1}$ in that order. Each of the two iterations can be viewed as moving a *pebble* through the relevant list. Because the lists may change dynamically, the following rules apply: If a currently pebbled container D is deleted from its list, the pebble is first moved to the successor of D, if any, in the relevant list. If a pebble reaches the end of its list, it waits there for new containers that may be inserted at the end of the list. One of the at most two pebbled containers is the *current container* D_{c}, whose gray vertices are enumerated as explained earlier. If D_{c} is regular, this involves first converting it to the loose representation. Once all gray vertices in D_{c} have been enumerated, D_{c} stops being the current container. If it is in the loose representation, we convert it to either the regular or the compact representation.

If the iteration happens in an exploration round, we always convert D_c to the regular representation, so that μ does not increase. If the iteration happens in a consolidation round, we attempt to convert it to the compact representation. If this fails because D_c contains more than $\lambda - 1$ gray vertices, we instead convert it to the regular representation; μ does not decrease. Then the pebble on D_c is moved to the list successor of D_c, and one of the at most two containers that are now pebbled is chosen to be the new current container. The iteration ends when both pebbles are at the end of their respective lists.

Since every container that is not gray-free belongs either to the iteration list or to the right list, it is clear that each round enumerates all vertices that are gray at the beginning of the round (and maybe some that become gray in the course of the round). A container and its gray vertices may be enumerated twice, namely once as part of the iteration list and once as part of the right list. The BFS algorithm can tolerate this, and no vertex is enumerated more than twice within one round because μ moves in only one direction within the round. A vertex can be gray for at most (part of) four consecutive rounds, so the total number of vertex enumerations is $O(n)$. Therefore the total time spent on enumeration is $O(n)$, except possibly for the following two contributions to the running time: (1) Containers that are enumerated but turn out to be gray-free; (2) Conversions of containers between different representations. As for (1), every container concerned is regular or on the right side, i.e., the number of such containers is bounded by $2(N - \mu)$. Since the iteration converts all $N - \mu$ regular containers to the loose representation, the contribution of (1) is dominated by that of (2). And as for (2), since the number of other conversions is within a constant factor of the number of conversions to the regular representation, it suffices to bound the latter by $O(n/\lambda)$. Call a conversion of a container D to the regular representation *proper* if D contains at least λ gray vertices at the time of the conversion, and *improper* otherwise. Improper conversions happen only in exploration rounds. Before the first conversion of a container D to the regular representation, λ vertices located in D must have become gray, and between two successive proper conversions of D at least λ vertices in D either change color or are enumerated. Moreover, between two consecutive proper conversions of D there can be at most one improper conversion of D (namely in an exploration round). Since the number of color changes and of vertex enumerations is $O(n)$, the bound follows.

Theorem 1. *The BFS problem can be solved on directed or undirected graphs with n vertices and m edges in $O(n+m)$ time with $n \log_2 3 + O((\log n)^2)$ bits of working memory.*

4.2 A Time-Space Tradeoff

In order to derive a time-space tradeoff from Theorem 1, we must take a slightly closer look at the data structure of Dodis et al. [5] behind Lemma 1. For a certain set S whose elements can be represented in $O(\log n)$ bits, a certain function $g : S \to S$ that can be evaluated in constant time and a certain start value $x_0 \in S$ that can be computed in constant time, the preprocessing of the data structure serves to compute and store a table Y of $x_0 = g^{(0)}(x_0), g^{(1)}(x_0), g^{(2)}(x_0), \ldots, g^{(\lfloor \log_2 N \rfloor)}(x_0)$, where $g^{(j)}$, for integer $j \geq 0$, denotes j-fold repeated application of g. In addition, we need the powers $3^0, 3^1, \ldots, 3^{q-1}$, which are also assumed to be stored in Y. If we carry out the preprocessing but store $g^{(j)}(x_0)$ and 3^j only for those values of j that are multiples of t for some given integer $t \geq 1$, the shortened table Y' occupies only $O(\lceil (\log n)/t \rceil \log n) = O((\log n)^2/t + \log n)$ bits, and the rest of the BFS algorithm works with $O(\log n)$ bits. Whenever the data structure of Sect. 3 is called upon to carry out an operation, it needs a constant number of entries of Y, which can be reconstructed from those in Y' in $O(t)$ time. This causes a slowdown of $O(t)$ compared to an algorithm that has the full table Y at its disposal. Thus Theorem 1 generalizes as follows:

Theorem 2. *For every given $t \geq 1$, the BFS problem can be solved on directed or undirected graphs with n vertices and m edges in $O((n+m)t)$ time with $n \log_2 3 + O((\log n)^2/t + \log n)$ bits of working memory.*

4.3 BFS with $n \log_2 3 + O(\log N)$ Bits

Suppose now that we are allowed only $O(\log n)$ extra bits. Then, with notation as in the previous subsection, we can no longer afford to store the table Y of $x_0, g(x_0), g^{(2)}(x_0), \ldots, g^{(\lfloor \log_2 N \rfloor)}(x_0)$ and $3^0, 3^1, \ldots, 3^{q-1}$. Instead we store only the two $O(\log n)$-bit quantities x_0 and 3 and compute $g^{(i)}(x_0)$ and 3^i from them as needed. Concerning the latter, 3^i can be computed in $O(\log q) = O(\lambda)$ time for arbitrary $i \in \{0, \ldots, q-1\}$ by a well-known method based on repeated squaring.

When the data structure of Dodis et al. [5] is used to represent an array A with index set $\{1, \ldots, N\}$, $A[j]$, for $j = 1, \ldots, N$, is associated with the node j in a complete N-node binary tree T whose nodes are numbered $1, \ldots, N$ in the manner of Heapsort, i.e., the root is 1, the parent of every nonroot node j is $\lfloor j/2 \rfloor$, and every left child is even. Suppose that a node $j \in \{1, \ldots, N\}$ is of height h in T. Then we can access (read or update) $A[j]$ in constant time after computing $g^{(h)}(x_0)$ from x_0, which takes $O(h+1)$ time. In the worst case $h = \Theta(\log n)$, so we can access A with a slowdown of $O(\log n)$ relative to an algorithm with access to the full table Y. This leads to the result of Theorem 2 for $t = \log n$, i.e., $O((n+m)\log n)$ time and $O(\log n)$ extra bits. However, for most j the height h is much smaller than $\log_2 n$, which hints at a possible improvement.

For $i = 1, \ldots, n$, let d_i be the (total) degree of the vertex i in the input graph. It is easy to see that the number of accesses to the color of i in the course of the execution of the BFS algorithm is $O(d_i + 1)$. The color of i is located in the container D_j, where $j = \lceil i/q \rceil$, or in a slave $D_{j'}$ with $j' > j$, and D_j is in fact a big digit stored in $A[j]$, where A is the array maintained with the data structure of Dodis et al. [5]. The depth of the node j in the corresponding binary tree T is exactly $\lfloor \log_2 j \rfloor = \lfloor \log_2 \lceil i/q \rceil \rfloor$, and it is not difficult to see that its height is at most $\lfloor \log_2 \lceil n/q \rceil \rfloor - \lfloor \log_2 \lceil i/q \rceil \rfloor \leq \log_2(n/i) + 2$. Therefore the running time of the complete BFS algorithm is $O((n+m)\log\log n + \sum_{i=1}^{n} d_i \log(2n/i))$.

If $d_1 \le \cdots \le d_n$, $\sum_{i=1}^{n} d_i \log_2(2n/i) \le (1/n)(\sum_{i=1}^{n} d_i)(\sum_{i=1}^{n} \log_2(2n/i)) = O(m)$. Thus if the vertex degrees form a nondecreasing sequence, the running time is $O((n+m)\log\log n)$. Since $\log_2(2n/i) \le 1 + r\log_2\log_2 n$ if $i \ge n/(\log_2 n)^r$ for some $r \ge 1$, the same is true if $\sum_{i=1}^{\lfloor n/(\log_2 n)^r \rfloor} d_i = O(m\log\log n/\log n)$ for some fixed $r \ge 1$. Informally, the latter condition is satisfied if G is approximately regular. In particular, it is satisfied if the ratio of the maximum degree in G to the average degree is (at most) polylogarithmic in n.

References

1. Angluin, D., Valiant, L.G.: Fast probabilistic algorithms for Hamiltonian circuits and matchings. J. Comput. Syst. Sci. **18**(2), 155–193 (1979)
2. Banerjee, N., Chakraborty, S., Raman, V.: Improved space efficient algorithms for BFS, DFS and applications. In: Dinh, T.N., Thai, M.T. (eds.) COCOON 2016. LNCS, vol. 9797, pp. 119–130. Springer, Cham (2016). https://doi.org/10.1007/978-3-319-42634-1_10
3. Banerjee, N., Chakraborty, S., Raman, V., Satti, S.R.: Space efficient linear time algorithms for BFS, DFS and applications. Theory Comput. Syst. **62**(8), 1736–1762 (2018)
4. Barnes, G., Buss, J.F., Ruzzo, W.L., Schieber, B.: A sublinear space, polynomial time algorithm for directed *s*-*t* connectivity. SIAM J. Comput. **27**(5), 1273–1282 (1998)
5. Dodis, Y., Pătraşcu, M., Thorup, M.: Changing base without losing space. In: Proceedings of the 42nd ACM Symposium on Theory of Computing (STOC 2010), pp. 593–602. ACM (2010)
6. Elmasry, A., Hagerup, T., Kammer, F.: Space-efficient basic graph algorithms. In: Proceedings of the 32nd International Symposium on Theoretical Aspects of Computer Science (STACS 2015). LIPIcs, vol. 30, pp. 288–301. Schloss Dagstuhl - Leibniz-Zentrum für Informatik (2015)
7. Hagerup, T.: Sorting and searching on the word RAM. In: Morvan, M., Meinel, C., Krob, D. (eds.) STACS 1998. LNCS, vol. 1373, pp. 366–398. Springer, Heidelberg (1998). https://doi.org/10.1007/BFb0028575
8. Hagerup, T.: Small uncolored and colored choice dictionaries. Computing Research Repository (CoRR), arXiv:1809.07661 [cs.DS] (2018)
9. Hagerup, T.: Highly succinct dynamic data structures. In: Gąsieniec, L.A., Jansson, J., Levcopoulos, C. (eds.) FCT 2019. LNCS, vol. 11651, pp. 29–45. Springer, Cham (2019). https://doi.org/10.1007/978-3-030-25027-0_3
10. Hagerup, T., Kammer, F.: Succinct choice dictionaries. Computing Research Repository (CoRR), arXiv:1604.06058 [cs.DS] (2016)
11. Kammer, F., Sajenko, A.: Simple 2^f-color choice dictionaries. In: Proceedings of the 29th International Symposium on Algorithms and Computation (ISAAC 2018). LIPIcs, vol. 123, pp. 66:1–66:12. Schloss Dagstuhl - Leibniz-Zentrum für Informatik (2018)
12. Katoh, T., Goto, K.: In-place initializable arrays. Computing Research Repository (CoRR), arXiv:1709.08900 [cs.DS] (2017)
13. Reingold, O.: Undirected connectivity in log-space. J. ACM **55**(4), 17:1–17:24 (2008)
14. Savitch, W.J.: Relationships between nondeterministic and deterministic tape complexities. J. Comput. Syst. Sci. **4**(2), 177–192 (1970)

A Turing Kernelization Dichotomy for Structural Parameterizations of $\mathcal{F}$-Minor-Free Deletion

Huib Donkers(✉) and Bart M. P. Jansen

Eindhoven University of Technology,
P.O. Box 513, 5600 MB Eindhoven, The Netherlands
{h.t.donkers,b.m.p.jansen}@tue.nl

Abstract. For a fixed finite family of graphs $\mathcal{F}$, the $\mathcal{F}$-MINOR-FREE DELETION problem takes as input a graph G and an integer ℓ and asks whether there exists a set $X \subseteq V(G)$ of size at most ℓ such that $G - X$ is $\mathcal{F}$-minor-free. For $\mathcal{F} = \{K_2\}$ and $\mathcal{F} = \{K_3\}$ this encodes VERTEX COVER and FEEDBACK VERTEX SET respectively. When parameterized by the feedback vertex number of G these two problems are known to admit a polynomial kernelization. Such a polynomial kernelization also exists for any $\mathcal{F}$ containing a planar graph but no forests.

In this paper we show that $\mathcal{F}$-MINOR-FREE DELETION parameterized by the feedback vertex number is MK[2]-hard for $\mathcal{F} = \{P_3\}$. This rules out the existence of a polynomial kernel assuming NP $\not\subseteq$ coNP/poly, and also gives evidence that the problem does not admit a polynomial Turing kernel. Our hardness result generalizes to any $\mathcal{F}$ not containing a P_3-subgraph-free graph, using as parameter the vertex-deletion distance to treewidth $\min \text{tw}(\mathcal{F})$, where $\min \text{tw}(\mathcal{F})$ denotes the minimum treewidth of the graphs in $\mathcal{F}$. For the other case, where $\mathcal{F}$ contains a P_3-subgraph-free graph, we present a polynomial Turing kernelization. Our results extend to $\mathcal{F}$-SUBGRAPH-FREE DELETION.

Keywords: Turing kernelization · Minor-free deletion · Subgraph-free deletion · Structural parameterization

1 Introduction

Background and Motivation. Kernelization is a framework for the scientific investigation of provably effective preprocessing procedures for NP-hard problems, framed in the language of parameterized complexity. A *kernelization* for a parameterized problem is a polynomial-time algorithm that transforms any parameterized instance (x, k) into an instance (x', k') with the same answer, such that $|x'|$ and k' are both bounded by $f(k)$ for some computable function f. The function f is the *size* of the kernel. Of particular interest are kernels

B. M. P. Jansen: Supported by NWO Gravitation grant "Networks".

I. Sau and D. M. Thilikos (Eds.): WG 2019, LNCS 11789, pp. 106–119, 2019.
https://doi.org/10.1007/978-3-030-30786-8_9

of polynomial size. Determining which parameterized problems admit kernels of polynomial size has become a rich area of algorithmic research [4,13,22].

A common approach in kernelization [1,12,18] is to take the solution size as the parameter k, with the aim of showing that large inputs that ask for a small solution can be efficiently reduced in size. However, this method does not give any nontrivial guarantees when the solution size is known to be proportional to the total size of the input. For that reason, there is an alternative line of research [6,7,11,16,19–21,25] that focuses on parameterizations based on a measure of nontriviality of the instance (cf. [23]). One formal way to capture nontriviality of a graph problem is to measure how many vertex-deletions are needed to reduce the input graph to a graph class in which the problem can be solved in polynomial time. Since many graph problems can be solved in polynomial time on trees and forests, the structural graph parameter *feedback vertex number* (the minimum number of vertex deletions needed to make the graph acyclic, i.e. a forest) is a relevant measure of nontriviality.

Previous research has shown that for the VERTEX COVER problem, there is a polynomial kernel parameterized by the feedback vertex number [19]. This preprocessing algorithm guarantees that inputs which are large with respect to their feedback vertex number, can be efficiently reduced. The VERTEX COVER problem is the simplest in a family of so-called minor-free deletion problems. For a fixed finite family of graphs $\mathcal{F}$, an input to $\mathcal{F}$-MINOR-FREE DELETION consists of a graph G and an integer ℓ. The question is whether there is a set X of at most ℓ vertices in G, such that the graph $G - X$ obtained by removing these vertices does not contain any graph from $\mathcal{F}$ as a minor. Motivated by the fact that VERTEX COVER and FEEDBACK VERTEX SET, arguably the simplest $\mathcal{F}$-MINOR-FREE DELETION problems, admit polynomial kernels when parameterized by the feedback vertex number, we set out to resolve the following question: Do *all* $\mathcal{F}$-MINOR-FREE DELETION problems admit a polynomial kernel when parameterized by the feedback vertex number?

Results. To our initial surprise, we prove that the answer to this question is *no.* While the parameterization by feedback vertex number admits polynomial kernels for $\mathcal{F} = \{K_2\}$ [19], for $\mathcal{F} = \{K_3\}$ [5,18,24], and for any set $\mathcal{F}$ containing a planar graph[1] but no forests [12], there are also cases that do not admit polynomial kernels (assuming $\mathsf{NP} \not\subseteq \mathsf{coNP/poly}$). For example, we will show that the case of $\mathcal{F}$ consisting of a single graph P_3 that forms a path on three vertices does not admit a polynomial kernel.

Recall that a graph is a forest if and only if its treewidth is one [3]. Hence the feedback vertex number is exactly the minimum number of vertex deletions needed to obtain a graph of treewidth one. Let $\mathrm{tw}(G)$ denote the treewidth of graph G, and define $\min \mathrm{tw}(\mathcal{F}) := \min_{H \in \mathcal{F}} \mathrm{tw}(H)$. Our lower bound also holds for $\mathcal{F}$-SUBGRAPH-FREE DELETION, which is the related problem that asks

[1] If $\mathcal{F}$ contains no forests, the size of an optimal solution is at most the size of a feedback vertex set: the kernel for the solution-size parameterization can be used.

whether there is a vertex set X of size at most k such that $G - X$ contains no graph $H \in \mathcal{F}$ as a *subgraph.* We prove the following.

Theorem 1. *Let $\mathcal{F}$ be a finite set of graphs, such that each graph in $\mathcal{F}$ has a connected component on at least three vertices. Then $\mathcal{F}$-*MINOR-FREE DELETION *and $\mathcal{F}$-*SUBGRAPH-FREE DELETION *do not admit polynomial kernels when parameterized by the vertex-deletion distance to a graph of treewidth* $\min \mathrm{tw}(\mathcal{F})$, *unless* $\mathsf{NP} \subseteq \mathsf{coNP/poly}$.

Theorem 1 implies the claimed lower bound for $\mathcal{F} = \{P_3\}$: when $\mathcal{F}$ contains an acyclic graph with at least one edge we have $\min \mathrm{tw}(\mathcal{F}) = 1$ and therefore the vertex-deletion distance to treewidth $\min \mathrm{tw}(\mathcal{F})$ equals the feedback vertex number. The theorem also generalizes earlier results of Cygan et al. [7, Theorem 13], who investigated the problem of *losing treewidth* by removing vertices.

Theorem 1 is obtained through a polynomial-parameter transformation from the CNF-SAT problem parameterized by the number of variables, for which a superpolynomial kernelization lower bound is known [8,14]. This transformation also rules out the existence of polynomial-size *Turing* kernelizations under a certain hardness assumption. Turing kernelization [10] is a relaxation of the traditional form of kernelization. Intuitively, it investigates whether inputs (x, k) can be solved efficiently using the answers to subproblems of size $f(k)$ which are provided by an oracle, which models an external computation cluster. Formally, a Turing kernelization of size f for a parameterized problem $\mathcal{Q}$ is an algorithm that can query an oracle to obtain the answer to any instance of problem $\mathcal{Q}$ of size and parameter bounded by $f(k)$ in a single step, and using this power solves any instance (x, k) in time polynomial in $|x| + k$. The reduction proving Theorem 1 also proves the non-existence of polynomial-size Turing kernelizations, unless all parameterized problems in the complexity class $\mathsf{MK}[2]$ defined by Hermelin et al. [17] have polynomial Turing kernels. (CNF-SAT with clauses of unbounded length, parameterized by the number of variables, is $\mathsf{MK}[2]$-complete [17, Theorem 1, cf. Theorem 10] and widely believed *not* to admit polynomial-size Turing kernels.)

Motivated by the general form of the lower bound in Theorem 1, we also investigate upper bounds and derive a complexity dichotomy. For any $\mathcal{F}$ that does not meet the criterion of Theorem 1, we obtain a polynomial Turing kernel.

Theorem 2. *Let $\mathcal{F}$ be a finite set of graphs, such that some $H \in \mathcal{F}$ has no connected component of three or more vertices. Then $\mathcal{F}$-*MINOR-FREE DELETION *and $\mathcal{F}$-*SUBGRAPH-FREE DELETION *admit polynomial Turing kernels when parameterized by the vertex-deletion distance to a graph of treewidth* $\min \mathrm{tw}(\mathcal{F})$.

Note that if some graph $H \in \mathcal{F}$ has no component of three or more vertices, then H does not contain P_3 as a subgraph and therefore consists of isolated vertices and edges. Hence $\min \mathrm{tw}(\mathcal{F}) = 1$ in nontrivial cases, so that the parameter is the feedback vertex number. Our Turing kernelization uses an adaptation of the Tutte-Berge formula to show that the $\mathcal{F}$-minor-free graphs that result after removing a solution, have a small witness structure that can be guessed by a Turing kernelization. After this guessing phase, we can reduce the problem to a

VERTEX COVER instance parameterized by feedback vertex set, which we can shrink using the existing kernelization [19] and query to the oracle.

Organization. We present preliminaries on graphs and kernelization in Sect. 2. Section 3 develops the lower bounds on (Turing) kernelization when all graphs in $\mathcal{F}$ have a connected component with at least three vertices. In Sect. 4 we show that in all other cases, a polynomial Turing kernelization exists.

2 Preliminaries

All graphs we consider in this paper are simple, finite and undirected. We denote the vertex set and edge set of a graph G by $V(G)$ and $E(G)$ respectively. For a vertex set $S \subseteq V(G)$ let $G[S]$ be the subgraph of G induced by S, and let $G - S$ denote the subgraph of G induced by $V(G) \setminus S$. For a vertex v we use $G - v$ as shorthand for $G - \{v\}$. For a non-negative integer n we use $n \cdot G$ to denote the graph consisting of n disjoint copies of G. Let $N_G(S)$ and $N_G(v)$ denote the open neighborhood in G of a vertex set S and a vertex v respectively. Let $\deg_G(v)$ denote the degree of v in G. The subscript may be omitted when G is clear from the context. We use $\text{FVS}(G)$ to denote the feedback vertex number of G.

A graph H is a minor of graph G, denoted by $H \preceq G$, if H can be obtained from G by a series of edge contractions, edge deletions, and vertex deletions. An H-model in G is a function $\varphi\colon V(H) \to 2^{V(G)}$ such that (i) for every vertex $v \in V(H)$, the graph $G[\varphi(v)]$ is connected, (ii) for every edge $\{u, v\} \in E(H)$ there exists an edge $\{u', v'\} \in E(G)$ with $u' \in \varphi(u)$ and $v' \in \varphi(v)$, and (iii) for distinct $u, v \in V(H)$ we have $\varphi(v) \cap \varphi(u) = \emptyset$. The sets $\varphi(v)$ are called *branch sets*. Clearly, $H \preceq G$ if and only if there is an H-model in G. For any function $f\colon A \to B$ and set $A' \subseteq A$ we use $f(A')$ as a shorthand for $\bigcup_{a \in A'} f(a)$. Specifically in the case of a minor-model φ and graph G, we use $\varphi(G)$ to denote $\bigcup_{v \in V(G)} \varphi(v)$. We say a graph H is a *component-wise minor* of a graph G, denoted as $H \precsim G$, when every connected component of H is a minor of G.

For *type* $\in$ {minor, subgraph} and a finite family of graphs $\mathcal{F}$, we define:

> $\mathcal{F}$-*type*-FREE DELETION
> **Input:** A graph G and an integer ℓ.
> **Parameter:** vertex-deletion distance to a graph of treewidth $\min \text{tw}(\mathcal{F})$.
> **Question:** Is there a set $X \subseteq V(G)$ of at most ℓ vertices such that $G - X$ does not contain any $H \in \mathcal{F}$ as a *type*?

A vertex $v \in V(G)$ is a *cut vertex* when its removal from G increases the number of connected components. A graph is called *biconnected* when it is connected and contains no cut vertex. A *biconnected component* of a graph G is a maximal biconnected subgraph of G. For any integer α, a graph G is called α-*robust* when $|V(G)| \geq \alpha$ and no vertex $v \in V(C)$ exists such that $G - v$ contains a connected component with less than $\alpha - 1$ vertices.

Proposition 1. *Any graph G has a unique maximal α-robust subgraph. Any α-robust subgraph of G is a subgraph of the maximal α-robust subgraph of G.*

Proof. The proposition follows straightforwardly from the fact that if $G[A]$ and $G[B]$ are α-robust, then so is $G[A \cup B]$. We now prove this fact.

Consider two vertex sets $A, B \subseteq V(G)$, such that $G[A]$ and $G[B]$ are α-robust. We show that $G[A \cup B]$ is α-robust. Since $G[A]$ is α-robust we have $|A| \geq \alpha$ so then $|A \cup B| \geq \alpha$. Suppose for contradiction that there exists a vertex $v \in A \cup B$ such that $G[A \cup B] - v$ contains a connected component of size smaller than $\alpha - 1$. Let C be the vertices of this connected component. We know C contains vertices of at least one of A and B. Assume w.l.o.g. $A \cap C \neq \emptyset$, then $G[A \cap C]$ is a connected component of size less than $\alpha - 1$ in $G[A] - v$. If $v \in A$ this directly contradicts α-robustness of $G[A]$, so assume $v \notin A$. Now $G[A]$ contains a connected component with less than $\alpha - 1$ vertices. Since $|C| < \alpha \leq |A|$ there exists a vertex $u \in A \setminus C$, so then $G[A] - u$ contains a connected component with less than $\alpha - 1$ vertices, which contradicts α-robustness of $G[A]$. □

For any graph G and integer α, let α-prune(G) denote the unique maximal α-robust subgraph of G, which may be empty. We define a *leaf-block* of a graph G as a biconnected component of G that contains at most one cut vertex of G. The size of a leaf-block H is $|V(H)|$. The size of the smallest leaf-block of a graph G is denoted as $\lambda(G)$. Observe that G is α-robust if and only if $\lambda(G) \geq \alpha$.

A *polynomial-parameter transformation* from parameterized problem $\mathcal{P}$ to parameterized problem $\mathcal{Q}$ is a polynomial-time algorithm that, given an instance (x, k) of $\mathcal{P}$, outputs an instance (x', k') of $\mathcal{Q}$ such that $(x, k) \in \mathcal{P} \Leftrightarrow (x', k') \in \mathcal{Q}$, and k' is upper-bounded by a polynomial in k.

Due to space restrictions, the proofs of some claims have been omitted and are given in the full version [9]. These claims are marked with a (★).

3 Lower Bound

In this section we consider the case where all graphs in $\mathcal{F}$ contain a connected component of at least three vertices and give a polynomial-parameter transformation from CNF-SAT parameterized by the number of variables. In this construction we make use of the way biconnected components of graphs G and H restrict the options for an H-model to exist in G.

Proposition 2. *Let H be an α-robust graph and let φ be a minimal H-model in a graph G, then $G[\varphi(H)]$ is α-robust.*

Proof. Take an arbitrary vertex $v \in \varphi(H)$ and let $u \in V(H)$ be such that $v \in \varphi(u)$. Since $H - u$ does not have connected components smaller than $\alpha - 1$, $G[\varphi(H)] - \varphi(u)$ cannot have connected components smaller than $\alpha - 1$. Consider a spanning tree of $G[\varphi(u)]$. Each leaf of this spanning tree must be connected to a vertex in a different branch set, otherwise φ is not minimal. We know every connected component in $G[\varphi(u)] - v$ contains at least one leaf of this spanning tree, hence every connected component of $G[\varphi(u)] - v$ is connected to $G[\varphi(H)] - \varphi(u)$. So $G[\varphi(H)] - v$ does not contain a connected component smaller than $\alpha - 1$. Since v was arbitrary, $G[\varphi(H)]$ is α-robust. □

Proposition 3. *Let φ be an H-model in G, and B a biconnected component of H. Then $G[\varphi(B)]$ contains a biconnected subgraph on at least $|B|$ vertices.*

Proof. Let φ' a minimal B-model in G such that $\varphi'(v) \subseteq \varphi(v)$ for all $v \in V(B)$. Hence $G[\varphi'(B)]$ is a subgraph of $G[\varphi(B)]$. It suffices to show that $G[\varphi'(B)]$ contains a biconnected component on at least $|V(B)|$ vertices. Since B is biconnected, it is $|V(B)|$-robust so by Proposition 2 we know $G[\varphi'(B)]$ is $|V(B)|$-robust. Hence $G[\varphi'(B)]$ contains a biconnected component on at least $|V(B)|$ vertices. □

Proposition 4 (★). *For any $\trianglelefteq \in \{\preceq, \precsim\}$, two integers $\alpha \geq \beta$, and graphs H and G we have that $H \trianglelefteq G \Rightarrow \alpha$-prune$(H) \trianglelefteq \beta$-prune$(G)$.*

We proceed to construct a clause gadget to be used in the polynomial-parameter transformation from CNF-SAT.

Lemma 1. *For any connected graph H with at least three vertices there exists a polynomial-time algorithm that, given an integer $n \geq 1$, outputs a graph G and a vertex set $S \subseteq V(G)$ of size n such that all of the following are true:*

1. $\text{tw}(G) \leq \text{tw}(H)$,
2. *G contains a packing of $3n-1$ vertex-disjoint H-subgraphs,*
3. *$G - S$ contains a packing of $3n-2$ vertex-disjoint H-subgraphs, and*
4. *$\forall v \in S$ there exists $X \subseteq V(G)$ of size $3n-1$ s.t. all of the following are true:*
   (a) $v \in X$,
   (b) *$G - X$ is H-minor-free,*
   (c) $\lambda(H)$-prune$(G - X) \precsim H$, *and*
   (d) *for all connected components G_c of $G - X$ that contain a vertex of S we have $|V(G_c)| < \lambda(H)$ and G_c contains exactly one vertex of S.*

Proof. Consider a subgraph L of H such that L is a smallest leaf-block of H. Let R be the graph obtained from H by removing all vertices of L that are not a cut vertex in H. Note that when H is biconnected, $L = H$ and R is an empty graph. We distinguish three distinct vertices a, b, c in H. Vertices c and b are both part of L, where c is the cut vertex (if there is one) and b is any other vertex in L. Finally vertex a is any vertex in H that is not c or b. See Fig. 1(a). In the construction of G we will combine copies of H such that a, b, and c form cut vertices in G and are part of two different H-subgraphs. Vertices b and c are chosen such that removing either one from a copy of H in G means no vertex from the L-subgraph of this copy of H can be used in a minimal H-model in G. In the remainder of this proof we use $f_{K \to K'} \colon V(K) \to V(K')$ for isomorphic graphs K and K' to denote a fixed isomorphism.

Take two copies of H, call them H_1 and H_2. Let R_1 and L_1 denote the subgraphs of H_1 related to R and L, respectively, by the isomorphism between H and H_1. Similarly let R_2 and L_2 denote the subgraphs of H_2. Take a copy of L which we call L_3. Let M be the graph obtained from the disjoint union of H_1, H_2, and L_3 by identifying the pair $f_{H \to H_1}(c)$ and $f_{H \to H_2}(b)$ into a single vertex

s, and identifying the pair $f_{H \to H_2}(c)$ and $f_{L \to L_3}(c)$ into a single vertex t. We label $f_{H \to H_1}(a)$, $f_{H \to H_1}(b)$, and $f_{L \to L_3}(b)$ as u, w, and v respectively.

This construction is motivated by the fact that the graphs $M - \{v, s\}$, $M - \{u, t\}$, and $M - \{w, t\}$ are all H-minor-free, which we will exploit in the formal correctness argument later. We will connect copies of M to each other via the vertices u, v, and w so that, although two vertices need to be removed in every copy of M, one such vertex can always be in two copies of M at the same time.

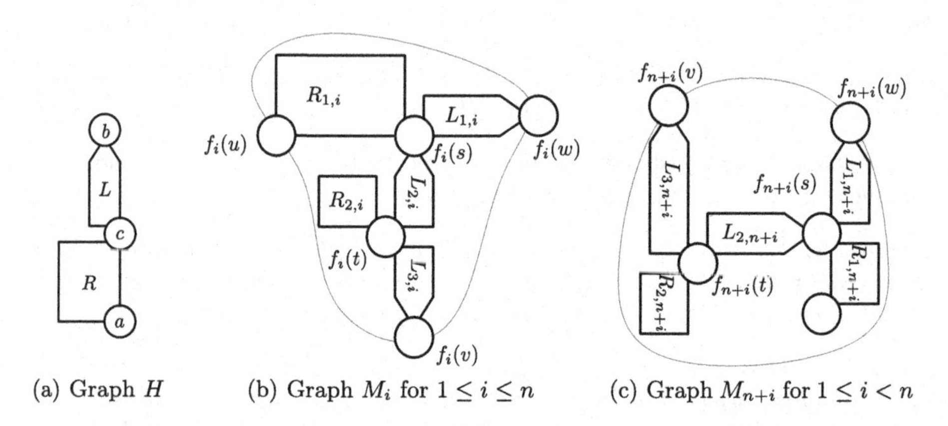

(a) Graph H (b) Graph M_i for $1 \leq i \leq n$ (c) Graph M_{n+i} for $1 \leq i < n$

Fig. 1. We show the situation where a is contained in R. Note that a can always be chosen such that it is contained in R when H is not biconnected. Note that the graphs in (b) and (c) are isomorphic but drawn differently.

Now take $2n - 1$ copies of M, call them $M_1, \ldots, M_{2n-1}$. For readability we denote $f_{M \to M_i}$ as f_i for all $1 \leq i \leq 2n - 1$. For all $1 \leq i < n$ we identify $f_i(w)$ and $f_{n+i}(v)$, and we identify $f_{n+i}(w)$ and $f_{i+1}(u)$. Let this graph be G, and let S be the set of vertices $f_i(v)$ for all $1 \leq i \leq n$. Let $H_{1,i}$, $H_{2,i}$, $R_{1,i}$, $R_{2,i}$, $L_{1,i}$, $L_{2,i}$, and $L_{3,i}$ denote the subgraphs in M_i that correspond to the subgraphs H_1, H_2, R_1, R_2, L_1, L_2, and L_3 in M. See Fig. 1(b) and (c).

This concludes the description of graph G and set S. It is easily seen that these can be constructed in polynomial time. It remains to verify that all conditions of the lemma statement are met.

(1) Since we connected copies of L and R in a treelike fashion along cut vertices, we did not introduce any new biconnected components. Since the treewidth of a graph is equal to the maximum treewidth over all its biconnected components we know that $\text{tw}(G) \leq \max\{\text{tw}(R), \text{tw}(L)\} = \text{tw}(H)$.

(2) For each $1 \leq i \leq n$ we can distinguish two H-subgraphs in M_i, namely $H_{1,i}$ and $L_{3,i} \cup R_{2,i}$. This gives us $2n$ H-subgraphs in G. Note that since all $M_1, \ldots, M_n$ are vertex-disjoint, these $2n$ H-subgraphs are also vertex-disjoint in G. For each $n < i \leq 2n - 1$ we distinguish one H-subgraph, namely $H_{2,i}$. Note that since $H_{2,i}$ is vertex-disjoint from all $M_1, \ldots, M_{i-1}, M_{i+1}, \ldots, M_{2n-1}$ we have a total of $2n + n - 1 = 3n - 1$ vertex-disjoint H-subgraphs in G. This packing is shown in Fig. 2(a).

(3) Alternatively, for each $1 \le i \le n$ we can distinguish one H-subgraph in M_i, namely $H_{2,i}$. For each $n < i \le 2n-1$ we distinguish two H-subgraphs in M_i, namely $H_{1,i}$ and $L_{3,i} \cup R_{2,i}$. Again these H-subgraphs are vertex-disjoint, and since they also do not contain any vertices of S, they form a packing of $n + 2(n-1) = 3n-2$ vertex-disjoint H-subgraphs in $G - S$. See Fig. 2(b).

(4) Let $f_j(v) \in S$ be an arbitrary vertex in S, implying $1 \le j \le n$, and take

$$X = \bigcup_{1 \le i < j} \{f_i(t), f_i(w), f_{i+n}(s)\} \cup \{f_j(v), f_j(s)\} \cup \bigcup_{j < i \le n} \{f_i(t), f_i(u), f_{i+n-1}(t)\}.$$

Observe that $|X| = 3n-1$ and $f_j(v) \in X$, so condition 4a of the lemma statement holds. Next, we give a proof sketch for Conditions 4b, 4c, and 4d for the case that $a \in V(R)$. A complete proof can be found in the full version [9].

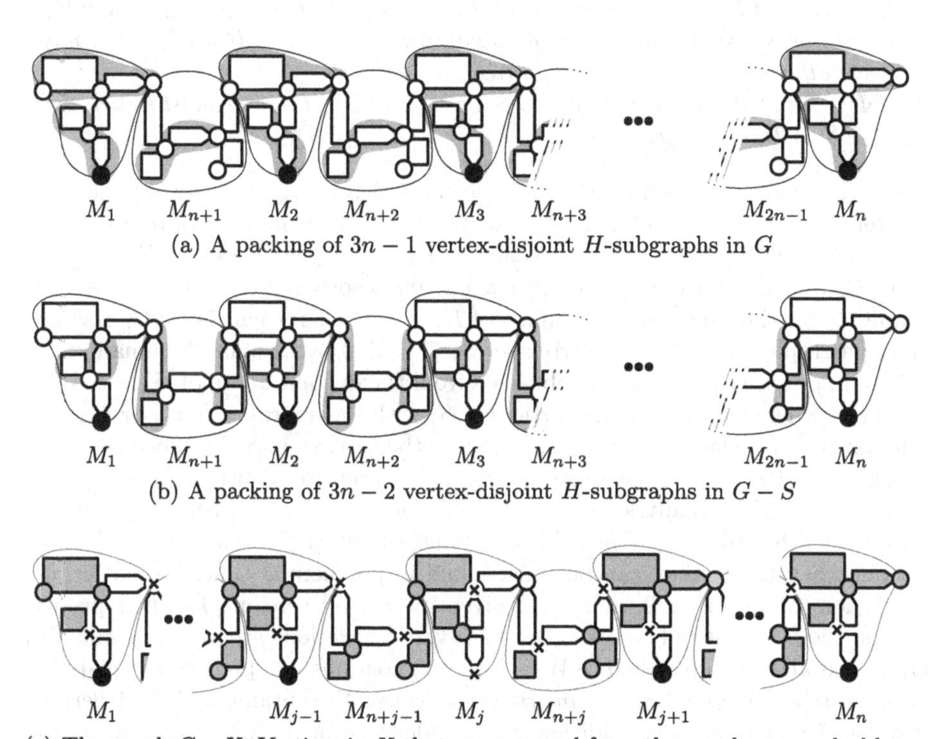

(a) A packing of $3n-1$ vertex-disjoint H-subgraphs in G

(b) A packing of $3n-2$ vertex-disjoint H-subgraphs in $G - S$

(c) The graph $G - X$. Vertices in X that are removed from the graph are marked by a cross. Vertices in S are marked black. A supergraph of $\lambda(H)$-prune$(G - X)$ is shown in gray. Note that when $|V(R)| = |V(L)|$, not all subgraphs and vertices marked gray are necessarily part of $\lambda(H)$-prune$(G - X)$.

Fig. 2. Illustrations of conditions 2, 3 and 4. Vertices in S are marked black.

To show condition 4b holds, by Proposition 4 it is sufficient to show that $\lambda(H)$-prune$(G - X)$ is H-minor-free, since $H = \lambda(H)$-prune(H). Figure 2(c) shows a super graph of $\lambda(H)$-prune$(G-X)$ in gray. It is easily verified from the figure that every connected component of $\lambda(H)$-prune$(G-X)$ contains insufficient vertices to contain H as a minor. It can also be seen that every connected component in $\lambda(H)$-prune$(G-X)$ is a subgraph of H, proving condition 4c. Similarly, condition 4d can also directly be seen to hold from the figure. □

Using the clause gadget described in Lemma 1 we give a polynomial-parameter transformation for the case where $\mathcal{F}$ contains a single, connected graph H.

Lemma 2. *For any connected graph H with at least three vertices there exists a polynomial-time algorithm that, given a* CNF*-formula Φ with k variables, outputs a graph G and an integer ℓ such that all of the following are true:*

1. *there is a set $S \subseteq V(G)$ of at most $2k$ vertices such that* $\text{tw}(G-S) \leq \text{tw}(H)$,
2. *if Φ is not satisfiable then there does not exist a set $X \subseteq V(G)$ of size at most ℓ such that $G-X$ is H-subgraph-free,*
3. *if Φ is satisfiable then there exists a set $X \subseteq V(G)$ of size at most ℓ such that $G-X$ is H-minor-free.*

Proof. Let $x_1, \ldots, x_k$ denote the variables of Φ, let $C_1, \ldots, C_m$ denote the sets of literals in each clause of Φ, and let n denote the total number of occurrences of literals in Φ, i.e. $n = \sum_{1 \leq j \leq m} |C_j|$. Let $H_1, \ldots, H_k$ be copies of H. In each copy H_i we arbitrarily label one vertex v_{x_i} and another $v_{\neg x_i}$. Let G_{var} be the graph obtained from the disjoint union of $H_1, \ldots, H_k$. For each clause C_j of Φ we create a graph called W_j and vertex set $S_j \subseteq V(W_j)$ by invoking Lemma 1 with H and $|C_j|$. Let G be the graph obtained from the disjoint union of $W_1, \ldots, W_m$ and G_{var} where we identify the vertices in S_j with the appropriate v_{x_i} or $v_{\neg x_i}$ as follows: For each clause C_j let $s_1, \ldots, s_{|C_j|}$ be the vertices in S_j in some arbitrary order, and let $c_1, \ldots, c_{|C_j|}$ be the literals in C_j, then we identify s_i and v_{c_i} for each $1 \leq i \leq |C_j|$. Finally let $\ell = k + 3n - 2m$ and $S = \bigcup_{1 \leq i \leq k} \{v_{x_i}, v_{\neg x_i}\}$. Note that $S_j \subseteq S$ for all $1 \leq j \leq m$. This concludes the description of G, ℓ, and S.

It is easy to see they can be constructed in polynomial time. Formal arguments for conditions 1, 2, and 3 are given in the full version [9]. Their proofs rely on the fact that only $\ell - k = 3n - 2m$ vertex-deletions are available outside G_{var}, and all are required since $W_i - V(G_{var})$ contains $3|C_i| - 2$ vertex disjoint H-subgraphs for each $1 \leq i \leq m$. However, since W_i contains $3|C_j| - 1$ vertex disjoint H-subgraphs, one additional vertex-deletion in W_i has to coincide with a vertex-deletion in G_{var}. This is possible if and only if a set $X' \subseteq V(G_{var})$ of k vertices can be chosen such that each W_i, $1 \leq i \leq m$, contains at least one vertex from X'. The choice of X' corresponds to a satisfying assignment for Φ. □

The construction from Lemma 2 can directly be used to give a polynomial-parameter transformation from CNF-SAT parameterized by the number of variables. Observe that if $G-X$ is $\mathcal{F}$-minor-free, then $G-X$ is also $\mathcal{F}$-subgraph-free. Similarly, if $G-X$ contains an H-subgraph for all $X \subseteq V(G)$ with $|X| \leq \ell$, then $G-X$ also contains an H-minor. Therefore, for any *type* $\in$ {minor, subgraph} and $\mathcal{F}$ consisting of one connected graph on at least three vertices, Lemma 2 gives a polynomial-parameter transformation from CNF-SAT parameterized by the number of variables to $\mathcal{F}$-*type*-FREE DELETION parameterized by deletion distance to $\min \text{tw}(\mathcal{F})$.

When $\mathcal{F}$ contains multiple graphs, each containing a connected component of at least three vertices, it is possible to select a connected component H of one of the graphs in $\mathcal{F}$ such that the construction described in Lemma 2 forms the main ingredient for a polynomial-parameter transformation. Selection of H and the remainder of the construction are described in the full version [9].

We conclude that a polynomial-parameter transformation exists for all *type* $\in$ {minor, subgraph} and $\mathcal{F}$ containing only graphs with a connected component on at least three vertices. Together with the fact that CNF-SAT is MK[2]-hard and does not admit a polynomial kernel unless NP $\subseteq$ coNP/poly (cf. [17, Lemma 9]), this proves the following generalization of Theorem 1.

Theorem 3 (★). *For type $\in$ {minor, subgraph} and a set $\mathcal{F}$ of graphs, all with a connected component of at least three vertices, $\mathcal{F}$-type-*FREE DELETION *parameterized by vertex-deletion distance to a graph of treewidth* $\min \text{tw}(\mathcal{F})$ *is* MK[2]*-hard and does not admit a polynomial kernel unless* NP $\subseteq$ coNP/poly.

4 A Polynomial Turing Kernelization

In this section we consider the case where $\mathcal{F}$ contains a graph with no connected component of more than two vertices; or in short $\mathcal{F}$ contains a P_3-subgraph-free graph. This graph consists of isolated vertices and disjoint edges. Let isol(G) denote the set of isolated vertices in a graph G, i.e. $\text{isol}(G) = \{v \in V(G) \mid \deg(v) = 0\}$. We first show that the removal of all isolated vertices from all graphs in $\mathcal{F}$ only changes the answer to $\mathcal{F}$-MINOR-FREE DELETION and $\mathcal{F}$-SUBGRAPH-FREE DELETION when the input is of constant size.

Lemma 3 (★). *For type $\in$ {minor, subgraph} and any family of graphs $\mathcal{F}$ containing a P_3-subgraph-free graph, let $\mathcal{F}' = \{F - \text{isol}(F) \mid F \in \mathcal{F}\}$. For any graph G, if G is $\mathcal{F}$-type-free but not $\mathcal{F}'$-type-free, then* $|V(G)| < \max_{F \in \mathcal{F}}(|V(F)| + 2|V(F)|^3)$.

After the removal of isolated vertices in $\mathcal{F}$ to obtain $\mathcal{F}'$, we know that $\mathcal{F}'$ contains a graph consisting entirely of disjoint edges, i.e. this graph is isomorphic to $c \cdot P_2$ for some integer $c \geq 0$. If $c = 0$ then $\mathcal{F}$-*type*-free graphs have constant size and the problem is polynomial-time solvable. We proceed assuming $c \geq 1$. Let the matching number of a graph G, denoted as $\nu(G)$, be the size of a maximum matching in G. We make the following observation.

Observation 1. *For all $c \geq 1$, graph G is $c \cdot P_2$-subgraph-free $\Leftrightarrow \nu(G) \leq c - 1$.*

We give a characterization of graphs with bounded matching number, based on an adaptation of the Tutte-Berge formula [2]. We use odd(G) to denote the number of connected components in G that consist of an odd number of vertices.

Lemma 4. (★). *For any graph G and integer m we have $\nu(G) \leq m$ if and only if $V(G)$ can be partitioned into three disjoint sets U, R, S such that all of the following are true:*

- *all connected components in $G[R]$ have an odd size of at least 3,*
- *$G[S]$ is independent,*
- *$N_G(S) \subseteq U$, and*
- *$|U| + \frac{1}{2}(|R| - \text{odd}(G[R])) \leq m$.*

Observe that for any partition U, R, S satisfying the first three conditions, we have $|R| - \text{odd}(G[R]) \geq \frac{2}{3}|R|$ since each component of $G[R]$ has at least three vertices. To satisfy the fourth condition therefore requires $|U| + \frac{1}{2}(\frac{2}{3}|R|) \leq m$, which will be a constant in our application. Since $N_G(S) \subseteq U$, the lemma guarantees that $N_G(R) \subseteq U$.

Let us showcase how Lemma 4 can be used to attack $\mathcal{F}$-Minor-Free Deletion when $\mathcal{F}$ consists of a single graph $c \cdot P_2$, so that the problem is to find a set $X \subseteq G$ of size at most ℓ such that $G - X$ has matching number less than c.

Theorem 4. *For any constant c, the $\{c \cdot P_2\}$-*Minor-Free Deletion *problem parameterized by the size k of a feedback vertex set, can be solved in polynomial time using an oracle that answers* Vertex Cover *instances with $\mathcal{O}(k^3)$ vertices.*

Proof. If an instance (G, ℓ) admits a solution X, then Lemma 4 guarantees that $V(G - X)$ can be partitioned into U, R, S satisfying the four conditions for $m = c - 1$. We try all relevant options for the sets U and R in the partition, of which there are only polynomially many since $|U| + \frac{1}{3}|R| \leq m \in \mathcal{O}(1)$.

For given sets $U, R \subseteq V(G)$, we can decide whether there is a solution X of size at most ℓ for which U, R, and $S := V(G) \setminus (U \cup R \cup X)$ form the partition witnessing that $G - X$ has matching number at most m, as follows. If some component of $G[R]$ has even size, or less than three vertices, we reject outright. Similarly, if $|U| + \frac{1}{2}(|R| - \text{odd}(G[R])) > m$, we reject. Now, if U and R were guessed correctly, then Lemma 4 guarantees that the only neighbors of R in the graph $G - X$ belong to U. Hence we infer that all vertices of $X' := N_G(R) \setminus U$ must belong to the solution X. Note that since S is an independent set in $G - X$, the solution X forms a vertex cover of $G - (U \cup R)$, so that $X'' := X \setminus X'$ is a vertex cover of $G' := G - (U \cup R \cup X')$. On the other hand, for every vertex cover X'' of G', the graph $G - (X' \cup X'')$ will have matching number at most m, as witnessed by the partition. Hence the problem of finding a minimum solution X whose corresponding graph $G - X$ has U and R as two of the classes in its witness partition, reduces to finding a minimum vertex cover of the graph G'. In terms of the decision problem, this means G has a solution of size at most ℓ with U and R as witness partite sets, if and only if G' has a vertex cover of size at most $\ell - |X'|$. Since $\text{FVS}(G') \leq \text{FVS}(G)$, we can apply the known [19] kernel for Vertex Cover parameterized by the feedback vertex number to reduce $(G', \ell - |X'|)$ to an equivalent instance with $\mathcal{O}(\text{FVS}(G)^3)$ vertices, which is queried to the oracle. If the oracle answers positively to any query, then (G, ℓ) has answer YES; otherwise the answer is NO. □

We remark that by using the polynomial-time reduction guaranteed by NP-completeness, the queries to the oracle can be posed as instances of the original $\mathcal{F}$-Minor-Free Deletion problem, rather than Vertex Cover. In the appendix we present our general (non-adaptive) Turing kernelization for the minor-free and subgraph-free deletion problems for all families $\mathcal{F}$ containing a P_3-subgraph-free graph, combining three ingredients. Lemma 3 allows us to focus on families whose graphs have no isolated vertices. The guessing strategy of Theorem 4 is the second ingredient. The final ingredient is required to deal with the fact that a solution subgraph $G - X$ that is $c \cdot P_2$-minor-free for some $c \cdot P_2 \in \mathcal{F}$, may still have one of the other graphs in $\mathcal{F}$ as a forbidden minor. To cope with this issue, we show that if $G - X$ has no matching of size c, but does contain a minor model of some graph in $\mathcal{F}$, then there is such a minor model of constant size. By employing a more expensive (but still polynomially bounded) guessing step, this allows us to complete the Turing kernelization and prove the following theorem.

Theorem 2 (★). *Let $\mathcal{F}$ be a finite set of graphs, such that some $H \in \mathcal{F}$ has no connected component of three or more vertices. Then $\mathcal{F}$-*Minor-Free Deletion *and $\mathcal{F}$-*Subgraph-Free Deletion*admit polynomial Turing kernels when parameterized by the vertex-deletion distance to a graph of treewidth* $\min \mathrm{tw}(\mathcal{F})$.

5 Conclusion

Earlier work [5,18,19,24] has shown that several $\mathcal{F}$-Minor-Free Deletion problems admit polynomial kernelizations when parameterized by the feedback vertex number. In this paper we showed that when $\mathcal{F}$ contains a forest and each graph in $\mathcal{F}$ has a connected component of at least three vertices, the $\mathcal{F}$-Minor-Free Deletion problem does *not* admit such a polynomial kernel unless $\mathsf{NP} \subseteq \mathsf{coNP/poly}$. This lower bound generalizes to any $\mathcal{F}$ where each graph has a connected component of at least three vertices, when we consider the vertex-deletion distance to treewidth $\min \mathrm{tw}(\mathcal{F})$ as parameter.

For all other choices of $\mathcal{F}$ we showed that a polynomial Turing kernelization exists for $\mathcal{F}$-Minor-Free Deletion parameterized by the feedback vertex number. The size of the Vertex Cover queries generated by the Turing kernelization does not depend on $\mathcal{F}$: the Turing kernelization can be shown to be *uniformly polynomial* (cf. [15]). However, it remains unknown whether the *running time* can be made uniformly polynomial, and whether the Turing kernelization can be improved to a traditional kernelization.

Our results leave open the possibility that all $\mathcal{F}$-Minor-Free Deletion problems admit a polynomial kernel when parameterized by the vertex-deletion distance to a *linear forest*, i.e. a collection of paths. Resolving this question may be an interesting direction for future work.

References

1. Agrawal, A., Lokshtanov, D., Misra, P., Saurabh, S., Zehavi, M.: Feedback vertex set inspired kernel for chordal vertex deletion. In: Proceedings of 28th SODA, pp. 1383–1398. SIAM (2017). https://doi.org/10.1137/1.9781611974782.90
2. Berge, C.: Sur le couplage maximum d'un graphe. Comptes rendus hebdomadaires des séances de l'Académie des sciences **247**, 258–259 (1958)
3. Bodlaender, H.L.: A partial k-arboretum of graphs with bounded treewidth. Theor. Comput. Sci. **209**(1–2), 1–45 (1998). https://doi.org/10.1016/S0304-3975(97)00228-4
4. Bodlaender, H.L.: Kernelization: new upper and lower bound techniques. In: Chen, J., Fomin, F.V. (eds.) IWPEC 2009. LNCS, vol. 5917, pp. 17–37. Springer, Heidelberg (2009). https://doi.org/10.1007/978-3-642-11269-0_2
5. Bodlaender, H.L., van Dijk, T.C.: A cubic kernel for feedback vertex set and loop cutset. Theory Comput. Syst. **46**(3), 566–597 (2010)
6. Bougeret, M., Sau, I.: How much does a treedepth modulator help to obtain polynomial kernels beyond sparse graphs? In: Proceedings of 12th IPEC. LIPIcs, vol. 89, pp. 10:1–10:13 (2017). https://doi.org/10.4230/LIPIcs.IPEC.2017.10
7. Cygan, M., Lokshtanov, D., Pilipczuk, M., Pilipczuk, M., Saurabh, S.: On the hardness of losing width. Theory Comput. Syst. **54**(1), 73–82 (2014). https://doi.org/10.1007/s00224-013-9480-1
8. Dell, H., van Melkebeek, D.: Satisfiability allows no nontrivial sparsification unless the polynomial-time hierarchy collapses. J. ACM **61**(4), 23:1–23:27 (2014). https://doi.org/10.1145/2629620
9. Donkers, H., Jansen, B.M.P.: A Turing kernelization dichotomy for structural parameterizations of $\mathcal{F}$-minor-free deletion. CoRR abs/1906.05565 (2019). http://arxiv.org/abs/1906.05565
10. Fernau, H.: Kernelization, Turing kernels. In: Kao, M.Y. (ed.) Encyclopedia of Algorithms, pp. 1043–1045. Springer, New York (2016). https://doi.org/10.1007/978-1-4939-2864-4_528
11. Fomin, F.V., Jansen, B.M.P., Pilipczuk, M.: Preprocessing subgraph and minor problems: When does a small vertex cover help? J. Comput. Syst. Sci. **80**(2), 468–495 (2014). https://doi.org/10.1016/j.jcss.2013.09.004
12. Fomin, F.V., Lokshtanov, D., Misra, N., Saurabh, S.: Planar $\mathcal{F}$-deletion: approximation, kernelization and optimal FPT algorithms. In: Proceedings of 53rd FOCS, pp. 470–479 (2012). https://doi.org/10.1109/FOCS.2012.62
13. Fomin, F.V., Lokshtanov, D., Saurabh, S., Zehavi, M.: Kernelization: Theory of Parameterized Preprocessing. Cambridge University Press, Cambridge (2019). https://doi.org/10.1017/9781107415157
14. Fortnow, L., Santhanam, R.: Infeasibility of instance compression and succinct PCPs for NP. J. Comput. Syst. Sci. **77**(1), 91–106 (2011). https://doi.org/10.1016/j.jcss.2010.06.007
15. Giannopoulou, A.C., Jansen, B.M.P., Lokshtanov, D., Saurabh, S.: Uniform kernelization complexity of hitting forbidden minors. ACM Trans. Algorithms **13**(3), 35:1–35:35 (2017). https://doi.org/10.1145/3029051
16. Guo, J., Hüffner, F., Niedermeier, R.: A structural view on parameterizing problems: distance from triviality. In: Proceedings of 1st IWPEC, pp. 162–173 (2004). https://doi.org/10.1007/978-3-540-28639-4_15
17. Hermelin, D., Kratsch, S., Soltys, K., Wahlström, M., Wu, X.: A completeness theory for polynomial (Turing) kernelization. Algorithmica **71**(3), 702–730 (2015). https://doi.org/10.1007/s00453-014-9910-8

18. Iwata, Y.: Linear-time kernelization for feedback vertex set. In: Proceedings of 44th ICALP. LIPIcs, vol. 80, pp. 68:1–68:14 (2017). https://doi.org/10.4230/LIPIcs.ICALP.2017.68
19. Jansen, B.M.P., Bodlaender, H.L.: Vertex cover kernelization revisited - upper and lower bounds for a refined parameter. Theory Comput. Syst. **53**(2), 263–299 (2013). https://doi.org/10.1007/s00224-012-9393-4
20. Jansen, B.M.P., Kratsch, S.: Data reduction for graph coloring problems. Inf. Comput. **231**, 70–88 (2013). https://doi.org/10.1016/j.ic.2013.08.005
21. Jansen, B.M.P., Pieterse, A.: Polynomial kernels for hitting forbidden minors under structural parameterizations. In: Proceedings of 26th ESA. LIPIcs, vol. 112, pp. 48:1–48:15 (2018). https://doi.org/10.4230/LIPIcs.ESA.2018.48
22. Lokshtanov, D., Misra, N., Saurabh, S.: Kernelization - preprocessing with a guarantee. In: The Multivariate Algorithmic Revolution and Beyond, pp. 129–161 (2012). https://doi.org/10.1007/978-3-642-30891-8_10
23. Niedermeier, R.: Reflections on multivariate algorithmics and problem parameterization. In: Proceedings of 27th STACS, pp. 17–32 (2010). https://doi.org/10.4230/LIPIcs.STACS.2010.2495
24. Thomassé, S.: A $4k^2$ kernel for feedback vertex set. ACM Trans. Algorithms **6**(2) (2010). https://doi.org/10.1145/1721837.1721848
25. Uhlmann, J., Weller, M.: Two-layer planarization parameterized by feedback edge set. Theor. Comput. Sci. **494**, 99–111 (2013). https://doi.org/10.1016/j.tcs.2013.01.029

Flip Distances Between Graph Orientations

Oswin Aichholzer[1], Jean Cardinal[2], Tony Huynh[2], Kolja Knauer[3], Torsten Mütze[4], Raphael Steiner[4(✉)], and Birgit Vogtenhuber[1]

[1] TU Graz, Graz, Austria
{oaich,bvogt}@ist.tugraz.at
[2] Université libre de Bruxelles (ULB), Brussels, Belgium
jcardin@ulb.ac.be,tony.bourbaki@gmail.com
[3] Université Aix-Marseille, Marseille, France
kolja.knauer@lis-lab.fr
[4] TU Berlin, Berlin, Germany
{muetze,steiner}@math.tu-berlin.de

Abstract. Flip graphs are a ubiquitous class of graphs, which encode relations on a set of combinatorial objects induced by elementary, local changes. A natural computational problem to consider is the flip distance: Given two objects, what is the minimum number of flips needed to transform one into the other?

We consider flip graphs on so-called α-orientations of a graph G, in which every vertex v has a specified outdegree $\alpha(v)$, and a flip consists of reversing all edges of a directed cycle. We prove that deciding whether the flip distance between two α-orientations of a planar graph G is at most 2 is NP-complete. This also holds in the special case of plane perfect matchings, where flips involve alternating cycles. We also consider the dual question of the flip distance between graph orientations in which every cycle has a specified number of forward edges, and a flip is the reversal of all edges in a minimal directed cut. In general, the problem remains hard, but if we only change sinks into sources, or vice-versa, then the problem can be solved in polynomial time.

Keywords: Flip distance · α-orientation · Graph orientation

T.H. supported by ERC Consolidator Grant 615640-ForEFront; K.K. partially supported by ANR grants GATO: ANR-16-CE40-0009-01, DISTANCIA: ANR-17-CE40-0015, and CAPPS: ANR-17-CE40-0018; T.M. is also affiliated with Charles University, Faculty of Mathematics and Physics, and was supported by GACR grant GA 19-08554S, and by DFG grant 413902284; R.S. funded by DFG-GRK 2434. B.V. partially supported by the Austrian Science Fund (FWF): I 3340-N35.
A full version of the paper, including further details and proofs, can be found on https://arxiv.org/abs/1902.06103.
This work was initiated during the workshop "Order & Geometry" 2018 in Ciążeń Palace. We thank the organizers and participants for the stimulating atmosphere.

I. Sau and D. M. Thilikos (Eds.): WG 2019, LNCS 11789, pp. 120–134, 2019.
https://doi.org/10.1007/978-3-030-30786-8_10

1 Introduction

The term flip is commonly used in combinatorics to refer to an elementary, local, reversible operation that transforms one combinatorial object into another. Such flip operations naturally yield a flip graph, whose vertices are the considered combinatorial objects, and two of them are adjacent if they differ by a single flip. A classical example is the flip graph of triangulations of a convex polygon [44]; see Fig. 1. The vertex set of this graph are all triangulations of the polygon, and two triangulations are adjacent if one can be obtained from the other by replacing the diagonal of a quadrilateral formed by two triangles by the other diagonal. Similar flip graphs have also been investigated for triangulations of general point sets in the plane [32], triangulations of topological surfaces [34], and planar graphs [7,8]. The flip distance between two combinatorial objects is the minimum number of flips needed to transform one into the other. It is known that computing the flip distance between two triangulations of a simple polygon [3] or of a point set [33] is NP-hard. The latter is known to be fixed-parameter tractable [28]. On the other hand, the NP-hardness of computing the flip distance between two triangulations of a convex polygon is a well-known open question [13,31,43]. Flip graphs involving other geometric configurations have also been studied, such as flip graphs of non-crossing perfect matchings of a point set in the plane, where flips are with respect to alternating 4-cycles [24], or alternating cycles of arbitrary length [25]. Very recently the distance computation between perfect matchings of a (non-geometric) graph with respect to alternating 4-cycles has been studied [6]. Other flip graphs include the flip graph on plane spanning trees [2], the flip graph of non-crossing partitions of a point set or dissections of a polygon [26], the mutation graph of simple pseudoline arrangements [41], the Eulerian tour graph of an Eulerian graph [46], and many others. There is also a vast collection of interesting flip graphs for non-geometric objects, such as bitstrings, permutations, combinations, and partitions [16].

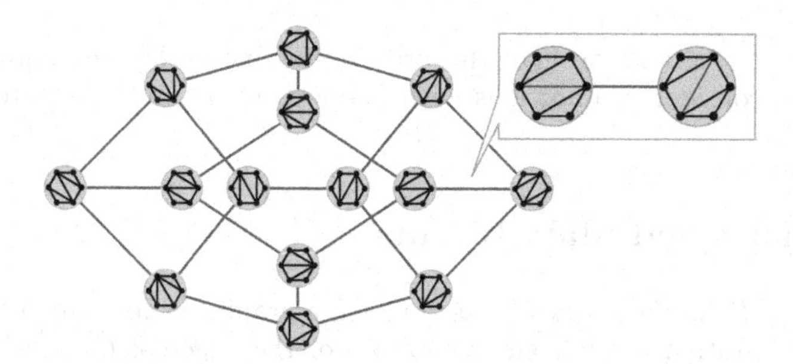

Fig. 1. The flip graph of triangulations of a convex polygon.

In essence, a flip graph provides the considered family of combinatorial objects with an underlying structure that reveals interesting properties about the objects. It can also be a useful tool for proving that a property holds for all objects, by proving that one particularly nice object has the property, and that the property is preserved under flips. Flip graphs are also an essential tool for solving fundamental algorithmic tasks such as random and exhaustive generation, see e.g. [4] and [39].

The focus of the present paper is on flip graphs for orientations of graphs satisfying some constraints. First, we consider so-called α-orientations, in which the outdegree of every vertex is specified by a function α, and the flip operation consists of reversing the orientation of all edges in a directed cycle. We study the complexity of computing the flip distance between two such orientations. An interesting special case of α-orientations corresponds to perfect matchings in bipartite graphs, where flips involve alternating cycles. We also consider the dual notion of c-orientations, in which the number of forward edges along each cycle is specified by a function c. Here a flip consists of reversing all edges in a directed cut. We also analyze the computational complexity of the flip distance problem in c-orientations.

There are several deep connections between flip graphs and polytopes. Specifically, many interesting flip graphs arise as the (1-)skeleton of a polytope. For instance, flip graphs of triangulations of a convex polygon are skeletons of associahedra [12], and flip graphs of regular triangulations of a point set in the plane are skeletons of secondary polytopes (see [32, Chapter 5]). Associahedra are generalized by quotientopes [36], whose skeletons yield flip graphs on rectangulations [11], bitstrings, permutations, and other combinatorial objects. Moreover, flip graphs of acyclic orientations or strongly connected orientations of a graph are skeletons of graphical and co-graphical zonotopes, respectively (see [37, Section 2]). Similarly, as we show below, flip graphs on α-orientations are skeletons of matroid intersection polytopes. We also consider vertex flips in c-orientations, inducing flip graphs that are distributive lattices and in particular subgraphs of skeletons of certain distributive polytopes. These polytopes specialize to flip polytopes of planar α-orientations, are generalized by the polytope of tensions of a digraph, and form part of the family of alcoved polytopes (see [18]).

In the next section, we give the precise statements of the computational problems we consider, connections with previous work, and the statements of our results.

2 Problems and Main Results

Flip Distance Between α-Orientations. Given a graph G and some $\alpha : V(G) \to \mathbb{N}_0$, an α-orientation of G is an orientation of the edges of G in which every vertex v has outdegree $\alpha(v)$. An example for a graph and two α-orientations for this graph is given in Fig. 2. A flip of a directed cycle C in some α-orientation X consists of the reversal of the orientation of all edges of C, as shown in the figure. Edges with distinct orientations in two given α-orientations X and Y induce an Eulerian subdigraph of both X and Y. They can therefore be partitioned into an edge-disjoint union of cycles in G which are directed in both X and Y. Hence the reversal of each such cycle in X gives rise to a flip sequence transforming X into Y and vice versa. We may thus define the flip distance between two α-orientations X and Y to be the minimum number of cycles in a flip sequence transforming X into Y. We are interested in the computational complexity of determining the flip distance between two given α-orientations.

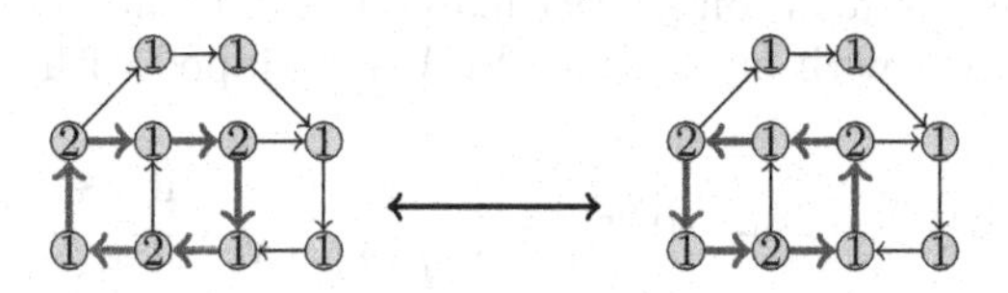

Fig. 2. Two α-orientations of a graph and a flip between them, where the values of α are depicted on the vertices.

Problem 1. Given a graph G, some $\alpha : V(G) \rightarrow \mathbb{N}_0$, a pair X, Y of α-orientations of G and an integer $k \geq 0$, decide whether the flip distance between X and Y is at most k.

The crucial difficulty of this problem is that a shortest flip sequence transforming X into Y may flip edges that are oriented the same in X and Y an even number of times, to reach Y with fewer flips compared to only flipping edges that are oriented differently in X and Y; see the example in Fig. 3. This motivates the following variant of the previous problem:

Problem 2. Given G, α, X, Y, k as in Problem 1, decide whether the flip distance between X and Y is at most k, where we only allow flipping edges that are oriented differently in X and Y.

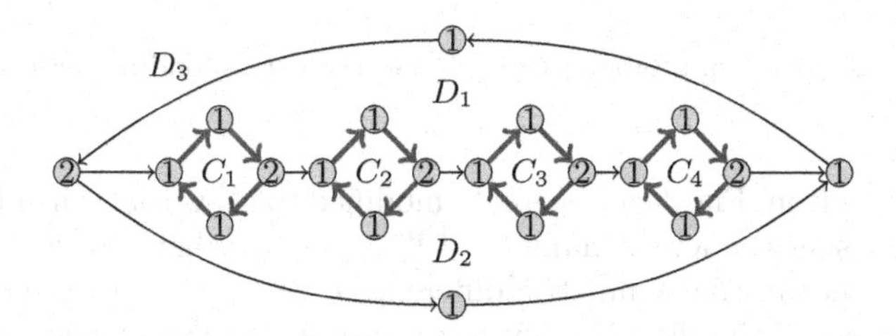

Fig. 3. An α-orientation X of a graph. The α-orientation Y obtained by flipping the four directed facial cycles $C_1, \ldots, C_4$ can be reached with fewer flips by flipping only the three directed facial cycles D_1, D_2, D_3 in this order.

From α-Orientations to Perfect Matchings. The flexibility in choosing a function α for a set of α-orientations on a graph allows us to capture numerous relevant combinatorial structures, including: domino and lozenge tilings of a plane region [40,45], planar spanning trees [22], (planar) bipartite perfect matchings [30], (planar) bipartite d-factors [14,38], Schnyder woods of a planar triangulation [9], Eulerian orientations of a (planar) graph [14], k-fractional orientations of a planar graph with specified outdegrees [5], and contact representations of planar graphs with homothetic triangles, rectangles, and k-gons [15,19,20,23].

In the following, we focus on perfect matchings of bipartite graphs. Consider any bipartite graph G with bipartition (V_1, V_2) equipped with

$$\alpha : V(G) \to \mathbb{N}_0, \quad \alpha(x) := \begin{cases} 1 & \text{if } x \in V_1, \\ d_G(x) - 1 & \text{if } x \in V_2. \end{cases}$$

With this definition, in each α-orientation of G, the edges directed from V_1 to V_2 form a perfect matching. This is illustrated in Fig. 4. Conversely, given a perfect matching M of G, orienting all edges of M from V_1 to V_2 and all the other edges from V_2 to V_1 yields an α-orientation of the above type. Furthermore, the directed cycles in any α-orientation of G correspond to the alternating cycles in the associated perfect matching. Flipping an alternating cycle in a perfect matching corresponds to exchanging matching and non-matching edges. An example of the flip graph of perfect matchings of a graph is given in Fig. 5. In this special case, Problem 1 boils down to:

Problem 3. Given a bipartite graph G, a pair X, Y of perfect matchings in G and an integer $k \geq 0$, decide whether the flip distance between X and Y is at most k.

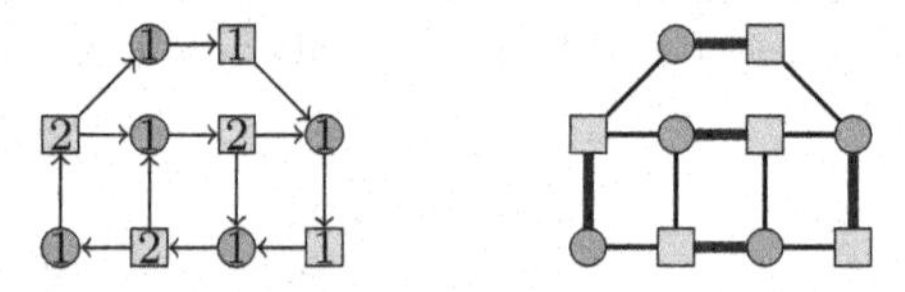

Fig. 4. An α-orientation of a bipartite graph and the corresponding perfect matching.

The example from Fig. 3 can be easily modified to show that when transforming X into Y using the fewest number of flips, we may have to flip alternating cycles that are not in the symmetric difference of X and Y; see the example in Fig. 6. If we restrict the flips to only use cycles in the symmetric difference of X and Y, then the problem of finding the flip distance becomes trivial, as the symmetric difference is a collection of disjoint cycles, and each of them has to be flipped, so Problem 2 is trivial for perfect matchings.

Flip Graphs and Matroid Intersection Polytopes. We give a geometric interpretation of the flip distance between α-orientations as the distance in the skeleton of a 0/1-polytope.

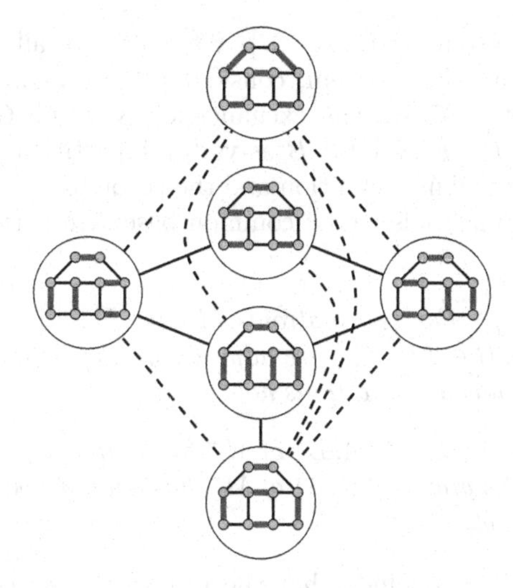

Fig. 5. The flip graph of perfect matchings of a graph. The solid edges indicate flips along facial cycles, and the dashed edges indicate flips along non-facial cycles.

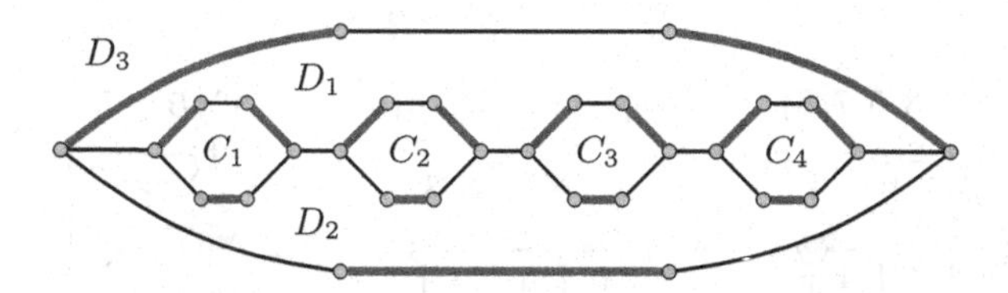

Fig. 6. A perfect matching X in a graph. The perfect matching Y obtained by flipping the four alternating facial cycles $C_1, \ldots, C_4$ can be reached with fewer flips by flipping only the three alternating facial cycles D_1, D_2, D_3 in this order.

Recall that a matroid is an abstract simplicial complex $(E, \mathcal{I})$, where $\mathcal{I} \subseteq 2^E$ satisfies the independent set augmentation property. The elements of $\mathcal{I}$ are called independent sets. A emphbase of the matroid is an inclusionwise maximal independent set.

It is well-known that perfect matchings in a bipartite graph $G = (V_1 \cup V_2, E)$ are common bases of two partition matroids $(E, \mathcal{I}_1)$ and $(E, \mathcal{I}_2)$, in which a set of edges is independent if no two share an endpoint in V_1, or, respectively, in V_2.

Similarly, α-orientations can be defined as common bases of two partition matroids. In this case, every edge of the graph G is replaced by a pair of parallel arcs, one for each possible orientation of the edge. One matroid encodes the constraint that in a base, for every edge exactly one orientation is chosen. The second matroid encodes the constraint that in a base, each vertex v has exactly $\alpha(v)$ outgoing arcs.

The common base polytope of two matroids is a 0/1-polytope obtained as the convex hull of the characteristic vectors of the common bases. Adjacency of two vertices of this polytope has been characterized by Frank and Tardos [21].

A shorter proof was given by Iwata [27]. We briefly recall their result in the next theorem. To state the theorem, consider a matroid $M = (E, \mathcal{I})$, a base $B \in \mathcal{I}$, and a subset $F \subseteq E$. The exchangeability graph $G(B, F)$ of M is a bipartite graph with $B \setminus F$ and $F \setminus B$ as vertex bipartition, and edge set $\{ij \mid B \setminus \{i\} \cup \{j\}$ is a basis$\}$. This definition and the theorem are illustrated in Fig. 7 for the two partition matroids whose common bases are perfect matchings of a graph.

Theorem 1 ([21,27]). *For two matroids $M^+ = (E, \mathcal{I}^+)$ and $M^- = (E, \mathcal{I}^-)$, two common bases $A, B \in \mathcal{I}^+ \cap \mathcal{I}^-$ are adjacent on the common base polytope if and only if all the following conditions hold:*

(i) the exchangeability graph $G(A, B)$ of M^+ has a unique perfect matching P^+,
(ii) the exchangeability graph $G(B, A)$ of M^- has a unique perfect matching P^-,
(iii) $P^+ \cup P^-$ is a single cycle.

From Theorem 1 we conclude that the flip graphs we consider on perfect matchings and α-orientations are precisely the skeletons of the corresponding polytopes of common bases.

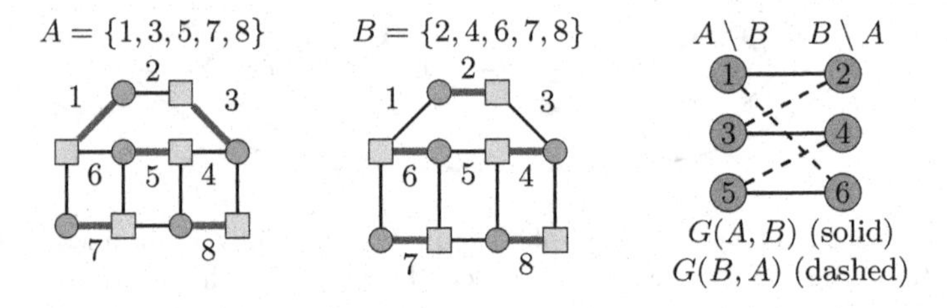

Fig. 7. Two common bases A and B (left and middle) of the matroids M^+ and M^-, where M^+ and M^- have as independent sets all subsets of edges of the graph where no two share an endpoint in the set of circled vertices, or the set of squared vertices, respectively. The right hand side shows the exchangeability graphs $G(A, B)$ of M^+ (solid edges) and $G(B, A)$ of M^- (dashed edges). As the conditions of Theorem 1 are met, the two bases are adjacent in the common base polytope, and adjacent in the flip graph shown in Fig. 5.

It is interesting to compare Problems 1 and 3 with the analogous problems for other families of matroid polytopes. For instance, it is known that for two bases A, B of a matroid, the exchangeability graph $G(A, B)$ has a perfect matching [10]. Hence A can be transformed into B by performing $|A \Delta B|/2$ exchanges of elements (where $A \Delta B$ is the symmetric difference of A and B), which is also the distance in the skeleton of the base polytope of the matroid. On the other hand, the problem of computing the flip distance between two triangulations of a convex polygon amounts to computing distances in skeletons of associahedra, which are known to be polymatroids (see [1] and references therein). This problem is neither known to be in P nor known to be NP-hard.

Also note that for other families of combinatorial polytopes, testing adjacency is already intractable. This is the case for instance for the polytope of the Traveling Salesman Problem (TSP) [35], whose skeleton is known to have diameter at most 4 [42]. On the other hand, the corresponding polytope is known to be the common base polytope of three matroids.

Hardness of Flip Distance Between Perfect Matchings and α-Orientations. We prove that Problem 3 is NP-complete, even for 2-connected bipartite subcubic planar graphs and $k = 2$. This implies that Problem 1 is NP-complete as well.

Theorem 2. *Given a 2-connected bipartite subcubic planar graph G and a pair X, Y of perfect matchings in G, deciding whether the flip distance between X and Y is at most two is* NP*-complete.*

We prove Theorem 2 by reduction from deciding directed Hamiltonicity of orientations of cubic planar graphs without sinks and sources. As direct consequences of this proof we get:

Corollary 1. *Unless* P = NP*, deciding whether the flip distance between two perfect matchings is at most k is not fixed-parameter tractable with respect to parameter k.*

Corollary 2. *Unless* P = NP*, the flip distance between two perfect matchings is not approximable within a multiplicative factor $3/2 - \epsilon$ in polynomial time, for any $\epsilon > 0$.*

We also prove that Problem 2 is NP-complete, even for 4-regular graphs and $k = 2$, by reduction from the following problem: Given a digraph D where each vertex has indegree and outdegree equal to 2, is $E(D)$ the union of two directed Hamiltonian cycles?

Theorem 3. *Given a 4-regular graph G and a pair X, Y of α-orientations of G, deciding whether the flip distance between X and Y is at most two is* NP*-complete. Moreover, the problem remains* NP*-complete if we only allow flipping edges that are oriented differently in X and Y.*

From α-Orientations in Planar Graphs to c-Orientations. In what follows, we generalize the problem, via planar duality, to flip distances in so-called *c*-orientations.

Consider an arbitrary 2-connected plane graph G and its planar dual G^*. Then for any orientation D of the edges of G, the directed dual D^* of D is obtained by orienting any dual edge forward if it crosses a left-to-right arc in D in a simultaneous plane embedding of G and G^*, and backward otherwise; see Fig. 8. Edge sets of directed cycles in D correspond to edge sets of minimal directed cuts in D^* and vice-versa. Hence D is acyclic (respectively, strongly connected) if and only if D^* is strongly connected (respectively, acyclic). A directed vertex cut is a cut consisting of all edges incident to a sink or a source vertex. Directed facial cycles in D are in bijection with the directed vertex cuts in D^*, and vice versa. The unbounded face in the plane embedding of D can be chosen such that it corresponds to a fixed vertex $\top$ in D^*.

Let D be an α-orientation of G. Given a minimal cut in D separating $U \subseteq V(D)$ from $\overline{U} := V(D) \setminus U$, we denote by $\delta^+(U)$ the edges pointing from U to $\overline{U}$ in D. We also let $d_D^+(v)$ denote the outdegree of vertex v in D. We have

$$|\delta^+(U)| = \sum_{v \in U} d_D^+(v) - |E(G[U])| = \sum_{v \in U} \alpha(v) - |E(G[U])|,$$

which only depends on α and G. Consequently, the set of orientations of G^* which are directed duals of α-orientations of G can be characterized by the property that for every cycle C in G^*, the number of edges in clockwise direction is fixed by a certain value $c(C)$ independent of the orientation. The flip operation between α-orientations of D consists of the reversal of a directed cycle. In the corresponding set of dual orientations of D^*, this translates to the reversal of the orientations of the edges in a minimal directed cut, as shown on Fig. 8.

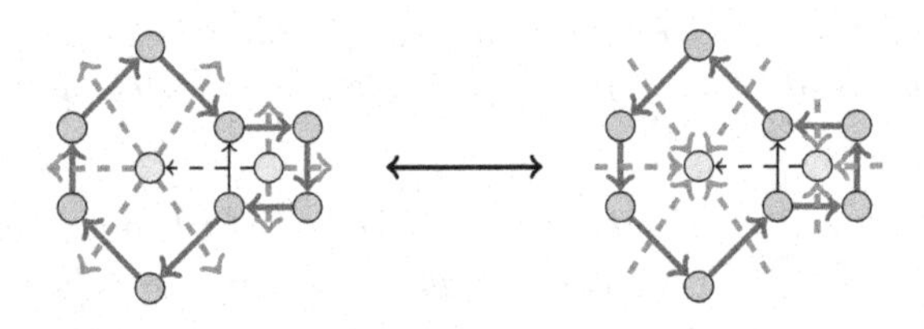

Fig. 8. Duality between flips in α-orientations (solid edges) and in c-orientations (dashed edges).

The same notion has been investigated more generally without planarity conditions under the name of c-orientations by Propp [38] and Knauer [29]. Given a graph G, we can fix an arbitrary direction of traversal for each cycle C. Given a graph and an assignment $c(C) \in \mathbb{N}_0$ to each cycle in G, one may define a c-orientation of G to be an orientation having exactly $c(C)$ edges in forward direction for every cycle C in G. Note that it is sufficient to define the function c on a cycle basis of G, which consists of linearly many cycles. The flip operation on the set $\mathcal{R}_c$ of such c-orientations of a graph is defined as the reversal of all edges in a minimal directed cut. It is not difficult to see that these flips make the set of c-orientations of a graph connected.

From the duality between planar α-orientations and planar c-orientations, determining flip distances between α-orientations of 2-connected planar graphs reduces to determining flip distances between the dual c-orientations. Note that planar duals of bipartite graphs are exactly the Eulerian planar graphs. Theorem 2 therefore directly yields:

Corollary 3. *Given an Eulerian planar graph G and c-orientations X, Y of G, deciding whether the flip distance between X and Y is at most two is* NP*-complete.*

c-Orientations and Distributive Lattices. A more local operation consists of flipping only directed vertex cuts, induced by sources and sinks, excluding a fixed vertex $\top$. We will refer to this special case as a vertex flip. Specifically, given a pair of c-orientations X, Y of a graph G with a fixed vertex $\top$, we aim to transform X into Y using only vertex flips at vertices distinct from $\top$.

A c-orientation X of G might contain a cycle C in G which is directed in X. According to the definition of a c-orientation, this means that C keeps the same orientation in every c-orientation of G. Consequently, any (minimal) directed cut in a c-orientation of G is disjoint from $E(C)$. Contracting the cycle C in G, we end up with a smaller graph G' containing the same (minimal) directed cuts, such that the c-orientations of G are determined by their corresponding orientations on G'. We can therefore safely assume that the c-orientations that we consider are all acyclic. Similarly, G will be assumed to be connected.

Problem 4. Given a connected graph G with a fixed vertex $\top$ and a pair X, Y of acyclic c-orientations, what is the length of a shortest vertex flip sequence transforming X into Y?

We now reason that every pair of c-orientations is reachable from each other by vertex flips. This property is provided in a much stronger way by a distributive lattice structure on the set $\mathcal{R}_c$; see Fig. 9. The next theorem is a special case of Theorem 1 in Propp [38] where the c-orientations are acyclic.

Theorem 4 ([29,38]). *Let G be a graph with fixed vertex $\top$ and $\mathcal{R}_c$ a set of acyclic c-orientations of G. Then the partial order $\leq_c$ on $\mathcal{R}_c$ in which Y covers X if and only if Y can be obtained from X by flipping a source defines a distributive lattice on $\mathcal{R}_c$.*

Hence Problem 4 consists of finding shortest paths in the cover graph of a distributive lattice, where the size of the lattice can be exponential in the size of the input G.

Facial Flips in Planar Graphs. When we consider Problem 4 on planar graphs, restricting to vertex flips and considering the dual plane graph amounts to considering only flips of directed facial cycles, excluding the outer face whose dual vertex is $\top$. We refer to these as facial flips. Felsner [14] considered distributive lattices induced by facial flips. The following computational problem is a special case of Problem 4.

Problem 5. Given a 2-connected plane graph G and a pair X, Y of strongly connected α-orientations, what is the length of a shortest facial flip sequence transforming X into Y?

Zhang, Qian, and Zhang [47] recently provided a closed formula for this flip distance, which can be turned into a polynomial-time algorithm. We prove the analogous stronger statement for Problem 4.

Theorem 5. *There is an algorithm that, given a graph G with a fixed vertex $\top$ and a pair X, Y of c-orientations of G, outputs a shortest vertex flip sequence between X and Y, and runs in time $\mathcal{O}(m^3)$ where m is the number of edges.*

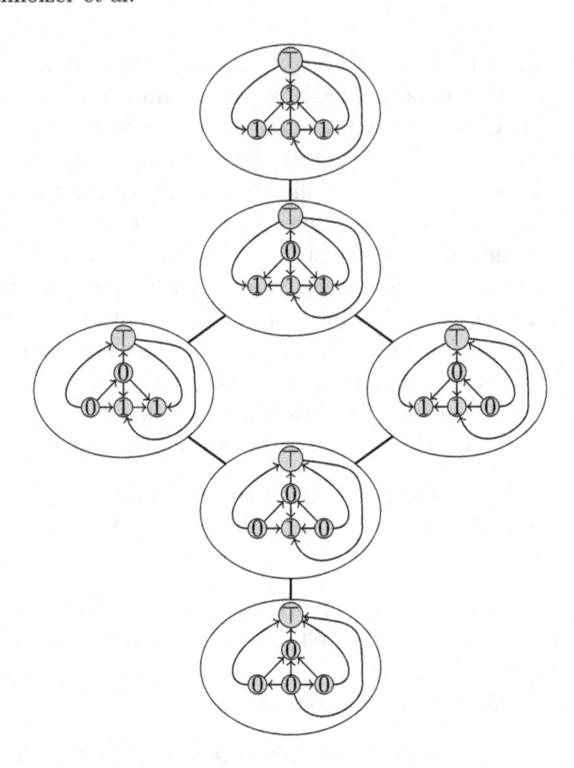

Fig. 9. The distributive lattice induced by vertex flips in c-orientations. The reference orientation at the bottom is the directed dual D^* of the orientation D of the graph G used in Figs. 4 and 5, where some parallel arcs incident with $\top$ are grouped together for simplicity. The numbers depicted at the vertices indicate the number of times that each vertex is flipped with respect to the reference orientation.

In the planar case, this directly translates to a polynomial-time algorithm for Problem 5. In [17], the distributive lattice structure on c-orientations is generalized to so-called Δ-bonds, also known as tensions. We believe that our proof of Theorem 5 can be generalized to these objects.

Flip Distance with Larger Cut Sets. While computing the cut flip distance between c-orientations is an NP-hard problem in general (Theorem 2), there is a polynomial-time-algorithm for computing the distance when only using vertex flips (Theorem 5). It is natural to ask for a threshold between the hard and easy cases of flip distance problems. Our proof for Theorem 2 involves very long directed cycles, which correspond to flips of directed cuts in the dual c-orientations with cut sets of large size. Consequently, one may hope that the problem gets easier when restricting the sizes of the cut sets involved in a flip sequence. Our last result destroys this hope:

Theorem 6. *Let X, Y be c-orientations of a connected graph G with fixed vertex $\top$. It is* NP*-hard to determine the length of a shortest cut flip sequence* *transforming X into Y, which consists only of minimal directed cuts with interiors of size at most two.*

The proof of Theorem 6 is by reduction from the problem of determining the jump number of a poset of height two, which is known to be NP-hard.

3 Open Problems

Problem 2 asks for a shortest flip sequence of directed cycles transforming one α-orientation X into another one Y, where we only allow flipping edges that are oriented differently in X and Y. Since these edges that are oriented differently in X and Y form an Eulerian subdigraph D of both X and Y, we get:

Question 1. What is the smallest number of directed cycles into which an Eulerian digraph can be decomposed?

We have seen in Theorem 3 that from a computational point of view, this problem is hard for general digraphs, but we wonder what happens when adding planarity constraints. Another interesting undirected variant of Question 1 is:

Question 2. Given a graph G with an Eulerian subgraph H, what is the smallest number of cycles of G such that their symmetric difference is H?

Concerning Theorem 6, we believe that for any bound on the size of the cuts, the corresponding flip distance will be NP-hard to compute. On the other hand, we use very particular graphs as gadgets, and we do not know the complexity of the corresponding problem for planar α-orientations. We think the following is an interesting special case which actually might be tractable:

Question 3. Let X, Y be perfect matchings of a planar bipartite 3-connected graph G. What is the complexity of determining the distance of X and Y with respect to alternating cycles that are either a face or the symmetric difference of two adjacent faces?

References

1. Aguiar, M., Ardila, F.: Hopf monoids and generalized permutahedra, September 2017. arXiv: 1709.07504
2. Aichholzer, O., Aurenhammer, F., Huemer, C., Vogtenhuber, B.: Gray code enumeration of plane straight-line graphs. Graphs Comb. **23**(5), 467–479 (2007). https://doi.org/10.1007/s00373-007-0750-z
3. Aichholzer, O., Mulzer, W., Pilz, A.: Flip distance between triangulations of asimple polygon is NP-complete. Discrete Comput. Geom. **54**(2), 368–389 (2015). https://doi.org/10.1007/s00454-015-9709-7
4. Avis, D., Fukuda, K.: Reverse search for enumeration. Discrete Appl. Math. **65**(1–3), 21–46 (1996). https://doi.org/10.1016/0166-218X(95)00026-N. First International Colloquium on Graphs and Optimization (GOI), Grimentz (1992)

5. Bernardi, O., Fusy, E.: A bijection for triangulations, quadrangulations, pentagulations, etc. J. Comb. Theory Ser. A **119**(1), 218–244 (2012). https://doi.org/10.1016/j.jcta.2011.08.006
6. Bonamy, M., et al.: The Perfect Matching Reconfiguration Problem, April 2019. arXiv:1904.06184
7. Bose, P., Hurtado, F.: Flips in planar graphs. Comput. Geom. **42**(1), 60–80 (2009). https://doi.org/10.1016/j.comgeo.2008.04.001
8. Bose, P., Verdonschot, S.: A history of flips in combinatorial triangulations. In: Márquez, A., Ramos, P., Urrutia, J. (eds.) EGC 2011. LNCS, vol. 7579, pp. 29–44. Springer, Heidelberg (2012). https://doi.org/10.1007/978-3-642-34191-5_3
9. Brehm, E.: 3-orientations and Schnyder-3-tree-decompositions, Diploma Thesis, Freie Universität Berlin (2000)
10. Brualdi, R.A.: Comments on bases in dependence structures. Bull. Australas. Math. Soc. **1**, 161–167 (1969). https://doi.org/10.1017/S000497270004140X
11. Cardinal, J., Sacristán, V., Silveira, R.I.: A note on flips in diagonal rectangulations. Discrete Math. Theor. Comput. Sci. **20**(2), 1–22 (2018)
12. Ceballos, C., Santos, F., Ziegler, G.M.: Many non-equivalent realizations of the associahedron. Combinatorica **35**(5), 513–551 (2015). https://doi.org/10.1007/s00493-014-2959-9
13. Cleary, S., St. John, K.: Rotation distance is fixed-parameter tractable. Inform. Process. Lett. **109**(16), 918–922 (2009). https://doi.org/10.1016/j.ipl.2009.04.023
14. Felsner, S.: Lattice structures from planar graphs. Electron. J. Comb. **11**(1), 24 (2004). Research Paper 15
15. Felsner, S.: Rectangle and square representations of planar graphs. In: Pach, J. (ed.) Thirty Essays on Geometric Graph Theory, pp. 213–248. Springer, New York (2013). https://doi.org/10.1007/978-1-4614-0110-0_12
16. Felsner, S., Kleist, L., Mütze, T., Sering, L.: Rainbow cycles in flip graphs. In: Symposium on Computational Geometry 2018, pp. 38:1–38:14 (2018). https://doi.org/10.4230/LIPIcs.SoCG.2018.38
17. Felsner, S., Knauer, K.: ULD-lattices and Δ-bonds. Comb. Probab. Comput. **18**(5), 707–724 (2009). https://doi.org/10.1017/S0963548309010001
18. Felsner, S., Knauer, K.: Distributive lattices, polyhedra, and generalized flows. Eur. J. Comb. **32**(1), 45–59 (2011). https://doi.org/10.1016/j.ejc.2010.07.011
19. Felsner, S., Schrezenmaier, H., Steiner, R.: Equiangular polygon contact representations. In: Brändstadt, A., Köhler, E., Meer, K. (eds.) WG 2018. LNCS, vol. 11159, pp. 203–215. Springer, Cham (2018). https://doi.org/10.1007/978-3-030-00256-5_17
20. Felsner, S., Schrezenmaier, H., Steiner, R.: Pentagon contact representations. Electron. J. Comb. **25**(3), 38 (2018). Paper 3.39
21. Frank, A., Tardos, E.: Generalized polymatroids and submodular flows. Math. Program. **42**(3, (Ser. B)), 489–563 (1988). https://doi.org/10.1007/BF01589418. Submodular optimization
22. Gilmer, P.M., Litherland, R.A.: The duality conjecture in formal knot theory. Osaka J. Math. **23**(1), 229–247 (1986)
23. Gonçalves, D., Lévêque, B., Pinlou, A.: Triangle contact representations and duality. Discrete Comput. Geom. **48**(1), 239–254 (2012). https://doi.org/10.1007/s00454-012-9400-1
24. Hernando, M.C., Hurtado, F., Noy, M.: Graphs of non-crossing perfect matchings. Graphs Comb. **18**(3), 517–532 (2002). https://doi.org/10.1007/s003730200038

25. Houle, M.E., Hurtado, F., Noy, M., Rivera-Campo, E.: Graphs of triangulations and perfect matchings. Graphs Comb. **21**(3), 325–331 (2005). https://doi.org/10.1007/s00373-005-0615-2
26. Huemer, C., Hurtado, F., Noy, M., Omaña-Pulido, E.: Gray codes for non-crossing partitions and dissections of a convex polygon. Discrete Appl. Math. **157**(7), 1509–1520 (2009). https://doi.org/10.1016/j.dam.2008.06.018
27. Iwata, S.: On matroid intersection adjacency. Discrete Math. **242**(1–3), 277–281 (2002). https://doi.org/10.1016/S0012-365X(01)00167-4
28. Kanj, I., Sedgwick, E., Xia, G.: Computing the flip distance between triangulations. Discrete Comput. Geom. **58**(2), 313–344 (2017). https://doi.org/10.1007/s00454-017-9867-x
29. Knauer, K.: Partial orders on orientations via cycle flips. Master's thesis, Faculty of Mathematics, TU Berlin (2007)
30. Lam, P.C.B., Zhang, H.: A distributive lattice on the set of perfect matchings of a plane bipartite graph. Order **20**(1), 13–29 (2003). https://doi.org/10.1023/A:1024483217354
31. Li, M., Zhang, L.: Better approximation of diagonal-flip transformation and rotation transformation. In: Hsu, W.-L., Kao, M.-Y. (eds.) COCOON 1998. LNCS, vol. 1449, pp. 85–94. Springer, Heidelberg (1998). https://doi.org/10.1007/3-540-68535-9_12
32. Loera, J.A.D., Rambau, J., Santos, F.: Triangulations: Structures for Algorithms and Applications. No. 25 in Algorithms and Computation in Mathematics. Springer, Heidelberg (2010). https://doi.org/10.1007/978-3-642-12971-1
33. Lubiw, A., Pathak, V.: Flip distance between two triangulations of a point set is NP-complete. Comput. Geom. **49**, 17–23 (2015). https://doi.org/10.1016/j.comgeo.2014.11.001
34. Negami, S.: Diagonal flips in triangulations of surfaces. Discrete Math. **135**(1–3), 225–232 (1994). https://doi.org/10.1016/0012-365X(93)E0101-9
35. Papadimitriou, C.H.: The adjacency relation on the traveling salesman polytope is NP-complete. Math. Program. **14**(3), 312–324 (1978). https://doi.org/10.1007/BF01588973
36. Pilaud, V., Santos, F.: Quotientopes, August 2018. arXiv: 1711.05353
37. Postnikov, A.: Permutohedra, associahedra, and beyond. Int. Math. Res. Not. IMRN **6**, 1026–1106 (2009). https://doi.org/10.1093/imrn/rnn153
38. Propp, J.: Lattice structure for orientations of graphs, September 2002. arXiv: math/0209005
39. Propp, J., Wilson, D.: Coupling from the past: a user's guide. In: Microsurveys in discrete probability, Princeton, NJ (1997). DIMACS Series Discrete Mathematics Theoretical Computer Science, vol. 41, pp. 181–192. American Mathematical Society, Providence, RI (1998)
40. Rémila, E.: The lattice structure of the set of domino tilings of a polygon. Theor. Comput. Sci. **322**(2), 409–422 (2004). https://doi.org/10.1016/j.tcs.2004.03.020
41. Ringel, G.: Über Geraden in allgemeiner lage. Elem. Math. **12**, 75–82 (1957)
42. Rispoli, F.J., Cosares, S.: A bound of 4 for the diameter of the symmetric traveling salesman polytope. SIAM J. Discrete Math. **11**(3), 373–380 (1998). https://doi.org/10.1137/S0895480196312462
43. Rogers, R.O.: On finding shortest paths in the rotation graph of binary trees. In: Proceedings of the Thirtieth Southeastern International Conference on Combinatorics, Graph Theory, and Computing, Boca Raton, FL, vol. 137, pp. 77–95 (1999)

44. Sleator, D.D., Tarjan, R.E., Thurston, W.P.: Rotation distance, triangulations, and hyperbolic geometry. J. Am. Math. Soc. **1**(3), 647–681 (1988). https://doi.org/10.2307/1990951
45. Thurston, W.P.: Conway's tiling groups. Am. Math. Mon. **97**(8), 757–773 (1990). https://doi.org/10.2307/2324578
46. Zhang, F.J., Guo, X.F.: Hamilton cycles in directed Euler tour graphs. Discrete Math. **64**(2–3), 289–298 (1987). https://doi.org/10.1016/0012-365X(87)90198-1
47. Zhang, W.J., Qian, J.G., Zhang, F.J.: Distance between α-orientations of plane graphs by facial cycle reversals. Acta Mathematica Sinica, English Series (2019). https://doi.org/10.1007/s10114-018-7403-4

Graph Functionality

Bogdan Alecu[1], Aistis Atminas[2], and Vadim Lozin[1](✉)

[1] Mathematics Institute, University of Warwick, Coventry, UK
{B.Alecu,V.Lozin}@warwick.ac.uk
[2] Department of Mathematics, London School of Economics, London, UK
A.Atminas@lse.ac.uk

Abstract. In the present paper, we introduce the notion of graph functionality, which generalizes simultaneously several other graph parameters, such as degeneracy or clique-width, in the sense that bounded degeneracy or bounded clique-width imply bounded functionality. Moreover, we show that this generalization is proper by revealing classes of graphs of unbounded degeneracy and clique-width, where functionality is bounded by a constant. We also prove that bounded functionality implies bounded VC-dimension, i.e. graphs of bounded VC-dimension extend graphs of bounded functionality, and this extension also is proper.

Keywords: Clique-width · Graph degeneracy · VC-dimension · Permutation graph · Graph representation

1 Introduction

Let $G = (V, E)$ be a simple graph, i.e. an undirected graph without loops and multiple edges. We denote by $A = A_G$ the adjacency matrix of G and by $A(x, y)$ the element of this matrix corresponding to vertices $x, y \in V$, i.e. $A(x, y) = 1$ if x and y are adjacent, and $A(x, y) = 0$ otherwise.

We say that a vertex $y \in V$ is a function of vertices $x_1, \ldots, x_k \in V$ if there exists a Boolean function f of k variables such that for any vertex $z \in V - \{y, x_1, \ldots, x_k\}$, we have $A(y, z) = f(A(x_1, z), \ldots, A(x_k, z))$.

The functionality $\text{fun}(y)$ of vertex y is the minimum k such that y is a function of k vertices. In particular, the functionality of an isolated vertex is 0, and the same is true for a dominating vertex, i.e. a vertex adjacent to all the other vertices in the graph. More generally, the functionality of a vertex y does not exceed the number of its neighbours (the degree of y) and the number of its non-neighbours. One more simple example of functional vertices is given by twins, i.e. vertices x and y that have the same set of neighbours different from x and y. Twins are functions of each other and their functionality is (at most) 1. The same is true for anti-twins, i.e. vertices whose neighbourhoods complement each other.

From a practical point of view, the notion of functional vertices is of interest in the area of graph learning and graph mining, since it makes graphs amenable

I. Sau and D. M. Thilikos (Eds.): WG 2019, LNCS 11789, pp. 135–147, 2019.
https://doi.org/10.1007/978-3-030-30786-8_11

to the techniques of Logical Analysis of Data [11], which is based on Boolean methods for pattern detection. This notion provides a tool for revealing dependencies that are hidden in the structure of the graph and for identifying alliances that are more complex than "friends" or "enemies".

From a theoretical point of view, the importance of this notion is due to the fact that it defines a new complexity measure, which we call *graph functionality.* The functionality $\text{fun}(G)$ of G is

$$\max_{H} \min_{y \in V(H)} \text{fun}(y),$$

where the maximum is taken over all induced subgraphs H of G. Similarly to many other graph parameters, this notion becomes valuable when its value is small, i.e. is bounded by a constant independent of the size of the graph. The purpose of this paper is to show that graphs of bounded functionality generalize simultaneously several other important graph properties, such as graphs of bounded vertex degree, degeneracy, arboricity, tree-width and clique-width. We prove this in Sect. 2. Moreover, in the same section we show that this generalization is proper by revealing classes of graphs where functionality is bounded but the other parameters are not. This includes permutation graphs, line graphs and, more generally, the intersection graphs of 3-uniform hypergraphs. On the other hand, in Sect. 3 we show that bounded functionality implies bounded VC-dimension, i.e. graphs of bounded VC-dimension extend graphs of bounded functionality, and this extension also is proper.

Throughout the paper, we consider only simple graphs and use standard terminology and notation. In particular, for a graph G, we denote by $V(G)$ and $E(G)$ the vertex set and the edge set of G, respectively. The neighbourhood $N(v)$ of a vertex $v \in V(G)$ is the set of vertices of G adjacent to v and the degree of v is $|N(v)|$. A vertex of degree 0 is called *isolated.* The closed neighbourhood of v is $N[v] = \{v\} \cup N(v)$. A chordless cycle of length n is denoted C_n. A graph H is an *induced subgraph* of a graph G if H can be obtained from G by vertex deletions. A class X of graphs is *hereditary* if it is closed under taking induced subgraphs.

2 Graphs of Small Functionality

From the discussion in the introduction, it follows that graphs of bounded functionality extend graphs of bounded vertex degree. More generally, they extend graphs of bounded degeneracy, where the *degeneracy* of G is the minimum k such that every induced subgraph of G has a vertex of degree at most k. A notion related to degeneracy is that of *arboricity,* which is the minimum number of forests into which the edges of G can be partitioned. The degeneracy of G is always between the arboricity and twice the arboricity of G and hence graphs of bounded functionality extend graphs of bounded arboricity too.

One more important graph parameter is *clique-width.* Many algorithmic problems that are generally NP-hard become polynomial-time solvable when restricted to graphs of bounded clique-width [6]. Clique-width is a relatively new notion and it generalizes another important graph parameter, tree-width, studied in the literature for decades. Clique-width is stronger than tree-width in the sense that graphs of bounded tree-width have bounded clique-width. In Sect. 2.1, we show that functionality is stronger than clique-width by proving that graphs of bounded clique-width have bounded functionality. Then in Sects. 2.2 and 2.3 we identify classes of graphs where functionality is bounded, but degeneracy and clique-width are not.

2.1 Graphs of Bounded Clique-Width

The notion of clique-width of a graph was introduced in [5]. The clique-width of a graph G is denoted $\mathrm{cwd}(G)$ and is defined as the minimum number of labels needed to construct G by means of the following four graph operations:

- creation of a new vertex v with label i (denoted $i(v)$),
- disjoint union of two labelled graphs G and H (denoted $G \oplus H$),
- connecting vertices with specified labels i and j (denoted $\eta_{i,j}$) and
- renaming label i to label j (denoted $\rho_{i\to j}$).

Every graph can be defined by an algebraic expression using the four operations above. This expression is called a k-expression if it uses k different labels. For instance, the cycle C_5 on vertices a, b, c, d, e (listed along the cycle) can be defined by the following 4-expression:

$$\eta_{4,1}(\eta_{4,3}(4(e) \oplus \rho_{4\to 3}(\rho_{3\to 2}(\eta_{4,3}(4(d) \oplus \eta_{3,2}(3(c) \oplus \eta_{2,1}(2(b) \oplus 1(a)))))))).$$

Alternatively, any algebraic expression defining G can be represented as a rooted binary tree, whose leaves correspond to the operations of vertex creation, the internal nodes correspond to the $\oplus$-operations, and the root is associated with G. The operations η and ρ are assigned to the respective edges of the tree. Figure 1 shows the tree representing the above expression defining a C_5.

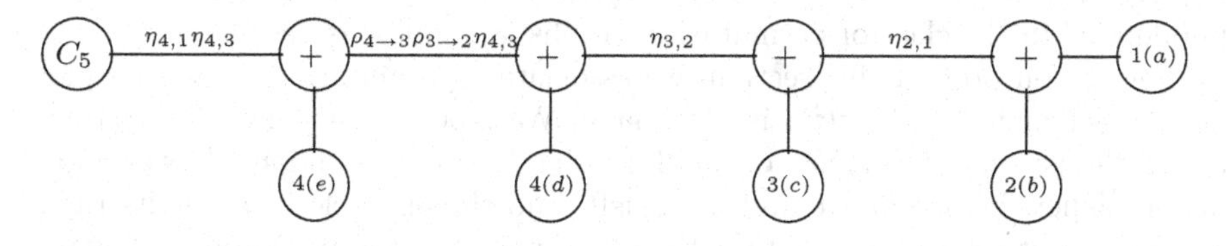

Fig. 1. The tree representing the expression defining a C_5

Among various examples of graphs of bounded clique-width we mention distance-hereditary graphs. These are graphs of clique-width at most 3 [8]. Every graph in this class can be constructed from a single vertex by successively adding either a pendant vertex or a twin (true or false) [4]. From this characterization we immediately conclude that the functionality of distance-hereditary graphs is at most one. More generally, in the next theorem we show that functionality is bounded for all classes of graphs of bounded clique-width.

Theorem 1. *For any graph G, $fun(G) \leq 2\text{cwd}(G) - 1$.*

Proof. Let G be a graph of clique-width k and let T be a tree corresponding to a k-expression that describes G. Consider a node v of the tree such that the tree T_v rooted at v has more than k leaves, and neither of the two children of v has this property (if no such v exists, we are done, since G has at most k vertices). Since T_v has more than k leaves, at least two of them, say x and y, have the same label at node v. On the other hand, T_v has at most $2k$ leaves by the choice of v. Therefore, G contains at most $2k - 2$ vertices that distinguish x and y, since x and y are not distinguished outside of T_v. As a result, the functionality of both x and y is at most $2k - 1$.

It is known (see e.g. [7]) that the clique-width of an induced subgraph of G cannot exceed the clique-width of G. Therefore, every induced subgraph of G has a vertex of functionality at most $2k - 1$. Thus, the functionality of G is at most $2k - 1$. □

In the proof of Theorem 1 the bound on functionality is achieved in a very specific way: in any graph of bounded clique width, there must exist two vertices whose neighbourhoods have small symmetric difference. The converse is generally not true. However, large symmetric difference for all pairs of vertices necessarily implies large clique-width, and this provides an alternative approach to identifying graphs of large clique-width. To illustrate this idea, in the next section we construct permutation graphs where the neighbourhoods of any pair of vertices have large symmetric difference. On the other hand, we show that the functionality of permutation graphs is bounded by a constant.

2.2 Permutation Graphs

Let π be a permutation of the elements in $\{1, 2, \ldots, n\}$. The permutation graph of π is a graph with vertex set $\{1, 2, \ldots, n\}$ in which two vertices i and j are adjacent if and only if $(i - j)(\pi(i) - \pi(j)) < 0$. Clique-width is known to be unbounded in the class of permutation graphs [8], and so is degeneracy.

For the purpose of this section, we associate a permutation π with its plot, i.e. the set of points $(i, \pi(i))$ in the plane. We label those points by $\pi(i)$ and define the *geometric neighbourhood* of a point k to be the union of two regions in the plane: the one above and to its left, and the one below and to its right. Then it is not difficult to see that the set of points of the permutation lying in the geometric neighbourhood of k is precisely the set of neighbours of vertex k in the permutation graph of π.

Theorem 2. *The functionality of permutation graphs is at most 8.*

Proof. Since the class of permutation graphs is hereditary, it suffices to show that every permutation graph contains a vertex of functionality at most 8. Let G be a permutation graph corresponding to a permutation π. The proof will be given in two steps: first, we show that if there is a vertex with a certain property in G (yet to be specified), then this vertex is a function of 4 other vertices.

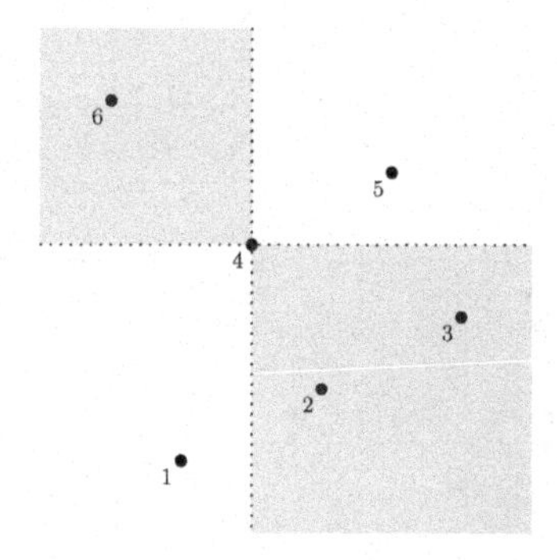

Fig. 2. Geometric representation of $\pi = 614253$, with the neighbourhood of 4 shaded

Second, we show how to find vertices that are "close enough" to having that property.

Step 1: Consider the plot of π. Among any 3 horizontally consecutive points, one is vertically between the two others. We call such a point *vertical middle* (in the permutation from Fig. 2, the vertical middle points are 4, 2 and 3). Similarly, among any 3 vertically consecutive points, one is horizontally between the two others, and we call this point *horizontal middle* (in Fig. 2, the horizontally middle points are 2, 5 and 4).

Now let us suppose that π has a point x that is simultaneously a horizontal and a vertical middle point. Then x is part of a triple x, b, t (not necessarily in that order) of horizontally consecutive points, where b is the bottom point (the lowest in the triple) and t is the top point (the highest in the triple). Also, x is part of a triple x, l, r (not necessarily in that order) of vertically consecutive points, where l is the leftmost and r is the rightmost point in the triple (see Fig. 3a for an illustration).

In general, x can be at any of the 9 intersection points of pairs of 3 consecutive vertical and horizontal lines, i.e. x is somewhere in X (see Fig. 3b). We also have $l \in L$, $r \in R$, $t \in T$ and $b \in B$ for the surrounding points (see Fig. 3b). The important thing to note is that, since the points are consecutive, those are the *only* points of the permutation lying in the shaded area $X \cup L \cup R \cup T \cup B$. Any point different from x, l, r, t, b lies in one of Q_1, Q_2, Q_3 or Q_4.

It is not difficult to see that the geometric neighbourhood corresponding to $(N(r) \cap N(b)) \cup (N(l) \cap N(t))$ (see Fig. 3a) will always contain Q_2 and Q_4, and will never intersect Q_1 or Q_3. Therefore, the function that describes how x depends on $\{l, r, t, b\}$ can be written as follows:

$$f(x_r, x_b, x_l, x_t) = x_r x_b \vee x_l x_t,$$

where x_r, x_b, x_l, x_t are Boolean variables corresponding to points r, b, l, t, respectively. In other words, a vertex $y \notin \{x, l, r, t, b\}$ is adjacent to x if and only if $f(A(y,r), A(y,b), A(y,l), A(y,t)) = 1$.

Step 2: Let us relax the simultaneous middle point condition to the following one: amongst every 5 vertically (respectively horizontally) consecutive points,

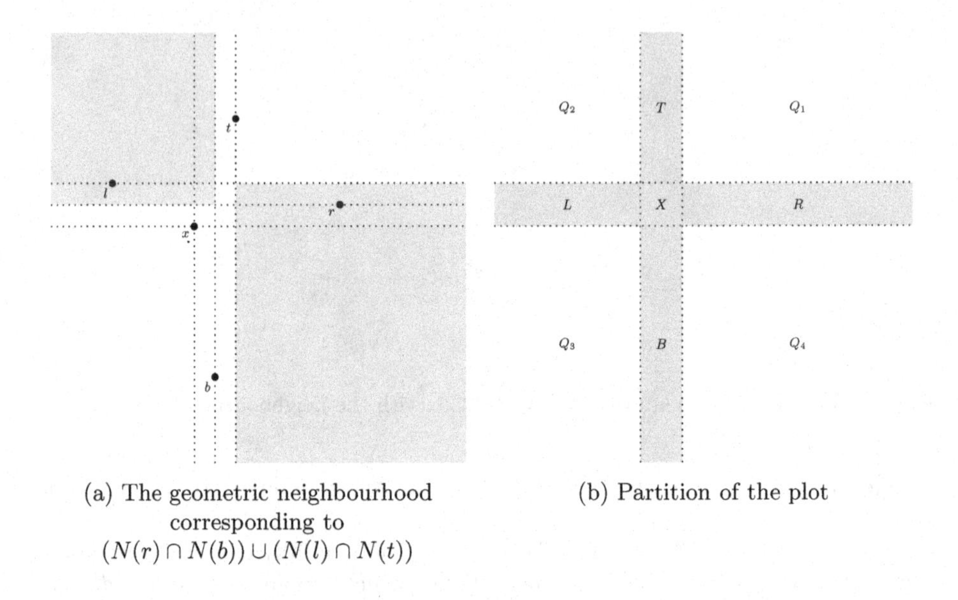

(a) The geometric neighbourhood corresponding to $(N(r) \cap N(b)) \cup (N(l) \cap N(t))$

(b) Partition of the plot

Fig. 3. A middle point x and its four surrounding points

call the middle three *weak horizontal* (respectively *vertical*) *middle points.* Note that if the number of points is divisible by 5, at least $\frac{3}{5}$ of them are weak vertical and at least $\frac{3}{5}$ of them are weak horizontal middle points. Using this observation it is not hard to deduce that if there are at least 13 points, then more than half of them are weak vertical and more than half of them are weak horizontal middle points. Therefore, there must exist a point x that is simultaneously both. We can deal with this case only, as the functionality of any graph on at most 12 vertices is at most 6, which is due to the fact that every vertex has at most 6 neighbours or non-neighbours. If x is simultaneously a weak vertical and weak horizontal middle point, then there must exist quintuples l, x, m_1, m_2, r and t, x, m_3, m_4, b (not necessarily in that order), where x is a simultaneous weak middle point in both directions, while m_1, m_2, m_3 and m_4 are the other weak middle points in their respective quintuples. By removing m_1, m_2, m_3 and m_4 from the graph, we find ourselves in the configuration of Step 1 and conclude that x is a function of $\{l, r, t, b\}$ in the reduced graph. Therefore, in the original graph x is a function of $\{l, r, t, b, m_1, m_2, m_3, m_4\}$, concluding the proof. □

In the rest of this section, we give a construction of permutation graphs where every pair of neighbourhoods has large symmetric difference. Together with Theorem 1, this construction gives an alternative proof of the known fact that permutation graphs have unbounded clique-width. For two distinct vertices x_1 and x_2, let $\mathrm{sd}(x_1, x_2)$ denote the number of vertices other than x_1 and x_2 that are adjacent to exactly one of them.

Theorem 3. *For any $t \in \mathbb{N}$, there is a permutation graph G such that for any distinct $x_1, x_2 \in V(G)$, $\mathrm{sd}(x_1, x_2) \geq t$.*

Proof. We will make use of the geometric representation of permutations discussed earlier. Given two vertices x_1 and x_2 of a permutation graph G, the symmetric difference of their neighbourhoods can be represented geometrically as an area in the plane (see Fig. 4). More precisely, a vertex different from x_1 and x_2 lies in the symmetric difference of their neighbourhoods if and only if the corresponding point of the permutation lies in the shaded area.

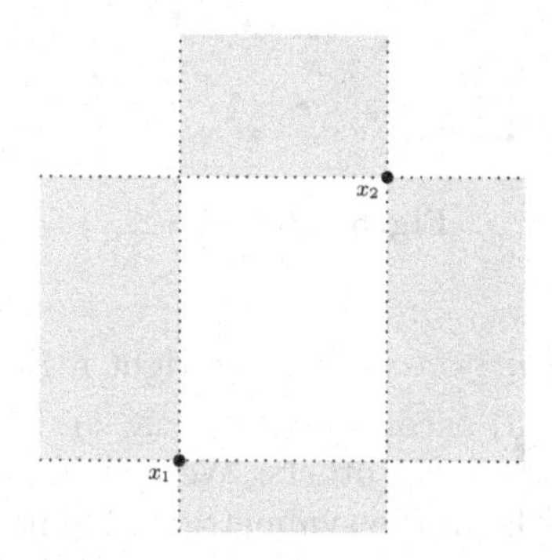

Fig. 4. Geometric symmetric difference of two points x_1 and x_2

In order to prove the theorem, it suffices, for each $t \in \mathbb{N}$, to exhibit a set S_t of points in the plane (with no two on the same vertical or horizontal line) such that for any pair $x_1, x_2 \in S_t$, there are at least t other points of S_t lying in the geometric symmetric difference of x_1 and x_2. Such a construction immediately gives rise to a permutation and thus a permutation graph where the symmetric difference of the neighbourhoods of any pair of vertices is at least t.

We construct sets S_t in the following way (see Fig. 5 for an example):

- start with all the points with integer coordinates between 0 and t inclusive;
- apply to the set the rotation sending $(1, 0)$ to $(1, \frac{1}{t+1})$ and $(0, 1)$ to $(-\frac{1}{t+1}, 1)$.

To see that these sets have indeed the desired property, let $x_1, x_2 \in S_t$. For simplicity, we will use the coordinates of the points before the rotation. Suppose $x_1 = (a_1, b_1)$ and $x_2 = (a_2, b_2)$. There are four possible cases (after switching x_1 and x_2 if necessary):

- If $a_1 = a_2$ and $b_1 < b_2$, then the t points (k, b_2), (l, b_1) with $k < a_1 < l$ are in the symmetric difference.
- Similarly, if $b_1 = b_2$ and $a_1 < a_2$, then the t points (a_1, k), (a_2, l) with $k < b_1 < l$ are in the symmetric difference.
- If $a_1 < a_2$ and $b_1 < b_2$, the following points all lie in the symmetric difference of x and y:
  (1) Points (a_1, k) with $k < b_1$ (in the bottom region).
  (2) Points (a_1, k) with $b_1 < k \leq b_2$ (in the left region).
  (3) Points (a_2, k) with $b_2 < k$ (in the top region).

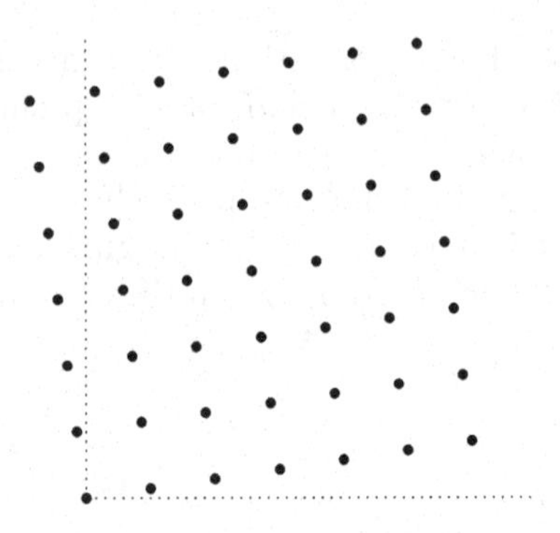

Fig. 5. The set S_6

(4) Points (a_2, k) with $b_1 \leq k < b_2$ (in the right region).

In particular, (1) and (3) account for at least $b_1 + t - b_2$ points, while (2) and (4) account for $2(b_2 - b_1)$ others. We conclude that in total, at least $t + (b_2 - b_1) > t$ points lie in the symmetric difference of x_1 and x_2.

- If $a_1 < a_2$ and $b_1 > b_2$, a similar index chasing argument exhibits at least t points in the symmetric difference of x_1 and x_2. □

2.3 Intersection Graphs

In this section, we show that unit interval graphs and line graphs have bounded functionality. It is known (see e.g. [9,12]) that clique-width is unbounded in both of those classes. The same is true for degeneracy, since line graphs and unit interval graphs contain arbitrarily large cliques.

Theorem 4. *The functionality of unit interval graphs is at most 2.*

Proof. Let G be a unit interval graph with n vertices and assume without loss of generality that G has no isolated vertices (by adding isolated vertices to a graph we increase neither its functionality not symmetric difference).

Take a unit interval representation for $G = (V, E)$ with the interval endpoints all distinct. We label the vertices $v_1, \ldots, v_n$ in the order in which they appear on the real line (from left to right), and denote the endpoints of interval I_i corresponding to vertex v_i by $a_i < b_i$.

For two distinct vertices v_i and v_j, write like before $\mathrm{sd}(v_i, v_j)$ to denote the number of vertices in $V \setminus \{v_i, v_j\}$ adjacent to exactly one of them. From the definition, it is immediate that $\mathrm{fun}(v_i) \leq \mathrm{sd}(v_i, v_j) + 1$.

We will bound

$$S = \sum_{i=1}^{n-1} \mathrm{sd}(v_i, v_{i+1}).$$

Note that any neighbour of v_i which is not a neighbour of v_{i+1} needs to have its right endpoint between a_i and a_{i+1}. Similarly, any neighbour of v_{i+1} but not of v_i needs to have its left endpoint between b_i and b_{i+1}. In other words, $\text{sd}(v_i, v_{i+1})$ is bounded above by the number of endpoints in $(a_i, a_{i+1}) \cup (b_i, b_{i+1})$ (we say bounded above and not equal, since it might happen that b_i lies between a_i and a_{i+1}, without contributing to the symmetric difference).

The key is now to note that any endpoint can be counted at most once in the whole sum S, since all (a_i, a_{i+1}) are disjoint (and the same applies to the (b_i, b_{i+1})), and the a's can only appear between b's (and vice-versa). In fact, a_1 and b_n are never counted in S, and if a_2 is between b_1 and b_2, then v_1 must be isolated, so a_2 is not counted either. The sum is thus at most $2n - 3$. Since it has $n-1$ terms, one of the terms, say $\text{sd}(v_t, v_{t+1})$, must be at most 1. Therefore, the functionality of both v_t and v_{t+1} is at most 2.

Since the class of unit interval graphs is hereditary, we conclude that the functionality of any unit interval graph is at most 2. □

Theorem 5. *The functionality of line graphs is at most 6.*

Proof. Let G be a graph and H be the line graph of G. Since the class of line graphs is hereditary, it suffices to prove that H has a vertex of functionality at most 6. We will prove a stronger result showing that *every* vertex of H has functionality at most 6.

Let x be a vertex in H, i.e. an edge in G. We denote the two endpoints of this edge in G by a and b. Assume first that both the degree of a and the degree of b are at least 4. Let $Y = \{y_1, y_2, y_3\}$ be a set of any three edges of G incident to a, and let $Z = \{z_1, z_2, z_3\}$ be a set of any three edges of G incident to b.

We claim that a vertex $v \notin \{x\} \cup Y \cup Z$ is adjacent to x in H if and only if it is adjacent to every vertex in Y or to every vertex in Z. Indeed, if v is adjacent to x in H, then the edge v intersects the edge x in G. If the intersection consists of a, then v is adjacent to every vertex in Y in the graph H, and if the intersection consists of b, then v is adjacent to every vertex in Z in the graph H. Conversely, let v be adjacent to every vertex in Y, then v must intersect the edges y_1, y_2, y_3 in G at vertex a, in which case v is adjacent to x in H. Similarly, if v is adjacent to every vertex in Z, then v intersects the edges z_1, z_2, z_3 in G at vertex b and hence v is adjacent to x in H.

Therefore, in the case when both a and b have degree at least 4 in G, the function that describes how x depends on $\{y_1, y_2, y_3, z_1, z_2, z_3\}$ in the graph H can be written as follows: $f(y_1, y_2, y_3, z_1, z_2, z_3) = y_1y_2y_3 \vee z_1z_2z_3$.

If the degree of a is less than 4, we include in Y all the edges of G distinct from x which are incident to a (if there are any) and remove the term $y_1y_2y_3$ from the function. Similarly, if the degree of b is less than 4, we include in Z all the edges of G distinct from x which are incident to b (if there are any) and remove the term $z_1z_2z_3$ from the function. If both terms have been removed, the function is defined to be identically 0, i.e. no vertices are adjacent to x in H, except for those in $Y \cup Z$. □

Having proved that the intersection graph of edges, i.e. the intersection graph of a family of 2-subsets, has bounded functionality, it is natural to ask whether the intersection graph of a family of k-subsets has bounded functionality for $k > 2$. This question is substantially harder and we answer it only for $k = 3$ (we omit the proof here).

Theorem 6. *Intersection graphs of 3-uniform hypergraphs have functionality bounded by 462.*

3 Graphs of Large Functionality

Knowing what is good without knowing what is bad is just half-knowledge. Therefore, in this section we turn to graphs of large functionality.

When we talk about graphs of large functionality we assume that we deal with an infinite family X of graphs, because in any finite collection of graphs functionality is bounded by a constant. Moreover, we can further assume that X is hereditary. Indeed, if X is not hereditary, we can extend it to a hereditary class by adding all induced subgraphs of graphs in X, and this extension has (un)bounded functionality if and only if X has, because by definition the functionality of an induced subgraph of a graph G is never larger than the functionality of G.

The notion of functional vertices was originally introduced in [1] for compact representation of graphs. This paper does not formally define the notion of graph functionality, but the results proved in [1] imply that any hereditary class of graphs of bounded functionality has $2^{O(n \log_2 n)}$ labelled graphs with n vertices. In the terminology of [3], these are classes with (at most) factorial speed of growth, or simply (at most) factorial classes. Therefore, in every superfactorial class functionality is unbounded. This is the case, for instance, for bipartite, co-bipartite and split graphs, since each of these classes contains at least $2^{n^2/4}$ labelled graphs with n vertices. This conclusion allows us to establish a relationship between functionality and one more important graph parameter known as VC-dimension.

A set system (X, S) consists of a set X and a family S of subsets of X. A subset $A \subseteq X$ is *shattered* if for every subset $B \subseteq A$ there is a set $C \in S$ such that $B = A \cap C$. The VC-dimension of (X, S) is the cardinality of a largest shattered subset of X.

The VC-dimension of a graph $G = (V, E)$ was defined in [2] as the VC-dimension of the set system (V, S), where S the family of closed neighbourhoods of vertices of G, i.e. $S = \{N[v] \ : \ v \in V(G)\}$. We denote the VC-dimension of G by $vc(G)$.

Theorem 7. *There exists a function f such that for any graph G, $vc(G) \leq f(fun(G))$.*

Proof. Fix a k and consider the class X_k of all graphs of functionality at most k. Clearly, X_k is hereditary. Assume X_k contains graphs of arbitrarily large

VC-dimension and let $G_1, G_2, \ldots$ be an infinite sequence of graphs from X_k with strictly increasing values of the VC-dimension. Let Y be the hereditary class containing all these graphs and all their induced subgraphs. Then Y is a hereditary subclass of X_k with unbounded VC-dimension. It is was shown in [13] that the only minimal hereditary classes of graph of unbounded VC-dimension are bipartite, co-bipartite and split graphs. But then Y and hence X_k contains one of these three classes, which is a contradiction to the fact that functionality is unbounded in these classes. Therefore, there is a constant $f(k)$ bounding the VC-dimension of graphs in X_k, which proves the result. □

Since large VC-dimension implies large functionality, it would be natural to construct graphs of large functionality through constructing graphs of large VC-dimension. The latter is an easy task. Indeed, consider the bipartite graph $D_n = (A, B, E)$ with two parts $|A| = n$ and $|B| = 2^n$. For each subset $C \subseteq A$ we create a vertex in B whose neighbourhood coincide with C. Clearly, the VC-dimension of D_n is n and hence with n growing the functionality of D_n grows as well.

However, this example is not very interesting in the sense that D_n contains vertices of low functionality (of low degree) and hence graphs of large functionality are hidden in D_n as proper induced subgraphs. A much more interesting task is constructing graphs where *all* vertices have large functionality. In what follows, we show that this is the case for hypercubes.

Let $V_n = \{0,1\}^n$ be the set of binary sequences of length n and let $v, w \in V_n$. The Hamming distance $d(v, w)$ between v and w is the number of positions in which the two sequences differ. A *hypercube* Q_n is the graph with vertex set $V_n = \{0,1\}^n$, in which two vertices are adjacent if and only if the Hamming distance between them equals 1.

Theorem 8. *Functionality of the hypercube* Q_n *is at least* $(n-1)/3$.

Proof. By symmetry, it suffices to show that the vertex $v = 00\ldots0 \in V_n$ has functionality at least $(n-1)/3$. Suppose v is a function of vertices in a set $S \subseteq V_n \backslash \{v\}$. To provide a lower bound on the size of S, and hence a lower bound on the functionality of v, for each $i = 1, 2, \ldots, n$ consider the set $S_i = \{w \in S : d(w, v) = i\}$, i.e. the set of all binary sequences in S that contain exactly i 1s. Also, consider the following set:

$$I = \{i \in \{1, 2, \ldots, n\} : \exists z = z_1 z_2 \ldots z_n \in S_1 \cup S_2 \cup S_3 \text{ with } z_i = 1\}.$$

Suppose $|I| \leq n-2$. Then there exist two positions i and j such that for any sequence $z = z_1 z_2 \ldots z_n \in S_1 \cup S_2 \cup S_3$, we have $z_i = 0$ and $z_j = 0$. Consider the following two vertices:

- $u = u_1 u_2 \ldots u_n$ with $u_k = 1$ if and only if $k = i$,
- $w = w_1 w_2 \ldots w_n$ with $w_k = 1$ if and only if $k = i$ or $k = j$.

We claim that u and w are not adjacent to any vertex $z \in S$. First, it is not hard to see that for any $z \in S_1 \cup S_2 \cup S_3$ we have $d(z, u) \geq 2$ and $d(z, w) \geq 2$.

Indeed, any $z \in S_1 \cup S_2 \cup S_3$ differs from u and w in position i, i.e. $z_i = 0$ and $u_i = w_i = 1$, and there must exist a $k \neq i, j$ with $z_k = 1$ and $u_k = w_k = 0$. Also, it is not difficult to see that $d(z, u) \geq 2$ and $d(z, w) \geq 2$ for any vertex $z \in S \backslash (S_1 \cup S_2 \cup S_3)$, because any such z has at least four 1s, while u and w have at most two 1s. Therefore, by definition, u and w are not adjacent to any vertex in S.

We see that the assumption that $|I| \leq n - 2$ leads to the conclusion that there are two vertices $u, w \in Q_n \backslash (S \cup \{v\})$ which are non-adjacent to any vertex in S, but have different adjacencies to v. This contradicts the fact that v is a function of the vertices in S. So, we must conclude that I has size at least $n-1$. As each vertex in $S_1 \cup S_2 \cup S_3$ has at most three 1s, we conclude that $S_1 \cup S_2 \cup S_3$ must contain at least $|I|/3 = (n-1)/3$ vertices. This completes the proof of the theorem. □

Theorem 7 shows that graphs of bounded VC-dimension constitute an extension of graphs of bounded functionality, while Theorem 8 shows that this extension is proper, since the hereditary closure of hypercubes constitutes a proper subclass of bipartite graphs.

4 Concluding Remarks and Open Problems

In this paper, we proved a number of results about graph functionality. However, many questions on this topic remain unanswered. Some of them are motivated by the results presented in the paper, for instance:

Problem 1. Is functionality bounded for the intersection graphs of k-uniform hypergraphs for $k > 3$? What about interval graphs?

Many other questions are motivated by related research. Of particular interest is the notion of *implicit representation* [10]. Similarly to bounded functionality, any hereditary class that admits an implicit representation is at most factorial. However, the question whether all factorial classes admit implicit representations, also known as the *implicit graph representation conjecture*, is widely open. Note that for non-hereditary classes the conjecture is not valid (see e.g. [14]). We ask whether there is any relationship between the two notions in the universe of hereditary classes.

Problem 2. Does implicit representation of graphs in a hereditary class imply bounded functionality in that class and/or vice versa?

One more open question is inspired by a result in [1] showing that if the family of prime (with respect to modular decomposition) graphs in a hereditary class X is factorial, then the entire class X is factorial.

Problem 3. Is it true that if prime (with respect to modular decomposition) graphs in a hereditary class X have bounded functionality, then all graphs in X have bounded functionality?

We note that a similar question for implicit representations is open too.

References

1. Atminas, A., Collins, A., Lozin, V., Zamaraev, V.: Implicit representations and factorial properties of graphs. Discrete Math. **338**, 164–179 (2015)
2. Alon, N., Brightwell, G., Kierstead, H., Kostochka, A., Winkler, P.: Dominating sets in k-majority tournaments. J. Comb. Theory Ser. B **96**, 374–387 (2006)
3. Balogh, J., Bollobás, B., Weinreich, D.: The speed of hereditary properties of graphs. J. Comb. Theory Ser. B **79**, 131–156 (2000)
4. Bandelt, H.-J., Mulder, H.M.: Distance-hereditary graphs. J. Comb. Theory Ser. B **41**, 182–208 (1986)
5. Courcelle, B., Engelfriet, J., Rozenberg, G.: Handle-rewriting hypergraph grammars. J. Comput. Syst. Sci. **46**, 218–270 (1993)
6. Courcelle, B., Makowsky, J.A., Rotics, U.: Linear time solvable optimization problems on graphs of bounded clique-width. Theory Comput. Syst. **33**, 125–150 (2000)
7. Courcelle, B., Olariu, S.: Upper bounds to the clique-width of a graph. Discrete Appl. Math. **101**, 77–114 (2000)
8. Golumbic, M.C., Rotics, U.: On the clique-width of some perfect graph classes. Int. J. Found. Comput. Sci. **11**(3), 423–443 (2000)
9. Gurski, F., Wanke, E.: Line graphs of bounded clique-width. Discrete Math. **307**, 2734–2754 (2007)
10. Kannan, S., Naor, M., Rudich, S.: Implicit representation of graphs. In: STOC 1988, pp. 334–343 (1988)
11. Lejeune, M., Lozin, V., Lozina, I., Ragab, A., Yacout, S.: Recent advances in the theory and practice of Logical Analysis of Data. Eur. J. Oper. Res. **275**, 1–15 (2019)
12. Lozin, V.: Minimal classes of graphs of unbounded clique-width. Ann. Comb. **15**, 707–722 (2011)
13. Lozin, V.: Graph parameters and Ramsey theory. Lect. Notes Comput. Sci. **10765**, 185–194 (2018)
14. Spinrad, J.P.: Efficient Graph Representations. Fields Institute Monographs, 19, xiii+342 pp. American Mathematical Society, Providence (2003)

On Happy Colorings, Cuts, and Structural Parameterizations

Ivan Bliznets[1,2] and Danil Sagunov[1](✉)

[1] St. Petersburg Department of Steklov Institute of Mathematics of the Russian Academy of Sciences, St. Petersburg, Russia
iabliznets@gmail.com, danilka.pro@gmail.com
[2] National Research University Higher School of Economics, St. Petersburg, Russia

Abstract. We study the Maximum Happy Vertices and Maximum Happy Edges problems. The former problem is a variant of clusterization, where some vertices have already been assigned to clusters. The second problem gives a natural generalization of Multiway Uncut, which is the complement of the classical Multiway Cut problem. Due to their fundamental role in theory and practice, clusterization and cut problems has always attracted a lot of attention. We establish a new connection between these two classes of problems by providing a reduction between Maximum Happy Vertices and Node Multiway Cut. Moreover, we study structural and distance to triviality parameterizations of Maximum Happy Vertices and Maximum Happy Edges. Obtained results in these directions answer questions explicitly asked in four works: Agrawal '17, Aravind et al. '16, Choudhari and Reddy '18, Misra and Reddy '17.

Keywords: Happy coloring · Maximum happy vertices · Maximum happy edges · Homophily law · Multiway cut · Distance to triviality · Treewidth · Clique-width · Parameterized complexity

1 Introduction

In this paper, we study Maximum Happy Vertices and Maximum Happy Edges. Both problems were recently introduced by Zhang and Li in [24], motivated by a study of algorithmic aspects of the homophyly law in large networks. Informally they paraphrase the law as "birds of a feather flock together". The law states that in social networks people are more likely to connect with people sharing similar interests with them. A social network is represented by a graph, where each vertex corresponds to a person in the network, and an edge between two vertices denotes that corresponding persons are connected within the network. Furthermore, we let vertices have colors assigned. The color of a vertex indicates type, character or affiliation of the corresponding person in the

This research was supported by the Russian Science Foundation (project 16-11-10123).

I. Sau and D. M. Thilikos (Eds.): WG 2019, LNCS 11789, pp. 148–161, 2019.
https://doi.org/10.1007/978-3-030-30786-8_12

network. An edge is called *happy* if its endpoints are colored with the same color. A vertex is called *happy* if all its neighbours are colored with the same color as the vertex itself. Equivalently, a vertex is happy if all edges incident to it are happy. The formal definitions of MAXIMUM HAPPY VERTICES and MAXIMUM HAPPY EDGES are the following:

MAXIMUM HAPPY VERTICES (MHV)
Input: A graph G, a partial coloring of vertices $p : S \to [\ell]$ for some $S \subseteq V(G)$ and an integer k.
Question: Is there a coloring $c : V(G) \to [\ell]$ extending the partial coloring p such that the number of happy vertices with respect to c is at least k?

MAXIMUM HAPPY EDGES (MHE)
Input: A graph G, a partial coloring of vertices $p : S \to [\ell]$ for some $S \subseteq V(G)$ and an integer k.
Question: Is there a coloring $c : V(G) \to [\ell]$ extending the partial coloring p such that the number of happy edges with respect to c is at least k?

MAXIMUM HAPPY EDGES has an immediate connection to MULTIWAY CUT. Precisely, if each color is used in precoloring exactly once, then MAXIMUM HAPPY EDGES is exactly the MULTIWAY UNCUT problem, i.e. the edge complement of MULTIWAY CUT. Thus, MAXIMUM HAPPY EDGES is a generalization of the MULTIWAY UNCUT problem. So, in this case the connection between clustering vertices by color and cutting edges in order to separate different colors is pretty obvious. However, this is not the case for vertex version of the problem, which we would like to connect with the vertex version of MULTIWAY CUT, NODE MULTIWAY CUT.

MAXIMUM HAPPY VERTICES can be seen as a sort of clusterization problem, in which some vertices already have prescribed color/cluster and the goal is to identify colors/clusters of initially uncolored/unassigned vertices. In some sense, we would like to clusterize the graph in such a way that overall boundary of clusters is minimized. Here, by a boundary of a cluster we understand vertices of the cluster that are connected to vertices outside the cluster. While it is possible to straightforwardly formulate the problem in terms of a special cutting problem, this kind of formalization will sound complicated and unnatural. We show that MHV can be easily transformed into NODE MULTIWAY CUT, thereby constructing an additional bridge between clusterization and cutting problems.

Recently, MHV and MHE have attracted a lot of attention and were studied from parameterized [1–3,7,19] and approximation [22–25] points of view as well as from experimental perspective [18]. Further, dozens of algorithms for the classical MULTIWAY CUT problem have been considered as well, which is the complement of a special case of MHE.

In 2015, Zhang and Li established that ℓ-MHE and ℓ-MHV are NP-hard for $\ell \geq 3$, where ℓ is the number of colors used. Later, Aravind et al. [2] showed that when the input graph is a tree, ℓ-MHV and ℓ-MHE can be solved in $\mathcal{O}(n\ell \log \ell)$ and in $\mathcal{O}(n\ell)$ time respectively. In [19], Misra and Reddy proved NP-hardness of both MHV and MHE on split and on bipartite graphs, and showed that MHV is polynomial time solvable on cographs.

From the approximation perspective, the currently best known results are the following. Zhang et al. [25] showed that MHV can be approximated within $\frac{1}{\Delta+1}$, where Δ is the maximum degree of the input graph, and MHE can be approximated within $\frac{1}{2} + \frac{\sqrt{2}}{4} f(\ell)$, where $f(\ell) = \frac{(1-1/\ell)\sqrt{\ell(\ell-1)}+1/\sqrt{2}}{\ell-1+1/2\ell} \leq 1$. They also claimed that a more careful analysis can improve the approximation ratio for MHV to $\frac{1}{\Delta+1/g(\Delta)}$, where $g(\Delta) = (\sqrt{\Delta} + \sqrt{\Delta+1})^2 \Delta > 4\Delta^2$.

The known results in parameterized complexity (not including kernelization) are summarized in Table 1. Results proved in the paper are marked by * in the table. Agrawal [1] provides $\mathcal{O}(k^2\ell^2)$-kernel for MHV, where ℓ is the number of used colors and k is the number of desired happy vertices. Independently, Gao and Gao [13] present a $(2^{k\ell+k} + k\ell + k + \ell)$-kernel for the general case and a $(7(k\ell + k) + \ell - 10)$-kernel in the case of planar graphs. We provide a kernel on $\mathcal{O}(d^3)$ vertices for MHV parameterized by the distance to clique, partially answering a question in [19]. Note that the kernel sizes mentioned in this paragraph correspond to the number of vertices in the kernels.

Table 1. Known and established results under distance-to-triviality and structural parameters. * marks results of this work. T indicates a result proven as a theorem. C indicates a result proven as a corollary. d denotes the distance parameter of the row.

<table>
<tr><th>Parameter</th><th>MHE</th><th>ℓ-MHE</th><th>MHV</th><th>ℓ-MHV</th></tr>
<tr><td>Distance to threshold graphs</td><td>?</td><td>?</td><td colspan="2">$d^{\mathcal{O}(d)} \cdot n^{\mathcal{O}(1)}$ [7]</td></tr>
<tr><td>Distance to clique</td><td colspan="4">$d^{\mathcal{O}(d)} \cdot n^{\mathcal{O}(1)}$ [19]</td></tr>
<tr><td>Distance to cluster</td><td>W[1]-hard C3*</td><td>$\ell^d \cdot n^{\mathcal{O}(1)}$</td><td colspan="2">$d^{\mathcal{O}(d)} \cdot n^{\mathcal{O}(1)}$ T3*</td></tr>
<tr><td>Distance to cographs</td><td rowspan="5">W[1]-hard C2*</td><td>?</td><td rowspan="5">W[1]-hard C3*</td><td>$(2\ell)^d \cdot n^{\mathcal{O}(1)}$</td></tr>
<tr><td>Treewidth</td><td>$\ell^{\text{tw}} \cdot n^{\mathcal{O}(1)}$ [1,3]</td><td>$\ell^{\text{tw}} \cdot n^{\mathcal{O}(1)}$ [3,19]</td></tr>
<tr><td>Pathwidth</td><td>$\ell^{\text{pw}} \cdot n^{\mathcal{O}(1)}$ [3,19]</td><td>$\ell^{\text{pw}} \cdot n^{\mathcal{O}(1)}$ [1,3]</td></tr>
<tr><td>Cliquewidth</td><td>?</td><td>?</td></tr>
<tr><td>Feedback vertex set number</td><td>$\ell^d \cdot n^{\mathcal{O}(1)}$</td><td>$(2\ell)^d \cdot n^{\mathcal{O}(1)}$</td></tr>
<tr><td>Vertex cover number</td><td colspan="4">$d^{\mathcal{O}(d)} \cdot n^{\mathcal{O}(1)}$ [19]</td></tr>
<tr><td>Split vertex deletion number</td><td colspan="4" rowspan="2">para-NP-hard [19]</td></tr>
<tr><td>Odd cycle transversal number</td></tr>
<tr><td>Neighbourhood diversity</td><td colspan="4">$2^{\text{nd}} \cdot n^{\mathcal{O}(1)}$ [3]</td></tr>
</table>

Our Results: The main contributions of our work are the following.

- We establish a natural connection between MAXIMUM HAPPY VERTICES on a graph G and NODE MULTIWAY CUT on a second power of a certain subgraph of G.
- We answer questions in [1,2] about existence of FPT-algorithm for MHV parameterized by the treewidth of the input graph only.
- Similarly, we answer one of the questions from Choudhari et al. [7] and Misra et al. [19] by showing W[1]-hardness of MHE parameterized by the cluster vertex deletion number. We show that MHV, in contrast to MHE, is in FPT when parameterized by the cluster vertex deletion number.
- We partially answer a question stated by Misra and Reddy in [19]. We provide a kernel of size $\mathcal{O}(d^3)$ for MHV, where d is the distance to cliques.
- Among other results, we also present the first algorithm for NODE MULTIWAY CUT parameterized by the clique-width of the input graph.

Organization of the Paper: Section 3 describes results under some structural and distance-to-triviality parameters. In Sect. 5 we provide results connecting NODE MULTIWAY CUT and MAXIMUM HAPPY VERTICES. In Sect. 6 we provide a polynomial kernel for MHV parameterized by the distance to clique.

2 Preliminaries

Basic Notation. We denote the set of positive integer numbers by $\mathbb{N}$. For each positive integer k, by $[k]$ we denote the set of all positive integers not exceeding k, $\{1, 2, \ldots, k\}$. We use ∞ to denote an infinitely large number, for which holds $n < \infty$ and $n + \infty = \infty + n = \infty$, where n is an arbitrary integer. We use $\sqcup$ for the disjoint union operator, i.e. $A \sqcup B$ equals $A \cup B$, with an additional constraint that A and B are disjoint.

We use the traditional $\mathcal{O}$-notation for asymptotical upper bounds. We additionally use the $\mathcal{O}^*$-notation that hides polynomial factors. Many of our results concern the parameterized complexity of the problems, including fixed-parameter tractable algorithms, kernelization algorithms, and some hardness results for certain parameters. For a detailed survey in parameterized algorithms we refer to the book of Cygan et al. [10]. In their book one may also find definitions of pathwidth and treewidth that are considered as parameters in some of our results.

Throughout the paper, we use standard graph notation and terminology, following the book of Diestel [12]. All graphs in our work are undirected simple graphs. We consider several graph classes in our work. *Interval graphs* are graphs whose vertices can be represented as intervals on the real line, so that a pair of vertices are connected by an edge if and only if their representative intervals intersect. *Cluster graphs* are graphs that are a disjoint union of cliques, or, equivalently, graphs that do not contain induced paths on three vertices.

We often refer to the distance to $\mathcal{G}$ parameter, where $\mathcal{G}$ is an arbitrary graph class. For a graph G, we say that a vertex subset $S \subseteq V(G)$ is a $\mathcal{G}$ *modulator* of G, if G becomes a member of $\mathcal{G}$ after deletion of S, i.e. $G \setminus S \in \mathcal{G}$. Then, the *distance to* $\mathcal{G}$ parameter of G is defined as the size of its smallest $\mathcal{G}$ modulator.

Graph Colorings. When dealing with instances of Maximum Happy Vertices or Maximum Happy Edges, we use a notion of colorings. A *coloring* of a graph G is a function that maps vertices of the graph to a set of colors. If this function is partial, we call such a coloring *partial.* If not stated otherwise, we use ℓ for the number of distinct colors, and assume that colors are integers in $[\ell]$. A partial coloring p is always given as a part of the input for both problems, along with graph G. We also call p a *precoloring* of the graph G, and use (G, p) to denote the graph along with the precoloring. The goal of both problems is to extend this partial coloring to a specific coloring c that maps each vertex to a color. We call c a *full coloring* (or simply, a coloring) of G that extends p. We may also say that c is a coloring of (G, p). For convenience, introduce the notion of potentially happy vertices, both for full and partial colorings.

Definition 1. We call a vertex v of (G, p) *potentially happy,* if there exists a coloring c of (G, p) such that v is happy with respect to c. In other words, if u and w are precolored neighbours of v, then $p(u) = p(w)$ (and $p(u) = p(v)$, if v is a precolored vertex). We denote the set of all potentially happy vertices in (G, p) by $\mathcal{H}(G, p)$.

By $\mathcal{H}_i(G, p)$ we denote the set of all potentially happy vertices in (G, p) such that they are either precolored with color i or have a neighbour precolored with color i:

$$\mathcal{H}_i(G, p) = \{v \in \mathcal{H}(G, p) \mid N[v] \cap p^{-1}(i) \neq \emptyset\}.$$

In other words, if a vertex $v \in \mathcal{H}_i(G, p)$ is happy with respect to some coloring c of (G, p), then necessarily $c(v) = i$.

Note that if c is a full coloring of a graph G, then $|\mathcal{H}(G, c)|$ is equal to the number of vertices in G that are happy with respect to c.

Clique-Width. Among other structural parameters, we consider clique-width in our work. We follow definitions presented by Lackner et al. in their work on Multicut parameterized by clique-width [17].

Due to the space restrictions, we omit the definition of clique-width and k-expressions to the full version of this paper. For more details on clique-width we refer to [14].

Omitted Proofs. Due to the space restricitons, we omit full proofs of some theorems, lemmata, corollaries or claims to the full version of this paper. For some of them we leave a proof sketch instead of a full proof, and for some of them we omit the proof completely. Such statements with omitted proofs are marked with the '$\star$' sign.

3 Structural and Distance-to-Triviality Parameters

In [1], Agrawal proved that Maximum Happy Vertices is W[1]-hard with respect to the standard parameter, the number of happy vertices. In [2,7,19] some structural parameters for MHV and MHE were studied. In [1], Agrawal also asked whether MHV admits an FPT algorithm when parameterized by the treewidth of the input graph alone. In this section, we show that both MHV and MHE are W[1]-hard with respect to certain distance-to-triviality and structural parameters, including treewidth, answering the question of Agrawal and some other questions. We start with the definition of a classical W[1]-complete (with respect to the solution size) problem.

> REGULAR MULTICOLORED INDEPENDENT SET
> **Input:** Graph G, with degree of every vertex in G equal to r, a partition of G into k cliques $V_1, V_2, \ldots, V_k$.
> **Parameter:** k
> **Question:** Is there a multicolored independent set in G of size k, i.e. a subset $S \subseteq V(G)$ of its vertices that is an independent set in G and $|S \cap V_i| = 1$ for every $i \in [k]$?

Theorem 1. MAXIMUM HAPPY VERTICES *is W[1]-hard when parameterized by the distance to graphs that are a disjoint union of paths consisting of three vertices.*

Proof. We reduce from REGULAR MULTICOLORED INDEPENDENT SET, that is W[1]-complete with respect to k due to [4].

Let $(G, k, V_1, V_2, \ldots, V_k)$ be an instance of REGULAR MULTICOLORED INDEPENDENT SET, and let r be the degree of every vertex in G, i.e. $r = |N(v)|$ for any $v \in V(G)$. We assume that $|V_i| \geq 2$ for each i, since otherwise the instance can be trivially reduced to an instance with a smaller k. We construct an instance (G', p, k') of MAXIMUM HAPPY VERTICES as follows.

We set $\ell = |V(G)|$, so each color corresponds to a unique vertex of G. For convenience, we use vertices of G as colors, instead of the numbers in $[\ell]$.

For each edge $uv \in E(G)$, we introduce a path on three vertices $t^u_{uv}, e_{uv}, t^v_{uv}$ in G', with e_{uv} being the middle vertex of the path. Endpoint vertices t^u_{uv} and t^v_{uv} are precolored in colors u and v respectively, i.e. $p(t^u_{uv}) = u$ and $p(t^v_{uv}) = v$, and the middle vertex is left uncolored.

We then introduce a selection gadget in G'. That is, we introduce k uncolored vertices $s_1, s_2, \ldots, s_k$. For each $i \in [k]$ and each color $v \in V_i$, we connect s_i with each vertex precolored in color v. Thus, a vertex t^u_{uv} becomes connected to exactly one vertex of the selection gadget s_i, where i is such that $u \in V_i$. The purpose of the selection gadget is that the color of s_i in the optimal coloring corresponds to a vertex that we should take in V_i in the initial instance of REGULAR MULTICOLORED INDEPENDENT SET.

We finally set $k' = kr$ and argue that $(G, k, V_1, V_2, \ldots, V_k)$ is a yes-instance of REGULAR MULTICOLORED INDEPENDENT SET if and only if (G', p, k') is a yes-instance of MAXIMUM HAPPY VERTICES.

Let $S \subseteq V(G)$ be a multicolored independent set of G, i.e. S is an independent set in G and $|S \cap V_i| = 1$ for each i. Let us construct a coloring c of $V(G')$ such that it extends p and at least $k' = kr$ vertices of G are happy with respect to c. For each i, set the color of s_i to v_i, i.e. $c(s_i) = v_i$, where $v_i \in S \cap V_i$. For each edge $uv \in E(G)$, set the color of e_{uv} to u, if $u \in S$, or to v, if $v \in S$, and to an arbitrary color otherwise. Formally, $c(e_{uv}) = u$, if $u \in S$, and $c(e_{uv}) = v$, if $v \in S$. If $u, v \notin S$, then $c(e_{uv})$ can be assigned an arbitrary color. Note that either $u \notin S$ or $v \notin S$, since S is an independent set.

G' has no other uncolored vertex, thus the construction of c is complete.

Claim 1. For each vertex $u \in S$ and each edge $uv \in E(G)$ incident to u, t^u_{uv} is happy with respect to c.

Proof of Claim 1. Indeed, t^u_{uv} is adjacent to exactly two vertices: s_i, where $u \in V_i$, and e_{uv}. Since $u \in S$, $c(s_i) = u$ and $c(e_{uv}) = u$ by construction of c. t^u_{uv} is a vertex precolored with color u, hence t^u_{uv} is happy with respect to c. ■

For each $u \in S$, there are exactly r edges adjacent to u, hence all r vertices precolored with color u are happy. $|S| = k$, hence at least kr vertices of G' are happy with respect to c.

It is left to prove that if (G', p, k') is a yes-instance of Maximum Happy Vertices, then $(G, k, V_1, V_2, \ldots, V_k)$ is a yes-instance of Regular Multicolored Independent Set.

Claim 2. Let c be an arbitrary coloring of $V(G')$ extending p. There are at most kr happy vertices in G' with respect to c. Moreover, all happy vertices are precolored vertices of at most k distinct colors.

Proof of Claim 2. Observe that for each $i \in [k]$, s_i is unhappy with respect to any coloring extending p, since neighbours of s_i are precolored with colors in V_i, and each color is presented exactly r times among its neighbours, and we assumed that V_i consists of at least two vertices.

For each $uv \in E(G)$, e_{uv} is adjacent to exactly two vertices t^u_{uv} and t^v_{uv}, which are precolored with two distinct colors u and v. Thus, e_{uv} is also unhappy with respect to any coloring extending p.

Hence, only precolored vertices of G' can be happy, i.e. vertices t^u_{uv} for $uv \in E(G)$. Each of them is adjacent to exactly one vertex of the selector gadget, i.e. vertex s_i for some $i \in [k]$. But for each $i \in [k]$, only the neighbours that share the same color as s_i can be happy. Thus, each happy vertex shares a color with one of k vertices of the selection gadget. Since each color is presented exactly r times in the partial coloring p, there can be at most kr such happy vertices. ■

Let c be a coloring of $V(G')$ extending p such that at least kr vertices of G' are happy with respect to c. According to Claim 2, exactly kr vertices of G' are happy with respect to c, and they are precolored with k different colors. Moreover, for each color, all r precolored vertices of this color are happy. Let S be the set of these k colors, i.e. $S = \{c(s_1), c(s_2), \ldots, c(s_k)\}$. We argue that S is an independent set in G. Note that $|S \cap V_i| = 1$ is then automatically satisfied, as V_i is a clique in G for each $i \in [k]$.

Claim 3. If there are kr happy vertices among the vertices of type t^v_{uv} in G' with respect to coloring c, that extends p, then $S = \{c(s_1), c(s_2), \ldots, c(s_k)\}$ is an independent set in G.

Proof of Claim 3. Indeed, suppose that S is not an independent set in G, i.e. there are vertices $u, v \in S$, such that $uv \in E(G)$. Then there is a path t^u_{uv}, e_{uv}, t^v_{uv} in G'. t^u_{uv} is a happy vertex of color $c(t^u_{uv}) = p(t^u_{uv}) = u$, hence $c(e_{uv}) = u$. Analogously, t^v_{uv} is a happy vertex of color v, hence $c(e_{uv}) = v$. We get that $u = c(e_{uv}) = v$, which contradicts our assumption. ■

We have shown that (G', p, k') is an instance equivalent to $(G, k, V_1, V_2, \dots, V_k)$; moreover, it can be constructed in polynomial time.

Note that the deletion of the selector gadget vertices in G' leads to G' being a disjoint union of paths consisting of three vertices. Thus, G' has the distance parameter being at most k, and if Maximum Happy Vertices is in FPT when parameterized by the distance to graphs being a disjoint union of path consisting of three vertices, then W[1]-complete Regular Multicolored Independent Set is also in FPT. Hence, MHV is W[1]-hard with respect to the distance parameter. □

The following corollary answers an open question posed in [1].

Corollary 1. Maximum Happy Vertices *is W[1]-hard with respect to parameters pathwidth, treewidth or clique-width, distance to cographs, feedback vertex set number.*

Proof Sketch. W[1]-hardness of MHV with respect to the parameters distance to cographs or feedback vertex set number is an immediate corollary of Theorem 1. Since graphs of type $n \times P_3$ (that is, graphs that are a disjoint union of paths consisting of three vertices) are simultaneously cographs and forests, respectively to the parameters. Proofs of bounded pathwidth, treewidth and clique-width for graphs with bounded distance to $n \times P_3$ can be found in the full version of the paper. □

Theorem 2. Maximum Happy Edges *is W[1]-hard when parameterized by the distance to graphs that are disjoint union of paths consisting of three vertices and is W[1]-hard when parameterized by the distance to graphs that are a disjoint union of cycles of length three.*

Proof Sketch. We adjust the reduction from Regular Multicolored Independent Set to MHV provided in the proof of Theorem 1.

Given an instance $(G, k, V_1, V_2, \dots, V_k)$ of Regular Multicolored Independent Set, we construct an instance (G', p, k') of Maximum Happy Edges as follows.

Let $n = |V(G)|$, $m = |E(G)|$. G' is constructed in the same way as in the proof of Theorem 1: for each edge $uv \in E(G)$, we introduce a path on three vertices t^u_{uv}, e_{uv}, t^v_{uv}, and set $p(t^u_{uv}) = u$, $p(t^v_{uv}) = v$, and e_{uv} is left uncolored; then we introduce the selection gadget vertices $s_1, s_2, \dots, s_k$, and introduce an edge between s_i and t^u_{uv} for each $i \subset [k]$, $u \in V_i$ and $uv \in E(G)$. For each $i \in [k]$, s_i is left uncolored.

Additionally, we introduce edges new to this construction: for each $i, j \in [k]$ and each edge $uv \in E(G)$, such that $u \in V_i$ and $v \in V_j$, we introduce edges between e_{uv} and s_i and between e_{uv} and s_j. In case $i = j$, we introduce only one edge.

We also need additional precolored vertices in order for this reduction to work. For each $i \in [k]$, and each $v \in V_i$, we introduce m new paths consisting of three vertices in G': for each $j \in [m]$, we introduce a path $a^1_{v,j}$, $a^2_{v,j}$, $a^3_{v,j}$. Every vertex in these new paths we precolor with color v, i.e. $p(a^1_{v,j}) = p(a^2_{v,j}) = p(a^3_{v,j}) = v$ for each j, and connect by a newly-introduced edge to exactly one vertex of the selector gadget s_i. These auxiliary vertices will ensure that for each $i \in [k]$, s_i is colored with one of the colors in V_i. Paths between these vertices are needed only to preserve the distance parameter.

We finally set $k' = kr + (m + kr) + (3k + 2n) \cdot m$ and argue that $(G, k, V_1, V_2, \ldots, V_k)$ is a yes-instance of Regular Multicolored Independent Set if and only if (G', p, k') is a yes-instance of Maximum Happy Edges. For the full proof of this fact, which is mostly by carefully counting each edge in G', we refer to the full version of our paper.

To prove the same for the distance to graphs being a disjoint union of cycles of length three, we note that in our construction of G', endpoints of the paths are precolored vertices. Hence, we can add an edge between endpoints of each path, i.e. between t^v_{uv} and t^u_{uv} for each $uv \in E(G)$ and between $a^1_{v,j}$ and $a^3_{v,j}$ for each $v \in V(G)$ and $j \in [m]$, and just increase the parameter k' by the number of newly-appeared happy edges. Namely, these are the edges between $a^1_{v,j}$ and $a^3_{v,j}$, thus we increase k' by $n \cdot m$, and the other parts of the construction remain the same. □

Corollary 2. Maximum Happy Edges *is W[1]-hard with respect to parameters pathwidth, treewidth or clique-width, distance to cographs, feedback vertex set number.*

The rest of the section focuses on the parameterized complexity of both MHV and MHE parameterized by the distance to cluster parameter. We separate MHE and MHV, showing that the former problem is W[1]-hard with respect to this parameter, but the latter admits an FPT-algorithm. This answers an open question posed in works of Choudhari and Reddy [7] and Misra and Reddy [19].

Corollary 3. Maximum Happy Edges *is W[1]-hard when parameterized by the cluster vertex deletion number.*

Proof. Observe that graph consisting of disjoint cycles of length three is a cluster graph. Then, by Theorem 2, MHE is W[1]-hard when parameterized by the distance to cluster graphs. □

Theorem 3. Maximum Happy Vertices *can be solved in $\mathcal{O}^*((2d)^d)$ time, where d is the distance to cluster parameter of the input graph.*

Proof Sketch. We adapt algorithms of Misra and Reddy presented in [19] in their proofs of FPT membership result for both MHV and MHE parameterized by the vertex cover number and by the distance to clique parameters. For a full proof we refer to the full version of our paper. □

4 Obtaining W[2]-Hardness

We are grateful for the anonymous reviewers of this paper for sharing ideas of how the statements of Theorems 1 and 2 can be changed to obtain W[2]-hardness with respect to structural parameters, strengthening Corrollaries 1, 2 and 3. This section is dedicated to these W[2]-hardness results.

Theorem 4 **($\star$).** MAXIMUM HAPPY VERTICES *is W[2]-hard when parameterized by the distance to graphs that are a disjoint union of stars.*

Theorem 5 **($\star$).** MAXIMUM HAPPY EDGES *is W[2]-hard when parameterized by the distance to graphs that are a disjoint union of stars and is W[2]-hard when parameterized by the distance to graphs that are a disjoin union of cliques.*

Corollary 4 **($\star$).** MAXIMUM HAPPY VERTICES *and* MAXIMUM HAPPY EDGES *are both W[2]-hard with respect to parameters pathwidth, treewidth or clique-width, distance to cographs, feedback vertex set number.*

5 Maximum Happy Vertices and Node Multiway Cut

This section reveals the connection between MAXIMUM HAPPY VERTICES and NODE MULTIWAY CUT. This connection is a natural supplement of the straightforward connection of the edge versions of the problems, MAXIMUM HAPPY EDGES and MULTIWAY CUT. It is more convenient for us to use a variation of NODE MULTIWAY CUT, called GROUP MULTIWAY CUT, where terminal groups are used instead of singleton terminals.

GROUP MULTIWAY CUT [6]
Input: A graph G and pairwise disjoint sets of terminals $\{T_1, T_2, \ldots, T_\ell\}$, and an integer k.
Question: Is there a set $S \subseteq V(G)$ of size at most k such that $G \setminus S$ has no $u - v$ path for any $u \in T_i, v \in T_j$ and $i \neq j$?

We start with the following crucial lemma.

Lemma 1. *Let G be a graph with precoloring p. Let $H \subseteq \mathcal{H}(G,p)$ be an arbitrary subset of its potentially happy vertices. Then a coloring c extending p, so that all vertices in H are happy with respect to c, exists if and only if there exists no path $u_1, u_2, \ldots, u_t$ in G, such that u_1 and u_t are precolored and $p(u_1) \neq p(u_t)$, and for each $i \in [t-1]$, either u_i or u_{i+1} is in H.*

Proof Sketch. The basic idea of the proof is that happy vertices are incident only to happy edges. Furthermore, two vertices of distinct colors cannot be connected by a path containing only of happy edges. For the full formal proof we refer to the full version of our paper. □

Theorem 6. *Let (G, p, k) be an instance of* Maximum Happy Vertices. *Then (G, p, k) is a yes-instance of* Maximum Happy Vertices *if and only if $(G^2[\mathcal{H}(G,p)], \{\mathcal{H}_1(G,p), \mathcal{H}_2(G,p), \ldots, \mathcal{H}_\ell(G,p)\}, |\mathcal{H}(G,p)| - k)$ is a yes-instance of* Group Multiway Cut.

Proof. Let (G, p, k) be a yes-instance of MHV. We show that $(G^2[\mathcal{H}(G,p)], \{\mathcal{H}_1(G,p), \mathcal{H}_2(G,p), \ldots, \mathcal{H}_\ell(G,p)\}, |\mathcal{H}(G,p)| - k)$ is a yes-instance of Group Multiway Cut. Since (G, p, k) is a yes-instance, there is a coloring c such that at least k vertices of (G, p) are happy with respect to c. Let H be a set of any k of these vertices, i.e. $|H| = k$ and all vertices in H are happy with respect to c in (G, p).

Observe that any path in G whose all edges are incident to at least one vertex of H, corresponds to a simple path in $G^2[H]$. Indeed, let $u_1, u_2, \ldots, u_t$ be a path in G such that $u_i \in H$ or $u_{i+1} \in H$ for each $i \in [t-1]$. Let $u_1^H, u_2^H, \ldots, u_{t_1}^H$ be the subsequence of $u_1, \ldots, u_t$ of vertices in H ($H \cap \{u_1, \ldots, u_t\} = \{u_1^H, \ldots, u_{t_1}^H\}$). Note that for each $i \in [t_1 - 1]$, u_i^H and u_{i+1}^H are either consequent in $u_1, \ldots, u_t$ or there is only one vertex between them in $u_1, \ldots, u_t$. That is, there is an edge between u_i^H and u_{i+1}^H in $G^2[H]$. Thus, $u_1^H, u_2^H, \ldots, u_{t_1}^H$ is a path in $G^2[H]$. Vice versa, any simple path in $G^2[H]$ corresponds to paths in G which edges are incident to vertices in H.

Since all vertices H are happy in (G, p), by Lemma 1, there is no path between differently precolored vertices with all edges incident to at least one vertex in H. Consider $G^2[H]$ and suppose that there exists a path between vertices v and w in $G^2[H]$, such that $v \in \mathcal{H}_i(G,p)$ and $w \in \mathcal{H}_j(G,p)$, and $i \neq j$. As shown above, this path corresponds to a path between v and w in G, and all edges of this path are incident to H. Since $v \in \mathcal{H}_i(G,p)$, there is a precolored vertex $v' \in N[v]$ with $p(v') = i$. Similarly, there is a $w' \in N[w]$ with $p(w') = j$. There is a path between v' and w' in G with all edges incident to H and $p(v') \neq p(w')$, a contradiction. Hence, no vertices in different sets of terminals $\mathcal{H}_i(G,p)$ and $\mathcal{H}_j(G,p)$ are connected in $G^2[H]$. Thus, $\mathcal{H}(G,p) \setminus H$ is an answer to $(G^2[\mathcal{H}(G,p)], \{\mathcal{H}_i(G,p)\}, |\mathcal{H}(G,p)| - k)$, so it is a yes-instance of Group Multiway Cut.

The proof in the other direction is similar: if S, ($|S| = |\mathcal{H}(G,p)| - k$), is a solution to the instance of Group Multiway Cut, then all k vertices in $\mathcal{H}(G,p) \setminus S$ can be happy simultaneously in (G, p). □

Theorem 6 shows the importance of potentially happy vertices in the input of MHV. Other vertices are playing role of common neighbours or precolored neighbours for potentially happy vertices. Note that the sets $\mathcal{H}(G,p)$ and $\mathcal{H}_i(G,p)$ are computable in polynomial time. Thus, an instance of MHV can be compressed in order to contain only useful information about potentially happy vertices. We formulate this in the following corollary.

Corollary 5 ($\star$). Maximum Happy Vertices, *parameterized by the number of potentially happy vertices h, (i) admits a polynomial compression into* Group Multiway Cut *with h vertices and (ii) admits a kernel with $\mathcal{O}(h^2)$ vertices and edges.*

Another interesting consequence of Theorem 6, along with Corollary 1, is a lower bound on algorithms for Group Multiway Cut parameterized by clique-width.

Corollary 6. Group Multiway Cut *is W[1]-hard when parameterized by the clique-width of the input graph.*

Proof. By Corollary 1, Maximum Happy Vertices is W[1]-hard when parameterized by the clique-width of the input graph. Take an instance (G, p, k) of MHV. As shown by Todinca in [21], if G has clique-width t, then the power G^c of G has clique-width at most $2tc^t$. Hence, G^2 has clique-width at most $2t2^t$. Then, as shown by Courcelle and Olariu in [9], every induced subgraph of a graph of clique-width t has clique-width at most t, so $G^2[\mathcal{H}(G,p)]$ has clique-width at most $2t2^t$ as well. So, in an instance $(G^2[\mathcal{H}(G,p)], \{\mathcal{H}_i(G,p)\}, |\mathcal{H}(G,p)| - k)$ of Group Multiway Cut equivalent to the instance (G, p, k), the clique-width of the input graph is bounded if the clique-width of G is bounded. Since the reduction from MHV to Group Multiway Cut is polynomial, the corollary statement follows. □

In contrast, we have that Node Multiway Cut is in FPT when parameterized by the clique-width of the input graph. We present an algorithm solving Node Multiway Cut using dynamic programming on a w-expression of G.

Theorem 7 (⋆). Node Multiway Cut *can be solved in* $(w+3)^{2w} \cdot n^{\mathcal{O}(1)}$, *if a w-expression of G is given.*

6 Polynomial Kernel for Maximum Happy Vertices

In this section, we present a polynomial kernel for MHV parameterized by the distance to clique. This partially answers a question of Misra and Reddy in [19], where they also showed FPT algorithms for both MHV and MHE parameterized by this parameter.

Theorem 8 (⋆). Maximum Happy Vertices *admits a kernel with* $\mathcal{O}(d^3)$ *vertices, where d is the distance to clique parameter, and the parameter and a clique modulator of G are not given explicitly.*

References

1. Agrawal, A.: On the parameterized complexity of happy vertex coloring. In: Brankovic, L., Ryan, J., Smyth, W.F. (eds.) IWOCA 2017. LNCS, vol. 10765, pp. 103–115. Springer, Cham (2018). https://doi.org/10.1007/978-3-319-78825-8_9
2. Aravind, N.R., Kalyanasundaram, S., Kare, A.S.: Linear time algorithms for happy vertex coloring problems for trees. In: Mäkinen, V., Puglisi, S.J., Salmela, L. (eds.) IWOCA 2016. LNCS, vol. 9843, pp. 281–292. Springer, Cham (2016). https://doi.org/10.1007/978-3-319-44543-4_22

3. Aravind, N., Kalyanasundaram, S., Kare, A.S., Lauri, J.: Algorithms and hardness results for happy coloring problems. arXiv preprint arXiv:1705.08282 (2017)
4. Belmonte, R., Golovach, P.A., van 't Hof, P., Paulusma, D.: Parameterized complexity of two edge contraction problems with degree constraints. In: Gutin, G., Szeider, S. (eds.) IPEC 2013. LNCS, vol. 8246, pp. 16–27. Springer, Cham (2013). https://doi.org/10.1007/978-3-319-03898-8_3
5. Boral, A., Cygan, M., Kociumaka, T., Pilipczuk, M.: A fast branching algorithm for cluster vertex deletion. Theory Comput. Syst. **58**(2), 357–376 (2015)
6. Chitnis, R., Fomin, F.V., Lokshtanov, D., Misra, P., Ramanujan, M.S., Saurabh, S.: Faster exact algorithms for some terminal set problems. In: Gutin, G., Szeider, S. (eds.) IPEC 2013. LNCS, vol. 8246, pp. 150–162. Springer, Cham (2013). https://doi.org/10.1007/978-3-319-03898-8_14
7. Choudhari, J., Reddy, I.V.: On structural parameterizations of happy coloring, empire coloring and boxicity. In: Rahman, M.S., Sung, W.-K., Uehara, R. (eds.) WALCOM 2018. LNCS, vol. 10755, pp. 228–239. Springer, Cham (2018). https://doi.org/10.1007/978-3-319-75172-6_20
8. Corneil, D.G., Rotics, U.: On the relationship between clique-width and treewidth. SIAM J. Comput. **34**(4), 825–847 (2005)
9. Courcelle, B., Olariu, S.: Upper bounds to the clique width of graphs. Discrete Appl. Math. **101**(1–3), 77–114 (2000)
10. Cygan, M., et al.: Parameterized Algorithms, vol. 3. Springer, Heidelberg (2015). https://doi.org/10.1007/978-3-319-21275-3
11. Cygan, M., Philip, G., Pilipczuk, M., Pilipczuk, M., Wojtaszczyk, J.O.: Dominating set is fixed parameter tractable in claw-free graphs. Theor. Comput. Sci. **412**(50), 6982–7000 (2011). https://doi.org/10.1016/j.tcs.2011.09.010
12. Diestel, R.: Graph Theory. Springer, Heidelberg (2018). https://doi.org/10.1007/978-3-662-53622-3
13. Gao, H., Gao, W.: Kernelization for maximum happy vertices problem. In: Bender, M.A., Farach-Colton, M., Mosteiro, M.A. (eds.) LATIN 2018. LNCS, vol. 10807, pp. 504–514. Springer, Cham (2018). https://doi.org/10.1007/978-3-319-77404-6_37
14. Hlineny, P., Oum, S.I., Seese, D., Gottlob, G.: Width parameters beyond tree-width and their applications. Comput. J. **51**(3), 326–362 (2007)
15. Hüffner, F., Komusiewicz, C., Moser, H., Niedermeier, R.: Fixed-parameter algorithms for cluster vertex deletion. Theory Comput. Syst. **47**(1), 196–217 (2008)
16. Jansen, K., Scheffler, P., Woeginger, G.: The disjoint cliques problem. RAIRO-Oper. Res. **31**(1), 45–66 (1997)
17. Lackner, M., Pichler, R., Rümmele, S., Woltran, S.: Multicut on graphs of bounded clique-width. In: Lin, G. (ed.) COCOA 2012. LNCS, vol. 7402, pp. 115–126. Springer, Heidelberg (2012). https://doi.org/10.1007/978-3-642-31770-5_11
18. Lewis, R., Thiruvady, D., Morgan, K.: Finding happiness: an analysis of the maximum happy vertices problem. Comput. Oper. Res. **103**, 265–276 (2019)
19. Misra, N., Reddy, I.V.: The parameterized complexity of happy colorings. In: Brankovic, L., Ryan, J., Smyth, W.F. (eds.) IWOCA 2017. LNCS, vol. 10765, pp. 142–153. Springer, Cham (2018). https://doi.org/10.1007/978-3-319-78825-8_12
20. Papadimitriou, C.H., Steiglitz, K.: Combinatorial Optimization: Algorithms and Complexity. Prentice Hall, Upper Saddle River (1981)
21. Todinca, I.: Coloring powers of graphs of bounded clique-width. In: Bodlaender, H.L. (ed.) WG 2003. LNCS, vol. 2880, pp. 370–382. Springer, Heidelberg (2003). https://doi.org/10.1007/978-3-540-39890-5_32

22. Xu, Y., Goebel, R., Lin, G.: Submodular and supermodular multi-labeling, and vertex happiness. CoRR (2016)
23. Zhang, P., Jiang, T., Li, A.: Improved approximation algorithms for the maximum happy vertices and edges problems. In: Xu, D., Du, D., Du, D. (eds.) COCOON 2015. LNCS, vol. 9198, pp. 159–170. Springer, Cham (2015). https://doi.org/10.1007/978-3-319-21398-9_13
24. Zhang, P., Li, A.: Algorithmic aspects of homophyly of networks. Theor. Comput. Sci. **593**, 117–131 (2015)
25. Zhang, P., Xu, Y., Jiang, T., Li, A., Lin, G., Miyano, E.: Improved approximation algorithms for the maximum happy vertices and edges problems. Algorithmica **80**(5), 1412–1438 (2018)

Shortest Reconfiguration of Matchings

Nicolas Bousquet[1], Tatsuhiko Hatanaka[2], Takehiro Ito[2], and Moritz Mühlenthaler[2](✉)

[1] CNRS, Laboratoire G-SCOP, Grenoble-INP, Univ. Grenoble-Alpes, Grenoble, France
nicolas.bousquet@grenoble-inp.fr
[2] Fakultät für Mathematik, TU Dortmund University, Dortmund, Germany
moritz.muehlenthaler@math.tu-dortmund.de

Abstract. Imagine that unlabelled tokens are placed on edges forming a matching of a graph. A token can be moved to another edge provided that the edges containing tokens remain a matching. The *distance* between two configurations of tokens is the minimum number of moves required to transform one into the other. We study the problem of computing the distance between two given configurations. We prove that if source and target configurations are maximal matchings, then the problem admits no polynomial-time sublogarithmic-factor approximation algorithm unless $\mathsf{P} = \mathsf{NP}$. On the positive side, we show that for matchings of bipartite graphs the problem is fixed-parameter tractable parameterized by the size d of the symmetric difference of the two given configurations. Furthermore, we obtain a d^{ε}-factor approximation algorithm for the distance of two maximum matchings of bipartite graphs for every $\varepsilon > 0$. The proofs of our positive results are constructive and can hence be turned into algorithms that output shortest transformations. Both algorithmic results rely on a close connection to the Directed Steiner Tree problem. Finally, we show that determining the exact distance between two configurations is complete for the class $\mathsf{D^P}$, and determining the maximum distance between any two configurations of a given graph is $\mathsf{D^P}$-hard.

Keywords: Matchings · Reconfiguration · Fixed-parameter tractability · Approximation hardness

1 Introduction

A reconfiguration problem asks for the existence of a step-by-step transformation between two given configurations, where in each step we apply some simple modification to the current configuration. The set of configurations may for instance be the set of k-colorings [3,10] or independent sets [15,16,18] of a graph, or the

N. Bousquet—This author was partially supported by ANR project GrR (ANR-18-CE40-0032).
T. Ito—Partially supported by JST CREST Grant Number JPMJCR1402, and JSPS KAKENHI Grant Numbers JP18H04091 and JP19K11814, Japan.

I. Sau and D. M. Thilikos (Eds.): WG 2019, LNCS 11789, pp. 162–174, 2019.
https://doi.org/10.1007/978-3-030-30786-8_13

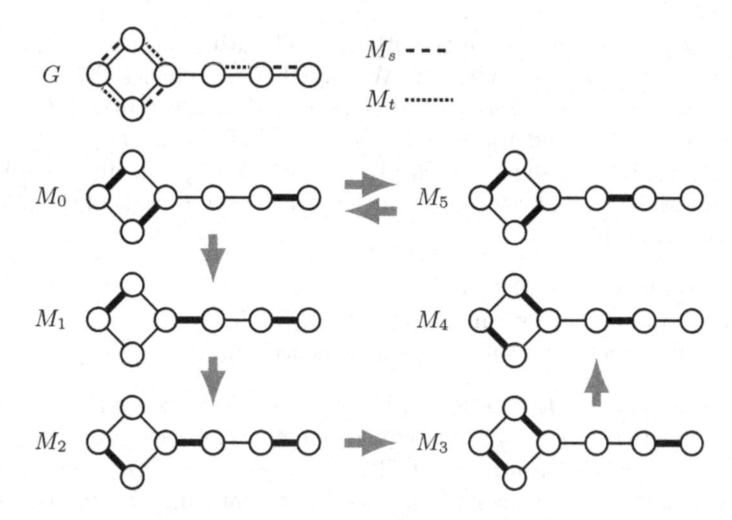

Fig. 1. A reconfiguration sequence $M_s = M_0, M_1, \ldots, M_4 = M_t$ of matchings in a graph G.

set of satisfying assignments of a Boolean formula [12]. A suitable modification may for example alter the color of a single vertex, or the truth value of a variable in a satisfying assignment. For recent surveys on reconfiguration problems the reader is referred to [14] or [20].

Recently, there has been considerable interest in the complexity of finding shortest transformations between configurations. Examples include finding a shortest transformation between triangulations of planar point sets [22] and simple polygons [1], configurations of the Rubik's cube [6], and satisfying assignments of Boolean formulas [19]. For all of these problems, except for the last one, we can decide efficiently if a transformation between two given configurations exists. However, deciding if there is a transformation of at most a given length is NP-complete. In particular, the flip distance of triangulations of planar point sets is known to be APX-hard [22] and, on the positive side, fixed-parameter tractable (FPT) in the length of the transformation [17]. Our reference problem is the task of computing the length of a shortest transformation between two matchings of a graph.

Reconfiguration of Matchings. A *matching* M of a graph is a set of pairwise independent edges. (Fig. 1 shows the six different matchings of the graph G.) A matching is *inclusion-wise maximal* if it is not properly contained in another matching. We may consider a matching as a placement of (unlabeled) *tokens* on independent edges. Then the *Token Jumping* (*TJ*) operation provides an adjacency relation on the set of matchings of a graph, all having the same cardinality[1]: Two matchings M and M' of a graph G are *adjacent* (under

[1] There is another well-studied operation called Token Sliding (TS). In this paper, we employ TJ as the default operation. However, some of our results apply also to TS.

TJ) if one can be obtained from the other by relocating a single token, that is, if $|M \setminus M'| = 1$ and $|M' \setminus M| = 1$. We say that a sequence $M_0, M_1, \ldots, M_\ell$ of matchings of G is a *reconfiguration sequence* of *length* ℓ from M to M', if $M_0 = M$, $M_\ell = M'$, and the matchings M_{i-1} and M_i are adjacent for each $i \in \{1, 2, \ldots, \ell\}$; see the sequence $M_0, M_1, \ldots, M_4$ in Fig. 1 as an example. The following question is often referred to as the *reachability variant* of the matching reconfiguration problem:

MATCHING RECONFIGURATION
Input: Graph G and two matchings M_s, M_t of G.
Question: Is there a reconfiguration sequence from M_s to M_t?

MATCHING RECONFIGURATION is known as an early example of a reconfiguration problem that admits a non-trivial polynomial-time algorithm [15]. For YES-instances, the algorithm from [15] gives a bound of $O(n^2)$ on the length of a transformation. The *distance* between two matchings is the length of a shortest transformation between them (under TJ). If there is no transformation between two matchings, we regard their distance as infinity. In this paper we study the complexity of the following optimization problem related to matching reconfiguration, which is also referred to as the *shortest variant.*

MATCHING DISTANCE
Input: Graph G and two matchings M_s, M_t of G.
Task: Compute the distance between M_s and M_t.

We also study two related *exact* problems. The first is the exact version of MATCHING DISTANCE, which takes as input also the supposed distance ℓ of the given matchings.

EXACT MATCHING DISTANCE
Input: Graph G, matchings M_s, M_t of G, and number $\ell \in \mathbb{N}$.
Question: Is ℓ equal to the distance between M_s and M_t?

We can similarly define EXACT MATCHING DIAMETER: Given a graph G and numbers $k, \ell \in \mathbb{N}$, decide if the maximum distance between any two matchings of G of cardinality k is equal to ℓ.

Related Results. Despite recent intensive studies on reconfiguration problems (see, e.g., [20]), most known algorithmic (positive) results are obtained for reachability variants. However, such algorithms sometimes also give answers to the corresponding shortest variants: if the algorithm constructs an actual reconfiguration sequence which, at any step, transforms an edge of the initial matching into an edge of the target one, then the transformation must indeed be a shortest one.

Generally speaking, finding shortest transformations is more difficult when we need a *detour*, a transformation that touches an element that is not in the symmetric difference of the source and target configurations. For such a detour-required case, only a few polynomial-time algorithms are known for shortest variants, e.g., satisfying assignments of a certain Boolean formulas [19], and independent sets under the TS operation for caterpillars [24]. Note that MATCHING RECONFIGURATION belongs to the detour-required case (in the example of Fig. 1, we need to use the edge in $E(G) \setminus (M_s \cup M_t)$ in any reconfiguration sequence).

The reconfiguration of matchings is a special case of the reconfiguration of independent sets of a graph. To see this, recall that matchings of a graph correspond to independent sets of its line graph. Therefore, by a result of Kamiński et al. [16], we can solve MATCHING DISTANCE in polynomial time if the line graph of a given graph is even-hole-free. Note that in this case no detour is required.

Our Results. Although the reconfiguration of independent sets is one of the most well-studied reconfiguration problems (see, e.g., a survey [20]), to the best of our knowledge, the shortest variant of independent sets under the TJ operation is known to be solvable only for even-hole-free graphs, as mentioned above. Thus, in this paper, we start a systematic study of the complexity of finding shortest reconfiguration sequences between matchings, and more generally, between independent sets of a graph.

Our first result shows that there is no polynomial-time algorithm that computes a sublogarithmic-factor approximation of the distance between two matchings unless $\mathsf{P} = \mathsf{NP}$. The result implies approximation hardness for the length of shortest transformations between b-matchings of a graph and shortest transformations between independent sets on any graph class containing line graphs of bipartite graphs. Note that maximal subclasses of line graphs of bipartite graphs that have been considered in the literature are either trivial (e.g., disjoint unions of cliques) or even-hole-free, so our result implies a sharp complexity bound.

Theorem 1. MATCHING DISTANCE *admits no polynomial-time $o(\log n)$-factor approximation algorithm, unless $\mathsf{P} = \mathsf{NP}$, even for maximal matchings of bipartite graphs of maximum degree three.*

The proof of Theorem 1 is provided in Sect. 2. Our main result is positive. We show that determining the distance between matchings of *bipartite* graphs is FPT, where the parameter is the size d of the symmetric difference of the two input matchings. An outline of the algorithm is provided in Sect. 3.1. We distinguish two main cases: either a shortest reconfiguration sequence contains a non-inclusion-wise maximal matching or not. For the former case we give a polynomial time algorithm. Note that this algorithm implies in particular a polynomial-time algorithm for finding a shortest transformation between two matchings if at least one of the two matchings is not inclusion-wise maximal. In order to deal with the latter case, we give a reduction from MATCHING DISTANCE to the problem DIRECTED STEINER TREE, where the number of terminals is of the order of the size of the symmetric difference of the source and target matchings. By putting everything together we obtain our main result.

Theorem 2. MATCHING DISTANCE *in bipartite graphs can be solved in time $2^d \cdot n^{O(1)}$, where d is the size of the symmetric difference of the two given matchings.*

Theorem 2 raises hopes for possible generalizations, e.g., an FPT algorithm for finding shortest transformations between matchings in general graphs or between independent sets of claw-free graphs. For maximum matchings, our reduction to DIRECTED STEINER TREE is approximation-preserving, which implies the following.

Corollary 1. MATCHING DISTANCE *restricted to maximum matchings in bipartite graphs admits a polynomial-time* d^{ε}*-factor approximation algorithm for every* $\varepsilon > 0$*, where* d *is the size of the symmetric difference of two given matchings.*

The proofs of Theorem 2 and Corollary 1 are given in Sect. 3.

We finally show that there is a polynomial-time algorithm that decides if the maximum distance between any two matchings of a graph is *finite*. In contrast, we also show that the problems EXACT MATCHING DISTANCE and EXACT MATCHING DIAMETER are both hard for the class $\mathsf{D^P}$, a class containing both NP and coNP.

Theorem 3. *The problem* EXACT MATCHING DISTANCE *is* $\mathsf{D^P}$*-complete and the problem* EXACT MATCHING DIAMETER *is* $\mathsf{D^P}$*-hard.*

The class $\mathsf{D^P}$ has been introduced by Papadimitriou and Yannakakis in [21] and is a natural complexity class for *exact* problems and *critical* problems. It was proved by Frieze and Teng that the related problem of deciding the diameter of the graph of a polyhedron is also $\mathsf{D^P}$-hard [11]. Section 4 is devoted to the proof of Theorem 3. Due to space restriction, proofs of statements marked by (∗) are not included in this extended abstract. We would like to remark that the NP-hardness of MATCHING DISTANCE has been established independently in [13].

Notation. Let $G = (V, E)$ be a graph. Unless stated otherwise, graphs are simple. For standard definitions and notation related to graphs, we refer the reader to [7]. We denote by $A \triangle B$, the symmetric difference of two sets A and B. That is, $A \triangle B := (A \setminus B) \cup (B \setminus A)$. Let $M \subseteq E$ be a matching of G. A vertex of G that is not incident to any edge in M is called M*-exposed* (or M*-free*), otherwise it is called *matched* or *covered.*

2 Approximation Hardness of Matching Distance

To prove Theorem 1, we first show the following slightly less general result.

Theorem 4. MATCHING DISTANCE *admits no* $o(\log n)$*-factor approximation unless* $\mathsf{P} = \mathsf{NP}$*, even when restricted to maximum matchings on bipartite graphs of maximum degree three.*

We show that a sublogarithmic-factor approximation for MATCHING DISTANCE yields a sublogarithmic-factor approximation of SET COVER. Since SET COVER is not approximable within a sublogarithmic factor, unless $\mathsf{P} = \mathsf{NP}$ [8],

Theorem 4 follows. Note that Theorem 4 also holds for the TS operation if M_1 and M_2 are maximum[2]. To obtain Theorem 1, we need to slightly alter the reduction used in the proof of Theorem 4 to transform maximum matchings into maximal matchings. (We refer the reader to the full version [4] for the detailed construction).

Let us briefly recall some definitions related to the SET COVER problem. An instance $I = (U, \mathcal{S})$ of SET COVER is given by a set U of *items* and a family $\mathcal{S}$ of subsets of U. The task is to find the minimum number of sets in $\mathcal{S}$ that are required to cover U. We denote this number by $\mathrm{OPT}(I)$ and let $n := |U|$ and $m := |\mathcal{S}|$. Furthermore, let $d := \max_{S \in \mathcal{S}}\{|S|\}$ be the maximum cardinality of a set in $\mathcal{S}$ and, for each $u \in U$, let $f_u := |\{S \in \mathcal{S} \mid u \in S\}|$ be the *frequency* of u, and let $f := \max_{u \in U}\{f_u\}$ be frequency of I.

Let us now give the reduction and the two main lemmas that guarantee the safeness of our construction.

Reduction. We construct from the SET COVER instance $I = (U, \mathcal{S})$ an instance $I' = (G, M_1, M_2)$ of MATCHING DISTANCE as follows. For each item $u \in U$, we create a cycle C_u of length four and label one of the vertices by c_u. Then, for each set $S \in \mathcal{S}$, we add a path P_S of length $L := |U|(2 + f + d)$ and label the end-points p_S and q_S. We may assume without loss of generality that L is even. Now, for each set $S \in \mathcal{S}$ and item $u \in U$, we join p_S to c_u by an edge if and only if $u \in S$. The two matchings M_1 and M_2 are constructed as follows. For each $S \in \mathcal{S}$, we leave q_S exposed and add every second other edge to both matchings. For each $u \in U$ there are two different perfect matchings of C_u and we add one to M_1 and the other to M_2. Note that the graph G is bipartite and, since L is even, only the vertices q_S are M_1- and M_2-exposed for $S \in \mathcal{S}$. In order to get the maximum degree down to three, we have to use a slightly more elaborate construction for the part of G that corresponds to the incidence graph of the SET COVER instance. This completes the construction of the instance I' of MATCHING DISTANCE.

Observe that the construction of I' performed in polynomial time. In order to change the matching on a cycle C_u, it is necessary to move each token on some path P_S such that $u \in S$. Intuitively, we think of L as a very large number, so in order to change the matching M_1 on each cycle C_u, $u \in U$, it is desirable to minimize the number of times we have to move the tokens on the long paths P_S, $S \in \mathcal{S}$. In order to obtain the approximation hardness result in Theorem 4, we establish the following correspondences between reconfiguration sequences from M_1 to M_2 and covers of U by sets in $\mathcal{S}$:

Lemma 1 (∗). *Let $C \subseteq \mathcal{S}$ be a cover of U. Then there is a reconfiguration sequence from M_1 to M_2 of length at most $2L|C| + 2|U|(2 + f + d)$.*

[2] If we delete an edge e of a maximum matching, we can only replace it by an edge e' sharing an endpoint with e. So for maximum matchings, TJ and TS rules are equivalent.

Lemma 2 (∗). *There is a polynomial-time algorithm A' that constructs from a reconfiguration sequence τ from M_1 to M_2 of length $|\tau|$ a cover $C \subseteq \mathcal{S}$ of U of cardinality at most $|\tau|/2L$.*

Combining Lemmas 1 and 2, we have that a $o(\log|I'|)$-factor approximation algorithm for Matching Distance yields a $o(\log n)$-factor approximation algorithm for Set Cover, which contradicts the approximation hardness result from [8].

3 Matching Distance in Bipartite Graphs is FPT

The goal of this section is Theorem 2, which states that the distance of two matchings of a bipartite graph is FPT, where the parameter is the size d of the symmetric difference of the source and target matchings. Let us fix an instance (G, M_s, M_t) of Matching Distance and let us assume that the graph $G = (V, E)$ is bipartite with bipartition $V = (U, W)$. According to [15, Proposition 1] we may check in polynomial time whether a transformation from M_s to M_t exists, so let us assume in the following such transformation exists. In the remainder of this section we denote by $\mathcal{C}$ (resp., $\mathcal{P}$) be the set of (M_s, M_t)-alternating cycles (resp., (M_s, M_t)-alternating paths) in $(V, M_s \triangle M_t)$.

3.1 Overview of the Algorithm

There are two distinct main cases we need to consider.

Case 1 (A shortest transformation from M_s to M_t visits a matching that is not inclusion-wise maximal).
We show that in this case we can find a shortest transformation from M_s to M_t in polynomial time, which may seem quite surprising in the light of the hardness result given in Theorem 1. We first observe that the following holds:

Lemma 3 (∗). *A shortest transformation between two matchings can be computed in polynomial time if at least one of them is not inclusion-wise maximal.*

To prove Lemma 3, we show that the distance of two matchings M_s and M_t, at least one of which is not inclusion-wise maximal, is either $|M_s \triangle M_t|/2$ or $|M_s \triangle M_t|/2 + 1$. We can check in polynomial time which case applies. Note that Lemma 3 also holds if we do not assume that the input graph is bipartite. The hard part of Case 1 is to prove the following lemma (which only holds for bipartite graphs):

Lemma 4 (∗). *There is a polynomial-time algorithm that outputs a shortest transformation from M_s to M_t via a matching M that is not inclusion-wise maximal, or indicates that no such transformation exists.*

Let us briefly summarize the algorithm. Since a shortest transformation from M_s to M_t visits a non-inclusion-wise maximal matching, we have that M_s and M_t cannot be maximum, so there is at least one M_s-augmenting path. Note that, given an M_s-augmenting path P, we may transform M_s into a non-inclusion-wise maximal matching by sliding tokens along P. We may then find a shortest transformation from the resulting matching to M_t in polynomial time according to Lemma 3. We show that it suffices to find an M_s-augmenting path that gives an overall shortest transformation. This task reduces to a shortest-path-computation in a suitable weighted digraph.

Case 2 (No shortest transformation from M_s to M_t visits a matching that is not inclusion-wise maximal).
Let us first assume that M_s and M_t are maximum. We reduce the task of finding a shortest transformation from M_s to M_t to the problem DIRECTED STEINER TREE, which is defined as follows.

DIRECTED STEINER TREE
Input: Directed graph $D = (V, A)$, integral arc weights $c \in \mathbb{Z}^A_{\geq 0}$, root vertex $r \in V$, and terminals $T \subseteq V$.
Task: Find a minimum-cost directed tree in D that connects the root r to each terminal.

The reduction to DIRECTED STEINER TREE and its correctness are sketched in Sect. 3.2. The main idea is to construct an instance of DIRECTED STEINER TREE such that each arc of positive cost corresponds to a token move and its cost corresponds to how many times the token has to be moved in a transformation from M_s to M_t. It is known that DIRECTED STEINER TREE parameterized by the number of terminals is FPT [2,9]. Our reduction gives at most $d/2$ terminals. We employ the FPT algorithm from [2] to obtain the following result.

Lemma 5. *Let M_s and M_t be maximum. Then there is an algorithm that finds in time $2^{d/2} \cdot n^{O(1)}$ a shortest transformation from M_s to M_t, or indicates that no such transformation exists.*

In order to deal with the case that M_s and M_t are not maximum, we first recall the following lemma from [15].

Lemma 6 ([15, Lemma 1]). *If M_s and M_t are maximum then there is a transformation from M_s to M_t if and only if, for each cycle $C \in \mathcal{C}$, there is an M_s-alternating path in G connecting an M_s-exposed vertex to C.*

Note that we assume that for each shortest transformation from M_s to M_t, each intermediate matching is inclusion-wise maximal. Therefore, in a shortest transformation from M_s to M_t, we *have* to use the algorithm from the constructive proof of Lemma 6 to transform the cycles in $\mathcal{C}$. Hence, in such a transformation, we have to consider for each cycle $C \in \mathcal{C}$ the two choices that C is reconfigured using either an M_s-exposed vertex from U or from W. Since each cycle has length at least four, we have that $\mathcal{C}$ contains at most $d/4$ cycles.

We branch over all of the at most $2^{d/4}$ possible choices. For each choice, we reduce the problem to the case where M_s and M_t are maximum as follows. We create two sub-instances: one for the cycles $\mathcal{C}_1$ that have to be reconfigured using exposed vertices in U and one for the cycles $\mathcal{C}_2$ that have to be reconfigured using exposed vertices in W. For the sub-instance of cycles in $\mathcal{C}_1$, we delete all the exposed vertices in W and the matching M_s then becomes maximum. (We perform a similar reduction in the other case). We finally show that no transformation maintaining maximal matchings all along is better than combining the optimal solutions of the two sub-instances and obtain the following result.

Lemma 7 (*). *Suppose no shortest transformation from M_s to M_t visits a matching that is not inclusion-wise maximal. Then there is an algorithm that finds in time $2^d \cdot n^{O(1)}$ a shortest transformation from M_s to M_t.*

Hence, Theorem 2 follows from Lemmas 4 and 7. Note that if M_s and M_t are maximum, then we may use the approximation algorithm for DIRECTED STEINER TREE from [5] instead of the exact algorithm from [2]. Since our reduction to DIRECTED STEINER TREE preserves costs, we obtain from an α-approximate solution of the DIRECTED STEINER TREE instance an α-approximate solution to MATCHING DISTANCE, which implies Corollary 1. Our techniques are not likely to generalize to matchings in non-bipartite graphs in a straight-forward way. We leave as an open problem whether finding a shortest transformation between two matchings in non-bipartite graphs is FPT in the size of the symmetric difference of source and target matchings.

3.2 Proof of Lemma 5: Reduction to Directed Steiner Tree

Let M_s and M_t be maximum matchings of G. We will reduce the task of finding a shortest transformation from M_s to M_t to the DIRECTED STEINER TREE problem. Note that if some edge e is not contained in any maximum matching of G, then we cannot move any token to e and e can be deleted from the graph. Therefore, we may assume that every edge of G is contained in some maximum matching. Let X_s be the set of M_s-exposed vertices of G. By the properties of the Edmonds-Gallai decomposition [23, Ch. 24.4b], we may assume the following[3].

Proposition 1 (*). *Without loss of generality we have $X_s \subseteq U$.*

Reduction. The main feature of the reduction is that it preserves costs. That is, an optimal Steiner tree of cost α corresponds to a shortest transformation from M_s to M_t of length at most α. We construct an instance $I' := (D, c, r, T)$ of DIRECTED STEINER TREE as follows. The digraph $D = (U', A)$ is given by

$$U' := \{v \in U \mid \exists \text{ an even-length } M_s\text{-alternating path from } X_s \text{ to } v\} \cup \{r\}$$
$$A := \{uw \mid u, w \in U, \exists v \in W : uv \in E \setminus M_s, vw \in M_s\} \cup R,$$

[3] Due to space restrictions, the definition of Edmonds-Gallai decomposition is not included in this extended abstract, see [4] for more details.

where r is a new vertex and $R := \{rv \mid v \in X_s\}$. For an arc $uw \in A$, let the weight c_{uw} be given by

$$c_{uw} := \begin{cases} 0 & \text{if } u = r, \\ 1 & \text{if there are two edges } uv \in M_t \text{ and } vw \text{ in } M_s. \\ 2 & \text{otherwise.} \end{cases}$$

The set T of terminals is given by $T := U' \cap \bigcup_{Z \in \mathcal{C} \cup \mathcal{P}} V(Z)$. Note that any two distinct items in $\mathcal{P} \cup \mathcal{C}$ are vertex-disjoint. The root of the Steiner tree is the vertex r. This completes the construction of the instance I'.

Let us now give an outline of the proof of Lemma 5, which states that finding a shortest transformation between two maximum matchings M_s and M_t of a bipartite graph is fixed parameter tractable, where the parameter is the size d of the symmetric difference of M_s and M_t.

We first observe that we may restrict our attention to Steiner trees with some structure on the paths $\mathcal{P}$ and cycles $\mathcal{C}$ of $(V, M_s \triangle M_t)$.

Proposition 2 (*). *Let F be a Steiner tree for I'. Then we can obtain in polynomial time a Steiner tree F' for I' of cost at most $c(F)$ with the following properties.*

(i) For each $P \in \mathcal{P}$, the tree F' contains all arcs in $A(P)$.
(ii) For each $C \in \mathcal{C}$, the tree F' misses exactly one arc of $A(C)$.
(iii) For each $P \in \mathcal{P}$, the root r is joined to the M_s-exposed vertex of P.

The proof of next lemma shows how to construct from a Steiner tree F' of cost $c(F')$ a reconfiguration sequence of length at most $c(F')$.

Lemma 8 (*). *Let F' be a Steiner tree for I'. Then we can construct in polynomial time a transformation from M_s to M_t of length at most $c(F')$.*

Let us sketch the proof of Lemma 8. If F' does not satisfy the properties of Proposition 2, then we may find in polynomial time a Steiner tree F for I' of cost at most $c(F')$ that does. We perform a DFS traversal of F giving a preference to visiting arcs with largest weight. Note that we visit each arc of F twice, once going "down" the tree and once going "up". Each arc of weight at least one corresponds to a token and the arc-weight specifies how often the corresponding token is moved during the traversal of F. When we traverse an arc of weight one going down the tree, then we move a token to its target destination, so we perform no token move when backtracking. On the other hand, an arc of weight two corresponds to moving a token away from its target position, so we have to undo the move when backtracking. The placement of the terminals on the items in $\mathcal{C} \cup \mathcal{P}$ ensures that after the traversal of F, all tokens have been moved to their target positions.

From Lemma 8 and the next lemma we may conclude that the shortest length of a transformation from M_s to M_t equals the optimal cost of a Steiner tree for I'.

Lemma 9 (*). *Let $M_0, M_1, \ldots, M_m$ be a transformation of length m from M_s to M_t. Then there is a Steiner tree F of I' such that $c(F) \leq m$.*

We combine our previous arguments to prove Lemma 5. The construction of I' can be performed in polynomial time. Moreover, the number of terminals of I' is at most $d/2$. So the Directed Steiner Tree algorithm from [2] computes an optimal Steiner tree F^* of I' in time $2^{d/2} \cdot n^{O(1)}$. Lemma 8 ensures that F^* can be turned into an transformation between M_s and M_t in polynomial time. Lemma 9 ensures that this transformation is of the shortest length. Finally, since the construction is polynomial, the approximation algorithm for Directed Steiner Tree from [5] yields Corollary 1.

4 Exact Distance and Diameter

We consider the problems Exact Matching Distance and Exact Matching Diameter. Before presenting our hardness results for these problems, we first prove that we can decide in polynomial if the maximal distance of any two matchings of a graph is finite. It will be convenient to consider the *reconfiguration graph* $\mathcal{M}_k(G)$ of matchings of a graph G, which is given as follows.

$$V(\mathcal{M}_k(G)) := \{M \subseteq E \mid M \text{ is a matching in } G, |M| = k\}$$
$$E(\mathcal{M}_k(G)) := \{MN \mid M, N \in V(\mathcal{M}_k(G)), |M \triangle N| = 2\}$$

We show that for $k \geq 0$ we can decide in polynomial time if $\mathcal{M}_k(G)$ is connected. First suppose that k is less than the size of a maximum matching of G. Then we can transform any matching of size k into one that is not inclusion-wise maximal by sliding tokens along an augmenting path. Hence, by the algorithm given in the proof of Lemma 3, the graph $\mathcal{M}_k(G)$ is connected. Now suppose that k is equal to the size of a maximum matching of G and consider the *Edmonds-Gallai decomposition* A, D, C of the vertex set of G, where C are the vertices that are covered by any maximum matching [23, Ch. 24.4b]. By showing that $\mathcal{M}_k(G)$ is connected if and only if the graph $G[C]$ has a unique perfect matching we obtain the following result.

Theorem 5 (*). *There is a polynomial-time algorithm that, given a graph G and a number $k \in \mathbb{N}$, decides if $\mathcal{M}_k(G)$ is connected.*

To obtain hardness results for Exact Matching Distance and Exact Matching Diameter it suffices to consider maximum matchings. By using a similar construction to the one from Sect. 2 we show that Exact Matching Distance and Exact Matching Diameter are hard for the class $\mathsf{D^P}$, which is given by $\mathsf{D^P} := \{L_1 \cap L_2 \mid L_1 \in \mathsf{NP}, L_2 \in \mathsf{coNP}\}$. We reduce from the problem Exact Vertex Cover, which asks whether the minimum size of a vertex cover of a graph is equal to a given number ℓ. Our construction guarantees that, if we can decide the size of a shortest transformation, then we can decide the size of a minimum vertex cover. Since Exact Matching Distance is in $\mathsf{D^P}$ (the question "is the distance of two matchings in $\mathcal{M}_k(G)$ at most ℓ" being in NP), Theorem 3 follows.

References

1. Aichholzer, O., Mulzer, W., Pilz, A.: Flip distance between triangulations of a simple polygon is NP-complete. Discret. Comput. Geom. **54**(2), 368–389 (2015). https://doi.org/10.1007/s00454-015-9709-7
2. Björklund, A., Husfeldt, T., Kaski, P., Koivisto, M.: Fourier meets Möbius: fast subset convolution. In: Proceedings of the Thirty-ninth Annual ACM Symposium on Theory of Computing, pp. 67–74. ACM (2007). https://doi.org/10.1145/1250790.1250801
3. Bonamy, M., Bousquet, N.: Recoloring graphs via tree decompositions. Eur. J. Comb. **69**, 200–213 (2018). https://doi.org/10.1016/j.ejc.2017.10.010
4. Bousquet, N., Hatanaka, T., Ito, T., Mühlenthaler, M.: Shortest reconfiguration of matchings. CoRR abs/1812.05419 (2018)
5. Charikar, M., et al.: Approximation algorithms for directed Steiner problems. J. Algorithms **33**(1), 73–91 (1999). https://doi.org/10.1006/jagm.1999.1042
6. Demaine, E.D., Eisenstat, S., Rudoy, M.: Solving the Rubik's cube optimally is NP-complete. In: 35th Symposium on Theoretical Aspects of Computer Science. STACS, vol. 96, pp. 24:1–24:13 (2018). https://doi.org/10.4230/LIPIcs.STACS.2018.24
7. Diestel, R.: Graph Theory, Graduate Texts in Mathematics, vol. 173, 3rd edn. Springer, Heidelberg (2005)
8. Dinur, I., Steurer, D.: Analytical approach to parallel repetition. In: Proceedings of the 46th Annual ACM Symposium on Theory of Computing, pp. 624–633. ACM, New York (2014). https://doi.org/10.1145/2591796.2591884
9. Dreyfus, S.E., Wagner, R.A.: The Steiner problem in graphs. Networks **1**(3), 195–207 (1971). https://doi.org/10.1002/net.3230010302
10. Feghali, C., Johnson, M., Paulusma, D.: A reconfigurations analogue of Brooks' theorem and its consequences. J. Graph Theory **83**(4), 340–358 (2016)
11. Frieze, A.M., Teng, S.H.: On the complexity of computing the diameter of a polytope. Comput. Complex. **4**(3), 207–219 (1994). https://doi.org/10.1007/BF01206636
12. Gopalan, P., Kolaitis, P.G., Maneva, E.N., Papadimitriou, C.H.: The connectivity of Boolean satisfiability: computational and structural dichotomies. SIAM J. Comput. 2330–2355 (2009). https://doi.org/10.1137/07070440X
13. Gupta, M., Kumar, H., Misra, N.: On the complexity of optimal matching reconfiguration. In: Catania, B., Královič, R., Nawrocki, J., Pighizzini, G. (eds.) SOFSEM 2019. LNCS, vol. 11376, pp. 221–233. Springer, Cham (2019). https://doi.org/10.1007/978-3-030-10801-4_18
14. van den Heuvel, J.: The complexity of change. In: Surveys in Combinatorics 2013, pp. 127–160. Cambridge University Press (2013)
15. Ito, T., et al.: On the complexity of reconfiguration problems. Theor. Comput. Sci. **412**(12–14), 1054–1065 (2011). https://doi.org/10.1016/j.tcs.2010.12.005
16. Kamiński, M., Medvedev, P., Milanič, M.: Complexity of independent set reconfigurability problems. Theor. Comput. Sci. **439**, 9–15 (2012). https://doi.org/10.1016/j.tcs.2012.03.004
17. Li, S., Feng, Q., Meng, X., Wang, J.: An improved FPT algorithm for the flip distance problem. In: 42nd International Symposium on Mathematical Foundations of Computer Science (MFCS 2017), vol. 83, pp. 65:1–65:13 (2017). https://doi.org/10.4230/LIPIcs.MFCS.2017.65

18. Lokshtanov, D., Mouawad, A.E.: The complexity of independent set reconfiguration on bipartite graphs. In: Proceedings of the Twenty-Ninth Annual ACM-SIAM Symposium on Discrete Algorithms (SODA), pp. 185–195. SIAM (2018)
19. Mouawad, A.E., Nishimura, N., Pathak, V., Raman, V.: Shortest reconfiguration paths in the solution space of Boolean formulas. SIAM J. Discret. Math. **31**(3), 2185–2200 (2017). https://doi.org/10.1137/16M1065288
20. Nishimura, N.: Introduction to reconfiguration. Algorithms **11**(4:52) (2018). https://doi.org/10.3390/a11040052
21. Papadimitriou, C.H., Yannakakis, M.: The complexity of facets (and some facets of complexity). J. Comput. Syst. Sci. **28**(2), 244–259 (1984). https://doi.org/10.1016/0022-0000(84)90068-0
22. Pilz, A.: Flip distance between triangulations of a planar point set is APX-hard. Comput. Geom. **47**(5), 589–604 (2014). https://doi.org/10.1016/j.comgeo.2014.01.001
23. Schrijver, A.: Combinatorial Optimization - Polyhedra and Efficiency, Algorithms and Combinatorics, vol. 24. Springer, Heidelberg (2003)
24. Yamada, T., Uehara, R.: Shortest reconfiguration of sliding tokens on a caterpillar. In: Kaykobad, M., Petreschi, R. (eds.) WALCOM 2016. LNCS, vol. 9627, pp. 236–248. Springer, Cham (2016). https://doi.org/10.1007/978-3-319-30139-6_19

Travelling on Graphs with Small Highway Dimension

Yann Disser[1], Andreas Emil Feldmann[2(✉)], Max Klimm[3], and Jochen Könemann[4]

[1] TU Darmstadt, Darmstadt, Germany
disser@mathematik.tu-darmstadt.de
[2] Charles University in Prague, Prague, Czechia
feldmann.a.e@gmail.com
[3] Humboldt-Universität zu Berlin, Berlin, Germany
max.klimm@hu-berlin.de
[4] University of Waterloo, Waterloo, Canada
jochen@uwaterloo.ca

Abstract. We study the Travelling Salesperson (TSP) and the Steiner Tree problem (STP) in graphs of low highway dimension. This graph parameter was introduced by Abraham et al. [SODA 2010] as a model for transportation networks, on which TSP and STP naturally occur for various applications in logistics. It was previously shown [Feldmann et al. ICALP 2015] that these problems admit a quasi-polynomial time approximation scheme (QPTAS) on graphs of constant highway dimension. We demonstrate that a significant improvement is possible in the special case when the highway dimension is 1, for which we present a fully-polynomial time approximation scheme (FPTAS). We also prove that STP is weakly NP-hard for these restricted graphs. For TSP we show NP-hardness for graphs of highway dimension 6, which answers an open problem posed in [Feldmann et al. ICALP 2015].

Keywords: Travelling Salesperson · Steiner Tree · Highway dimension · Approximation scheme · NP-hardness

1 Introduction

Two fundamental optimization problems already included in Karp's initial list of 21 NP-complete problems [33] are the Travelling Salesperson

Y. Disser—Supported by the 'Excellence Initiative' of the German Federal and State Governments and the Graduate School CE at TU Darmstadt.
A. E. Feldmann—Supported by the Czech Science Foundation GAČR (grant #17-10090Y), and by the Center for Foundations of Modern Computer Science (Charles Univ. project UNCE/SCI/004).
M. Klimm—Supported by the German Research Foundation (DFG) as part of Math$^+$ (project AA3-4).
J. Könemann—Supported by the Discovery Grant Program of the Natural Sciences and Engineering Research Council of Canada.

I. Sau and D. M. Thilikos (Eds.): WG 2019, LNCS 11789, pp. 175–189, 2019.
https://doi.org/10.1007/978-3-030-30786-8_14

problem (TSP) and the STEINER TREE problem (STP). Given an undirected graph $G = (V, E)$ with non-negative edge weights $w : E \to \mathbb{R}^+$, the TSP asks to find the shortest closed walk in G visiting all nodes of V. Besides its fundamental role in computational complexity and combinatorial optimization, this problem has a variety of applications ranging from circuit manufacturing [29,41] and scientific imaging [14] to vehicle routing problems [40] in transportation networks. For the STP, a subset $R \subseteq V$ of nodes is marked as *terminals*. The task is to find a weight-minimal connected subgraph of G containing the terminals. It has plenty of fundamental applications in network design including telecommunication networks [42], computer vision [20], circuit design [30], and computational biology [22,43], but also lies at the heart of line planning in public transportation [17].

Both TSP and STP are APX-hard in general [6,13,21,34,39,45] implying that, unless P = NP, none of these problems admit a *polynomial-time approximation scheme (PTAS)*, i.e., an algorithm that computes a $(1+\varepsilon)$-approximation in polynomial time for any given constant $\varepsilon > 0$. On the other hand, for restricted inputs PTASs do exist, e.g., for planar graphs [5,18,28,36], Euclidean and Manhattan metrics [7,44], and more generally low doubling[1] metrics [8].

We study another class of graphs captured by the notion of *highway dimension*, which was proposed by Abraham et al. [3]. This graph parameter models transportation networks and is thus of particular importance in terms of applications for both TSP and STP. On a high level, the highway dimension is based on the empirical observation of Bast et al. [9,10] that travelling from a point in a network to a sufficiently distant point on a shortest path always passes through a sparse set of "hubs". The following formal definition is taken from [25] and follows the lines of Abraham et al. [3].[2] Here the *distance* between two vertices is the length of the shortest path between them, according to the edge weights. The *ball* $B_v(r)$ of radius r around a vertex v contains all vertices with distance at most r from v.

Definition 1. *For a scale* $r \in \mathbb{R}_{>0}$, *let* $\mathcal{P}_{(r,2r]}$ *denote the set of all vertex sets of shortest paths with length in* $(r, 2r]$. *A* shortest path cover *for scale* r *is a hitting set for* $\mathcal{P}_{(r,2r]}$, *i.e., a set* $\textsc{spc}(r) \subseteq V$ *such that* $|\textsc{spc}(r) \cap P| \neq \emptyset$ *for all* $P \in \mathcal{P}_{(r,2r]}$. *The vertices of* $\textsc{spc}(r)$ *are the* hubs *for scale* r. *A shortest path cover* $\textsc{spc}(r)$ *is* locally *h-sparse, if* $|\textsc{spc}(r) \cap B_v(2r)| \leq h$ *for all vertices* $v \in V$. *The* highway dimension *of* G *is the smallest integer* h *such that there is a locally* h*-sparse shortest path cover* $\textsc{spc}(r)$ *for every scale* $r \in \mathbb{R}_{>0}$ *in* G.

The algorithmic consequences of this graph parameter were originally studied in the context of road networks [1–3], which are conjectured to have fairly small

[1] A metric is said to have *doubling dimension* d if for all $r > 0$ every ball of radius r can be covered by at most 2^d balls of half the radius $r/2$.

[2] It is often assumed that all shortest paths are unique when defining the highway dimension, since this allows good polynomial approximations of this graph parameter [2]. In this work however, we do not rely on these approximations, and thus do not require uniqueness of shortest paths.

highway dimension. Road networks are generally non-planar due to overpasses and tunnels, and are also not Euclidean due to different driving or transmission speeds. This is even more pronounced in public transportation networks, where large stations have many incoming connections and plenty of crossing links, making Euclidean (or more generally low doubling) and planar metrics unsuitable as models. Here the highway dimension is better suited, since longer connections are serviced by larger and sparser stations (such as train stations and airports) that can act as hubs.

The main question posed in this paper is whether the structure of graphs with low highway dimension admits PTASs for problems such as TSP and STP, similar to Euclidean or planar instances. It was shown that *quasi-polynomial time approximation schemes (QPTASs)* exist for these problems [24], i.e., $(1+\varepsilon)$-approximation algorithms with runtime $2^{\mathrm{polylog}(n)}$ assuming that ε and the highway dimension of the input graph are constants. However it was left open whether this can be improved to polynomial time.

1.1 Our Results

Our main result concerns graphs of the smallest possible highway dimension, and shows that for these *fully polynomial time approximation schemes (FPTASs)* exist, i.e., a $(1+\varepsilon)$-approximation can be computed in time polynomial in both the input size and $1/\varepsilon$. Thus at least for this restricted case we obtain a significant improvement over the previously known QPTAS [24].

Theorem 2. *Both* Travelling Salesperson *and* Steiner Tree *admit an FPTAS on graphs with highway dimension* 1.

From an application point of view, so-called hub-and-spoke networks that can typically be seen in air traffic networks can be argued to have very small highway dimension close to 1: their star-like structure implies that hubs are needed at the centers of stars only, where all shortest paths converge. From a more theoretical viewpoint, we show that surprisingly the STP problem is non-trivial on graphs highway dimension 1, since it is still NP-hard even on this very restricted case. Interestingly, together with Theorem 2 this implies [49] that STP is *weakly* NP-hard on graphs of highway dimension 1. This is in contrast to planar graphs or Euclidean metrics, for which the problem is strongly NP-hard.

Theorem 3. *The* Steiner Tree *problem is weakly* NP*-hard on graphs with highway dimension* 1.[3]

It was in fact left as an open problem in [24] to determine the hardness of STP and also TSP on graphs of constant highway dimension. Theorem 3 settles this question for STP. We also answer the question for TSP, but in this case we are not able to bring down the highway dimension to 1 so that the following theorem does not complement Theorem 2 tightly.

Theorem 4. *The* Travelling Salesperson *problem is* NP*-hard on graphs with highway dimension* 6.

[3] The proofs of Theorems 3 and 4 are deferred to the full version of the paper.

1.2 Techniques

We present a step towards a better understanding of low highway dimension graphs by giving new structural insights on graphs of highway dimension 1. It is not hard to find examples of (weighted) complete graphs with highway dimension 1 (cf. [24]), and thus such graphs are not minor-closed. Nevertheless, it was suggested in [24] that the *treewidth* of low highway dimension graphs might be bounded polylogarithmically in terms of the *aspect ratio* α, which is the maximum distance divided by the minimum distance between any two vertices of the input graph.

Definition 5. *A* tree decomposition *of a graph* $G = (V, E)$ *is a tree* D *where each node* v *is labelled with a bag* $X_v \subseteq V$ *of vertices of* G*, such that the following holds: (a)* $\bigcup_{v \in V(D)} X_v = V$*, (b) for every edge* $\{u, w\} \in E$ *there is a node* $v \in V(D)$ *such that* X_v *contains both* u *and* w*, and (c) for every* $v \in V$ *the set* $\{u \in V(D) \mid v \in X_u\}$ *induces a connected subtree of* D*. The* width *of the tree decomposition is* $\max\{|X_v| - 1 \mid v \in V(D)\}$*. The* treewidth *of a graph* G *is the minimum width among all tree decompositions for* G*.*

As suggested in [24], one may hope to prove that the treewidth of any graph of highway dimension h is, say, $O(h \,\mathrm{polylog}(\alpha))$. As argued in Sect. 4, it unfortunately is unlikely that such a bound is generally possible. In contrast to this, our main structural insight on graphs of highway dimension 1 is that they have treewidth $O(\log \alpha)$. This implies FPTASs for TSP and STP, since we may reduce the aspect ratio of any graph with n vertices to $O(n/\varepsilon)$ and then use algorithms by Bodlaender et al. [16] to compute optimum solutions to TSP and STP in graphs of treewidth t in $2^{O(t)}n$ time. Since reducing the aspect ratio distorts the solution by a factor of $1 + \varepsilon$, this results in an approximation scheme. Although these are fairly standard techniques for metrics (cf. [24]), in our case we need to take special care, since we need to bound the treewidth of the graphs resulting from this reduction, which the standard techniques do not guarantee.

It remains an intriguing open problem to understand the complexity and structure of graphs of constant highway dimension larger than 1.

1.3 Related Work

The Travelling Salesperson problem (TSP) is among Karp's initial list of 21 NP-complete problems [33]. For general metric instances, the best known approximation algorithm is due to Christofides [23] and computes a solution with cost at most 3/2 times the LP value. For unweighted instances, the best known approximation guarantee is 7/5 and is due to Sebő and Vygen [47]. In general the problem is APX-hard [34,39,45]. For geometric instances where the nodes are points in $\mathbb{R}^d$ and distances are given by some l_p-norm, there exists a PTAS [4,44] for fixed d. When $d = \log n$, the problem is APX-hard [48]. Krauthgamer and Lee [38] generalized the PTAS to hyperbolic space. Grigni et al. [28] gave a PTAS for unweighted planar graphs which was later generalized by Arora et al. [5] to the weighted case. For improvements of the running time see Klein [36].

The STEINER TREE problem (STP) is contained in Karp's list of NP-complete problems as well [33]. The best approximation algorithm for general metric instances is due to Byrka et al. [19] and computes a solution with cost at most $\ln(4) + \epsilon < 1.39$ times that of an LP relaxation. Their algorithm improved upon previous results by, e.g., Robins and Zelikovsky [46] and Hougardy and Prömel [32]. Also the STP is APX-hard [21] in general. For Euclidean distances and nodes in $\mathbb{R}^d$ with d constant there is a PTAS due to Arora [4]. For $d = \log|R|/\log\log|R|$ where R is the terminal set, the problem is APX-hard [48]. For planar graphs, there is a PTAS for STP [18], and even for the more general STEINER FOREST problem for graphs with bounded genus [11]. Note that STP remains NP-complete for planar graphs [27].

It is worth mentioning that alternate definitions of the highway dimension exist.[4] In particular, in a follow-up paper to [3], Abraham et al. [1] define a version of the highway dimension, which implies that the graphs also have bounded doubling dimension. A related model for transportation networks was given by Kosowski and Viennot [37] via the so-called *skeleton dimension*, which also implies bounded doubling dimension. Hence for these definitions, Bartal et al. [8] already provide a PTAS for TSP. The highway dimension definition used here (cf. Definition 1) on the other hand allows for metrics of large doubling dimension as noted by Abraham et al. [3]: a star has highway dimension 1 (by using the center vertex to hit all paths), but its doubling dimension is unbounded. While it may be reasonable to assume that road networks (which are the main concern in the works of Abraham et al. [1–3]) have low doubling dimension, there are metrics modelling transportation networks for which it can be argued that the doubling dimension is large, while the highway dimension should be small. These settings are better captured by Definition 1. For instance, the so-called hub-and-spoke networks that can typically be seen in air traffic networks are star-like networks and are unlikely to have small doubling dimension while still having very small highway dimension close to 1. Thus in these examples it is reasonable to assume that the doubling dimension is a lot larger than the highway dimension.

Feldmann et al. [24] showed that graphs with low highway dimension can be embedded into graphs with low treewidth. This embedding gives rise to a QPTAS for both TSP and STP but also other problems. However, the result in [24] is only valid for a less general definition of the highway dimension from [2], i.e., there are graphs which have constant highway dimension according to Definition 1 but for which the algorithm of [24] cannot be applied. For the less general definition from [2], Becker et al. [12] give a PTAS for BOUNDED-CAPACITY VEHICLE ROUTING in graphs of bounded highway dimension. Also the k-CENTER problem has been studied on graphs of bounded highway dimension, both for the less general definition [12] and the more general one used here [25,26].

[4] See [24, Section 9] and [15] for detailed discussions on different definitions of the highway dimension.

2 Structure of Graphs with Highway Dimension 1

In this section, we analyse the structure of graphs with highway dimension 1. To this end, let us fix a graph G with highway dimension 1 and a shortest path cover $\text{SPC}(r)$ for each scale $r \in \mathbb{R}^+$. As a preprocessing, we remove edges that are longer than the shortest path between their endpoints, so that the triangle inequality holds.

We begin by analysing the structure of the graph $G_{\leq 2r}$, which is spanned by all edges of the input graph G of length at most $2r$. If G has highway dimension 1 it exhibits the following key property.

Lemma 6. *Let G be a metric graph with highway dimension 1, $r \in \mathbb{R}^+$ a scale, and $\text{SPC}(r)$ a shortest path cover for scale r. Then, every connected component of $G_{\leq 2r}$ contains at most one hub.*

Proof. For the sake of contradiction, let $r \in \mathbb{R}^+$ and let $x, y \in \text{SPC}(r)$ be a closest pair of distinct hubs in some component of $G_{\leq 2r}$. Let further P be a shortest path in $G_{\leq 2r}$ between x and y using only edges of length at most $2r$. (Note that P need not be a shortest path between x and y in G.) In particular, there is no other hub from $\text{SPC}(r) \setminus \{x, y\}$ along P. This implies that every edge of P that is not incident to either x or y must be of length at most r, since otherwise the edge would be a shortest path of length $(r, 2r]$ between its endpoints (using that G is metric) contradicting the fact that $\text{SPC}(r)$ is a shortest path cover for scale r.

Since the highway dimension of G is 1, any ball $B_w(2r)$ around a vertex $w \in V(P)$ contains at most one of the hubs $x, y \in \text{SPC}(r)$. Let $x', y' \in P$ be the vertices indicent to x and y along P, respectively. Since the length of the edge $\{x, x'\}$ is at most $2r$, the ball $B_{x'}(2r)$ must contain x and, by the observation above, it cannot contain y (in particular $\{x, y\}$ is not an edge). Symmetrically, the ball $B_{y'}(2r)$ contains y but not x. Consequently, $x' \neq y'$ and neither of these two vertices can be a hub of scale r, i.e., the path P contains at least two vertices different from x and y.

Let $V_x = \{w \in V : \text{dist}(x, w) < \text{dist}(y, w)\}$ contain all vertices closer to x than to y, where $\text{dist}(\cdot, \cdot)$ refers to the distance in the original graph G. As all edge weights are strictly positive, we have that $\text{dist}(x, y) > 0$ and thus $y \notin V_x$. Since P starts with vertex $x \in V_x$ and ends with vertex $y \notin V_x$ we deduce that there is an edge $\{u, v\}$ of P such that $u \in V_x$ and $v \notin V_x$. In particular, $\text{dist}(x, u) < \text{dist}(y, u)$ and $\text{dist}(y, v) \leq \text{dist}(x, v)$. We must have $\{u, v\} \neq \{y', y\}$, since otherwise $\text{dist}(x, y') < \text{dist}(y, y') \leq 2r$ and hence $B_{y'}(2r)$ would contain x. Similarly, we have $\{u, v\} \neq \{x, x'\}$, since otherwise $B_{x'}(2r)$ would contain y. Note that, by definition, $u \neq y$ and $v \neq x$, and hence $x, y \notin \{u, v\}$. Consequently, since every edge of P not incident to either x or y must have length at most r, we conclude that $\{u, v\}$ has length at most r.

Finally, consider the scale $r' \in \mathbb{R}^+$, defined such that $2r' = \text{dist}(x, u) + \text{dist}(u, v)$. Let Q and Q' denote shortest paths between x, u and v, y in G, respectively. Then the ball $B_v(2r')$ around v contains Q by definition of r'. From $\text{dist}(y, v) \leq \text{dist}(x, v) \leq \text{dist}(x, u) + \text{dist}(u, v) = 2r'$ it follows that

$B_v(2r')$ contains Q' as well. Also, $\text{dist}(y, v) \leq \text{dist}(x, v)$ means that $B_v(2r)$ cannot contain x, and hence $2r' = \text{dist}(x, u) + \text{dist}(u, v) \geq \text{dist}(x, v) > 2r$, which implies $r' > r$. W.l.o.g., assume that $\text{dist}(x, u) \leq \text{dist}(v, y)$ (otherwise consider scale $2r' = \text{dist}(y, v) + \text{dist}(u, v)$ and the ball $B_u(2r')$). Our earlier observation that $\text{dist}(u, v) \leq r$ with $r < r'$ then yields $\text{dist}(v, y) \geq \text{dist}(x, u) = 2r' - \text{dist}(u, v) > r'$. In other words, the lengths of both paths Q and Q' are in $(r', 2r']$, and so they both need to contain a hub of $\text{SPC}(r')$. However, by definition of u, v, the paths Q and Q' are vertex disjoint, which means that the ball $B_v(2r')$, which contains Q and Q', also contains at least two hubs from $\text{SPC}(r')$. This is a contradiction with G having highway dimension 1. □

Given a graph G, we now consider graphs $G_{\leq 2r}$ for exponentially growing scales. In particular, for any integer $i \geq 0$ we define the scale $r_i = 2^i$ and call a connected component of $G_{\leq 2r_i}$ a *level-i component.* Note that the level-i components partition the graph G, and that the level-i components are a *refinement* of the level-$(i+1)$ components, i.e., every level-i component is contained in some level-$(i+1)$ component. W.l.o.g., we scale the edge weights of the graph such that $\min_{e \in E} w(e) = 3$, so that there are no edges on level 0, and every level-0 component is a singleton. Let $\alpha = \frac{\max_{u \neq v} \text{dist}(u,v)}{\min_{u \neq v} \text{dist}(u,v)} = \frac{\max_{u \neq v} \text{dist}(u,v)}{3}$ be the aspect ratio of G. In our applications we may assume that G is connected, so that there is exactly one level-$(1 + \lceil \log_2(\alpha) \rceil)$ component containing all of G.

Since every edge is a shortest path between its endpoints, every edge $e = \{u, v\}$ that connects a vertex u of a level-i component C with a vertex v outside C is hit by a hub of $\text{SPC}(r_j)$, where j is the level for which $w(e) \in (r_j, 2r_j]$. Moreover, since v lies outside C, we have $w(e) > 2r_i$ and, thus, $j \geq i + 1$. The following definition captures the set of the hubs through which edges can possibly leave C.

Definition 7. *Let C be a level-i component of G. We define the set of* interface points *of C as $I_C := \bigcup_{j \geq i} \{u \in \text{SPC}(r_j) : \text{dist}_C(u) \leq 2r_j\}$, where $\text{dist}_C(u)$ denotes the minimum distance from u to a vertex in C (if $u \in C$, $\text{dist}_C(u) = 0$).*

Note that, for technical reasons, we explicitly add every hub at level i of a component to its set of interface points as well, even if such a hub does not connect the component with any vertex outside at distance more than $2r_i$.

Lemma 8. *If G has highway dimension 1, then each interface I_C of a level-i component C contains at most one hub for each level $j \geq i$.*

Proof. Assume that there are two hubs $u, v \in \text{SPC}(r_j)$ in I_C, and recall that we preprocessed the graph so that the triangle inequality holds. Then u and v must be contained in the same level-j component C', since u and v are connected to C with edges of length at most $2r_j$ (or are contained in C) and $C \subseteq C'$. This contradicts Lemma 6. □

Using level-i components and their interface points we can prove that the treewidth of a graph with highway dimension 1 is bounded in terms of the aspect ratio.

Lemma 9. *If a graph G has highway dimension 1 and aspect ratio α, its treewidth is at most $1 + \lceil \log_2(\alpha) \rceil$.*

Proof. The tree decomposition of G is given by the refinement property of level-i components. That is, let D be a tree that contains a node v_C for every level-i component C for all levels $0 \leq i \leq 1 + \lceil \log_2(\alpha) \rceil$. For every node v_C we add an edge in D to node $v_{C'}$, if C is a level-i component, C' is a level-$(i+1)$ component, and $C \subseteq C'$. The bag X_C for node v_C contains the interface points I_C. For a level-0 component C the bag X_C additionally contains the single vertex u contained in C.

Clearly, the tree decomposition has Property (a) of Definition 5, since the level-0 components partition the vertices of G and every vertex of G is contained in a bag X_C corresponding to a level-0 component C. Also, Property (b) is given by the bags X_C for level-0 components C, since for every edge e of G one of its endpoints u is a hub of $\text{SPC}(r_i)$ where i is such that $w(e) \in (r_i, 2r_i]$, and the other endpoint w is contained in a level-0 component C, for which X_C contains u and w.

For Property (c), first consider a vertex u of G, which is not contained in any set of interface points for any level-i component and any $0 \leq i \leq \log_2(\alpha)$. Such a vertex only appears in the bag X_C for the level-0 component C containing u, and thus the node v_C for which the bag contains u trivially induces a connected subtree of D.

Any other vertex u of G is an interface point. Let i be the highest level for which $u \in I_C$ for some level-i component C. We claim that $u \in C$, which implies that C is the unique level-i component containing u in its interface. To show our claim, assume $u \notin C$. Then, by definition, I_C contains u because $u \in \text{SPC}(r_j)$ for some $j \geq i$ and u has some neighbour at distance at most $2r_j$ in C. Since we preprocessed the graph such that every edge is a shortest path between its endpoints, this means that there must be an edge $e = \{u, v\}$ with $w(e) \in (r_j, 2r_j]$ and $v \in C$. Since $u \notin C$, we have $i < j$. Let C' be the unique level-j component with $C \subseteq C'$. Then, by definition, $u \in I_{C'}$, which contradicts the maximality of i. This proves our claim and shows that the highest level component C with $u \in X_C$ is uniquely defined. Moreover, we obtain $u \in \text{SPC}(r_i)$.

Now consider a level-i' component C' with $i' < i$, such that $u \in X_{C'}$, and let C'' be the unique level-$(i'+1)$ component containing C'. We claim that $u \in X_{C''}$. If $u \in C' \subseteq C''$, then $u \in X_{C''}$, since $u \in \text{SPC}(r_i)$, $\text{dist}_{C''}(u) = 0 \leq 2r_i$ and $i' + 1 \leq i$. If $u \notin C'$, then $u \in X_{C'}$ implies $u \in I_{C'}$, which means that there must be a vertex $w \in C'$ with $\text{dist}(u, w) \leq 2r_i$. But then $w \in C''$ and thus $\text{dist}_{C''}(u) \leq 2r_i$. Together with $u \in \text{SPC}(r_i)$, this implies $u \in X_{C''}$, as claimed. Since $v_{C'}$ is a child of $v_{C''}$ in the tree D, it follows inductively that the nodes of D with bags containing u induce a subtree of D with root v_C, which establishes Property (c).

By Lemma 8 each set of interface points contains at most one hub of each level. Since all edges have length at least 3, there are no hubs in $\text{SPC}(r_0)$ on level 0. This means that each bag of the tree decomposition contains at most $1+\lceil \log_2(\alpha) \rceil$

interface points. The bags for level-0 components contain one additional vertex. Thus the treewidth of G is at most $1 + \lceil \log_2(\alpha) \rceil$, as claimed. □

An additional property that we will exploit for our algorithms is the following. A (μ, δ)-*net* $N \subseteq V$ is a subset of vertices such that (a) the distance between any two distinct *net points* $u, w \in N$ is more than μ, and (b) for every vertex $v \in V$ there is some net point $w \in N$ at distance at most δ. For graphs of highway dimension 1 however, we can obtain nets with additional favourable properties, as the next lemma shows.

Lemma 10. *For any graph G of highway dimension* 1 *and any $r > 0$, there is an $(r, 3r)$-net such that every connected component of $G_{\leq r}$ contains exactly one net point. Moreover this net can be computed in polynomial time.*

Proof. We first derive an upper bound of $3r$ for the diameter of any connected component of $G_{\leq r}$. Lemma 6 implies that a connected component C contains at most one hub x of $\text{SPC}(r/2)$. By definition, any shortest path in C of length in $(r/2, r]$ must pass through x. We also know that every edge of C has length at most r. Consequently, every edge in C not incident to x must have length at most $r/2$, since each edge constitutes a shortest path between its endpoints. This implies that any shortest path in C that is not hit by x must have length at most $r/2$: if C contains a shortest path P with length more than $r/2$ not containing x we could repeatedly remove edges of length at most $r/2$ from P until we obtain a shortest path of length in $(r/2, r]$ not hit by x, a contradiction. Now consider a shortest path P in G of length more than $r/2$ from some vertex $v \in C$ to x (note that this path may not be entirely contained in C). Let $\{u, w\}$ be the unique edge of P such that $\text{dist}(v, u) \leq r/2$ and $\text{dist}(v, w) > r/2$. If the length of the edge $\{u, w\}$ is at most $r/2$ then $\text{dist}(v, w) \leq r$, and thus $w = x$, since the part of the path from v to w is a shortest path of length in $(r/2, r]$ and thus needs to pass through x. Otherwise the length of the edge $\{u, w\}$ is in the interval $(r/2, r]$, which again implies $w = x$, since the edge must contain x. In either case, $\text{dist}(v, x) \leq 3r/2$. This implies that every vertex in C is at distance at most $3r/2$ from x, and thus the diameter of C is at most $3r$.

To compute the $(r, 3r)$-net, we greedily pick an arbitrary vertex of each connected component of $G_{\leq r}$. As the distances between components of $G_{\leq r}$ is greater than r, and every vertex lies in some component containing a net point, we get the desired distance bounds. Clearly this net can be computed in polynomial time. □

3 Approximation Schemes

In general the aspect ratio of a graph may be exponential in the input size. A key ingredient of our algorithms is to reduce the aspect ratio α of the input graph $G = (V, E)$ to a polynomial. For STP and TSP, standard techniques can be used to reduce the aspect ratio to $O(n/\varepsilon)$ when aiming for a $(1+\varepsilon)$-approximation. This was for instance also used in [24] for low highway dimension graphs, but here we need to take special care not to destroy the structural properties given by Lemma 9 in this process. In particular, we need to reduce the aspect ratio and maintain the fact that the treewidth is bounded.

Therefore, we reduce the aspect ratio of our graphs by the following preprocessing. Both metric TSP and STP admit constant factor approximations in polynomial time using well-known algorithms [19,23]. We first compute a solution of cost c using a β-approximation algorithm for the problem at hand (TSP or STP). For TSP, the diameter of the graph G clearly is at most $c/2$. For STP we remove every vertex of V that is at distance more than c from any terminal, since such a vertex cannot be part of the optimum solution. After having removed all such vertices in this way, we obtain a graph G of diameter at most $3c$. Thus, in the following, we may assume that our graph G has diameter at most $3c$. We then set $r = \frac{\varepsilon c}{3n}$ in Lemma 10 to obtain a $(\frac{\varepsilon c}{3n}, \frac{\varepsilon c}{n})$-net $N \subseteq V$. As a consequence the metric induced by N (with distances of G) has aspect ratio at most $\frac{3c}{\varepsilon c/(3n)} = O(n/\varepsilon)$, since the minimum distance between any two net points of N is at least $\frac{\varepsilon c}{3n}$ and the maximum distance is at most $3c$. We will exploit this property in the following.

By Lemma 10, each connected component of $G_{\leq \frac{\varepsilon c}{3n}}$ contains exactly one net point of N. Let $\eta\colon V \mapsto N$ map each vertex of G to the unique net point in the same connected component of $G_{\leq \frac{\varepsilon c}{3n}}$. We define a new graph G' with vertex set $N \subseteq V$ and edge set $\{\{\eta(u), \eta(v)\} : \{u, v\} \in E \wedge \eta(u) \neq \eta(v)\}$. The length of each edge $\{w, w'\}$ of G' is the shortest path distance between w and w' in G. This new graph G' may not have bounded highway dimension, but we claim that it has treewidth $O(\log(n/\varepsilon))$.

Lemma 11. *If G has highway dimension 1, the graph G' with vertex set N has treewidth $O(\log(n/\varepsilon))$. Moreover, a tree decomposition for G' of width $O(\log(n/\varepsilon))$ can be computed in polynomial time.*

Proof. We construct a tree decomposition D' of G' as follows. Following Lemma 9 we can compute a tree decomposition D of width at most $1 + \lceil \log_2(\alpha) \rceil$, where α is the aspect ratio of G: for this we need to compute a locally 1-sparse shortest path cover $\text{SPC}(r_i)$ for each level i, which can be done in polynomial time via an XP algorithm [24] if the highway dimension is 1. We then find the level-i components and their interface points, from which the tree decomposition D and its bags can be constructed. Since there are $O(\log \alpha)$ levels and α is at most exponential in the input size (which includes the encoding length of the edge weights), we can compute D in polynomial time.

We construct D' from D by replacing every bag X of D by a new bag $X' = \{\eta(v) : v \in X\}$ containing the net points for the vertices in X. It is not hard to see that Properties (a) and (b) of Definition 5 are fulfilled by D', since they are true for D. For Property (c), note that for any edge $\{u, v\}$ of G, the set of all bags of D that contain u or v form a connected subtree of D. This is because the bags containing u form a connected subtree (Property (c)), the same is true for v, and both these subtrees share at least one node labelled by a bag containing the edge $\{u, v\}$ (Property (b)). Consequently, the set of all bags containing vertices of any connected subgraph of G form a connected subtree. In particular, for any connected component A of $G_{\leq \frac{\varepsilon c}{3n}}$, the set of bags of D containing at least one vertex of A form a connected subtree. This implies Property (c) for D'. Thus, D' is indeed a tree decomposition of G' according to Definition 5. Note that D' can be computed in polynomial time.

To bound the width of D', recall that a bag X of the tree decomposition D of G contains the interface points I_C of a level-i component C, in addition to one more vertex of C on the lowest level $i = 0$. Each interface point is a hub from $\text{SPC}(r_j)$ at some level $j \geq i$ and is at distance at most $2r_j$ from C. In particular, if $2r_i \leq \frac{\varepsilon c}{3n}$ then C is a component of $G_{\leq 2r_i} \subseteq G_{\leq \frac{\varepsilon c}{3n}}$, and all hubs of $I_C \cap \text{SPC}(r_j)$ for which $2r_j \leq \frac{\varepsilon c}{3n}$ lie in the same connected component A of $G_{\leq \frac{\varepsilon c}{3n}}$ as C. These hubs are therefore all mapped to the same net point w in A by η. In addition to w, the bag $X' = \{\eta(v) : v \in X\}$ resulting from X and η contains at most one vertex for every level j such that $2r_j > \frac{\varepsilon c}{3n}$. As $r_j = 2^j$, this condition is equivalent to $j > \log_2(\frac{\varepsilon c}{3n}) - 1$. As there are $1 + \lceil \log_2(\alpha) \rceil$ levels in total, there are $O(\log(\frac{\alpha n}{\varepsilon c}))$ hubs in X'. This bound is obviously also valid in case $2r_i > \frac{\varepsilon c}{3n}$. We preprocessed the graph G so that its diameter is at most $3c$ and its minimum distance is 3, which implies an aspect ratio α of at most c for G. This means that every bag X' contains $O(\log(n/\varepsilon))$ vertices, and thus the claimed treewidth bound for G' follows. □

We are now ready to prove our main result.

Proof (of Theorem 2). To solve TSP or STP on G we first use the above reduction to obtain G' and its tree decomposition D', and then compute an optimum solution for G'. For TSP, G' is already a valid input instance, but for STP we need to define a terminal set, which simply is $R' = \{\eta(v) \mid v \in R\}$ if R is the terminal set of G. Bodlaender et al. [16] proved that for both TSP and STP there are deterministic algorithms to solve these problems exactly in time $2^{O(t)}n$, given a tree decomposition of the input graph of width t. By Lemma 11 we can thus compute the optimum to G' in time $2^{O(\log(n/\varepsilon))} \cdot n = (n/\varepsilon)^{O(1)}$. Afterwards, we convert the solution for G' back to a solution for G, as follows.

For TSP we may greedily add vertices of V to the tour on N by connecting every vertex $v \in V$ to the net point $\eta(v)$. As the vertices N of G' form a $(\frac{\varepsilon c}{3n}, \frac{\varepsilon c}{n})$-net of V, this incurs an additional cost of at most $2\frac{\varepsilon c}{n}$ per vertex, which sums up to at most $2\varepsilon c$. Let OPT and OPT' denote the costs of the optimum tours in G and G', respectively. We know that $c \leq \beta \cdot \text{OPT}$, since we used a β-approximation algorithm to compute c. Furthermore, the optimum tour in G can be converted to a tour in G' of cost at most OPT by short-cutting, due to the triangle inequality. Thus $\text{OPT}' \leq \text{OPT}$, which means that the cost of the computed tour in G is at most $\text{OPT}' + 2\varepsilon c \leq (1 + 2\beta\varepsilon)\text{OPT}$.

Similarly, for STP we may greedily connect a terminal v of G to the terminal $\eta(v)$ of G' in the computed Steiner tree in G'. This adds an additional cost of at most $\frac{\varepsilon c}{n}$, which sums up to at most εc. Let now OPT and OPT' be the costs of the optimum Steiner trees in G and G', respectively. We may convert a Steiner tree T in G into a tree T' in G' by using edge $\{\eta(u), \eta(v)\}$ for each edge $\{u, v\}$ of T. Note that the resulting tree T' contains all terminals of G', since $R' = \{\eta(v) \mid v \in R\}$. As the vertices N of G' form a $(\frac{\varepsilon c}{3n}, \frac{\varepsilon c}{n})$-net of V, the cost of T' is at most $\text{OPT} + 2\varepsilon c$ if the cost of T is OPT (by the same argument as used for the proof of Lemma 11). As before, we know that $c \leq \beta \cdot \text{OPT}$, and thus the cost of the computed Steiner tree in G is at most $\text{OPT}' + \varepsilon c \leq \text{OPT} + 3\varepsilon c \leq (1 + 3\beta\varepsilon)\text{OPT}$.

Hence we obtain FPTASs for both TSP and STP, which compute $(1 + \varepsilon)$-approximations within a runtime that is polynomial in the input size and $1/\varepsilon$. □

4 Conclusions

We showed that, somewhat surprisingly, graphs of highway dimension 1 exhibit a rich combinatorial structure. On one hand, it was already known [24] that these graphs are not minor-closed and thus their treewidth is unbounded. Here we additionally showed that STP is weakly NP-hard on such graphs, further confirming that these graphs have non-trivial properties. On the other hand, we proved in Lemma 9 that the treewidth of a graph of highway dimension 1 is logarithmically bounded in the aspect ratio α. This in turn can be exploited to obtain a very efficient FPTAS for both STP and TSP.

At this point one may wonder whether it is possible to generalize Lemma 9 to larger values of the highway dimension. In particular, in [24] it was suggested that the treewidth of a graph of highway dimension h might be bounded by, say, $O(h \operatorname{polylog}(\alpha))$. However such a bound is highly unlikely in general, since it would have the following consequence for the k-CENTER problem, for which k vertices (centers) need to be selected in a graph such that the maximum distance of any vertex to its closest center is minimized. It was shown in [25] that it is NP-hard to compute a $(2 - \varepsilon)$-approximation for k-CENTER on graphs of highway dimension $O(\log^2 n)$, for any $\varepsilon > 0$. Given such a graph, the same preprocessing of Sect. 3 could be used to derive an analogue of Lemma 11, i.e., a graph G' of treewidth $O(\operatorname{polylog}(n/\varepsilon))$ could be computed for the net N. Moreover, a 2-approximation for k-CENTER can be computed in polynomial time on any graph [31], and if the input has treewidth t a $(1+\varepsilon)$-approximation can be computed in $(t/\varepsilon)^{O(t)} n^{O(1)}$ time [35]. Using the same arguments to prove Theorem 2 for STP and TSP, it would now be possible to compute a $(1 + \varepsilon)$-approximation for k-CENTER in quasi-polynomial time (cf. [26]). That is, we would obtain a QPTAS for graphs of highway dimension $O(\log^2 n)$, which is highly unlikely given that computing a $(2 - \varepsilon)$-approximation is NP-hard on such graphs.

The above argument rules out any bound of $(h \log \alpha)^{O(1)}$ for graphs of highway dimension h and aspect ratio α, unless NP-hard problems admit quasi-polynomial time algorithms. In fact, we conjecture that the k-CENTER problem is NP-hard to approximate within a factor of $2-\varepsilon$ for graphs of constant highway dimension (for some constant larger than 1). If this is true, then the above argument even rules out a treewidth bound of $f(h) \operatorname{polylog}(\alpha)$ for any function f. Thus, in order to answer the open problem of [24] and obtain a PTAS for graphs of constant highway dimension, a different approach seems to be needed.

References

1. Abraham, I., Delling, D., Fiat, A., Goldberg, A.V., Werneck, R.F.: Highway dimension and provably efficient shortest path algorithms. J. ACM **63**(5), 41 (2016)
2. Abraham, I., Delling, D., Fiat, A., Goldberg, A.V., Werneck, R.F.: VC-dimension and shortest path algorithms. In: Aceto, L., Henzinger, M., Sgall, J. (eds.) ICALP 2011. LNCS, vol. 6755, pp. 690–699. Springer, Heidelberg (2011). https://doi.org/10.1007/978-3-642-22006-7_58
3. Abraham, I., Fiat, A., Goldberg, A.V., Werneck, R.F.: Highway dimension, shortest paths, and provably efficient algorithms. In: Proceedings of the 21st Annual ACM-SIAM Symposium Discrete Algorithms (SODA), pp. 782–793 (2010)
4. Arora, S.: Polynomial time approximation schemes for Euclidean traveling salesman and other geometric problems. J. ACM **45**(5), 753–782 (1998)
5. Arora, S., Grigni, M., Karger, D.R., Klein, P.N., Woloszyn, A.: A polynomial-time approximation scheme for weighted planar graph TSP. In: Proceedings of the 9th Annual ACM-SIAM Symposium Discrete Algorithms (SODA), pp. 33–41 (1998)
6. Arora, S., Lund, C., Motwani, R., Sudan, M., Szegedy, M.: Proof verification and hardness of approximation problems. In: Proceedings of the 33rd Annual IEEE Symposium Foundations Computer Science (FOCS), pp. 14–23 (1992)
7. Arora, S., Raghavan, P., Rao, S.: Approximation schemes for Euclidean k-medians and related problems. In: Proceedings of the 30th Annual ACM Symposium Theory Computer (STOC), pp. 106–113 (1998)
8. Bartal, Y., Gottlieb, L.-A., Krauthgamer, R.: The traveling salesman problem: low-dimensionality implies a polynomial time approximation scheme. In: Proceedings of the 44th Annual ACM Symposium Theory Computer (STOC), pp. 663–672 (2012)
9. Bast, H., Funke, S., Matijevic, D.: Ultrafast shortest-path queries via transit nodes. In: The Shortest Path Problem: Ninth DIMACS Implementation Challenge, vol. 74, pp. 175–192 (2009)
10. Bast, H., Funke, S., Matijevic, D., Sanders, P., Schultes, D.: In transit to constant time shortest-path queries in road networks. In: Proceedings of the 9th Workshop Algorithm Engineering and Experiments (ALENEX) (2007)
11. Bateni, M., Hajiaghayi, M.T., Marx, D.: Approximation schemes for Steiner forest on planar graphs and graphs of bounded treewidth. J. ACM **58**, 21:1–21:37 (2011)
12. Becker, A., Klein, P.N., Saulpic, D.: Polynomial-time approximation schemes for k-center, k-median, and capacitated vehicle routing in bounded highway dimension. In: Proceedings of the 26th Annual European Symposium on Algorithms (ESA 2018), pp. 8:1–8:15 (2018)
13. Bern, M., Plassmann, P.: The Steiner problem with edge lengths 1 and 2. Inform. Process. Lett. **32**, 171–176 (1989)
14. Bland, R., Shallcross, D.: Large traveling salesman problems arising from experiments in X-ray crystallography: a preliminary report on computation. Oper. Res. Lett. **8**, 125–128 (1989)
15. Blum, J.: Hierarchy of transportation network parameters and hardness results (2019). arXiv: 1905.11166 [cs.DM]
16. Bodlaender, H.L., Cygan, M., Kratsch, S., Nederlof, J.: Deterministic single exponential time algorithms for connectivity problems parameterized by treewidth. In: Fomin, F.V., Freivalds, R., Kwiatkowska, M., Peleg, D. (eds.) ICALP 2013. LNCS, vol. 7965, pp. 196–207. Springer, Heidelberg (2013). https://doi.org/10.1007/978-3-642-39206-1_17

17. Borndörfer, R., Neumann, M., Pfetsch, M.E.: The line connectivity problem. In: Fleischmann, B., Borgwardt, K.H., Klein, R., Tuma, A. (eds.) Operations Research Proceedings, pp. 557–562. Springer, Heidelberg (2008). https://doi.org/10.1007/978-3-642-00142-0_90
18. Borradaile, G., Kenyon-Mathieu, C., Klein, P.: A polynomial-time approximation scheme for Steiner tree in planar graphs. In: Proceedings of the 18th Annual ACM-SIAM Symposium Discrete Algorithms (SODA), pp. 1285–1294 (2007)
19. Byrka, J., Grandoni, F., Rothvoß, T., Sanità, L.: An improved LP-based approximation for Steiner tree. In: Proceedings of the 42nd Annual ACM Symposium Theory Computer (STOC), pp. 583–592 (2010)
20. Chen, C.Y., Grauman, K.: Efficient activity detection in untrimmed video with max-subgraph search. IEEE Trans. Pattern Anal. Mach. Intell. **39**, 908–921 (2018)
21. Chlebík, M., Chlebíková, J.: The Steiner tree problem on graphs: inapproximability results. Theor. Comput. Sci. **406**, 207–214 (2008)
22. Chowdhury, S.A., Shackney, S.E., Heselmeyer-Haddad, K., Ried, T., Schäffer, A.A., Schwartz, R.: Phylogenetic analysis of multiprobe uorescence in situ hybridization data from tumor cell populations. Bioinformatics **29**, i189–i198 (2013)
23. Christofides, N.: Worst-case analysis of a new heuristic for the travelling salesman problem. Technical report 388. Graduate School of Industrial Administration, Carnegie Mellon University (1976)
24. Feldmann, A.E., Fung, W.S., Könemann, J., Post, I.: A $(1+\epsilon)$-embedding of low highway dimension graphs into bounded treewidth graphs. SIAM J. Comput. **41**, 1667–1704 (2018)
25. Feldmann, A.E.: Fixed parameter approximations for k-center problems in low highway dimension graphs. Algorithmica (2018)
26. Feldmann, A.E., Marx, D.: The parameterized hardness of the k-center problem in transportation networks. In: Proceedings of the 16th Scandinavian Symposium and Workshop Algorithm Theory (SWAT), pp. 19:1–19:13 (2018)
27. Garey, M.R., Johnson, D.S.: The rectilinear Steiner tree problem is NP-complete. SIAM J. Appl. Math. **32**, 826–834 (1977)
28. Grigni, M., Koutsoupias, E., Papadimitriou, C.H.: An approximation scheme for planar graph TSP. In: Proceedings of the 36th Annual IEEE Symposium Foundations Computer Science (FOCS), pp. 640–645 (1995)
29. Grötschel, M., Holland, O.: Solution of large-scale symmetric travelling salesman problems. Math. Program. **51**, 141–202 (1991)
30. Held, S., Korte, B., Rautenbach, D., Vygen, J.: Combinatorial optimization in VLSI design. In: Chvatal, V. (ed.) Combinatorial Optimization: Methods and Applications, pp. 33–96. IOS Press, Amsterdam (2011)
31. Hochbaum, D.S., Shmoys, D.B.: A unified approach to approximation algorithms for bottleneck problems. J. ACM **33**(3), 533–550 (1986)
32. Hougardy, S., Prömel, H.J.: A 1.598 approximation algorithm for the Steiner problem in graphs. In: Proceedings of the 10th Annual ACM-SIAM Symposium Discrete Algorithms (SODA), pp. 448–453 (1999)
33. Karp, R.M.: Reducibility among combinatorial problems. In: Miller, R.E., Thatcher, J.W., Bohlinger, J.D. (eds.) Complexity of Computer Computations. IRSS, pp. 85–103. Springer, Boston (1972). https://doi.org/10.1007/978-1-4684-2001-2_9
34. Karpinski, M., Lampis, M., Schmied, R.: New inapproximability bounds for TSP. J. Comput. Syst. Sci. **81**, 1665–1677 (2015)

35. Katsikarelis, I., Lampis, M., Paschos, V.T.: Structural parameters, tight bounds, and approximation for (k, r)-center. In: Proceedings of the 28th International Symposium Algorithms Computer (ISAAC), pp. 50:1–50:13 (2017)
36. Klein, P.: A linear-time approximation scheme for TSP in undirected planar graphs with edge-weights. SIAM J. Comput. **37**(6), 1926–1952 (2008)
37. Kosowski, A., Viennot, L.: Beyond highway dimension: small distance labels using tree skeletons. In: Proceedings of the 28th Annual ACM-SIAM Symposium Discrete Algorithms (SODA), pp. 1462–1478 (2017)
38. Krauthgamer, R., Lee, J.R.: Algorithms on negatively curved spaces. In: Proceedings of the 47th Annual IEEE Symposium Foundations Computer Science (FOCS), pp. 119–132 (2006)
39. Lampis, M.: Improved inapproximability for TSP. Theory Comput. **10**, 217–236 (2014)
40. Laporte, G., Nobert, Y., Desrochers, M.: Optimal routing under capacity and distance restrictions. Oper. Res. **33**, 1050–1073 (1985)
41. Lenstra, J., Rinnooy Kan, A.: Some simple applications of the traveling salesman problem. Oper. Res. Quart. **26**, 717–733 (1975)
42. Ljubić, I., Weiskirchner, R., Pferschy, U., Klau, G.W., Mutzel, P., Fischetti, M.: An algorithmic framework for the exact solution of the prizecollecting Steiner tree problem. Math. Program. **105**, 427–449 (2006)
43. Loboda, A.A., Artyomov, M.N., Sergushichev, A.A.: Solving generalized maximum-weight connected subgraph problem for network enrichment analysis. In: Frith, M., Storm Pedersen, C.N. (eds.) WABI 2016. LNCS, vol. 9838, pp. 210–221. Springer, Cham (2016). https://doi.org/10.1007/978-3-319-43681-4_17
44. Mitchell, J.S.B.: Guillotine subdivisions approximate polygonal subdivisions: a simple polynomial-time approximation scheme for geometric TSP, k-MST, and related problems. SIAM J. Comput. **28**(4), 1298–1309 (1999)
45. Papadimitriou, C.H., Vempala, S.: On the approximability of the traveling salesman problem. Combinatorica **26**, 101–120 (2006)
46. Robins, G., Zelikovsky, A.: Tighter bounds for graph Steiner tree approximation. SIAM J. Discret. Math. **19**, 122–134 (2005)
47. Sebő, A., Vygen, J.: Shorter tours by nicer ears: 7/5-approximation for the graph-TSP, 3/2 for the path version, and 4/3 for two-edge-connected subgraphs. Combinatorica **34**, 1–34 (2014)
48. Trevisan, L.: When Hamming meets Euclid: the approximability of geometric TSP and Steiner tree. SIAM J. Comput. **30**, 475–485 (2000)
49. Vazirani, V.V.: Approximation Algorithms. Springer, New York (2001)

The Power of Cut-Based Parameters for Computing Edge Disjoint Paths

Robert Ganian[1] and Sebastian Ordyniak[2(✉)]

[1] Algorithms and Complexity Group, Vienna University of Technology, Vienna, Austria

[2] Algorithms Group, University of Sheffield, Sheffield, UK

sordyniak@gmail.com

Abstract. This paper revisits the classical Edge Disjoint Paths (EDP) problem, where one is given an undirected graph G and a set of terminal pairs P and asks whether G contains a set of pairwise edge-disjoint paths connecting every terminal pair in P. Our aim is to identify structural properties (parameters) of graphs which allow the efficient solution of EDP without restricting the placement of terminals in P in any way. In this setting, EDP is known to remain NP-hard even on extremely restricted graph classes, such as graphs with a vertex cover of size 3.

We present three results which use edge-separator based parameters to chart new islands of tractability in the complexity landscape of EDP. Our first and main result utilizes the fairly recent structural parameter treecut width (a parameter with fundamental ties to graph immersions and graph cuts): we obtain a polynomial-time algorithm for EDP on every graph class of bounded treecut width. Our second result shows that EDP parameterized by treecut width is unlikely to be fixed-parameter tractable. Our final, third result is a polynomial kernel for EDP parameterized by the size of a minimum feedback edge set in the graph.

Keywords: Edge disjoint path problem · Feedback edge set · Treecut width · Parameterized complexity

1 Introduction

Edge Disjoint Paths (EDP) is a fundamental routing graph problem: we are given a graph G and a set P containing pairs of vertices (*terminals*), and are asked to decide whether there is a set of $|P|$ pairwise edge disjoint paths in G connecting each pair in P. Similarly to its counterpart, the Vertex Disjoint Paths (VDP) problem, EDP has been at the center of numerous results in structural graph theory, approximation algorithms, and parameterized algorithms [1,7,8,12,14,16,18,19,23].

Robert Ganian acknowledges support by the Austrian Science Fund (FWF, Project P31336), and is also affiliated with FI MUNI, Czech Republic.

I. Sau and D. M. Thilikos (Eds.): WG 2019, LNCS 11789, pp. 190–204, 2019.
https://doi.org/10.1007/978-3-030-30786-8_15

Both EDP and VDP are NP-complete in general [13], and a significant amount of research has focused on identifying structural properties which make these problems tractable. For instance, Robertson and Seymour's seminal work in the Graph Minors project [19] provides an $\mathcal{O}(n^3)$ time algorithm for both problems for every fixed value of $|P|$. Such results are often viewed through the more refined lens of the *parameterized complexity* paradigm [4,6]; there, each problem is associated with a numerical parameter k (capturing some structural property of the instance), and the goal is to obtain algorithms which are efficient when the parameter is small. Ideally, the aim is then to obtain so-called *fixed-parameter* algorithms for the problem, i.e., algorithms which run in time $f(k) \cdot n^{\mathcal{O}(1)}$ where f is a computable function and n the input size; the aforementioned result of Robertson and Seymour is hence an example of a fixed-parameter algorithm where $k = |P|$, and we say that the problem is FPT (w.r.t. this particular parameterization). In cases where fixed-parameter algorithms are unlikely to exist, one can instead aim for so-called XP algorithms, i.e., algorithms which run in polynomial time for every fixed value of k.

Naturally, one prominent question that arises is whether we can use the structure of the input graph itself (captured via a *structural parameter*) to solve EDP and VDP. Here, we find a stark contrast in the difficulty between these two, otherwise closely related, problems. Indeed, while VDP is known to be FPT with respect to the well-established structural parameter *treewidth* [21], EDP is NP-hard even on graphs of treewidth 3 [8]. What's worse, the same reduction shows that EDP remains NP-hard even on graphs with a vertex cover of size 3 [8], which rules out fixed-parameter and XP algorithms for the vast majority of studied graph parameters (including, e.g., *treedepth* and the *size of a minimum feedback vertex set*).

We note that previous research on the problem has found ways of circumventing these negative results by imposing additional restrictions. Zhou et al. [23] introduced the notion of an augmented graph, which contains information about how terminal pairs need to be connected, and used the treewidth of this graph to solve EDP. Recent work [11], which primarily focused on the complexity of EDP on near-forests and with respect to parameterizations of the augmented graphs, has also observed that EDP admits a fixed-parameter algorithm when parameterized by treewidth and the maximum degree of the graph.

Our Contribution. The aim of this paper is to provide new algorithms and matching lower bounds for solving the EDGE DISJOINT PATHS problem *without imposing any restrictions on the number and placement of terminals.* In other words, our aim is to be able to identify structural properties of the graph which guarantee tractability of the problem without knowing any information about the placement of terminals. The only positive result known so far in this setting requires us to restrict the degree of the input graph; however, in the bounded-degree setting there is a simple treewidth-preserving reduction from EDP to VDP (see Proposition 1), and so the problem only becomes truly interesting when the input graphs can contain vertices of higher degree.

Our main result is an XP algorithm for EDP when parameterized by the structural parameter *treecut width* [17,22]. Treecut width is inherently tied to the theory of graph immersions; in particular, it has a similar relationship to graph immersions and cuts as treewidth has to graph minors and separators. Since its introduction, treecut width has been successfully used to obtain fixed-parameter algorithms for problems which are W[1]-hard w.r.t. treewidth [9,10]; however, this is the first time that it has been used to obtain an algorithm for a problem that is NP-hard on graphs of bounded treewidth.

One "feature" of algorithmically exploiting treecut width is that it requires the solution of a non-trivial dynamic programming step. In previous works, this was carried out mostly by direct translations into INTEGER LINEAR PROGRAMMING instances with few integer variables [9] or by using network flows [10]. In the case of EDP, the dynamic programming step requires us to solve an instance of EDP with a vertex cover of size k where every vertex outside of the vertex cover has a degree of 2; we call this problem SIMPLE EDP and solve it in the dedicated Sect. 3. It is worth noting that there is only a very small gap between SIMPLE EDP (for which we provide an XP algorithm) and graphs with a vertex cover of size 3 (where EDP is known to be NP-hard).

In view of our main result, it is natural to ask whether the algorithm can be improved to a fixed-parameter one. After all, given the parallels between EDP parameterized by treecut width (an edge-separator based parameter) and VDP parameterized by treewidth (a vertex-separator based parameter), one would rightfully expect that the fixed-parameter tractability result on the latter [21] would be mirrored in the former case. Surprisingly, we rule this out by showing that EDP parameterized by treecut width is W[1]-hard [4,6] and hence unlikely to be fixed-parameter tractable; in fact, we obtain this lower-bound result even in the more restrictive setting of SIMPLE EDP. The proof is based on an involved reduction from an adapted variant of the MULTIDIMENSIONAL SUBSET SUM problem [10,11] and forms our second main contribution.

Having ruled out fixed-parameter algorithms for EDP parameterized by treecut width and in view of previous lower-bound results, one may ask whether it is even possible to obtain such an algorithm for any reasonable parameterization. We answer this question positively by using the size of a minimum feedback edge set as a parameter. In fact, we show an even stronger result: as our final contribution, we obtain a so-called *linear kernel* [4,6] for EDP parameterized by the size of a minimum feedback edge set.

Organization of the Paper. After introducing the required preliminaries in Sect. 2, we proceed to introducing SIMPLE EDP, solving it via an XP algorithm and establishing our lower-bound result (Sect. 3). Section 4 then contains our algorithm for EDP parameterized by treecut width. Finally, in Sect. 5 we obtain a polynomial kernel for EDP parameterized by the size of a minimum feedback edge set.

2 Preliminaries

We use standard terminology for graph theory, see for instance [5]. Given a graph G, we let $V(G)$ denote its vertex set and $E(G)$ its edge set. The (open) neighborhood of a vertex $x \in V(G)$ is the set $\{y \in V(G) : xy \in E(G)\}$ and is denoted by $N_G(x)$. For a vertex subset X, the neighborhood of X is defined as $\bigcup_{x \in X} N_G(x) \setminus X$ and denoted by $N_G(X)$; we drop the subscript if the graph is clear from the context. *Contracting* an edge $\{a, b\}$ is the operation of replacing vertices a, b by a new vertex whose neighborhood is $(N(a) \cup N(b)) \setminus \{a, b\}$. For a vertex set A (or edge set B), we use $G - A$ ($G - B$) to denote the graph obtained from G by deleting all vertices in A (edges in B), and we use $G[A]$ to denote the *subgraph induced on* A, i.e., $G - (V(G) \setminus A)$.

A *forest* is a graph without cycles, and an edge set X is a *feedback edge set* if $G - X$ is a forest. The *feedback edge set number* of a graph G, denoted by $\mathbf{fes}(G)$, is the smallest integer k such that G has a feedback edge set of size k. We use $[i]$ to denote the set $\{0, 1, \ldots, i\}$.

We assume that readers are familiar with basic notions in the area of parameterized complexity [3,6], such as the classes FPT and W[1] and the notion of *kernelization.*

2.1 Edge Disjoint Path Problem

Throughout the paper we consider the following problem.

Edge Disjoint Paths (EDP)	
Input:	A graph G and a set P of *terminal pairs*, i.e., a set of subsets of $V(G)$ of size two.
Question:	Is there a set of pairwise edge disjoint paths connecting every set of terminal pairs in P?

A vertex which occurs in a terminal pair is called a *terminal*, and a set of pairwise edge disjoint paths connecting every set of terminal pairs in P is called a *solution.* Without loss of generality, we assume that G is connected. The Vertex Disjoint Paths (VDP) problem is defined analogously as EDP, with the sole distinction being that the paths must be vertex-disjoint.

The following proposition establishes a link between EDP and VDP on graphs of bounded degree. Since we will not need the notion of *treewidth* [20] for any other result presented in the paper, we refer to the standard textbooks [3,6] for its definition.

Proposition 1. *There exists a linear-time reduction from* EDP *to* VDP *with the following property: if the input graph has treewidth k and maximum degree d, then the output graph has treewidth at most $k \cdot d + 1$.*

We remark that Proposition 1 in combination with the known fixed-parameter algorithm for VDP parameterized by treewidth [21] provides an alternative proof for the fixed-parameter tractability of EDP parameterized by degree and treewidth [11].

2.2 Treecut Width

The notion of treecut decompositions was introduced by Wollan [22], see also [17]. A family of subsets $X_1, \ldots, X_k$ of X is a *near-partition* of X if they are pairwise disjoint and $\bigcup_{i=1}^{k} X_i = X$, allowing the possibility of $X_i = \emptyset$.

Definition 1. *A* treecut decomposition *of G is a pair $(T, \mathcal{X})$ which consists of a rooted tree T and a near-partition $\mathcal{X} = \{X_t \subseteq V(G) : t \in V(T)\}$ of $V(G)$. A set in the family $\mathcal{X}$ is called a* bag *of the treecut decomposition.*

For any node t of T other than the root r, let $e(t) = ut$ be the unique edge incident to t on the path to r. Let T_u and T_t be the two connected components in $T - e(t)$ which contain u and t, respectively. Note that $(\bigcup_{q\in T_u} X_q, \bigcup_{q\in T_t} X_q)$ is a near-partition of $V(G)$, and we use E_t to denote the set of edges with one endpoint in each part. We define the *adhesion* of t ($\mathbf{adh}(t)$) as $|E_t|$; we explicitly set $\mathbf{adh}(r) = 0$ and $E(r) = \emptyset$.

The *torso* of a treecut decomposition $(T, \mathcal{X})$ at a node t, written as H_t, is the graph obtained from G as follows. If T consists of a single node t, then the torso of $(T, \mathcal{X})$ at t is G. Otherwise let $T_1, \ldots, T_\ell$ be the connected components of $T - t$. For each $i = 1, \ldots, \ell$, the vertex set $Z_i \subseteq V(G)$ is defined as the set $\bigcup_{b\in V(T_i)} X_b$. The torso H_t at t is obtained from G by *consolidating* each vertex set Z_i into a single vertex z_i (this is also called *shrinking* in the literature). Here, the operation of consolidating a vertex set Z into z is to substitute Z by z in G, and for each edge e between Z and $v \in V(G) \setminus Z$, adding an edge zv in the new graph. We note that this may create parallel edges.

The operation of *suppressing* (also called *dissolving* in the literature) a vertex v of degree at most 2 consists of deleting v, and when the degree is two, adding an edge between the neighbors of v. Given a connected graph G and $X \subseteq V(G)$, let the *3-center* of (G, X) be the unique graph obtained from G by exhaustively suppressing vertices in $V(G) \setminus X$ of degree at most two. Finally, for a node t of T, we denote by $\tilde{H}_t$ the 3-center of (H_t, X_t), where H_t is the torso of $(T, \mathcal{X})$ at t. Let the *torso-size* $\mathbf{tor}(t)$ denote $|\tilde{H}_t|$.

Definition 2. *The width of a treecut decomposition $(T, \mathcal{X})$ of G is $\max_{t\in V(T)}\{\mathbf{adh}(t), \mathbf{tor}(t)\}$. The treecut width of G, or $\mathbf{tcw}(G)$ in short, is the minimum width of $(T, \mathcal{X})$ over all treecut decompositions $(T, \mathcal{X})$ of G.*

Without loss of generality, we shall assume that $X_r = \emptyset$. We conclude this subsection with some notation related to treecut decompositions. Given a tree node t, let T_t be the subtree of T rooted at t. Let $Y_t = \bigcup_{b\in V(T_t)} X_b$, and let G_t denote the induced subgraph $G[Y_t]$. A node $t \neq r$ in a rooted treecut decomposition is *thin* if $\mathbf{adh}(t) \leq 2$ and *bold* otherwise (Fig. 1).

While it is not known how to compute optimal treecut decompositions efficiently, there exists a fixed-parameter 2-approximation algorithm which we can use instead.

Theorem 1 ([15]). *There exists an algorithm that takes as input an n-vertex graph G and integer k, runs in time $2^{\mathcal{O}(k^2 \log k)} n^2$, and either outputs a treecut decomposition of G of width at most $2k$ or correctly reports that $\mathbf{tcw}(G) > k$.*

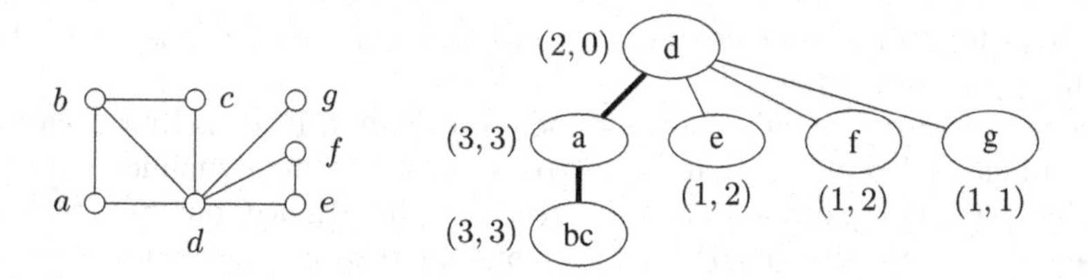

Fig. 1. A graph G and a width-3 treecut decomposition of G, including the torso-size (left value) and adhesion (right value) of each node.

A treecut decomposition $(T, \mathcal{X})$ is *nice* if it satisfies the following condition for every thin node $t \in V(T)$: $N(Y_t) \cap (\bigcup_{b \text{ is a sibling of } t} Y_b) = \emptyset$. The intuition behind nice treecut decompositions is that we restrict the neighborhood of thin nodes in a way which facilitates dynamic programming. Every treecut decomposition can be transformed into a nice treecut decomposition of the same width in cubic time [9].

For a node t, we let $B_t = \{ b \text{ is a child of } t \mid |N(Y_b)| \leq 2 \wedge N(Y_b) \subseteq X_t \}$ denote the set of thin children of t whose neighborhood is a subset of X_t, and we let $A_t = \{ a \text{ is a child of } t \mid a \notin B_t \}$ be the set of all other children of t. Then $|A_t| \leq 2k + 1$ for every node t in a nice treecut decomposition [9].

We refer to previous work [9,15,17,22] for a more detailed comparison of treecut width to other parameters. Here, we mention only that treecut width lies "between" treewidth and treewidth plus maximum degree.

3 The Simple Edge Disjoint Paths Problem

Before we start working towards our algorithm for solving EDP parameterized by treecut width, we will first deal with a simpler (but crucial) setting for the problem. We call this the Simple Edge Disjoint Paths problem (Simple EDP) and define it below.

Simple EDP	
Input:	An EDP instance (G, P) such that $V(G) = A \cup B$ where B is an independent set containing vertices of degree at most 2.
Parameter:	$k = \|A\|$
Question:	Is (G, P) a **YES**-instance of EDP?

Notice that every instance of Simple EDP has treecut width at most k, and so it forms a special case of EDP parameterized by treecut width. Indeed, the treecut decomposition where T is a star, the center bag contains A, and each leaf bag contains a vertex from B (except for the root r, where $X_r = \emptyset$), has treecut width at most k. This contrasts to the setting where G has a vertex cover of size 3 and all vertices outside the vertex cover have degree 3; the treecut width of such graphs is not bounded by any constant, and EDP is known to be NP-complete in this setting [8].

The main reason we introduce and focus on SIMPLE EDP is that it captures the combinatorial problem that needs to be solved in the dynamic step of the algorithm for EDP parameterized by treecut width. Hence, our first task here will be to solve SIMPLE EDP by an algorithm that can later be called as a subroutine.

Lemma 1. SIMPLE EDP *can be solved in time* $\mathcal{O}(|P|^{\binom{k}{2}+1}(k+1)!)$.

Sketch of Proof. Let (G, P) with partition A and B and $k = |A|$ be an instance of SIMPLE EDP. Let the *terminal graph* of G, denoted by G^T, as the graph with vertex set V and edge set P. Our first course of action will be to simplify the instance by removing all vertices in B that are not part of any terminal pair; to this end, we add multi-edges into $G[A]$ which represent removed degree-2 vertices. We now make the following two observations:

O1 Consider a path H connecting a terminal pair $p \in P$ in a solution. Because B is an independent set and every vertex in B has degree at most two and is contained in at least one terminal pair in P, we obtain that all inner vertices of H are from A. Hence, H contains at most $k + 2$ vertices and all inner vertices of H are contained in A. It follows that H is completely characterized by the sequence of vertices it uses in A. Consequently, there are at most $\sum_{\ell=1}^{k} \binom{k}{\ell}\ell! \leq (k+1)!$ different types of paths that need to be considered for the connection of any terminal pair.

O2 $G^T[B]$ is a disjoint union of paths and cycles. This is because every vertex v of G can be contained in at most $|N_G(v)|$ terminal pairs in P (otherwise we immediately reject) and all vertices in B have degree at most two.

Let u and v be two distinct vertices in A. Because $|A| \leq k$, we can enumerate all possible paths between u and v in $G[A]$ in time $\mathcal{O}((k+1)!)$. We will represent each such path H as a binary vector E_H, whose entries are indexed by all sets of two distinct vertices in A, such that $E_H[e] = 1$ if H uses the edge e and $E_H[e] = 0$ otherwise. Moreover, we will denote by $E_{u,v}$ the set $\{ E_H \mid H \text{ is a path between } u \text{ and } v \text{ in } G[A] \}$; intuitively, $E_{u,v}$ captures all possible sets of edges that need to be used in order to connect u to v.

Let S be a solution for (G, P). The algorithm represents every solution S for (G, P) as a solution vector E_S of natural numbers whose entries are indexed by all sets $\{u, v\}$ of two distinct vertices in A. More specifically, for two distinct vertices u and v in A, $E_S[\{u, v\}]$ is equal to the number of edges between u and v used by the paths in S. The algorithm uses dynamic programming to compute the set $\mathcal{L}$ of all solution vectors; clearly, $\mathcal{L} \neq \emptyset$ if and only if (G, P) is a **YES**-instance. We compute $\mathcal{L}$ in two main steps:

(S1) the algorithm computes the set $\mathcal{L}_A$ of all solution vectors for the subinstance $(G[A], P')$ of (G, P), where P' is the subset of P containing all terminal pairs $\{p, q\}$ with $p, q \in A$.

(S2) the algorithm computes the set of all solution vectors for the sub-instance $(G, P \setminus P')$. Note that every terminal pair p in $P \setminus P'$ is either completely contained in B, in which case it forms an edge of a path or a cycle in $G^T[B]$, or p has one vertex in A and the other vertex in B, which is the endpoint of a path in $G^T[B]$. The algorithm now computes the set of all solution vectors for the sub-instance $(G, P \setminus P')$ in two steps:

(S2A) For every cycle C in $G^T[B]$, the algorithm computes the set $\mathcal{L}_C$ of all solution vectors for the sub-instance $(G[A \cup V(C)], P_C)$, where P_C is the subset of P containing all terminal pairs $\{p, q\}$ such that $p, q \in C$.

(S2B) For every path H in $G^T[B]$, the algorithm computes the set $\mathcal{L}_H$ of all solution vectors for the sub-instance $(G[A \cup V(H)], P_H)$, where P_H is the subset of P containing all terminal pairs $\{p, q\}$ with $\{p, q\} \cap V(H) \neq \emptyset$.

In the end, the set of all *hypothetical solution vectors* $\mathcal{L}'$ for (G, P) is obtained as $\mathcal{L}_A \oplus (\oplus_{C \text{ is a cycle of } G^T[B]} \mathcal{L}_C) \oplus (\oplus_{H \text{ is a path of } G^T[B]} \mathcal{L}_H)$, where $\mathcal{P} \oplus \mathcal{P}'$ for two sets $\mathcal{P}$ and $\mathcal{P}'$ of solution vectors is equal to $\{ R + R' \mid R \in \mathcal{P} \wedge R' \in \mathcal{P}' \}$. Each vector in $\mathcal{L}'$ describes one possible set of multi-edges in $G[A]$ that can be used to connect all terminal pairs in P. In order to compute $\mathcal{L}$, one simply needs to remove all vectors from $\mathcal{L}'$ which require more multi-edges than are available in $G[A]$; in particular, to obtain $\mathcal{L}$ we delete each S from $\mathcal{L}'$ such that there exist $u, v \in A$ where $E_S[\{u, v\}]$ exceeds the number of multi-edges between u and v in G. The algorithm then returns **YES** if $\mathcal{L}$ is non-empty and otherwise the algorithm returns **NO**. ∎

Notice that Lemma 1 does not provide a fixed-parameter algorithm for SIMPLE EDP. Our second task for this section will be to rule out the existence of such algorithms (hence also ruling out the fixed-parameter tractability of EDP parameterized by treecut width).

Before we proceed, we would like note that this outcome was highly surprising for the authors. Indeed, not only does this "break" the parallel between {VDP, treewidth} and {EDP, treecut width}, but inspecting the dynamic programming algorithm for EDP parameterized by treecut width presented in Sect. 4 reveals that solving SIMPLE EDP is the only step which requires more than "FPT-time". In particular, if SIMPLE EDP were FPT, then EDP parameterized by treecut width would also be FPT. This situation contrasts the vast majority of dynamic programming algorithms for parameters such as treewidth and clique-width [2], where the complexity bottleneck is usually tied to the size of the records used and not to the computation of the dynamic step.

Our lower-bound result is based on a parameterized reduction from the following problem, whose W[1]-hardness follows from recent work of the authors [10,11]:

Multidimensional Subset Sum (MSS)	
Input:	An integer k, a set $S = \{s_1, \ldots, s_n\}$ of item-vectors with $s_i \in \mathbb{N}^k$ for every i with $1 \leq i \leq n$, a target vector $t \in \mathbb{N}^k$, and an integer ℓ.
Parameter:	k
Question:	Is there a subset $S' \subseteq S$ with $\|S'\| \geq \ell$ such that $\sum_{s \in S'} s \leq t$?

Lemma 2. Simple EDP *is* W[1]*-hard.*

4 An Algorithm for EDP on Graphs of Bounded Treecut Width

The goal of this section is to provide an XP algorithm for EDP parameterized by treecut-width. The core of the algorithm is a dynamic programming procedure which runs on a nice treecut decomposition $(T, \mathcal{X})$ of the input graph G.

Overview. Our first aim is to define the data table the algorithm is going to dynamically compute for individual nodes of the treecut decomposition; to this end, we introduce two additional notions. For a node t, we say that Y_t (or G_t) contains an *unmatched* terminal s if $\{s, e\} \in P$, $s \in Y_t$ and $e \notin Y_t$; let U_t be the multiset containing all unmatched terminals Y_t (one entry in U_t per tuple in P which contains an unmatched terminal). For a subgraph H of G, let $P_H \subseteq P$ denote the subset of terminal pairs whose both endpoints lie in H.

Let a *record* for node t be a tuple (δ, I, F, L) where:

- δ is a partitioning of E_t into *internal* (I'), *leaving* (L'), *foreign* (F') and *unused* (U');
- I is a set of subsets of size 2 that is a perfect matching between the edges in I';
- F is a set of subsets of size 2 that is a perfect matching between the edges in F';
- L is a perfect matching between U_t and the edges in L'.

Intuitively, a record captures all the information we need about one possible interaction between a solution to EDP and the edges in E_t. In particular, unmatched terminals need to cross between Y_t and G_t using an edge in E_t and L captures the first edge used by a path from an unmatched terminal in the solution, while I and F capture information about paths which intersect with E_t but whose terminals both lie in Y_t and $V(G_t) \setminus Y_t$, respectively. We formalize this intuition below through the notion of a *valid record.*

Definition 3. *A record* $\lambda = (\delta, I, F, L)$ *is* valid *for* t *if* (G^λ, P^λ) *is a* ***YES****-instance of* EDP*, where* (G^λ, P^λ) *is constructed from* (G_t, P_{G_t}) *as follows:*

1. *For each* $\{\{a, b\}, \{c, d\}\} \in I$ *where* $a, c \in Y_t$*, add a new vertex into* G_t *and connect it to* a *and* c *by edges (note that if* $a = c$ *then this simply creates a new leaf and hence this operation can be ignored).*

2. *For each* $\{s, \{a, b\}\} \in L$ *where* $a \in Y_t$*, add a new tuple* $\{s, e'\}$ *into* P_{G_t} *and a new leaf* e' *into* G_t *adjacent to* a*.*
3. *For each* $\{\{a, b\}, \{c, d\}\} \in F$ *where* $a, c \in Y_t$*, add two new leaves* b', d' *into* G_t*, make them adjacent to* a *and* c *respectively, and add* $\{b', d'\}$ *into* P_{G_t}*.*

We are now ready to define our data tables: for a node $t \in V(T)$, let $D(t)$ be the set of all valid records for t. We now make two observations. First, for any node t in a nice treecut decomposition of width k, it holds that there exist at most $4^k \cdot k!$ distinct records and hence $|D(t)| \leq 4^k \cdot k!$; indeed, there are 4^k possible choices for δ, and for each such choice and each edge e in E_t one has at most k options of what to match with e. Second, if r is the root of T, then either $D(r) = \emptyset$ or $D(r) = \{(\emptyset, \emptyset, \emptyset, \emptyset)\}$; furthermore, (G, P) is a **YES**-instance if and only if the latter holds. Hence it suffices to compute $D(r)$ in order to solve EDP.

The next lemma shows that $D(t)$ can be computed efficiently for all leaves of t.

Lemma 3. *Given* (G, P)*, a width-*k *treecut decomposition* $(T, \mathcal{X})$ *of* G *and a leaf* $t \in V(T)$ *as the input, it is possible to compute* $D(t)$ *in time* $k^{\mathcal{O}(k^2)}$*.*

At this point, all that is left to obtain a dynamic leaves-to-root algorithm which solves EDP is the dynamic step, i.e., computing the data table for a node $t \in V(t)$ from the data tables of its children. Unfortunately, that is where all the difficulty of the problem lies, and our first step towards handling this task will be the introduction of two additional notions related to records. The first is *correspondence*, which allows us to associate each solution to (G, P) with a specific record for t; on an intuitive level, a solution corresponds to a particular record if that record precisely captures the "behavior" of that solution on E_t. Correspondence will, among others, be used to later argue the correctness of our algorithm.

Definition 4. *A solution* $\mathcal{S}$ *to* (G, P) corresponds *to a record* $\lambda = (\delta, I, F, L)$ *for* t *if the conditions* **1.**–**4.** *stated below hold for every* a*-*b *path* $S \in \mathcal{S}$ *such that* $S \cap E_t \neq \emptyset$*. We let* $s = |S \cap E_t|$ *and we denote individual edges in* $S \cap E_t$ *by* $e_1, e_2, \dots e_s$*, ordered from the edge nearest to* a *along* S*.*

1. *If* $a, b \notin Y_t$*, then for each odd* $i \in [s]$*,* F *contains* (e_i, e_{i+1})*.*
2. *If* $a, b \in Y_t$*, then for each odd* $i \in [s]$*,* I *contains* (e_i, e_{i+1})*.*
3. *If* $\{a, b\} \cap Y_t = \{a\}$*, then* L *contains* (a, e_1)*, and for each even* $i \in [s]$ F *contains* (e_i, e_{i+1})*.*
4. *There are no elements in* I, F, L *other than those specified above.*

Note that "restricting" the solution $\mathcal{S}$ to the instance (G^λ, P^λ) used in Definition 3 yields also a solution to (G^λ, P^λ); in particular, for each path $S \in \mathcal{S}$ that intersects E_t, one replaces the path segments of S in $G \setminus Y_t$ by the newly created vertices to obtain a solution to (G^λ, P^λ). Consequently, if $\mathcal{S}$ corresponds to λ then λ must be valid (however, it is clearly not true that every valid record has a solution to the whole instance that corresponds to it). Moreover, since Definition 4 is constructive and deterministic, for each solution $\mathcal{S}$ and node t there exists precisely one corresponding valid record λ.

The second notion that we will need is that of *simplification*. This is an operation which takes a valid record λ for a node t and replaces G_t by a "small representative" so that the resulting graph retains the existence of a solution corresponding to λ. Simplification can also be seen as being complementary to the construction of (G^λ, P^λ) used in Definition 3 (instead of modeling the implications of a record on G_t, we model its implications on $G - Y_t$), and will later form an integral part of our procedure for computing valid records for nodes.

Definition 5. *The* simplification *of a node t in accordance with $\lambda = (\delta, I, F, L)$ is an operation which transforms the instance (G, P) into a new instance (G', P') obtained from $(G - Y_t, P_{G-Y_t})$ as follows. (1) For each $\{s, \{a, b\}\} \in L$ where $(s, e) \in P$ and $b \notin Y_t$, we add (s, e) to P' and create a vertex s adjacent to b. (2) For each $\{\{a, b\}, \{c, d\}\} \in I$ where $a, c \in Y_t$ and $a \neq c$, we add (a, c) into P', add vertices a and c into G', and add edges $\{a, b\}$ and $\{c, d\}$. (3) For each $\{\{a, b\}, \{c, d\}\} \in F$ where $a, c \in Y_t$ and $b \neq d$, we create a vertex x and set $N(x) = \{b, d\}$.*

Observation 1. *If there exists a solution to (G, P) which* corresponds *to a record $\lambda = (\delta, I, F, L)$ for t, and if (G', P') is the result of simplification of t in accordance with λ, then (G', P') admits a solution. On the other hand, if (G', P') is the result of simplification of t in accordance with a record λ and if (G', P') admits a solution, then (G, P) also admits a solution.*

The Dynamic Step. The following crucial lemma represents the tool that allows us to deal with the dynamic step of our leaf-to-root computation along the decomposition.

Lemma 4. *There is an algorithm which takes as input (G, P) along with a width-k treecut decomposition $(T, \mathcal{X})$ of G and a non-leaf node $t \in V(T)$ and $D(t')$ for every child t' of t, runs in time $(k|P|)^{\mathcal{O}(k^2)}$, and outputs $D(t)$.*

Sketch of Proof. We begin by looping through all of the at most $4^k \cdot k!$ distinct records for t; for each such record λ, our task is to decide whether it is valid, i.e., whether (G^λ, P^λ) is a **YES**-instance. On an intuitive level, our aim will now be to use branching and simplification in order to reduce the question of checking whether λ is valid to an instance of Simple EDP.

In our first layer of branching, we will select a record from the data tables of each node in A_t. Formally, we say that a *record-set* is a mapping $\tau : t' \in A_t \mapsto \lambda_{t'} \in D(t')$. Note that the number of record-sets is upper-bounded by $(4^k \cdot k!)^{3(2k+1)}$, and we will loop over all possible record-sets.

Next, for each record-set τ, we will apply simplification to each node $t' \in A_t$ in accordance with $\tau(t')$, and recall that each vertex v created by this sequence of simplifications has degree at most 2. We then apply a reduction rule to ensure that each such vertex is only adjacent to $(V(G) \setminus Y_t) \cup X_t$. At this point, every vertex contained in a bag $X_{t'}$ for $t' \in A_t$ has degree at most 2 and is only adjacent to $X_t \cup (V(G) \setminus Y_t)$. Now, we apply one additional reduction rule which allows us to replace every thin node by vertices of degree at most 2 adjacent to X_t (there are only a few possible kinds of records that thin nodes can have, and we use simple replacement rules for each individual case).

At this point, every vertex in $V(G^\lambda) \setminus X_t$ is of degree at most 2 and only adjacent to X_t, and so (G^λ, P^λ) is an instance of SIMPLE EDP. All that is left is to invoke Lemma 1; if it is a **YES**-instance then we add λ to $D(t)$, and otherwise we do not.

We conclude the proof by arguing correctness. Assume λ is a valid record. By Definition 3, this implies that (G^λ, P^λ) admits a solution S. For each child $t' \in A_t$, S corresponds to some record $\lambda^S_{t'}$ for t; consider now the branch in our algorithm which sets $\tau(t') = \lambda^S_{t'}$. Then by Observation 1 it follows that each simplification carried out by the algorithm preserves the existence of a solution to (G^λ, P^λ). Hence the instance of SIMPLE EDP we obtain at the end of this branch must also be a **YES**-instance. ■

Theorem 2. EDP *can be solved in time at most* $\mathcal{O}(n^3) + k^{\mathcal{O}(k^2)}n^2 + (k|P|)^{\mathcal{O}(k^2)}n$, *where* k *is the treecut width of the input graph and* n *is the number of its vertices.*

Proof. We begin by invoking Theorem 1 to compute a treecut decomposition of G of width at most $2k$ and then converting it into a nice treecut decomposition (this takes time $k^{\mathcal{O}(k^2)}n^2$ and $\mathcal{O}(n^3)$, respectively). Afterwards, we use Lemma 3 to compute $D(t)$ for each leaf of T, followed by a recursive leaf-to-root application of Lemma 4. Once we compute $D(r)$ for the root r of T, we output **YES** if and only if $D(r) = \{(\emptyset, \emptyset, \emptyset, \emptyset)\}$. ■

5 Kernelizing EDP Parameterized by Feedback Edge Set

The goal of this section is to provide a fixed-parameter algorithm for EDP which exploits the structure of the input graph exclusively. While treecut width cannot be used to obtain such an algorithm, here we show that the feedback edge set number can. More specifically, we obtain a linear kernel for EDP parameterized by the feedback edge set number. Our kernel relies on the following two facts:

Fact 1. *A minimum feedback edge set of a graph* G *can be obtained by deleting the edges of minimum spanning trees of all connected components of* G, *and hence can be computed in time* $\mathcal{O}(|E(G)| \cdot \log |V(G)|)$.

Fact 2 ([12]). EDP *can be solved in polynomial time when* G *is a forest.*

For the purposes of this section, it will be useful to assume that each vertex $v \in V(G)$ occurs in at most one terminal pair, each vertex in a terminal pair has degree 1 in G, and each terminal pair is not adjacent to each other. Note that for any instance without these properties, we can add a new leaf vertex for each terminal, attach it to the original terminal, and replace the original terminal in P with the leaf vertex [11,23].

Consider an instance (G, P) of EDP and let $X \subseteq E(G)$ be a minimum feedback edge set X. Let Y be the set of all vertices incident to at least one edge from X, and let $Q = G - X$. Similarly as before, given a subgraph H of G, we say that H contains an *unmatched* terminal s if $\{s,t\} \in P$, $s \in V(H)$ and $t \notin V(H)$. We begin with two simple reduction rules which allow us to remove degree 2 vertices and leaves not containing a terminal.

Reduction Rule 1. *Let $v \in V(G)$ be such that $|N_G(v)| = 1$. If v is not a terminal, then delete v from G.*

Reduction Rule 2. *Let $v, a, b \in V(G)$ be such that $N_G(v) = \{a, b\}$ and $\{a, b\} \notin E$. Then delete v and add the edge ab into E. Furthermore, if $\{a, v\}$ or $\{v, b\}$ were in X then add $\{a, b\}$ in X.*

Of crucial importance is our third rule, which allows us to prune the instance of subtrees with a single edge to Y. For a subgraph H of G, recall that $P_H \subseteq P$ denotes the subset of terminal pairs whose both endpoints lie in H.

Reduction Rule 3. *Let L be a connected component of $G - Y$ such that there exists a single edge $\{\ell \in L, y \in Y\}$ between L and Y.*

*a. If L contains no unmatched terminal and (L, P_L) is a **YES**-instance of* EDP*, then set $P := P \setminus P_L$ and $G := G \setminus V(L)$.*
*b. If L contains precisely one unmatched terminal s where $\{s, t\} \in P$ and the instance $(L, P_L \cup \{s, \ell\})$ is a **YES**-instance of* EDP*, then set $P := P \setminus P_L$ and $G := ((V(G) \setminus V(L)) \cup \{s\}, (E(G) \setminus (E(L) \cup \{\{\ell, y\}\}) \cup \{y, s\})$.*
*c. In all other cases, (G, P) is a **NO**-instance of* EDP.

After exhaustive application of Reduction Rules 1, 2 and 3 we observe that each leaf in Q is either in Y or adjacent to a vertex in Y. The simple rule below is required to obtain a bound on the number of leaves in Q in the subsequent step.

Reduction Rule 4. *If $\{a, b\} \in P$ and a, b are leaves in G such that $N(a) = N(b)$, then remove a and b from G and P.*

After exhaustive application of Reduction Rules 1, 2, 3 and 4, we can prove:

Lemma 5. *If Q contains more than $4|X|$ leaves, then (G, P) is a **NO**-instance.*

Finally, we put everything together in the proof of the desired theorem.

Theorem 3. EDP *admits a linear kernel parameterized by the feedback edge set number of the input graph.*

Proof. Let (G, P) be an instance of EDP; w.l.o.g. we assume that G is and remains connected (note that if G becomes disconnected due to a later application of a reduction rule, one can simply kernelize each connected component separately). We begin by computing a minimum feedback edge set X of G using Fact 1. We then exhaustively apply Reduction Rules 1, 2, 3 and 4; since EDP is polynomial-time tractable by Fact 2, the time required to apply each rule is easily seen to be polynomial.

After no more rules can be applied, we compare the number of leaves in $Q = G - X$ to $|X|$ (note that $V(G) = V(X)$). If Q contains more than $4|X|$ leaves, then we reject in view of Lemma 5. On the other hand, if Q contains at most $4|X|$ leaves, then we claim that Q contains at most $11|X| - 2$ vertices. Indeed, the number of vertices of degree at least 3 in a forest is at most equal to the number of leaves minus two and in particular Q has at most $4|X| - 2$ vertices of degree at least 3. Moreover, due to the exhaustive application of Reduction Rule 2 it follows that the number of degree two vertices is at most $|X|$. And so, by putting together the bounds on $|Y|$ along with the number of vertices of degree 1 and 2 and 3, we obtain $|V(G)| = |V(Q)| \leq 2|X| + 4|X| + |X| + 4|X| - 2$, as claimed. ■

References

1. Chekuri, C., Khanna, S., Bruce Shepherd, F.: An O(sqrt(n)) approximation and integrality gap for disjoint paths and unsplittable flow. Theory Comput. **2**(7), 137–146 (2006)
2. Courcelle, B., Makowsky, J.A., Rotics, U.: Linear time solvable optimization problems on graphs of bounded clique-width. Theory Comput. Syst. **33**(2), 125–150 (2000)
3. Cygan, M., et al.: Parameterized Algorithms. Springer, Heidelberg (2015)
4. Cygan, M., et al.: Parameterized Algorithms. Springer, Berlin (2014)
5. Diestel, R.: Graph Theory, 4th edn. Springer, Heidelberg (2010)
6. Downey, R.G., Fellows, M.R.: Fundamentals of Parameterized Complexity. Texts in Computer Science. Springer, Heidelberg (2013)
7. Ene, A., Mnich, M., Pilipczuk, M., Risteski, A.: On routing disjoint paths in bounded treewidth graphs. In: Proceedings of the SWAT 2016. LIPIcs, vol. 53, pp. 15:1–15:15. Schloss Dagstuhl (2016)
8. Fleszar, K., Mnich, M., Spoerhase, J.: New algorithms for maximum disjoint paths based on tree-likeness. In: Proceedings of the ESA 2016, pp. 42:1–42:17 (2016)
9. Ganian, R., Kim, E.J., Szeider, S.: Algorithmic applications of tree-cut width. In: Italiano, G.F., Pighizzini, G., Sannella, D.T. (eds.) MFCS 2015. LNCS, vol. 9235, pp. 348–360. Springer, Heidelberg (2015). https://doi.org/10.1007/978-3-662-48054-0_29
10. Ganian, R., Klute, F., Ordyniak, S.: On structural parameterizations of the bounded-degree vertex deletion problem. In: Proceedings of the STACS 2018, pp. 33:1–33:14 (2018)
11. Ganian, R., Ordyniak, S., Sridharan, R.: On structural parameterizations of the edge disjoint paths problem. In Proceedings of the ISAAC 2017. LIPIcs, vol. 92, pp. 36:1–36:13. Schloss Dagstuhl - Leibniz-Zentrum fuer Informatik (2017)

12. Garg, N., Vazirani, V.V., Yannakakis, M.: Primal-dual approximation algorithms for integral flow and multicut in trees. Algorithmica **18**(1), 3–20 (1997)
13. Karp, R.M.: On the computational complexity of combinatorial problems. Networks **5**(1), 45–68 (1975)
14. Kawarabayashi, K., Kobayashi, Y., Kreutzer, S.: An excluded half-integral grid theorem for digraphs and the directed disjoint paths problem. In: Proceedings of the STOC 2014, pp. 70–78. ACM (2014)
15. Kim, E., Oum, S., Paul, C., Sau, I., Thilikos, D.M.: An FPT 2-approximation for tree-cut decomposition. In: Sanità, L., Skutella, M. (eds.) WAOA 2015. LNCS, vol. 9499, pp. 35–46. Springer, Cham (2015). https://doi.org/10.1007/978-3-319-28684-6_4
16. Kolliopoulos, S.G., Stein, C.: Approximating disjoint-path problems using packing integer programs. Math. Program. **99**(1), 63–87 (2004)
17. Marx, D., Wollan, P.: Immersions in highly edge connected graphs. SIAM J. Discret. Math. **28**(1), 503–520 (2014)
18. Nishizeki, T., Vygen, J., Zhou, X.: The edge-disjoint paths problem is NP-complete for series-parallel graphs. Discret. Appl. Math. **115**(1–3), 177–186 (2001)
19. Robertson, N., Seymour, P.D.: Graph minors XIII. The disjoint paths problem. J. Comb. Theory Ser. B **63**(1), 65–110 (1995)
20. Robertson, N., Seymour, P.D.: Graph minors. XVIII. tree-decompositions and well-quasi-ordering. J. Comb. Theory Ser. B **89**(1), 77–108 (2003)
21. Scheffler, P.: Practical linear time algorithm for disjoint paths in graphs with bounded tree-width. In: Technical report TR 396/1994. FU Berlin, Fachbereich 3 Mathematik (1994)
22. Wollan, P.: The structure of graphs not admitting a fixed immersion. J. Comb. Theory Ser. B **110**, 47–66 (2015)
23. Zhou, X., Tamura, S., Nishizeki, T.: Finding edge-disjoint paths in partial k-trees. Algorithmica **26**(1), 3–30 (2000)

Geometric Representations of Dichotomous Ordinal Data

Patrizio Angelini[1(✉)], Michael A. Bekos[1], Martin Gronemann[2], and Antonios Symvonis[3]

[1] Wilhelm-Schickhard-Institut für Informatik, Universität Tübingen, Tübingen, Germany
{angelini,bekos}@informatik.uni-tuebingen.de
[2] Institut für Informatik, Universität zu Köln, Köln, Germany
gronemann@informatik.uni-koeln.de
[3] School of Applied Mathematical and Physical Sciences, NTUA, Athens, Greece
symvonis@math.ntua.gr

Abstract. Motivated by the study of ordinal embeddings in machine learning and by the recognition of Euclidean preferences in computational social science, we study the following problem. Given a graph G, together with a set of relationships between pairs of edges, each specifying that an edge must be longer than another edge, is it possible to construct a straight-line drawing of G satisfying all these relationships?

We mainly consider a dichotomous setting, in which edges are partitioned into *short* and *long*, as otherwise there are simple (planar) instances that do not admit a solution. Since the problem is NP-hard even in this setting, we study under which conditions a solution always exists. We prove that degeneracy-2 graphs, subcubic graphs, double-wheels, and 4-colorable graphs in which the short edges induce a caterpillar always admit a realization. These positive results are complemented by negative instances, even when the input graph is composed of a maximal planar graph, namely a double-wheel graph, and an edge. We conjecture that planar graphs always admit a (not necessarily planar) realization in the dichotomous setting.

Keywords: Geometric representations · Ordinal data · Graph drawing

1 Introduction

When modeling an application domain by means of a graph, it is sometimes necessary to associate weights to the edges, in order to represent the strength of the binary relationship among the actors. When a *quantitative* information is available, it is possible to assign specific values for the weights; this is however not always the case, as many application domains only carry an *ordinal* information.

I. Sau and D. M. Thilikos (Eds.): WG 2019, LNCS 11789, pp. 205–217, 2019.
https://doi.org/10.1007/978-3-030-30786-8_16

A typical example, which first triggered our work, comes from automatic classification systems (say, of images), whose training set often consists of a set of answers to questions of the type: *Given three images A, B, and C; is A more similar to B or to C?* This setting is modeled in machine learning by the notion of *ordinal embedding* [21]. Consider a set of objects $x_1, \ldots, x_n$ in an abstract space, and suppose that the only information about their displacement is a set of *ordinal constraints* of the form $dist(x_i, x_j) < dist(x_k, x_l)$. The goal is to find a point configuration $p_1, \ldots, p_n$ in the d-dimensional Euclidean space $\mathbb{R}^d$ that, by preserving as many ordinal constraints as possible, returns a good approximation of the displacement of $x_1, \ldots, x_n$.

Ordinal embeddings are also known as *non-metric multi-dimensional scaling*, as opposed to the classical *multi-dimensional scaling* [7], where the goal is to preserve the actual (possibly weighted) graph distances. Ordinal embeddings were first applied in the 60's by Shepard [19,20] and Kruskal [14,15] to the analysis of psychometric data. They have been later studied in several domains, mainly from a heuristic point of view [1,3,13,17,22]. Alon et al. [5] also provided approximation algorithms. A major contribution is due to Terada and von Luxburg [21], who proved that, if n is large enough, it suffices to know the h nearest neighbors of each point, for some parameter h, to approximately reconstruct the point set.

Another example comes from the recognition of Euclidean multidimensional preferences [6,9,10,16] in computational social science. In this setting, two sets of objects are given, the *decision makers* and the *alternatives*, and the goal is to place them in $\mathbb{R}^d$ so that their distances represent the preferences of each decision maker. One can see this problem as an ordinal embedding one in which the graph is bipartite; however, the goal here is to test if all constraints can be satisfied rather than to find an approximation. Efficient algorithms exist when $d = 1$ [9,10], while for any $d \geq 2$ the problem is as hard as the existential theory of the reals (that is, it is $\exists\mathbb{R}$-hard, and thus NP-hard) [16].

We approach these problems from a graph drawing perspective. Given a graph G and a set R of relationships, each specifying that an edge must be longer than another, the goal is to construct a *realization* of $\langle G, R \rangle$, i.e., a straight-line drawing of G in $\mathbb{R}^2$ satisfying R. We consider the special case in which R induces a partition of the edges of $G = (V, E)$ into k sets $E_1, \ldots, E_k$ and the goal is to find a straight-line drawing of G in which all the edges in E_i are (strictly) shorter than those in E_j, for $1 \leq i < j \leq k$. Aichholzer et al. [2] studied this problem when R describes a total order on the edges (i.e., $k = |E|$) with the extra requirement that the drawing of G has to be planar. Another variant of this problem, which has been proven to be NP-hard with and without the planarity requirement [8,11,18], requires all edges to have the same length.

We mainly focus on the case $k = 2$, where the partition consists of *short* and *long* edges; in this case, R is *dichotomous*. In image classification, the dichotomous setting may originate from answers to the question: *Given two images A and B, are they similar to each other?* Also, the setting in which only the h nearest neighbors of each vertex are known [21] is equivalent to the one in which h of the edges incident to a vertex must be shorter than the remaining ones; note that in this case the short/long dichotomy is defined on a local scale. Finally, the recognition of d-dimensional Euclidean preferences has also been studied when a decision maker either likes (short edge) an alternative or not (long edge) [12,16], both for global and local dichotomies. Since the $\exists\mathbb{R}$-hardness [16] extends to these relaxations, our problem is $\exists\mathbb{R}$-hard, and thus NP-hard, even when G is bipartite.

Alam et al. [4] independently proved NP-hardness, when G is complete, and that degree-2 contractible graphs always admit realizations in the dichotomous setting; a graph is *degree-2 contractible* if there is a sequence of edge-contractions, each involving an edge with a vertex of degree at most 2, that results in the removal of all edges. This includes 2-trees, and thus series-parallel graphs.

Our Contribution. In Sect. 2, we extend the result by Alam et al. [4] to any set of (not necessarily dichotomous) relationships. Given that both hardness results [4,16] hold when G is a non-planar graph, we study the role of planarity in our problem. As noted in [2], one cannot hope for planar realizations, even for small graphs like K_4. Thus, we investigate whether planar graphs admit not-necessarily planar realizations. While we could construct a negative instance for $k = 3$ in which G is planar, even if we allow non-planar realizations (Sect. 2), we are not aware of any planar instance for the dichotomous setting. For this reason, we focus the rest of our study on the dichotomous setting. In Sect. 3, we present negative instances when G is a maximal planar graph plus an edge (including K_5), and give algorithms to construct realizations for any set of dichotomous relationships for several graph classes: *degeneracy-2 graphs* (i.e., graphs where every subgraph has at least one vertex of degree at most 2), subcubic graphs, and *double wheels* (i.e., maximal planar graphs composed of a cycle and two vertices, each connected to all vertices of the cycle). Since degree-2 contractible graphs have degeneracy 2, we obtain another strengthening of the result by Alam et al. [4]. We finally prove that every 4-colorable graph admits a realization, if the subgraph induced by the short edges is a *caterpillar*, namely a tree such that the removal of all the leaves yields a path. This includes planar graphs, and (even non-planar) bipartite graphs, for which the problem is NP-hard if we neglect the additional condition on the short edges. We conclude with open problems in Sect. 4.

Preliminaries. We start with an observation, based on the fact that, if there exists a set of long edges whose removal creates different connected components, we can draw such components separately and place them far away from each other.

Observation 1. *Let G be a graph and let R be a set of relationships partitioning its edges into $k \geq 2$ sets $E_1, \ldots, E_k$. Then, it is not a loss of generality to assume that the subgraph obtained by removing the longest edges (that is, E_k) is connected and spanning. In particular, if R is dichotomous, the subgraph induced by the short edges is connected and spanning.*

Next, we describe a configuration that enforces a crossing between specific edges, which we use for constructing instances that do not admit any realization.

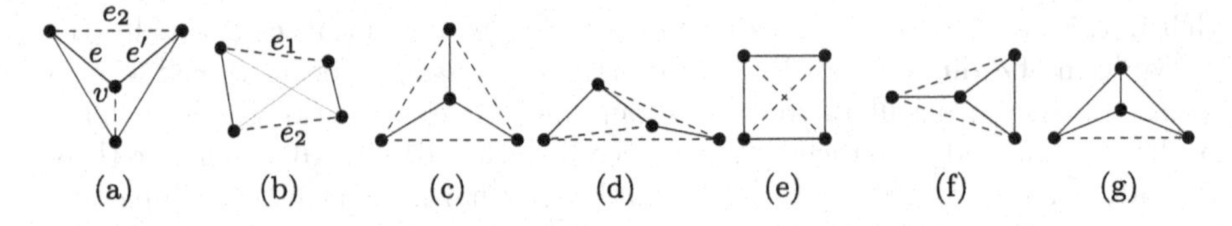

Fig. 1. (a)–(b) Illustrations for the proof of Lemma 1; (c)–(g) realizations of K_4 for any combination of short (solid) and long (dashed) edges that complies with Observation 1.

Lemma 1. *Let G be a graph containing K_4 as a subgraph, and let R be a set of relationships partitioning its edges into $k \geq 2$ sets $E_1, \ldots, E_k$. Let $e_1 \in E_i$ and $e_2 \in E_j$ be two independent edges of the K_4, with $i \leq j$. Suppose that each of the remaining edges of the K_4 belongs to one of the sets $E_1, \ldots, E_{i-1}$. Then, in any realization of $\langle G, R \rangle$, edges e_1 and e_2 cross.*

Proof. Assume there is a realization of $\langle G, R \rangle$ in which e_1 and e_2 do not cross. We distinguish two cases, based on whether the four vertices of the K_4 are in convex position or not. In the latter case, one of them, say v, lies inside the triangle created by the remaining three. Note that in this realization, no two edges of the K_4 cross each other. However, v is incident to an edge, say e_1, that is longer than its other two incident edges, say e and e' (see Fig. 1a). Notice that the angle between e_1 and e (e_1 and e') must be smaller than 90°, as otherwise the opposite edge would be longer than e_1. Hence, the angle between e and e' is larger than 180°, a contradiction to the fact that v lies inside the triangle.

We now consider the case in which the endpoints of e_1 and e_2 are in convex position. Thus, together with two shorter edges, they form a planar convex quadrilateral (see Fig. 1b). Notice that every interior angle of the quadrilateral is formed by exactly an edge in $E_1, \ldots, E_{i-1}$ and one in $E_i, \ldots, E_k$. Hence, all interior angles are smaller than 90°, as otherwise the edges corresponding to the two diagonals of the quadrilateral would be longer than e_1 or e_2. As a result, their sum is smaller than 360°, a contradiction. □

The fact that a realization of K_4 may require a crossing was already observed by Aichholzer et al. [2] in the dichotomous setting. Lemma 1 restates this result in a slightly stronger form, since it enforces a specific pair of edges to cross, and holds for any $k \geq 2$. On the other hand, we observe that when crossings are allowed, then K_4 is realizable for any set of dichotomous relationships; see Fig. 1.

Another observation for the dichotomous setting is that it is not a loss of generality to assume that G is biconnected. To see this, consider any cut-vertex v of G, which defines several subgraphs of G that are all incident to v. Suppose that each of these subgraphs admits a realization with respect to R. By scaling up or down each of these realizations uniformly, we can ensure that each short edge of R is shorter than 1 and each long edge is longer than 1. This results in a realization for $\langle G, R \rangle$. On the other hand, any realization of $\langle G, R \rangle$ clearly contains a realization of each of the subgraphs defined by v.

We conclude this section with a positive result for the dichotomous setting when the subgraph induced by the short edges has some special structure.

Lemma 2. *Let G be a graph and R a set of dichotomous relationships. Then, $\langle G, R\rangle$ admits a realization, if the subgraph induced by the short edges is (i) a rooted spanning tree T and there is no long edge connecting two vertices whose depths in T differ by at most 1, or (ii) a Hamiltonian cycle of G.*

Proof. For (i), we place the vertices on the x-axis, such that each vertex lies in a small neighborhood of the point with x-coordinate equal to its depth in T. Hence, the minimum (maximum) distance between any two vertices joined by a long (short) edge is almost 2 (slightly more than 1). For (ii), we place the vertices of G at the corners of a regular n-gon Q_n, as they appear along the Hamiltonian cycle. Hence, the short edges are drawn as the boundary edges of Q_n, while the long edges are drawn as chords of Q_n. Consequently, the length of each of the short edges of R is smaller than the length of the long edges of R. □

2 Realizability When $k \geq 3$

In this section, we consider instances $\langle G, R\rangle$ in which the set of relationships R is not dichotomous, that is, it defines a partition $E_1, \ldots, E_k$ of the edges of G such that $k \geq 3$. We first prove that a negative instance $\langle G, R\rangle$ exists even if $k = 3$ (that is, R is *trichotomous*) and G is a planar 3-tree with seven vertices. On the positive side we show that, if G is a degree-2 contractible graph, then $\langle G, R\rangle$ can always be realized for any value of k, which generalizes the result by Alam et al. [4] from the dichotomous setting to the general one.

Theorem 1. *For every $n \geq 7$, there exists an n-vertex planar 3-tree G and a set of trichotomous relationships R such that $\langle G, R\rangle$ does not admit any realization.*

Proof. Let G be the planar 3-tree on seven vertices of Fig. 2a. The set R of trichotomous relationships is such that $E_1 = \{(u_1, u_6), (u_1, u_7), (u_2, u_5), (u_2, u_6), (u_3, u_5), (u_3, u_7)\}$, $E_2 = \{(u_1, u_4), (u_2, u_4), (u_3, u_4)\}$, and $E_3 = \{(u_1, u_2), (u_2, u_3), (u_3, u_1), (u_4, u_5), (u_4, u_6), (u_4, u_7)\}$. Suppose, for a contradiction, that

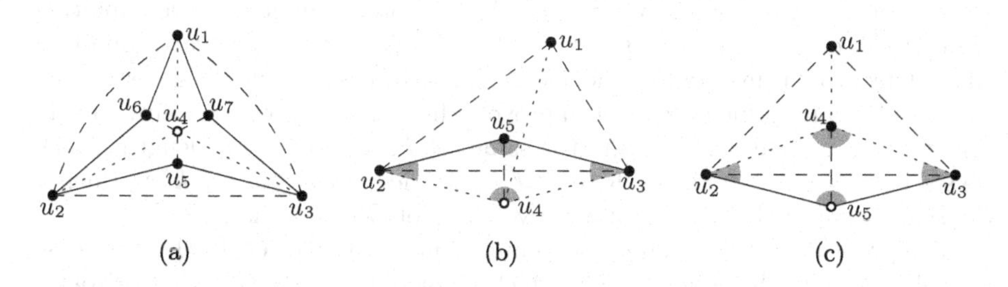

Fig. 2. Illustration for the proof of Theorem 1. The edges of E_1, E_2 and E_3 are illustrated solid, dotted and dashed, respectively.

$\langle G, R \rangle$ admits a realization. We distinguish two cases: (i) at least one of the edges of E_2 crosses an outer edge of G, and (ii) none of the edges of E_2 crosses an outer edge of G. Note that the outer edges of G belong to E_3.

Consider Case (i) and assume w.l.o.g. that (u_1, u_4) crosses (u_2, u_3); see Fig. 2b. By applying Lemma 1 to the K_4 formed by the vertices $\langle u_2, u_3, u_4, u_5 \rangle$, we have that edges (u_2, u_3) and (u_4, u_5) cross. We claim that the gray-shaded angle at u_4 is at least 120°. To see this, observe that vertices $\langle u_1, u_2, u_4 \rangle$ and $\langle u_1, u_3, u_4 \rangle$ form two triangles, whose longest edges are (u_1, u_2) and (u_1, u_3), respectively. Thus, the angles opposite to these edges are at least 60° each. Since these two angles together form the gray-shaded angle at u_4, the claim holds. Since the edges (u_2, u_5) and (u_3, u_5) are both shorter than the edges (u_2, u_4) and (u_3, u_4), the gray-shaded angle at u_5 is larger than the one at u_4; hence, it is larger than 120°. Also, in both triangles formed by $\langle u_2, u_4, u_5 \rangle$ and $\langle u_3, u_4, u_5 \rangle$, the longest edge is (u_4, u_5). Thus, each of the gray-shaded angles at u_2 and u_3 is larger than 60°. Since the sum of the angles of the quadrilateral formed by $\langle u_2, u_4, u_3, u_5 \rangle$ is larger than 360°, we have a contradiction.

Consider now Case (ii). As in the previous case, edges (u_2, u_3) and (u_4, u_5) cross, by Lemma 1. Analogously, (u_1, u_3) and (u_4, u_7) cross, and (u_1, u_2) and (u_4, u_6) cross, by Lemma 1. These observations together with the assumption of Case (ii) imply that vertex u_4 must lie inside the triangle formed by $\langle u_1, u_2, u_3 \rangle$; see Fig. 2c. In this case, the edges (u_1, u_4), (u_2, u_4), and (u_3, u_4) form three angles around u_4, one of which is at least 120°; say w.l.o.g. the angle between (u_2, u_4) and (u_3, u_4). Since the edges (u_2, u_5) and (u_3, u_5) are both shorter than the edges (u_2, u_4) and (u_3, u_4), the gray-shaded angle at u_5 is larger than the one at u_4; hence, it is larger than 120°. Since in both triangles formed by $\langle u_2, u_4, u_5 \rangle$ and $\langle u_3, u_4, u_5 \rangle$ the longest edge is (u_4, u_5), each of the gray-shaded angles at u_2 and u_3 is larger than 60°, and thus the sum of the angles of the quadrilateral formed by $\langle u_2, u_4, u_3, u_5 \rangle$ is larger than 360°; a contradiction.

By Cases (i)–(ii), $\langle G, R \rangle$ does not admit a realization. Since any planar graph that contains G as a subgraph has the same property, the statement follows. □

We now present our positive result for degree-2 contractible graphs.

Theorem 2. *Let G be a degree-2 contractible graph and let R be a set of relationships partitioning its edges into $k \geq 2$ sets. Then, $\langle G, R \rangle$ admits a realization.*

Proof. Let $v_1, \dots, v_n$ be the vertices of G in the order implied by the definition of degree-2 contractible graphs, and let $E_1, \dots, E_k$ be the given partition of its edges. Our proof is inspired by the one in [4] for the case $k = 2$. Beside extending this result to any value of k, we also prove a slightly stronger statement, namely that we can draw all the edges in the same set E_i with the same length, which we set to $\ell_i = k + i$. As a result, $\frac{\ell_i}{2} < \ell_j < 2\ell_i$, for any $1 \leq i, j \leq k$, and hence the ratio between the lengths of any two edges of G is less than 2.

The proof is by induction on the number m of edges of G. In the base case $m = 1$, we draw the unique edge, which belongs to a set E_j, as a horizontal line segment of length ℓ_j. In the inductive case, in which G has $m > 0$ edges, we consider the last vertex v_n in the order $v_1, \dots, v_n$ of the vertices. Let G^- be the graph obtained from G by contracting one of the edges incident to v_n. Note that, if v_n has degree 2 in G, this contraction may introduce an edge in G^-, between the two neighbors of v_n, that does not belong to G. In this case, we add this edge to set E_1.

By induction, G^- admits a drawing Γ^- in which every edge in set E_i, with $1 \leq i \leq k$, has length ℓ_i. If v_n has degree 1 in G, we construct a drawing Γ of G in which every edge in set E_i, with $1 \leq i \leq k$, has length ℓ_i by placing v_n in Γ^- on any point at distance ℓ_j from the unique neighbor of v_n. If v_n has degree 2 in G, let v_l and v_r be the two neighbors of v_j in G. Since v_n has been contracted to one of v_l and v_r in order to obtain G^-, edge (v_l, v_r) exists in G^- and belongs to some set E_j, with $1 \leq j \leq k$ (recall that $j = 1$, if (v_l, v_r) does not belong to G). Also, let E_x and E_y, with $1 \leq x, y \leq k$, be the sets that contain edges (v_n, v_l) and (v_n, v_r), respectively. Consider two circles C_l and C_r with radii ℓ_x and ℓ_y, centered at v_l and v_r in Γ^-, respectively. Since v_l and v_r have distance ℓ_j in Γ^-, and since $\ell_x + \ell_y > \ell_j$ (given that $\ell_x, \ell_y > \frac{\ell_j}{2}$), circles C_l and C_r intersect in two distinct points. Hence, we can obtain Γ by placing v_n in Γ^- on one of these two points, which ensures that edges (v_n, v_l) and (v_n, v_r) have length ℓ_x and ℓ_y, respectively. Since v_n has degree at most 2, by definition of degree-2 contractible graph, this concludes the proof. □

3 Realizability in the Dichotomous Setting

In this section we study whether certain graph classes admit realizations for any set of dichotomous relationships, and provide positive and negative examples. While $\langle K_4, R \rangle$ admits a realization for any set R of dichotomous relationships, we show that for the complete graph K_5 there exists a set of dichotomous relationships making it not realizable. We prove this statement in the next theorem for a more general class of graphs, whose members are almost planar, in the sense that the removal of a single edge is sufficient to ensure planarity.

Theorem 3. *For every odd $n \geq 5$, there exists an n-vertex graph G, composed of a maximal planar graph and of a single edge, and a set of dichotomous relationships R such that $\langle G, R \rangle$ does not admit any realization.*

Proof. Graph G is initialized as a *double wheel* on n vertices, i.e., starting from two *central* vertices, s and t, and a cycle $C = v_1, \ldots, v_{n-2}$, we connect every v_i to both s and t; see Fig. 3a. All the edges of the cycle C are long, while all the edges incident to s and t, are short. Finally, we add a long edge between s and t.

Note that any two consecutive vertices v_i and v_{i+1} along C form a K_4 together with s and t, in which (s, t) and (v_i, v_{i+1}) form an independent pair of long edges. Hence, by Lemma 1, edge (s, t) must cross every edge (v_i, v_{i+1}) of C. Since n is odd, this implies that v_1 and v_{n-2} are in the same half-plane defined by the line through s and t, but then (v_1, v_{n-2}) cannot cross (s, t), a contradiction. Note that, when n is even, a realization exists; see Fig. 3b. □

Theorem 3 implies that there exist negative instances of the problem in which the input graph G is composed of a double wheel and of an edge between its two

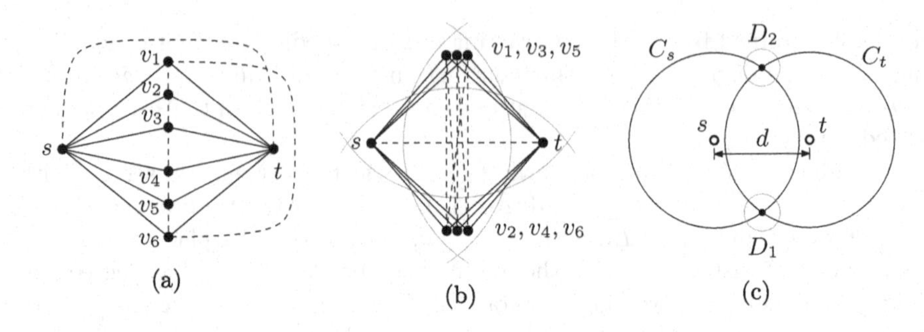

Fig. 3. Illustrations for the proofs of (a–b) Theorem 3 and of (c) Theorem 4.

central vertices; also, for $n = 5$ the obtained graph is the complete graph K_5, which has degree at most 4. In the following, we prove that this result is tight in two senses. First, we show in Theorem 4 that, if G is a double wheel (without the edge between its central vertices), then $\langle G, R \rangle$ is a positive instance for every set R of dichotomous relationships. Second, we show in Theorem 6 that, if G is a *subcubic* graph (i.e., its vertices have degree at most 3), then $\langle G, R \rangle$ is a positive instance for every set R of dichotomous relationships.

Note that a subcubic graph is not necessarily degree-2 contractible, even if the graph is not *cubic* (i.e., its vertices have degree exactly 3). For example, any subdivision of a cubic graph is not degree-2 contractible. On the other hand, if a subcubic graph has a vertex of degree at most 2, then it has degeneracy at most 2. Thus, as an auxiliary result for Theorem 6, we prove in Theorem 5 that a degeneracy-2 graph always admits a realization in the dichotomous setting. This result is interesting per sé, as it provides another strengthening of the result by Alam et al. [4], since a degree-2 contractible graph has degeneracy at most 2.

Theorem 4. *Let G be a double wheel graph on $n \geq 4$ vertices and let R be any set of dichotomous relationships. Then, $\langle G, R \rangle$ admits a realization.*

Proof sketch. We place the two central vertices s and t of G at a distance $d < \sqrt{3}$ to be defined later. Consider two circles C_s and C_t with radius 1, centered at s and t, respectively; see Fig. 3c. Since $d < \sqrt{3}$, C_s and C_t intersect at two points p_1 and p_2 whose distance is larger than 1. Thus, there exist two small disks D_1 and D_2, centered at p_1 and p_2, respectively, with radius $\epsilon > 0$ such that the distance between any point inside D_1 and any point inside D_2 is larger than 1.

Our strategy is to place every vertex v_i of the cycle $C = v_1, \ldots, v_{n-2}$ of G, except for at most two of them, in the interior of either D_1 or D_2. Note that each of these disks is split by C_s and C_t into four subregions, which realize all possible combinations of distances, larger or smaller than 1, from s and t. Thus, when we place a vertex v_i in a disk, we choose the subregion that realizes the required lengths of edges (s, v_i) and (t, v_i).

We first place v_1 in D_1. Then, for each $i = 2, \ldots, n-2$, we place v_i inside D_1 or D_2 based on the placement of v_{i-1} and on the required length of edge (v_{i-1}, v_i). Namely, if this edge is short, we place v_i in the same disk as v_{i-1}, otherwise in the other disk. This strategy works well when the number of long edges in C is even (including the case in which all edges of C are short), since the placement of v_{n-2} inside D_1 or D_2 determined by our strategy is coherent with the required length of edge (v_1, v_{n-2}). When the number of long edges along C is odd, we have to find a way to "break the parity".

With this in mind we distinguish two cases, based on whether there exist two consecutive long edges (v_{i-1}, v_i) and (v_i, v_{i+1}) along C, or not. In the former case, we place vertex v_i in a suitable position outside the two disks that respects the required distances to s and t, and allows v_{i-1} and v_{i+1} to be placed in different disks, hence breaking the parity. Note that, if we choose the value of the distance d between s and t to be smaller than 1, then we can guarantee a position for v_i. In the latter case, we consider three consecutive edges along C such that (v_{i-1}, v_i) is short, (v_i, v_{i+1}) is long, and (v_{i+1}, v_{i+2}) is short, and we break the parity by placing v_{i-1} and v_{i+2} in the same disk. This is done via an additional case analysis, based on the required lengths of the edges connecting v_i and v_{i+1} to s and t. Note that, in one of the cases, we need the value of the distance d between s and t to be larger than 1 (in particular, we set it equal to $1.5 < \sqrt{3}$). Since the distance between D_1 and D_2 is larger than 1 for any value $d < \sqrt{3}$, and since only one of the cases needs to be applied in order to break the parity, we can set the required value at the beginning of the construction. □

Theorem 5. *Let G be a degeneracy-2 graph and let R be any set of dichotomous relationships. Then, $\langle G, R\rangle$ admits a realization s.t. all vertices of G lie on a circle $\mathcal{C}$ with radius $\sqrt{2}/2$ and each short (long) edge in R is smaller (larger) than 1.*

Proof. Let $v_1, \ldots, v_n$ be an ordering of the vertices of G such that vertex v_i has degree at most 2 in the subgraph H_i of G induced by vertices $v_1, \ldots, v_i$, for each $i = 1, \ldots, n$. In the base of the recursion, we place vertex v_1 at any point of $\mathcal{C}$. For the recursive step of our algorithm, assume that we have computed a realization of the subgraph H_{i-1} of G induced by the vertices $v_1, \ldots, v_{i-1}$, for some $i = 2, \ldots, n-1$ such that: (I.1) all vertices of H_{i-1} lie on $\mathcal{C}$, (I.2) no two vertices of H_{i-1} are at antipodal points of $\mathcal{C}$, (I.3) each short (long) edge of H_{i-1} in R has length smaller (larger) than 1. In the following, we describe how to determine a position of v_i, such that (I.1)–(I.3) are satisfied for graph H_i.

Recall that vertex v_i has at most two neighbors in H_{i-1}. Assume first that v_i has exactly two neightbors in H_{i-1}, say v_j and v_k; see Fig. 4a. By (I.1) and (I.2), vertices v_j and v_k lie on two points, say p_j and p_k of $\mathcal{C}$, which are not antipodal. Let $\mathcal{C}_j$ and $\mathcal{C}_k$ be two circles of unit radius centered at p_j and p_k, respectively. Since the sum of the radii of $\mathcal{C}_j$ and $\mathcal{C}_k$ is larger than the diameter of $\mathcal{C}$, circles $\mathcal{C}_j$ and $\mathcal{C}_k$ overlap. Further, the overlap of $\mathcal{C}_j$ and $\mathcal{C}_k$ contains an arc of $\mathcal{C}$ of positive length, which we denote by $S(j,k)$. In particular, $S(j,k)$ lies on the shorter of the two arcs of $\mathcal{C}$ defined by p_j and p_k; also, $S(j,k)$ would degenerate to a single point only if p_j and p_k were antipodal. Symmetrically, on the longer of the two arcs of $\mathcal{C}$ defined by p_j and p_k, there is an arc of $\mathcal{C}$ with positive length, which is contained neither in $\mathcal{C}_j$ nor in $\mathcal{C}_k$; we denote this arc by $L(j,k)$. Again, $L(j,k)$ would degenerate to a single point, only if p_j and p_k were antipodal.

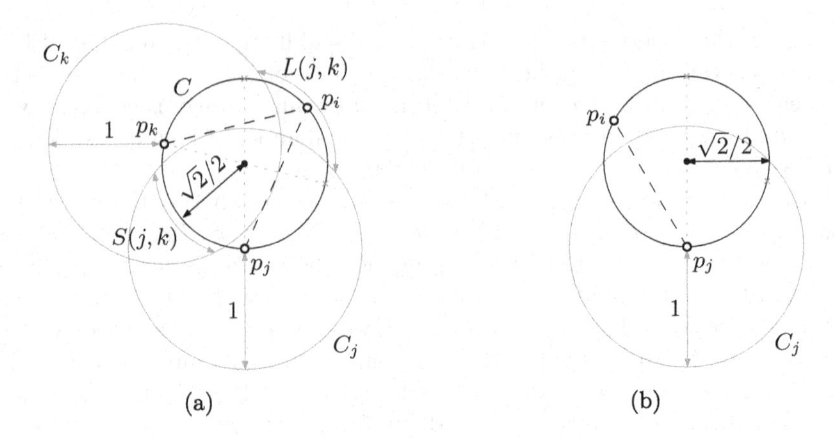

Fig. 4. Illustrations for the proofs of Theorem 5.

If the edges from v_i to v_j and v_k are both short (long), then we place v_i at any point p_i of $S(j,k)$ ($L(j,k)$, respectively), that is antipodal to none of the points where $v_1, \ldots, v_{i-1}$ reside. Clearly, (I.1) and (I.2) are satisfied. If $p_i \in S(j,k)$, then p_i lies in the interior of both $\mathcal{C}_j$ and $\mathcal{C}_k$. Otherwise, p_i is contained neither in $\mathcal{C}_j$ nor in $\mathcal{C}_k$. In both cases (I.3) is also satisfied. Consider now the case where the edge from v_i to v_j is short, while the edge from v_i to v_k is long; the other case is symmetric. We place v_i at any point on the arc of $\mathcal{C}$ that lies in the interior of $\mathcal{C}_j$, but not along $S(j,k)$, and is antipodal to none of the points where $v_1, \ldots, v_{i-1}$ reside. Note that this arc exists since $\mathcal{C}_j$ and $\mathcal{C}_k$ have the same radius and are centered in different points of $\mathcal{C}$. By construction, (I.1)–(I.3) are clearly satisfied.

If v_i has only one neighbor in H_{i-1}, say v_j, placed at point p_j, we consider a circle $\mathcal{C}_j$ with unit radius centered at p_j; see Fig. 4b. We place v_i on a point of $\mathcal{C}$ that is antipodal to none of the points where $v_1, \ldots, v_{i-1}$ reside, and is either in the interior or in the exterior of $\mathcal{C}_j$, based on whether edge (v_i, v_j) is short or long, respectively. Thus, (I.1)–(I.3) are satisfied. □

Theorem 6. *Let G be a subcubic graph and let R be any set of dichotomous relationships. Then, $\langle G, R\rangle$ admits a realization.*

Proof. Let H be the subgraph of G induced by the short edges of R. Suppose first that there exists no vertex of G that is incident to three short edges of R. In this case, graph H forms either a Hamiltonian path or a Hamiltonian cycle of G, since H is connected and spanning by Observation 1, and since G is subcubic. Hence, $\langle G, R\rangle$ admits a realization by Lemma 2.

Suppose now that G has a vertex v that is incident to three short edges of R. If we remove v, then the resulting graph G' becomes 2-degenerate. Note that G' is connected, since G can be assumed biconnected (as observed in Preliminaries). Let $R' \subset R$ be the set of dichotomous relationships for G'. By Theorem 5, $\langle G', R'\rangle$ admits a realization such that all vertices of G' lie on a circle $\mathcal{C}$ with radius $\sqrt{2}/2$ and each short (long) edge of G' in R' has length smaller (larger) than 1. Since

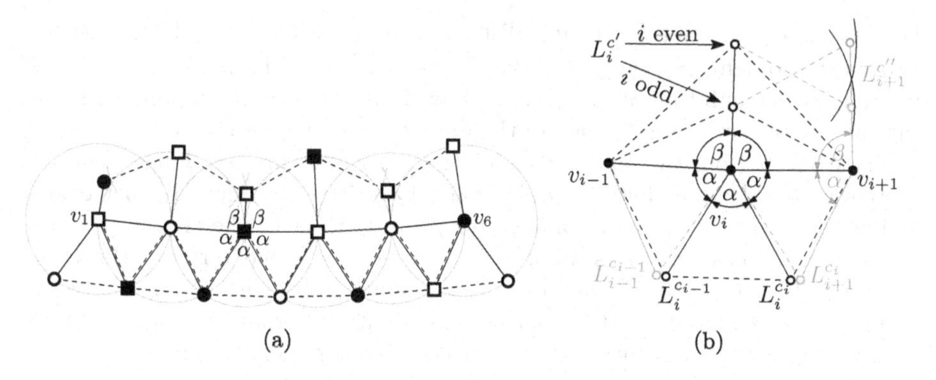

Fig. 5. Illustrations for the proof of Theorem 7.

v is incident to three short edges of R in G, and since the radius of $\mathcal{C}$ is smaller than 1, if we place v at the center of $\mathcal{C}$, then all the edges incident to v are shorter than 1. □

We conclude by extending the previous two results to 4-colorable graphs, which include all planar graphs, all bipartite graphs, and all graphs of maximum degree 4 (except for K_5). For this, however, we have to impose some constraints on the structure of the subgraph induced by the short edges.

Theorem 7. *Let G be a 4-colorable graph and let R be a set of dichotomous relationships such that the subgraph induced by the short edges is a spanning caterpillar. Then, $\langle G, R\rangle$ admits a realization.*

Proof. Recall that a caterpillar is a tree such that the removal of all the leaves yields a path $v_1, \ldots, v_m$, called *spine*. Let $f : V \to C$ be a 4-coloring of G, where $C = \{1, \ldots, 4\}$. Denote by L_i^j the leaves adjacent to v_i with color j. For each pair of consecutive spine vertices v_i and v_{i+1}, with $1 \leq i < m$, we identify a color $c_i \in C$, which we call *common color* of v_i and v_{i+1} with the following properties. The common color c_i is different from the colors of v_i and v_{i+1} in the 4-coloring f; also, no three consecutive vertices share the same common color, i.e., for all $1 < i < m$, it holds that $c_{i-1} \neq c_i$. To compute the common colors, we first set $c_1 \in C \backslash \{f(v_1), f(v_2)\}$; then, for each $i > 1$, we set $c_i = C \backslash \{f(v_i), f(v_{i+1}), c_{i-1}\}$.

We place the spine vertices at unit distances along an arc such that the spine is slightly bent (see Fig. 5a) by setting $\alpha = 60° + \gamma$ and $\beta = 90° - \frac{3}{2}\gamma$ for some small $\gamma > 0$. The leaves of every v_i are placed according to their color. Consider the three sets $L_i^{c_{i-1}}$, $L_i^{c_i}$ and $L_i^{c'}$, where $c' = C \setminus \{c_{i-1}, c_i, f(v_i)\}$. We will say that one of these sets is placed at a *spot*, meaning that all its vertices lie in a small neighborhood around a point.

The first two sets $L_i^{c_{i-1}}$, $L_i^{c_i}$ are placed at unit distance from v_i at the two spots indicated in Fig. 5b. The interior angles $\alpha > 60°$ ensure that $L_i^{c_{i-1}}$ and $L_i^{c_i}$ are at a distance greater than 1 from each other and also from v_{i-1} and v_{i+1} (dashed). The third set $L_i^{c'}$ is placed on the inner side of the arc (see Fig. 5b).

For i even, we place set $L_i^{c'}$ at unit distance from v_i, while for i odd at distance $\frac{1}{2}$ from v_i. For any $0 < \gamma \leq \min\{3.386°, 60°/n\}$, it can be proved that $L_i^{c'}$ is not only far enough from v_{i-1} and v_{i+1}, but also from their corresponding spots that are placed on the inner side of the arc (see, e.g., $L_i^{c''}$ in Fig. 5b).

By construction, the short edges of the caterpillar are not longer than 1. Moreover, the distance between every pair of vertices that have distinct colors and are adjacent to the same spine vertex is larger than 1. The critical part are the distances between leaves adjacent to consecutive spine vertices. However, since $L_i^{c_i}$, $L_{i+1}^{c_i}$ have the common color c_i, there cannot exist an edge between them. The sets placed on the inner side of the arc alternate in their distance from the spine vertex ensuring that their distance is larger than 1. □

4 Conclusions

We studied the problem of constructing geometric realizations of graphs accompanied with ordinal relationships on their edge lengths. We derived constructive algorithms for some graph classes, mainly in the dichotomous case, and presented some negative instances. Our work raises several open problems.

- The most intriguing question is whether every instance $\langle G, R\rangle$ such that G is a planar graph admits a realization in the dichotomous setting. We conjecture an affirmative answer to this question. Towards settling this conjecture, a possible direction is to consider restrictions on the subgraph induced by the short edges that make planar graphs realizable, as in Theorem 7, or to focus on special families of planar graphs, e.g., triangle-free planar graphs, planar graphs with maximum degree 4, or planar 3-trees.
- As an extension of Theorem 5, another interesting class in the dichotomous setting is the one of degeneracy-3 graphs, which include (even non-planar) 3-trees. Note that the negative instances from Theorem 3 have degeneracy 4.
- Another possible direction is to consider instances in which the subgraph induced by the long edges (rather than by the short edges) has a special structure, e.g., it forms a Hamiltonian cycle.
- Finally, from the ordinal embeddings point of view, the variant in which the edge relationships are specified locally to each vertex is equally important.

Acknowledgement. The authors would like to thank Michael Kaufmann and Ulrike von Luxburg for useful discussions.

References

1. Agarwal, S., Wills, J., Cayton, L., Lanckriet, G.R.G., Kriegman, D.J., Belongie, S.J.: Generalized non-metric multidimensional scaling. In: Meila, M., Shen, X. (eds.) AISTATS. JMLR Proceedings, vol. 2, pp. 11–18. JMLR.org (2007)
2. Aichholzer, O., Hoffmann, M., van Kreveld, M.J., Rote, G.: Graph drawings with relative edge length specifications. In: CCCG (2014)

3. Ailon, N.: An active learning algorithm for ranking from pairwise preferences with an almost optimal query complexity. J. Mach. Learn. Res. **13**, 137–164 (2012)
4. Alam, M.J., Kobourov, S.G., Pupyrev, S., Toeniskoetter, J.: Weak unit disk and interval representation of graphs. In: Mayr, E.W. (ed.) WG 2015. LNCS, vol. 9224, pp. 237–251. Springer, Heidelberg (2016). https://doi.org/10.1007/978-3-662-53174-7_17
5. Alon, N., Badoiu, M., Demaine, E.D., Farach-Colton, M., Hajiaghayi, M.T., Sidiropoulos, A.: Ordinal embeddings of minimum relaxation: general properties, trees, and ultrametrics. ACM Trans. Algorithms **4**(4), 46:1–46:21 (2008)
6. Bennett, J.F., Hays, W.L.: Multidimensional unfolding: determining the dimensionality of ranked preference data. Psychometrika **25**(1), 27–43 (1960)
7. Borg, I., Groenen, P.: Modern Multidimensional Scaling: Theory and Applications. Springer, Heidelberg (2005)
8. Cabello, S., Demaine, E.D., Rote, G.: Planar embeddings of graphs with specified edge lengths. J. Graph Algorithms Appl. **11**(1), 259–276 (2007)
9. Chen, J., Pruhs, K., Woeginger, G.J.: The one-dimensional Euclidean domain: finitely many obstructions are not enough. Soc. Choice Welf. **48**(2), 409–432 (2017)
10. Doignon, J., Falmagne, J.: A polynomial time algorithm for unidimensional unfolding representations. J. Algorithms **16**(2), 218–233 (1994)
11. Eades, P., Wormald, N.C.: Fixed edge-length graph drawing is NP-hard. Discret. Appl. Math. **28**(2), 111–134 (1990)
12. Elkind, E., Lackner, M.: Structure in dichotomous preferences. In: Yang, Q., Wooldridge, M. (eds.) IJCAI, pp. 2019–2025. AAAI Press (2015)
13. Jamieson, K.G., Nowak, R.D.: Low-dimensional embedding using adaptively selected ordinal data. In: Allerton Conference on Communication, Control, and Computer, pp. 1077–1084. IEEE (2011)
14. Kruskal, J.: Multidimensional scaling by optimizing goodness of fit to a nonmetric hypothesis. Psychometrika **29**(1), 1–27 (1964)
15. Kruskal, J.: Nonmetric multidimensional scaling: a numerical method. Psychometrika **29**(2), 115–129 (1964)
16. Peters, D.: Recognising multidimensional Euclidean preferences. In: Singh, S.P., Markovitch, S. (eds.) AAAI, pp. 642–648. AAAI Press (2017)
17. Quist, M., Yona, G., Yu, B.: Distributional scaling: an algorithm for structure-preserving embedding of metric and nonmetric spaces. J. Mach. Learn. Res. **5**, 399–420 (2004)
18. Saxe, J.B.: Embeddability of weighted graphs in k-space is strongly NP-hard. In: 17th Allerton Conference on Communication, Control, and Computer, pp. 480–489. IEEE (1979)
19. Shepard, R.N.: The analysis of proximities: multidimensional scaling with an unknown distance function. I. Psychometrika **27**(2), 125–140 (1962)
20. Shepard, R.N.: The analysis of proximities: multidimensional scaling with an unknown distance function. II. Psychometrika **27**(3), 219–246 (1962)
21. Terada, Y., von Luxburg, U.: Local ordinal embedding. In: ICML. JMLR Workshop and Conference Proceedings, vol. 32, pp. 847–855. JMLR.org (2014)
22. Vo, D., Vo, N., Challa, S.: Weighted MDS for Sensor Localization. In: Gervasi, O., Murgante, B., Laganà, A., Taniar, D., Mun, Y., Gavrilova, M.L. (eds.) ICCSA 2008. LNCS, vol. 5073, pp. 409–418. Springer, Heidelberg (2008). https://doi.org/10.1007/978-3-540-69848-7_34

Linear MIM-Width of Trees

Svein Høgemo(✉), Jan Arne Telle, and Erlend Raa Vågset

Department of Informatics, University of Bergen, Bergen, Norway
svein.hogemo@student.uib.no, {jan.arne.telle,erlend.vagset}@uib.no

Abstract. We provide an $O(n \log n)$ algorithm computing the linear maximum induced matching width of a tree and an optimal layout.

Keywords: Width parameters · Exact algorithms · Linear MIM-width · Acyclic graphs

1 Introduction

The study of structural graph width parameters like tree-width, clique-width and rank-width has been ongoing for a long time, and their algorithmic use has been steadily increasing [11,18]. The maximum induced matching width, denoted MIM-width, and the linear variant linear MIM-width, introduced by Vatshelle in 2012 [21], are graph parameters having very strong modelling power. The linear MIM-width parameter asks for a linear layout of vertices such that the bipartite graph induced by edges crossing any vertex cut has a maximum induced matching of bounded size. Belmonte and Vatshelle [2] showed that INTERVAL graphs, CONVEX graphs and PERMUTATION graphs, where clique-width can be proportional to the square root of the number of vertices [10], all have linear MIM-width 1, and that an optimal layout can be found in polynomial time.

Since many well-known classes of graphs have bounded MIM-width or linear MIM-width, algorithms that run in XP time in these parameters will yield polynomial-time algorithms on several interesting graph classes at once. Such algorithms have been developed for many problems: by Bui-Xuan et al. [4] for the class of LCVS-VP - Locally Checkable Vertex Subset and Vertex Partitioning - problems, by Jaffke et al. for non-local problems like FEEDBACK VERTEX SET [14,15] and also for GENERALIZED DISTANCE DOMINATION [13], by Golovach et al. [9] for output-polynomial ENUMERATION OF MINIMAL DOMINATING sets, by Bergougnoux and Kanté [3] for several Connectivity problems, and by Galby et al. for SEMITOTAL DOMINATION [8]. These results give a common explanation for many classical results in the field of algorithms on special graph classes and extends them to the field of parameterized complexity.

Note that very low MIM-width or linear MIM-width still allows quite complex cuts compared to similarly defined graph parameters. For example, carving-width 1 allows just a single edge, maximum matching-width 1 a star graph,

Long version with extra figures and full proofs is published on arxiv [12].

I. Sau and D. M. Thilikos (Eds.): WG 2019, LNCS 11789, pp. 218–231, 2019.
https://doi.org/10.1007/978-3-030-30786-8_17

and rank-width 1 a complete bipartite graph. In contrast, linear MIM-width 1 allows any cut where the neighborhoods of the vertices in a color class can be ordered linearly w.r.t. inclusion. In fact, it is an open problem whether the class of graphs having linear MIM-width 1 can be recognized in polynomial-time or if this is NP-complete. Sæther et al. [19] showed that computing the exact MIM-width and linear MIM-width of general graphs is W[1]-hard and not in APX unless NP = ZPP, while Yamazaki [22] shows that under the small set expansion hypothesis it is not in APX unless P = NP. The only graph classes where we know an exact polynomial-time algorithm computing linear MIM-width are the above-mentioned classes INTERVAL, BI-INTERVAL, CONVEX and PERMUTATION that all have structured neighborhoods implying linear MIM-width 1 [2]. Belmonte and Vatshelle also gave polynomial-time algorithms showing that CIRCULAR ARC and CIRCULAR PERMUTATION graphs have linear MIM-width at most 2, while DILWORTH k and k-TRAPEZOID have linear MIM-width at most k [2]. Recently, Fomin et al. [7] showed that linear MIM-width for the very general class of H-GRAPHS is bounded by $2|E(H)|$, and that a layout can be found in polynomial time if given an H-representation of the input graph. However, none of these results compute the exact linear MIM-width. On the negative side, Mengel [16] has shown that STRONGLY CHORDAL SPLIT graphs, CO-COMPARABILITY graphs and CIRCLE graphs all can have MIM-width, and linear MIM-width, linear in the number of vertices.

Just as linear MIM-width can be seen as the linear variant of MIM-width, path-width can be seen as the linear variant of tree-width. Linear variants of other well-known parameters like clique-width and rank-width have also been studied. Arguably, the linear variant of MIM-width commands a more noteworthy position, since in contrast to these other linear parameters, for almost all well-known graph classes where the original parameter (MIM-width) is bounded, but clique-width is unbounded, then also the linear variant (linear MIM-width) is bounded.

In this paper we give an $O(n \log n)$ algorithm computing the linear MIM-width of an n-node tree. This is the first graph class of linear MIM-width larger than 1 having a polynomial-time algorithm computing linear MIM-width and thus constitutes an important step towards a better understanding of this parameter. The path-width of trees was first studied in the early 1990s by Möhring [17], with Ellis et al. [6] giving an $O(n \log n)$ algorithm computing an optimal path-decomposition, and Skodinis [20] an $O(n)$ algorithm. In 2013 Adler and Kanté [1] gave linear-time algorithms computing the linear rank-width of trees and also the linear clique-width of trees, by reduction to the path-width algorithm. Even though linear MIM-width is very different from path-width, the basic framework of our algorithm is similar to the path-width algorithm in [6].

In Sect. 2 we give some standard definitions and prove the Path Layout Lemma, that if a tree T has a path P such that all components of $T \setminus N[P]$ have linear MIM-width at most k then T itself has a linear layout with linear MIM-width at most $k+1$. We use this to prove a classification theorem stating that a tree T has linear MIM-width at least $k+1$ if and only if there is a node

v such that after rooting T in v, at least three children of v *themselves have at least one child* whose rooted subtree has linear MIM-width at least k. From this it follows that the linear MIM-width of an n-node tree is no more than $\log n$. Our $O(n \log n)$ algorithm computing linear MIM-width of a tree T picks an arbitrary root r and proceeds bottom-up on the rooted tree T_r. In Sect. 3 we show how to assign labels to the rooted subtrees encountered in this process giving their linear MIM-width. However, as with the algorithm computing pathwidth of a tree, the label is sometimes complex, consisting of linear MIM-width of a sequence of subgraphs of decreasing linear MIM-width, that are not themselves full rooted subtrees.

Proposition 1 gives a 7-way case analysis giving a subroutine used to update the label at a node given the labels at all children. In Sect. 4 we give our bottom-up algorithm, which will make calls to the subroutine underlying Proposition 1 in order to compute the complex labels and the linear MIM-width.

Finally, we use all the computed labels to lay out the tree in an optimal manner.

2 Classifying Linear MIM-Width of Trees

We use standard graph theoretic notation, see e.g. [5]. For a graph $G = (V, E)$ and subset of its nodes $S \subseteq V$ we denote by $N[S]$ the *closed neighborhood* of S, by $N(S) = N[S] \setminus S$ its *open neighborhood*, and by $G[S]$ the graph induced by S. For a graph G we denote by $\mathrm{mim}(G)$ the size of its *maximum induced matching* (MIM), the largest number of edges whose endpoints induce a matching. Let σ be a total order corresponding to the enumeration $v_1, \ldots, v_n$ of the nodes of G; this will also be called a *linear layout* of G. For any index $1 \leq i < n$ we have a cut of σ that defines the bipartite graph on edges "crossing the cut" i.e. edges with one endpoint in $\{v_1, \ldots, v_i\}$ and the other endpoint in $\{v_{i+1}, \ldots, v_n\}$. The *maximum induced matching width* of G under layout σ is denoted $\mathrm{mw}(\sigma, G)$, and is defined as the maximum, over all cuts of σ, of the value attained by the MIM of the cut, i.e. of the bipartite graph defined by the cut.

The *linear maximum induced matching width* – *linear MIM-width* – of G is denoted $\mathrm{lmw}(G)$, and is the minimum value of $\mathrm{mw}(\sigma, G)$ over all possible layouts σ of the vertices of G.

We start by showing that if we have a path P in a tree T then the linear MIM-width of T is no larger than the maximum linear MIM-width of any component of $T \setminus N[P]$, plus 1. To discuss these components, the following notion is useful.

Definition 1 (Dangling Tree). *Let T be a tree containing the adjacent nodes v and u. The* dangling tree *from v in u, $T\langle v, u\rangle$, is the component of $T \setminus (u, v)$ containing u.*

Given a node $x \in T$ with neighbors $\{v_1, \ldots, v_d\}$, the forest obtained by removing $N[x]$ from T is a collection of dangling trees $\{T\langle v_i, u_{i,j}\rangle\}$, where $u_{i,j} \neq x$ is some neighbor of v_i. We can generalize this to a path $P = (x_1, \ldots, x_p)$ in place of x, such that $T \setminus N[P] = \{T\langle v_{i,j}, u_{i,j,m}\rangle\}$, where $v_{i,j} \in N(P)$ is a neighbor of x_i and $u_{i,j,m} \notin N[P]$. See top part of Fig. 1. This naming convention will be used in the following sections.

Lemma 1 (Path Layout Lemma). *Let T be a tree. If there exists a path $P = (x_1, \ldots, x_p)$ in T such that every connected component of $T \setminus N[P]$ has linear MIM-width $\leq k$ then $\text{lmw}(T) \leq k+1$. Moreover, given the layouts for the components, we can in linear time compute the layout for T.*

Proof. Given the optimal linear layouts of the connected components of $T \setminus N[P]$, we give the below algorithm LinOrd constructing a linear layout σ_T on the nodes of T showing that linear MIM-width of T is $\leq k+1$. The layout σ_T starts out empty and the algorithm has an outer loop going through vertices in the path $P = (x_1, \ldots, x_p)$. When arriving at x_i it uses the concatenation operator $\oplus$ to add the path node x_i before looping over all neighbors $v_{i,j}$ of x_i adding the linear layouts of each dangling tree from $v_{i,j}$ and then $v_{i,j}$ itself. See Fig. 1 for an illustration.

function LinOrd(T: tree, $P = (x_1, \ldots, x_p)$: path, $\{\sigma_{T\langle v_{i,j}, u_{i,j,m}\rangle}\}$: lin-ords)
  $\sigma_T \leftarrow \emptyset$ ▷ The list starts out empty
  for $i \leftarrow 1, p$ **do** ▷ For all nodes on path $(x_1, \ldots, x_p)$
    $\sigma_T \leftarrow \sigma_T \oplus x_i$ ▷ Append path node
    for $j \leftarrow 1, |N(x_i) \setminus P|$ **do** ▷ For all nbs of x_i not on path: $v_{i,j}$
      for $m \leftarrow 1, |N(v_{i,j}) \setminus x_i|$ **do** ▷ For all dangling trees from $v_{i,j}$
        $\sigma_T \leftarrow \sigma_T \oplus \sigma_{T\langle v_{i,j}, u_{i,j,m}\rangle}$ ▷ Append given order of $T\langle v_{i,j}, u_{i,j,m}\rangle$
      $\sigma_T \leftarrow \sigma_T \oplus v_{i,j}$ ▷ Append $v_{i,j}$

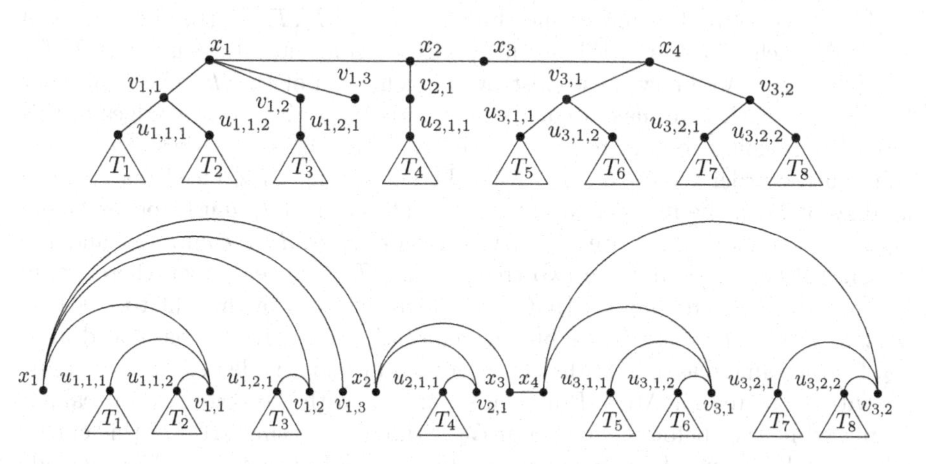

Fig. 1. A tree with a path $P = (x_1, x_2, x_3, x_4)$, with nodes in $N[N[P]]$ and dangling trees featured, and below it the layout given by the Path Layout Lemma.

Firstly, from the algorithm it should be clear that each node of T is added exactly once to σ_T, that it runs in linear time, and that there is no cut containing two crossing edges from two separate dangling trees. Now we must show that σ_T does not contain cuts with MIM larger than $k+1$. By assumption the layout of each dangling tree has no cut with MIM larger than k, and since these layouts can be found as subsequences of σ_T it followes that then also σ_T has no cut with more than k edges from a single dangling tree $T\langle v_{i,j}, u_{i,j,m}\rangle$. Also, we know that edges from two separate dangling trees cannot both cross the same cut. The only edges of T left to account for, i.e. not belonging to one of the dangling trees, are those with both endpoints in $N[N[P]]$, the nodes at distance at most 2 from a node in P. For every cut of σ_T that contains more than a single crossing edge (x_i, x_{i+1}), there is a unique $x_i \in P$ and a unique $v_{i,j} \in N(x_i)$ such that every edge with both endpoints in $N[N[P]]$ that crosses the cut is incident on either x_i or $v_{i,j}$. Since the edge connecting x_i and $v_{i,j}$ also crosses the cut, at most one of these edges can be taken into an induced matching. With these observations in mind, it is clear that $\text{lmw}(T) \leq \text{mw}(\sigma_T, T) \leq k+1$.

Definition 2 (k-neighbor and k-component Index). *Let x be a node in the tree T and v a neighbor of x. If v has a neighbor $u \neq x$ such that* $\text{lmw}(T\langle v, u\rangle) \geq k$*, then we call v a* k-neighbor *of x. The* k-component index *of x is equal to the number of k-neighbors of x and is denoted* $\text{D}_T(x,k)$*, or shortened to* $\text{D}(x,k)$.

Theorem 1 (Classification of the Linear MIM-width of Trees). *For a tree T and $k \geq 1$ we have* $\text{lmw}(T) \geq k+1$ *if and only if* $\text{D}(x,k) \geq 3$ *for some node x.*

Proof. We first prove the backward direction by contradiction. Thus we assume $\text{D}(x,k) \geq 3$ for a node x and there is a linear layout σ such that $\text{mw}(\sigma, T) \leq k$.

Let v_1, v_2, v_3 be the three k-neighbors of x and T_1, T_2, T_3 the three trees of $T \setminus N[x]$ each of linear MIM-width k, with v_i connected to a node of T_i for $i = 1, 2, 3$, that we know must exist by the definition of $\text{D}(x,k)$. We know that for each $i = 1, 2, 3$ we have a cut C_i in σ with MIM$=k$ and all k edges of this induced matching coming from the tree T_i. Wlog we assume these three cuts come in the order C_1, C_2, C_3, i.e. with the cut having an induced matching of k edges of T_2 in the middle. Note that in σ all nodes of T_1 must appear before C_2 and all nodes of T_3 after C_2, as otherwise, since T is connected and the distance between T_2 and the two trees T_1 and T_3 is at least two, there would be an extra edge crossing C_2 that would increase MIM of this cut to $k+1$. It is also clear that v_1 has to be placed before C_2 and v_3 has to be placed after C_2, for the same reason, e.g. the edge between v_1 and a node of T_1 cannot cross C_2 without increasing MIM. But then we are left with the vertex x that cannot be placed neither before C_2 nor after C_2 without increasing MIM of this cut by adding at least one of (v_1, x) or (v_3, x) to the induced matching. We conclude that $\text{D}(x,k) \geq 3$ for a node x implies linear MIM-width at least $k+1$.

For the full proof of the forward direction, please see the full paper [12], here we give a sketch. We assume that every node in T has $\text{D}(x,k) < 3$ and show that then $\text{lmw}(T) \leq k$. We define the following node subsets: $D_{=2} = \{x \in V(T) \mid \text{D}(x,k) = 2\}$ and $D_{=1} = \{x \in V(T) \mid \text{D}(x,k) = 1\}$. We show that there is always a path P in T such that all the connected components in $T \setminus N[P]$ have linear MIM-width $\leq k-1$ in the following way:

If $D_{=2} \neq \emptyset$*:* Then we show that the nodes in $D_{=2}$ induce a path $P = (x_1, \ldots, x_{|P|})$ in T. This path, plus two extra nodes x_0 and $x_{|P|+1}$, the second k-neighbor of x_1 and $x_{|P|}$ respectively, constitute a path as described above.

If $D_{=2} = \emptyset$*, but* $D_{=1} \neq \emptyset$*:* Then we show that starting out with an arbitrary node in $D_{=1}$, and adding the k-neighbor of the previously added node as long as it has such a neighbor, constitutes a path as described above.

If $D_{=2} = D_{=1} = \emptyset$*:* Then any node in T obviously constitutes a path as described above.

By the Path Layout Lemma, we then get that $\text{lmw}(T) \leq k$.

By Theorem 1, every tree with linear MIM-width $k \geq 2$ must be at least 3 times bigger than the smallest tree with linear MIM-width $k-1$, which implies the following.

Remark 1. The linear MIM-width of an n-node tree is $\mathcal{O}(\log n)$.

3 Rooted Trees, k-critical Nodes and Labels

Our algorithm computing linear MIM-width will work on a rooted tree, processing it bottom-up. We will choose an arbitrary node r of the tree T and denote by T_r the tree rooted in r. For any node x we denote by $T_r[x]$ the standard *complete subtree* of T_r rooted in x. During the bottom-up processing of T_r we will compute a label for various subtrees. The notion of a k-critical node is crucial for the definition of labels.

Definition 3 (k-critical Node). *Let T_r be a rooted tree with* $\text{lmw}(T_r) = k$*. We call a node x in T_r* k-critical *if it has exactly two children v_1 and v_2 that each has at least one child, u_1 and u_2 respectively, such that* $\text{lmw}(T_r[u_1]) = \text{lmw}(T_r[u_2]) = k$*. Thus x is k-critical in T if and only if* $\text{lmw}(T) = k$ *and* $\text{D}_{T_r[x]}(x, k) = 2$*.*

If $\text{lmw}(T_r) = k$, then T_r cannot have two k-critical nodes, as by Theorem 1, T_r would else have linear MIM-width $k+1$. For a detailed proof, see the full paper [12].

Remark 2. If T_r has linear MIM-width k it has at most one k-critical node.

Definition 4 (Label). *Let rooted tree T_r have* $\text{lmw}(T_r) = k$*. Then* $\text{label}(T_r)$ *consists of a list of decreasing numbers, $(a_1, \ldots, a_p)$, where $a_1 = k$, appended with a string called last_type, which tells us where in the tree an a_p-critical node lies, if it exists at all. If $p = 1$, then the label is* simple, *otherwise it is* complex*. The label is defined recursively, with type 0 being a basc case for singletons and for stars, and with type 4 being the only one defining a complex label.*

- *Type 0: r is a leaf, i.e. T_r is a singleton, then* $\text{label}(T_r) = (0, t.0)$*; or all children of r are leaves, then* $\text{label}(T_r) = (1, t.0)$

- *Type 1: No k-critical node in T_r, then* $\text{label}(T_r) = (k, t.1)$
- *Type 2: r is the k-critical node in T_r, then* $\text{label}(T_r) = (k, t.2)$
- *Type 3: A child of r is k-critical in T_r, then* $\text{label}(T_r) = (k, t.3)$
- *Type 4: There is a k-critical node u^k in T_r that is neither r nor a child of r. Let w be the parent of u^k. Then* $\text{label}(T_r) = k \oplus \text{label}(T_r \setminus T_r[w])$

In type 4 we note that $\text{lmw}(T_r \setminus T_r[w]) < k$, since otherwise u^k would have three k-neighbors (two children in the tree and also its parent), and then by Theorem 1 $\text{lmw}(T_r) = k + 1$. Therefore, all numbers in $\text{label}(T_r \setminus T_r[w])$ are smaller than k, and a complex label is thus a list of decreasing numbers followed by *last_type* $\in \{t.0, t.1, t.2, t.3\}$.

We now give a Proposition that, for any node x in T_r, will be used to compute $\text{label}(T_r[x])$ based on labels of subtrees rooted at descendants of x. The subroutine underlying this Proposition (see the decision tree in Fig. 2) will be used when reaching node x in the bottom-up processing of T_r.

Proposition 1. *Let x be a node of T_r with children $Child(x)$, and given* $\text{label}(T_r[v])$ *for all $v \in Child(x)$. Let $k = max_{v \in Child(x)}$ $\{\text{lmw}(T_r[v])\}$ and $N_k = \{v \in Child(x) \mid \text{lmw}(T[v]) = k\}$ and denote by $N_k = \{v_1, \ldots, v_q\}$ and by $l_i = \text{label}(T_r[v_i])$. Let $t_k = \text{D}_{T_r[x]}(x, k)$ by noting that $t_k = |\{v_i \in N_k \mid v_i \text{ has child } u_j \text{ with } \text{lmw}(T_r[u_j]) = k\}|$. Given this, we find* $\text{label}(T_r[x])$ *as follows:*

- ***Case 0:*** *if $|Child(x)| = 0$ then* $\text{label}(T_r[x]) = (0, t.0)$*;*
  else if $k = 0$, then $\text{label}(T_r[x]) = (1, t.0)$
- ***Case 1:*** *Every label in N_k is simple and has last_type equal to $t.1$ or $t.0$, and $t_k \leq 1$. Then,* $\text{label}(T_r[x]) = (k, t.1)$
- ***Case 2:*** *Every label in N_k is simple and has last_type equal to $t.1$ or $t.0$, but $t_k = 2$. Then,* $\text{label}(T_r[x]) = (k, t.2)$
- ***Case 3:*** *Every label in N_k is simple and has last_type equal to $t.1$ or $t.0$, but $t_k \geq 3$. Then,* $\text{label}(T_r[x]) = (k + 1, t.1)$
- ***Case 4:*** *$|N_k| \geq 2$ and for some $v_i \in N_k$, either l_i is a complex label, or l_i has last_type equal to either $t.2$ or $t.3$. Then,* $\text{label}(T_r[x]) = (k + 1, t.1)$
- ***Case 5:*** *$|N_k| = 1$, l_1 is a simple label and l_1 has last_type equal to $t.2$. Then,* $\text{label}(T_r[x]) = (k, t.3)$
- ***Case 6:*** *$|N_k| = 1$, l_1 is either complex or has last_type equal to $t.3$, and $k \notin \text{label}(T_r[x] \setminus T_r[w])$, where w is the parent of the k-critical node in $T_r[v_1]$. Then,* $\text{label}(T_r[x]) = k \oplus \text{label}(T_r[x] \setminus T_r[w])$
- ***Case 7:*** *$|N_k| = 1$, l_1 is either complex or has last_type equal to $t.3$, and $k \in \text{label}(T_r[x] \setminus T_r[w])$, where w is the parent of the k-critical node in $T_r[v_1]$. Then,* $\text{label}(T_r[x]) = (k + 1, t.1)$

Proof. For a full proof see the full paper [12], here we only give a sketch. Observe the decision tree in Fig. 2 which takes care of all cases, 1 up to 7, apart from the base cases. It follows from the definition of labels, k, N_k and t_k that cases 1 up to 7 of Proposition 1 corresponds to cases 1 up to 7 in the decision tree, and this shows that exactly one case applies to every possible rooted tree. To prove that labels are assigned correctly a case analysis is made based on Definition 4 and position of k-critical nodes. We argue for the two most complicated cases only.

Case 6: x has only one child v with $\text{lmw}(T_r[v]) = k$, and there is a k-critical node u^k with parent w – neither of which are equal to x – in $T_r[v]$, i.e. $T_r[v]$ is a type 3 or type 4 tree. Moreover, no tree rooted in another child of w, apart from u^k, can have linear MIM-width $\geq k$, since this would imply $\text{D}_{T_r[v]}(u^k, k) = 3$ and thus $\text{lmw}(T_r[v]) > k$; nor can $T_r[x] \setminus T_r[w]$ have linear MIM-width $= k$, since then we would have k in label$(T_r[x] \setminus T_r[w])$ disagreeing with the condition of Case 6. Therefore $\text{D}_{T_r[x]}(u, k) = 2$, and $\text{lmw}(T_r[x]) = k$. $T_r[x]$ is thus a type 4 tree and the label is assigned according to the definition.

Case 7: $T_r[v]$, u^k and w are as described in Case 6. But here, $\text{lmw}(T_r[x] \setminus T_r[w]) = k$ (since the condition says that k is in its label), and thus w is a k-neighbor of its child u^k. By Theorem 1 $\text{lmw}(T_r[x]) = k + 1$, and $T_r[x]$ is a type 1 tree.

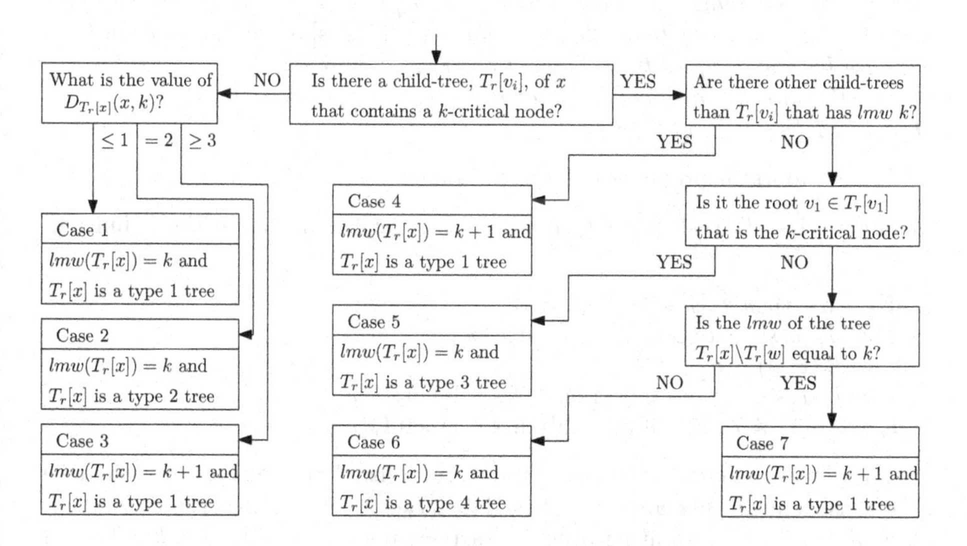

Fig. 2. A decision tree corresponding to the case analysis of Proposition 1

4 Computing the Linear MIM-Width of Trees and Finding an Optimal Layout

The subroutine underlying Proposition 1 will be used in a bottom-up algorithm that starts out at the leaves and works its way up to the root, computing labels of subtrees $T_r[x]$. However, in two cases (Case 6 and 7) we need the label of $T_r[x] \setminus T_r[w]$, which is not a complete subtree rooted in any node of T_r. Note that the label of $T_r[x] \setminus T_r[w]$ is again given by a (recursive) call to Proposition 1 and is then stored as a suffix of the complex label of $T_r[x]$. We will compute these labels by iteratively calling Proposition 1 (substituting the recursion by iteration). We first need to carefully define the subtrees involved when dealing with complex labels.

From the definition of labels it is clear that only type 4 trees lead to a complex label. In that case we have a tree $T_r[x]$ of linear MIM-width k and a k-critical node u^k that is neither x nor a child of x, and the recursive definition gives $\text{label}(T_r[x]) = k \oplus \text{label}(T_r[x] \setminus T_r[w])$ for w the parent of u^k. Unravelling this recursive definition, this means that if $\text{label}(T_r[x]) = (a_1, \ldots, a_p, last\_type)$, we can define a list of nodes $(w_1, \ldots, w_{p-1})$ where w_i is the parent of an a_i-critical node in $T_r[x] \setminus (T_r[w_1] \cup \ldots \cup T_r[w_{i-1}])$. We expand this list with $w_p = x$, such that there is a unique node in $T_r[x]$ corresponding to each number in $\text{label}(T_r[x])$, and $T_r[x] \setminus (T_r[w_1] \cup \ldots \cup T_r[w_p]) = \emptyset$.

Now, in the first level of a recursive call to Proposition 1 the role of $T_r[x]$ is taken by $T_r[x] \setminus T_r[w_1]$, and in the next level it is taken by $(T_r[x] \setminus T_r[w_1]) \setminus T_r[w_2]$ etc. The following definition gives a shorthand for denoting these trees.

Definition 5. *Let x be a node in T_r, $\text{label}(T_r[x]) = (a_1, a_2, \ldots, a_p, last\_type)$ and the corresponding list of vertices $(w_1, \ldots, w_p)$ is as we describe in the above text. For any non-negative integer s, the tree* $\mathbf{T_r[x, s]}$ *is the subtree of $T_r[x]$ obtained by removing all trees $T_r[w_i]$ from $T_r[x]$, where $a_i \geq s$.*

In other words, if q is such that $a_q \geq s > a_{q+1}$, then $T_r[x, s] = T_r[x] \setminus (T_r[w_1] \cup T_r[w_2] \cup \ldots \cup T_r[w_q])$.

Some important properties of $T_r[x, s]$ follow:

Remark 3. Let $T_r[x, s]$, $\text{label}(T_r[x, s])$, $(w_1, \ldots, w_p)$ and q as in the definition. Then

1. if $s > a_1$, then $T_r[x, s] = T_r[x]$
2. $\text{label}(T_r[x, s]) = (a_{q+1}, \ldots, a_p, last\_type)$
3. $\text{lmw}(T_r[x, s]) = a_{q+1} < s$
4. $\text{lmw}(T_r[x, s+1]) = s$ if and only if $s \in \text{label}(T_r[x])$
5. $T_r[x, s+1] \neq T_r[x, s]$ if and only if $s \in \text{label}(T_r[x])$

The above Remarks follow from the definitions. Note that for any s the tree $T_r[x, s]$ is defined only after we know $\text{label}(T_r[x])$. In the algorithm, we compute $\text{label}(T_r[x])$ by iterating over increasing values of s (until $s > \text{lmw}(T_r[x])$ since by Remark 3.1 we then have $T_r[x, s] = T_r[x]$) and we could hope for a loop invariant saying that we have correctly computed $\text{label}(T_r[x, s])$. However, $T_r[x, s]$ is only known once we are done. Instead, each iteration of the loop will correctly compute the label of the following subtree called $T_{union}[x, s]$, which is not always equal to $T_r[x, s]$, but importantly for $s > \text{lmw}(T_r[x])$, we will have $T_{union}[x, s] = T_r[x, s] = T_r[x]$.

Definition 6. *Let x be a node in T_r with children $v_1, \ldots, v_d$. $T_{union}[x, s]$ is then equal to the tree induced by x and the union of all $T_r[v_i, s]$ for $1 \leq i \leq d$. More technically, $T_{union}[x, s] = T_r[V_{union}]$ where $V_{union} = x \cup V(T_r[v_1, s]) \cup \ldots \cup V(T_r[v_d, s])$.*

Given a tree T, we find its linear MIM-width by rooting it in an arbitrary node r, and computing labels by processing T_r bottom-up. The answer is given by the first element of label$(T_r[r])$, which by definition is equal to lmw(T). At a leaf x of T_r we initialize by label$(T_r[x]) \leftarrow (0, t.0)$, and at a node x for which all children are leaves we initialize by label$(T_r[x]) \leftarrow (1, t.0)$, according to Definition 4. When reaching a higher node x we compute label of $T_r[x]$ by calling function MakeLabel(T_r, x).

function MakeLabel(T_r, x) ▷ finds *cur_label* = label$(T_r[x])$
  cur_label $\leftarrow (0, t.0)$ ▷ This is label$(T_{union}[x, 0])$
  $\{v_1, \ldots, v_d\}$ = children of x
  if $0 \in$ label$(T_r[v_i])$ for some i **then**
    cur_label $\leftarrow (1, t.0)$ ▷ This is then label$(T_{union}[x, 1])$
  for $s \leftarrow 1, \max_{i=1}^{d}$\{first element of label$(T_r[v_i])$\} **do**
    $\{l'_1, \ldots, l'_d\} = \{\text{label}(T_r[v_i, s+1]) \mid 1 \le i \le d\}$
    $N_s = \{v_i \mid 1 \le i \le d,\ s \in l'_i\}$
    $t_s = |\{v_i \mid v_i \in N_s,\ v_i \text{ has child } u_j \text{ s.t. } s \in \text{label}(T_r[u_j, s+1])\}|$
    if $|N_s| > 0$ **then**
      case $\leftarrow$ the case from Prop. 1 applying to s, $\{l'_1, \ldots, l'_d\}$, N_s and t_s
      cur_label $\leftarrow$ as given by *case* in Prop. 1 ($s \oplus$ *cur_label* if Case 6)

Lemma 2. *Given labels at descendants of node x in T_r,* MakeLabel*(T_r, x) computes* label$(T_r[x])$ *as the value of cur_label.*

Proof. See the full paper [12] for a full proof, here we give only a sketch. The crucial issue is to prove the loop invariant: "At the end of the s'th iteration of the for loop the value of *cur_label* is equal to label$(T_{union}[x, s+1])$." This invariant suffices since for $s >$ lmw$(T_r[x])$, we have $T_{union}[x, s] = T_r[x]$.

That this invariant holds for every iteration of the for loop, is proven by induction. The base case states that before the first iteration (i.e. in the first five lines of MakeLabel), the value of *cur_label* is set to be equal to label$(T_{union}[x, 1]$. This is true because $T_{union}[x, 1]$ is equal to a star if and only if x has a child v such that $0 \in$ label$(T_r[v])$, and equal to a singleton otherwise.

To prove the induction step, we assume that the value of *cur_label* is equal to label$(T_{union}[x, s]$ at the beginning of the s'th iteration. To show that *cur_label* = label$(T_{union}[x, s+1]$ at the end of the s'th iteration, we argue for the correspondence given by the below Table, between parameters used in Proposition 1 and parameters used in the for loop of MakeLabel.

Table 1. Correspondences between variables

Proposition 1	for loop iteration s	Explanation
$T_r[x], k$	$T_{union}[x, s+1], s$	Tree needing label, max lmw of children
$T_r[v_1], ..., T_r[v_d]$	$T_r[v_i, s], ..., T_r[v_d, s]$	Subtrees of children
$l_1, ..., l_d, N_k, t_k$	$l'_1, ..., l'_d, N_s, t_s$	Child labels, those with max, root comp. index
label$(T_r[x] \setminus T_r[w])$	*cur_label*	This is also label$(T_{union}[x, s+1] \setminus T_r[w, s+1])$

Let us here give only the two most complicated of these arguments, showing that t_s computed in iteration s of the for loop corresponds to $t_k = \mathrm{D}_{T_r[x]}(x,k)$ in Proposition 1 – meaning we need to show that $t_s = \mathrm{D}_{T_{union}[x,s+1]}(x,s)$. Consider v_i, a child of x. In accordance with MAKELABEL we say that v_i *contributes to* t_s if $v_i \in N_s$ and v_i has a child u_j with s in its label. We thus need to show that v_i contributes to t_s if and only if v_i is an s-neighbor of x in $T_{union}[x,s+1]$. Observe that by Remark 3.4, $\mathrm{lmw}(T_r[v_i,s+1]) = \mathrm{lmw}(T_r[u_j,s+1]) = s$ if and only if s is in the labels of both $T_r[v_i]$ and $T_r[u_j]$. If $s \notin \mathrm{label}(T_r[u_j,s+1])$, then $\mathrm{lmw}(T_r[u_j,s+1]) < s$, and if this is true for all children of v_i, then v_i is not an s-neighbor of x in $T_{union}[x,s+1]$. If $s \notin \mathrm{label}(T_r[v_i,s+1])$, then $\mathrm{lmw}(T_r[v_i,s+1]) < s$ and no subtree of $T_r[v_i,s+1]$ can have linear MIM-width s. However, if $s \in \mathrm{label}(T_r[u_j,s+1])$ and $s \in \mathrm{label}(T_r[v_i,s+1])$ (this is when v_i contributes to t_s), then $T_r[v_i,s+1] \cap T_r[u_j]$ must be equal to $T_r[u_j,s+1]$ and $T_r[u_j,s+1] \subseteq T_{union}[x,s+1]$. We thus conclude that v_i is an s-neighbor of x in $T_{union}[x,s+1]$ if and only if v_i contributes to t_s, so $t_s = \mathrm{D}_{T_{union}[x,s+1]}(x,s)$.

We then show that if $T_{union}[x,s+1]$ is a Case 6 or Case 7 tree – that is, $|N_s| = 1$, and $T_r[v_1,s+1]$ is a type 3 or type 4 tree, with w being the parent of an s-critical node – then the algorithm has $\mathrm{label}(T_{union}[x,s+1] \setminus T_r[w,s+1])$ available for computation, indeed that this is the value of *cur_label*. We know, by definition of label and Remark 3.5 that $T_r[v_i,s+1] \setminus T_r[v_i,s] = T_r[w,s+1]$. But since $|N_s| = 1$, for every $j \neq i$, $T_r[v_j,s+1] \setminus T_r[v_j,s] = \emptyset$. Therefore $T_{union}[x,s+1] \setminus T_{union}[x,s] = T_r[w,s+1]$ and $T_{union}[x,s+1] \setminus T_r[w,s+1] = T_{union}[x,s]$. But by the induction assumption, *cur_label* $= \mathrm{label}(T_{union}[x,s])$. Thus *cur_label* corresponds to $\mathrm{label}(T_r[x] \setminus T_r[w])$ in Proposition 1.

By the correspondences in Table 1, we conclude from Proposition 1, Definition 6 and the inductive assumption, that *cur_label* $= \mathrm{label}(T_{union}[x,s+1])$ at the end of the s'th iteration of the for loop in MAKELABEL. The loop runs for k iterations, with k the biggest number in any label of the children of x. At the end, *cur_label* is thus equal to $\mathrm{label}(T_{union}[x,k+1])$. Since $k \geq \mathrm{lmw}(T_r[v_i])$ for all i, by definition $T_r[v_i,k+1] = T_r[v_i]$ for all i, and thus $T_{union}[x,k+1] = T_r[x]$. Therefore, when MAKELABEL finishes, *cur_label* $= \mathrm{label}(T_r[x])$.

Theorem 2. *Given any tree T, $\mathrm{lmw}(T)$ can be computed in $\mathcal{O}(n \log n)$-time.*

Proof. We find $\mathrm{lmw}(T)$ by bottom-up processing of T_r and returning the first element of $\mathrm{label}(T_r)$. After correctly initializating at leaves and nodes whose children are all leaves, we make a call to MAKELABEL for each of the remaining nodes. Correctness follows by Lemma 2 and induction on the structure of the rooted tree.

For the timing we show that each call runs in $\mathcal{O}(\log n)$ time. m is given as the biggest number in any label of children of x, which is $\mathcal{O}(\log n)$ by Remark 1. For every integer s from 1 to m, the algorithm checks how many labels of children of x contain s (to compute N_s), and how many labels of grandchildren of x contain s (to compute t_s). The labels are sorted in descending order, therefore the whole loop goes only once through each of these labels, each of length $O(\log n)$. Other than this, MAKELABEL only does a constant amount of work. Therefore, MAKELABEL(T_r, x), if x has a children and b grandchildren, takes time proportional to $O(\log n)(a+b)$. As the sum of the number of children and grandchildren over all nodes of T_r is $O(n)$, we conclude that the total runtime to compute $\text{lmw}(T)$ is $\mathcal{O}(n \log n)$.

Theorem 3. *A layout of linear MIM-width* $\text{lmw}(T)$ *of a tree* T *can be found in* $\mathcal{O}(n \log n)$*-time.*

Proof. For a detailed proof, see the full paper [12]. Given T we first run the algorithm computing $\text{lmw}(T)$ finding the label of every full rooted subtree in T_r. We give a recursive layout-algorithm that uses these labels in tandem with LinOrd presented in the Path Layout Lemma. We call it on a rooted tree where labels of all subtrees are known. For simplicity we call this rooted tree T_r even though in recursive calls this is not the original root r and tree T:

(1) Let $\text{lmw}(T_r) = k$ and find a path P in T_r such that all trees in $T_r \setminus N[P]$ have linear MIM-width $< k$. The path depends on the type of T_r as explained below.
(2) Call this layout-algorithm recursively on every rooted tree in $T_r \setminus N[P]$ to obtain optimal linear layouts; for this, we need the correct labels for these trees, but they are easy to obtain.
(3) Call LinOrd on T_r, P and the layouts provided in step 2.

The path P is found for every Type of tree as follows:

Type 0 trees: Choose $P = (r)$.
Type 1 trees: Choose P to start at the root r, and as long as the last node in P has a k-neighbor $v \notin P$, v is appended to P.
Type 2 trees: We look at the trees rooted in the two k-neighbors of r, $T_r[v_1]$ and $T_r[v_2]$. These are Type 1 trees. Choose paths P_1, P_2 for $T_r[v_1]$ and $T_r[v_2]$ as described above. Gluing these paths together at r we get the path for T_r.
Type 3 trees: r has exactly one child v such that $T_r[v]$ is of type 2. Choose P as described above for $T_r[v]$.
Type 4 trees: In these trees, T_r contains precisely one node $w \neq r$ such that w is the parent of a k-critical node, x. $T_r[w]$ is a type 3 tree. Choose P for $T_r[w]$ as described above which will be the path for T_r.

For all paths chosen above, the trees in $T \setminus N[P]$ have linear MIM-width strictly less than k since no node in P has a k-neighbor that is not in P. For the recursive calls we need labels for all subtrees in $T \setminus N[P]$. In every case except Type 4 trees, all these subtrees are full rooted subtrees of T_r, and the label is clearly known. In Type 4 trees, the subtree $T_r \setminus T_r[w]$, where w the parent of a k-critical node, is not a full rooted subtree. In this case we must update the label of every ancestor y of w, but this is simple, since $\text{label}(T_r[y] \setminus T_r[w]) = \text{label}(T_r[y, k])$ which we get by removing the first element from $\text{label}(T_r[y])$.

References

1. Adler, I., Kanté, M.M.: Linear rank-width and linear clique-width of trees. Theor. Comput. Sci. **589**, 87–98 (2015)
2. Belmonte, R., Vatshelle, M.: Graph classes with structured neighborhoods and algorithmic applications. Theor. Comput. Sci. **511**, 54–65 (2013)
3. Bergougnoux, B., Kanté, M.M.: Rank based approach on graphs with structured neighborhood. CoRR, abs/1805.11275 (2018)
4. Bui-Xuan, B.-M., Telle, J.A., Vatshelle, M.: Fast dynamic programming for locally checkable vertex subset and vertex partitioning problems. Theor. Comput. Sci. **511**, 66–76 (2013)
5. Diestel, R.: Graph Theory. Graduate Texts in Mathematics, vol. 173, 4th edn. Springer, Heidelberg (2012)
6. Ellis, J.A., Sudborough, I.H., Turner, J.S.: The vertex separation and search number of a graph. Inf. Comput. **113**(1), 50–79 (1994)
7. Fomin, F.V., Golovach, P.A., Raymond, J.-F.: On the tractability of optimization problems on H-graphs. In: Proceedings of the ESA 2018, pp. 30:1–30:14 (2018)
8. Galby, E., Munaro, A., Ries, B.: Semitotal domination: new hardness results and a polynomial-time algorithm for graphs of bounded MIM-width. CoRR, abs/1810.06872 (2018)
9. Golovach, P.A., Heggernes, P., Kanté, M.M., Kratsch, D., Sæther, S.H., Villanger, Y.: Output-polynomial enumeration on graphs of bounded (local) linear MIM-width. Algorithmica **80**(2), 714–741 (2018)
10. Golumbic, M.C., Rotics, U.: On the clique-width of some perfect graph classes. Int. J. Found. Comput. Sci. **11**(3), 423–443 (2000)
11. Hlinený, P., Oum, S., Seese, D., Gottlob, G.: Width parameters beyond tree-width and their applications. Comput. J. **51**(3), 326–362 (2008)
12. Høgemo, S., Telle, J.A., Raa Vågset, E.: Linear MIM-width of trees. CoRR, arXiv:1907.04132 (2019)
13. Jaffke, L., Kwon, O., Strømme, T.J.F., Telle, J.A.: Generalized distance domination problems and their complexity on graphs of bounded MIM-width. In 13th International Symposium on Parameterized and Exact Computation, IPEC 2018, Helsinki, Finland, 20–24 August 2018, pp. 6:1–6:14 (2018)
14. Jaffke, L., Kwon, O., Telle, J.A.: Polynomial-time algorithms for the longest induced path and induced disjoint paths problems on graphs of bounded MIM-width. In 12th International Symposium on Parameterized and Exact Computation, IPEC 2017, Vienna, Austria, 6–8 September 2017, pp. 21:1–21:13 (2017)
15. Jaffke, L., Kwon, O., Telle, J.A.: A unified polynomial-time algorithm for feedback vertex set on graphs of bounded MIM-width. In 35th Symposium on Theoretical Aspects of Computer Science, STACS 2018, Caen, France, 28 February–3 March 2018, pp. 42:1–42:14 (2018)
16. Mengel, S.: Lower bounds on the MIM-width of some graph classes. Discrete Appl. Math. **248**, 28–32 (2018)
17. Möhring, R.H.: Graph problems related to gate matrix layout and PLA folding. In: Tinhofer, G., Mayr, E., Noltemeier, H., Syslo, M.M. (eds.) Computational Graph Theory. COMPUTING, vol. 7, pp. 17–51. Springer, Vienna (1990). https://doi.org/10.1007/978-3-7091-9076-0_2
18. Oum, S.: Rank-width: algorithmic and structural results. Discrete Appl. Math. **231**, 15–24 (2017)

19. Sæther, S.H., Vatshelle, M.: Hardness of computing width parameters based on branch decompositions over the vertex set. Theor. Comput. Sci. **615**, 120–125 (2016)
20. Skodinis, K.: Construction of linear tree-layouts which are optimal with espect to vertex separation in linear time. J. Algorithms **47**(1), 40–59 (2003)
21. Vatshelle, M.: New width parameters of graphs. Ph.D. thesis, University of Bergen, Norway (2012)
22. Yamazaki, K.: Inapproximability of rank, clique, Boolean, and maximum induced matching-widths under small set expansion hypothesis. Algorithms **11**(11), 173 (2018)

Approximating Minimum Dominating Set on String Graphs

Dibyayan Chakraborty(✉), Sandip Das, and Joydeep Mukherjee

Indian Statistical Institute, Kolkata, India
dibyayancg@gmail.com, sandipdas@isical.ac.in, joydeep.m1981@gmail.com

Abstract. A *string* graph is an intersection graph of simple curves on the plane. For $k \geq 0$, *B_k-VPG graphs* are intersection graphs of simple rectilinear curves having at most k cusps (bends). It is well-known that any string graph is a B_k-VPG graph for some value of k. For $k \geq 0$, *unit B_k-VPG graphs* are intersection graphs of simple rectilinear curves having at most k cusps (bends) and each segment of the curve being unit length. Any string graph is a unit-B_k-VPG graph for some value of k.

In this article, we show that the Minimum Dominating Set (MDS) problem for unit B_k-VPG graphs is NP-Hard for all $k \geq 1$ and provide an $O(k^4)$-approximation algorithm for all $k \geq 0$. Furthermore, we also provide an 8-approximation for the MDS problem for the *vertically-stabbed* L-*graphs*, intersection graphs of L-paths intersecting a common vertical line. The same problem is known to be APX-Hard (MFCS, 2018). As a by-product of our proof, we obtained a 2-approximation algorithm for the *stabbing segment with rays* (SSR) problem introduced and studied by Katz et al. (Comput. Geom. 2005).

Keywords: String graph · Dominating set · Approximation algorithm

1 Introduction

A *string graph* is a graph with simple curves on a plane as vertices, and two vertices are adjacent if they intersect. String graphs are important as it contains all intersection graphs of connected sets in $\mathbb{R}^2$. Asinowski et al. [1] introduced the concept of *B_k-VPG graphs* to initiate a systematic study of string graphs and its subclasses. A *path* is a simple rectilinear curve and a *k-bend path* is a path having k cusps (bend). The *B_k-VPG graphs* are intersection graphs of k-bend paths. Any string graph has a B_k-VPG representation for some k [1]. A *unit k-bend path* is a k-bend path with each segments being of unit length. The *unit B_k-VPG graphs* are intersection graphs of unit k-bend paths. Any string graph has a unit $B_{k'}$-VPG representation for some k'.

A *dominating set* of a graph $G = (V, E)$ is a subset D of vertices V such that each vertex in $V \setminus D$ is adjacent to some vertex in D. The Minimum Dominating Set (MDS) problem is to find a minimum cardinality dominating set of a graph G. It is not possible to approximate the MDS problem on string graphs with n

I. Sau and D. M. Thilikos (Eds.): WG 2019, LNCS 11789, pp. 232–243, 2019.
https://doi.org/10.1007/978-3-030-30786-8_18

vertices to within $(1-\alpha)\ln n$ unless $NP \subseteq DTIME(n^{O(\log\log n)})$ [5]. Researchers have studied the MDS problem extensively on important graph classes like *planar* graphs, *permutation* graphs, *cocomparability graphs* all of which are subclasses of string graphs [1,13,14,16] and hence unit B_k-VPG graphs for some k. The MDS problem admits constant factor approximation algorithms on intersection graphs of *unit disks*, *rectangles with bounded aspect ratio*, *r-regular polygons* etc. all of which are also subclasses of string graphs [10,12]. Mehrabi [18] gave constant factor approximation algorithms for the MDS problem on restricted subclasses of B_1-VPG graphs. However, for $k \geq 1$, it is not known if there is an $f(k)$-approximation algorithm for the MDS problem on B_k-VPG graphs. We show that the MDS problem remains NP-Hard on unit B_k-VPG graphs with $k \geq 1$.

Theorem 1. *It is NP-Hard to solve the* MDS *problem on unit B_k-VPG graphs with $k \geq 1$.*

We achieved a constant factor approximation algorithm for the MDS problem on unit B_k-VPG graphs for a given unit B_k-VPG representation and fixed k.

Theorem 2. *Given a unit B_k-VPG representation of a graph G with n vertices, there is an $O(k^2n^5)$-time $O(k^4)$-approximation algorithm to solve the* MDS *problem on G.*

The MDS problem remains difficult in restricted families of B_1-VPG graphs. An L-path is a 1-bend path having the shape 'L'. A *vertically-stabbed*-L-*representation* of a graph is a collection of L-paths along with a vertical line intersecting all the paths such that each path in the collection represents a vertex of the graph and two paths intersect if and only if the vertices they represent are adjacent in the graph. A graph is a *vertically-stabbed*-L *graph* if it has a vertically-stabbed-L-representation. Bandyapadhyay et al. [2] proved APX-hardness for the MDS problem on vertically-stabbed-L graphs. The class of vertically-stabbed-L graphs was introduced by McGuinness [17] and it contains many important graph classes like *interval graphs*, *outerplanar graphs*, *permutation graphs*, *circle graphs* as subclasses. Researchers have studied the MDS problem on these classes of graphs [4,7–9,11]. An ϵ-net based algorithm of Mehrabi [18] gives an $O(1)$-approximation algorithm for the MDS problem on vertically-stabbed-L-graphs. The specific value of the constant (which is at least 32) was not reported by the author. In this paper, we prove the following.

Theorem 3. *Given a vertically-stabbed*-L-*representation of a graph G with n vertices, there is an $O(n^5)$-time 8-approximation algorithm to solve the* MDS *problem on G.*

To prove Theorems 2 and 3, we needed to prove a lemma about the *stabbing segment with rays (*SSR*)* problem introduced by Katz et al. [15]. In this problem, the inputs are a set of disjoint leftward-directed horizontal rays and a set of disjoint vertical segments. The objective is to select a minimum number of leftward-directed horizontal rays that intersect all vertical segments. Throughout this article, we let $\mathcal{SSR}(R, V)$ denote an SSR instance where R is a given set of disjoint leftward-directed horizontal rays and V is a given set of disjoint vertical segments. We observe the following lemma.

Lemma 1. *Let R be a set of leftward-directed horizontal rays and V be a set of vertical segments. The cost of an optimal solution for the Integer Linear Program (ILP) of $\mathcal{SSR}(R, V)$ is at most twice the cost of an optimal solution for the relaxed linear program (LP) of $\mathcal{SSR}(R, V)$.*

As a consequence of the above Lemma 1, we have a subquadratic 2-approximation algorithm for the SSR problem.

Theorem 4. *There is an $O((n + m)\log(n + m))$-time 2-approximation algorithm for SSR problem where n and m are the number of rays and segments, respectively.*

We prove both Lemma 1 and Theorem 4 in Sect. 2 and prove Theorem 3 in Sect. 3. In Sect. 4, we prove Theorems 1 and 2.

2 Proof of Lemma 1 and Theorem 4

In this section, we represent a *leftward-directed horizontal ray* by simply a *ray* and a *vertical segment* by a *segment* in short. Let R be a set of disjoint rays and V be a set of disjoint vertical segments. We assume each segment intersects at least one ray in R and no two segments in V has the same x-coordinate.

To prove Lemma 1, first we present an iterative algorithm consisting of three main steps. The first step is to include all rays $r \in R$ in heuristic solution S whenever some segments in V intersect precisely a single ray r in that iterative step. In the next step, delete all segments intersecting any ray in S from V. In the final step, find a ray in $R \setminus S$ whose x-coordinate of the right endpoint is the smallest among all rays in $R \setminus S$ and delete it from R (when there are multiple such rays, choose anyone arbitrarily). We repeat the above three steps until V is empty. The above algorithm takes $O((|R| + |V|)\log(|R| + |V|))$ time (using segment trees [3]) and outputs a set S of rays such that all segments in V intersect at least one ray in S.

We describe the above algorithm formally in Algorithm 1. Below we introduce some notations used to describe the algorithm. We assign *token* $T_r = \{r\}$ for each $r \in R$ initially. For $i \geq 1$, let R_i, V_i, S_i be the set of rays, the set of segments and the heuristic solution constructed by this Algorithm 1, respectively at the *end* of ith iteration. A ray $r \in R_i$ is *critical* if there is a segment $v \in V_i$ such that r is the only ray in R_i that intersects v. We describe a *discharging technique* below.

Let D be a subset of R. A ray $r \in D$ lies *in between* two rays $r', r'' \in D$ if the y-coordinate of r lies in between those of r', r''. A ray $r \in D$ lies *just above* (resp. *just below*) a ray $r' \in D$ if y-coordinate of r is greater (resp. smaller) than that of r' and no other ray lies in between r, r' in D. Two rays $r, r' \in D$ are *neighbours* of each other if r lies just above or below r'.

Discharging Method: Let $r \in R_{i-1} \backslash S_i$ be a ray whose x-coordinate of the right endpoint is the smallest. The phrase "r discharges the token to its neighbours" in the ith iteration means the following operations in the given order.

(i) Let r' lie just above r and r'' lie just below r in $R_{i-1} \setminus S_i$. For all $x \in T_r$ (x and r not necessarily distinct) do the following. If there is a segment in V_i that intersects x, r' and r then assign $T_{r'} = T_{r'} \cup \{x\}$ and if there is a segment in V_i that intersects x, r'' and r then $T_{r''} = T_{r''} \cup \{x\}$.
(ii) Make $T_r = \emptyset$ after performing the above step.

Algorithm 1. SSR-Algorithm

Input: A set R of leftward-directed rays and a set V of vertical segments.
Output: A subset of R that intersects all segments in V.

1: $T_r = \{r\}$ for each $r \in R$ and $i \leftarrow 1, V_0 \leftarrow V, R_0 \leftarrow R, S \leftarrow \emptyset, S_0 \leftarrow \emptyset$ ▷ Initialisation.
2: **while** $V_{i-1} \neq \emptyset$ **do**
3: $S \leftarrow S \cup \{r \colon r \in R_{i-1}, r$ is critical after $(i-1)^{th}$ iteration$\}$ and $S_i \leftarrow S$. ▷ Critical ray collection.
4: $V_i \leftarrow$ the set obtained by deleting all segments from V_{i-1} that intersect a ray in S_i.
5: Find a $r \in R_{i-1} \setminus S_i$ whose x-coordinate of the right endpoint is the smallest.
6: r discharges the token to its neighbours.
7: $R_i \leftarrow$ The set obtained by deleting $\{r\} \cup S_i$ from R_{i-1}. ▷ Discharging token step.
8: $i \leftarrow i + 1$;
9: **end while**
10: **return** S

We have the following observation.

Observation A. *For some $v \in V_k, k \geq 1$, if some ray $r \in R_0$ intersects v, then either $r \in R_k$ or there exists some ray $r' \in R_k$ such that $r \in T_{r'}$.*

Proof. Assume $r \notin R_k$. Let $<r_1, r_2, \ldots, r_k>$ be a sorted order of the rays such that for $i < j$, r_i discharged the token to the neighbours before r_j. Due to step 5 of SSR-algorithm, $X = <r_1, r_2, \ldots, r_k>$ is an increasing sequence based on the x-coordinate of their right endpoint. Observe that, whenever a ray $r_i \in X$ discharged its token to its neighbours in the ith iteration, all the vertical segments in V_i intersected by r_i also intersects one of the immediate neighbours of r_i. Again as $v \in V_k$, v is not intersected by critical ray within k iteration. Hence the result follows. □

Lemma 2. *For a ray r, there are at most two tokens containing r.*

Proof. If r never discharged its token to its neighbours, then the statement is true. Let r discharged the token to its neighbours at iteration i. Note that, r discharged tokens to at most two of its neighbours. Since r gets deleted after the discharging step, the rays whose token contain r become neighbours of each other.

Let j be the minimum integer with $i < j$ such that at the end of $(j-1)^{th}$ iteration, there is a ray $p \in R_{j-1}$ which is critical and $r \in T_p$. Note that iteration of SSR-Algorithm may stop before encountering such events. However, within iteration i to $j-1$, there may exist some rays which discharged their tokens containing r due to step 5 of SSR-Algorithm.

To prove the lemma, we use induction to show that there are at most two tokens containing r in any iteration from i upto $j-1$. Consider some k, $i < k < j$, such that $x_1, x_2 \in R_{k-1}$ be only two rays where $r \in T_{x_1}$ and $r \in T_{x_2}$. Notice that, x_1 and x_2 are neighbours of each other and without loss of generality assume x_1 lies just above x_2 in V_{k-1}. Assume x_1 discharged its token at k^{th} iteration. If there exists a neighbour of x_1 (say x_3) which is different from x_2, then due to the discharging step of kth iteration, x_1 passes the token to its neighbours (i.e x_2 and x_3) and gets deleted from R_{k-1} to create R_k. If x_3 does not exist, then x_1 shall pass the token only to x_2. Therefore x_2 becomes the top-most ray among those rays in R_k which intersect some segment intersecting r.

Moreover, if x was the only ray in R_{k-1} such that $r \in T_x$, then x was the top-most (or bottom-most) ray among those rays in R_{k-1} which intersect some segment intersecting r. Therefore, at the end of kth iteration there is exactly one ray $x' \in R_k$ such that $r \in T_{x'}$ and x' must be the top-most (resp. bottom-most) ray among those rays in R_k which intersect some segment intersecting r.

Hence we conclude that for each k with $i \leq k < j$, there is at most two rays $r', r'' \in R_k$ such that $r \in T_{r'} \cap T_{r''}$ and they are neighbours. If there is exactly one ray $r''' \in R_k$ such that $r \in T_{r'''}$ then r''' must be the top-most or bottom-most ray among those rays in R_k which intersect some segment intersecting r.

In iteration j, ray p is critical and $r \in T_p$ and p is put in heuristic solution. If p is the only ray whose token contained r, only T_p will contain r after the termination of Algorithm 1. Let $r', p \in R_{j-1}$ be the rays whose token contained r. They must be neighbours. Without loss of generality assume that p lies just above r'. If both r', p are selected in S_j, then there is nothing to prove. Now consider the set A of segments in V_j that intersects r but not p. Note that, no ray above p intersects any segment in A. Hence r' becomes the only ray in next iterative step whose token contains r and r' turns to be bottom most ray among those rays in R_{j-1} which intersect some segment intersecting r. Now consider any iteration $k > j$. By similar arguments as above, there would be at most one ray in R_k that contains the token r. Hence the lemma follows. □

For a segment $v \in V$, let $N(v) \subseteq R$ be the set of rays that intersect v. Let $r \in S$ be a ray, i be the minimum integer such that $r \in S_i$. There must exist a segment $\nu_r \in V_{i-1}$ such that r is the only ray in R_{i-1} that intersects ν_r and all rays in $N(\nu_r) \setminus \{r\}$ must have passed the token to its neighbours. So, for each ray $r \in S$, there exists a segment ν_r such that for all $x \in N(\nu_r) \setminus \{r\}$ we have $T_x = \emptyset$. We call ν_r a *critical segment with respect to* r.

Observation B. *For a ray* $r \in S$ *let* ν_r *be a critical segment with respect to* r. *Then* $N(\nu_r) \subseteq T_r$.

Proof. Consider any arbitrary but fixed deleted ray $y \in N(\nu_r) \setminus \{r\}$ which was deleted at some j^{th} iteration. By Observation A, there exists a ray $y' \in R_j$ such that y' intersects v and $y \in T_{y'}$. Applying the above argument for all rays in $N(\nu_r) \setminus \{r\}$, we have the proof. □

Lemma 3. *If* S *is the set returned by* SSR-*algorithm with rays* R *and segments* V, *then* $|S| \leq 2|OPT|$, *where* OPT *is an optimum solution of* $\mathcal{SSR}(R, V)$.

Proof. Let R be the set of rays and V be the set of segments with $|R| = n, |V| = m$. Consider the ILP formulation Q of $\mathcal{SSR}(R, V)$. For each ray $r \in R$, let $x_r \in \{0, 1\}$ denote the variable corresponding to r. Objective is to minimize $\sum_{r \in R} x_r$ with constraints $\sum_{r \in N(v)} x_r \geq 1$ for all $v \in V$. Let the corresponding relaxed LP formulation be Q_l.

Let $\mathbf{Q}_l = \{x_r\}_{r \in R}$ be an optimal solution of Q_l. Consider SSR-algorithm. Here, define $y_r = 1$ if $r \in S$, $y_r = 0$ if $r \notin S$ and $\mathbf{Q}' = \{y_r\}_{r \in R}$, obtained by the algorithm. This is a feasible solution of Q as SSR-algorithm terminates only when no segments are left in V_i. Now we fix any arbitrary $r \in S$ and ν_r be a critical segment with respect to r. Then due to Observation B, we know that for all $z \in N(\nu_r) \setminus \{r\}$ we have $T_z = \emptyset$ and $N(\nu_r) \subseteq T_r$. Therefore, for the constraint corresponding to ν_r in Q_l, we have that

$$\sum_{z \in N(\nu_r)} y_z = 1 \leq \sum_{z \in N(\nu_r)} x_z \leq \sum_{z \in T_r} x_z \qquad \text{[since } N(\nu_r) \subseteq T_r \text{ by Observation B]}$$

Therefore, from above argument and from Lemma 2 we conclude that

$$|S| = \sum_{r \in S} y_r = \sum_{r \in S} \sum_{z \in N(\nu_r)} y_z \leq \sum_{r \in S} \sum_{z \in T_r} x_z \leq 2 \sum_{z \in R} x_z \leq 2|OPT|.$$

Hence we have the proof. □

The proofs of Lemma 1 and Theorem 4 follows directly from the proof of Lemma 3.

3 Proof of Theorem 3

In this section, we shall give an $O(n^5)$-time 8-approximation algorithm to solve the MDS problem on vertically-stabbed-L graphs. In the rest of the paper, $OPT(Q)$ and $OPT(Q_l)$ denote the cost of the optimum solution of an ILP Q and LP Q_l respectively.

Overview of the Algorithm: First, we solve the relaxed LP formulation of the ILP of the MDS problem on the input vertically-stabbed-L graph G and create two subproblems. We shall show that one of those two subproblems is equivalent to the SSR problem and the other is equivalent to a *Stabbing Rays with Segments* problem (defined below) introduced by Katz et al. [15]. We solve these two subproblems individually.

In the ***Stabbing Rays with Segments*** (SRS) problem, the input is a set R of disjoint leftward-directed horizontal rays and a set V of disjoint vertical segments. The objective is to select a minimum cardinality subset of V that intersects all rays in R. We shall propose a heuristic algorithm that gives a feasible solution, and its cost is at most twice the cost of the optimum solution of the ILP of *SRS*.

2-Approximation Algorithm for SRS Problem: With each segment $v \in V$, we associate a token T_v which is a subset of V. Initialise $T_v = \emptyset$ for each $v \in V$. Let r_i be the ray whose right-endpoint, (x_i, y_i), has the smallest x-coordinate. We assume without loss of generality that x- and y-coordinates of the endpoints of the rays are all distinct. Assuming that there is a feasible solution to the SRS instance, there must exist a segment of V that intersects r_i. Let $N(r_i) \subseteq V$ be the set of segments that intersect r_i. Let v_{top} (resp. v_{bot}) be a segment in $N(r_i)$ whose top endpoint is top-most (resp., bottom endpoint is bottom-most); it may be that $v_{top} = v_{bot}$. We add both v_{top} and v_{bot} to our heuristic solution set S. Also we set $T_{v_{top}} = T_{v_{bot}} = N(r_i)$. We remove from R all of the rays that intersect v_{top} or v_{bot}, delete all segments in $N(r_i)$ and then repeat the above steps untill $R = \emptyset$. Observe that for each ray r, there is a segment $v \in S$ that intersects r. Also observe that for each segment $v \in V$, there are at most two tokens such that both of them contains v.

Lemma 4. *Let Q be the ILP of the SRS instance with a set of rays R and set of segments V as input and Q_l be the corresponding relaxed LP. Then $OPT(Q) \leq 2 \cdot OPT(Q_l)$.*

Proof. Let $\mathbf{X} = \{x_v\}_{v \in V}$ be an optimal solution of Q_l where x_v denotes the value of the variable in Q_l corresponding to $v \in V$. Let S be the solution returned by the above algorithm with R, V as input. Now define for each $v \in V$, $y_v = 1$ if $v \in S$, $y_v = 0$ if $v \notin S$ and let $\mathbf{Y} = \{y_v\}_{v \in V}$. Observe that $\mathbf{Y}$ is a feasible solution of Q. For each $z \in S$, there is a ray r_i such that $T_z = N(r_i)$. Therefore, $y_z = 1 \leq \sum\limits_{v \in N(r_i)} x_v = \sum\limits_{v \in T_z} x_v$.

As a segment v is contained in at most two tokens, using the above inequality we have

$$|S| = \sum_{v \in S} y_v \leq \sum_{v \in S} \sum_{v' \in T_v} x_{v'} \leq 2 \sum_{v' \in V} x_{v'} = 2 \cdot OPT(Q_l)$$

Hence the result follows. □

Now we describe our approximation algorithm for MDS problem on vertically-stabbed-L graphs. Let $\mathcal{R} = \{\mathrm{L}_u\}_{u \in V}$ be a vertically-stabbed-L-representation of a graph $G = (V, E)$. Without loss of generality, we assume that (i) the vertical line $x = 0$ intersects all the L-paths in $\mathcal{R}$ and the x-coordinate of the corner point of each L-path in $\mathcal{R}$ is strictly less than 0, and (ii) whenever two distinct L-paths intersect in $\mathcal{R}$, they intersect at exactly one point.

For a vertex $u \in V$, let $N[u]$ denote the closed neighbourhood of u in G, $H_u = \{c \in N[u] \colon \mathrm{L}_c$ intersects the horizontal segment of $\mathrm{L}_u\}$ and let V_u denote the set $N(u) \setminus H_u$. Based on these we have the following ILP (say Q) of the problem of finding a minimum dominating set of G.

$$\begin{array}{ll} \text{minimize} & \sum_{v \in V} x_v \\ \text{subject to} & \sum_{v \in H_u} x_v + \sum_{v \in V_u} x_v \geq 1, \ \forall u \in V \\ x_v \in \{0,1\}, \ \forall v \in V & \\ & Q \end{array}$$

Let Q_l be the the relaxed LP formulation of Q and $\mathbf{Q}_l = \{x_v \colon v \in V\}$ be an optimal solution of Q_l. Now we define the following sets.

$$A_1 = \left\{ u \in V \colon \sum_{v \in H_u} x_v \geq \frac{1}{2} \right\}, A_2 = \left\{ u \in V \colon \sum_{v \in V_u} x_v \geq \frac{1}{2} \right\}$$

$$H = \bigcup_{u \in A_1} H_u, V = \bigcup_{u \in A_2} V_u$$

Based on these, we consider the following two integer programs Q' and Q''.

$\begin{array}{ll} \text{minimize} & \sum_{v \in H} x'_v \\ \text{subject to} & \sum_{v \in H_u} x'_v \geq 1, \forall u \in A_1 \\ x'_v \in \{0,1\}, \ v \in H & \\ & Q' \end{array}$	$\begin{array}{ll} \text{minimize} & \sum_{v \in V} x''_v \\ \text{subject to} & \sum_{v \in V_u} x''_v \geq 1, \forall u \in A_2 \\ x''_v \in \{0,1\}, \ v \in V & \\ & Q'' \end{array}$

Let Q'_l and Q''_l be the relaxed LP of Q' and Q'' respectively. Clearly, the solutions of Q' and Q'' gives a feasible solution for Q. Hence $OPT(Q) \leq OPT(Q') + OPT(Q'')$. For each $x_v \in \mathbf{Q}_l$, define $y_v = \min\{1, 2x_v\}$ and define $\mathbf{Y}_l = \{y_v\}_{x_v \in \mathbf{Q}_l}$. Notice that $\mathbf{Y}_l$ gives a solution to Q'_l and Q''_l. Therefore, $OPT(Q'_l) + OPT(Q''_l) \leq 4 \cdot OPT(Q_l)$. We have the following lemma.

Lemma 5. *$OPT(Q') \leq 2 \cdot OPT(Q'_l)$ and $OPT(Q'') \leq 2 \cdot OPT(Q''_l)$.*

Proof. Note that for each vertex $u \in A_1$, H_u is non-empty and for each $v \in H_u$, L_v intersects the horizontal segment of L_u. Let R be the set of horizontal segments of the L-paths representing the vertices in A_1 and S be the set of vertical segments of the L-paths representing the vertices in H. Since all horizontal segments in R intersect the vertical line $x = 0$ and the x-coordinates of the vertical segments in S is strictly less than 0, we can consider the horizontal segments in R as rightward directed rays. Hence, solving Q' is equivalent to solving the ILP, say $\mathcal{E}$, of the problem of finding a minimum cardinality subset of vertical segments S that intersects all rays in the set R of rightward-directed rays. Hence solving $\mathcal{E}$ is equivalent to solving an SRS instance with R and S as input. By Lemma 4, we have that

$$OPT(Q') = OPT(\mathcal{E}) \leq 2 \cdot OPT(\mathcal{E}_l) \leq 2 \cdot OPT(Q'_l)$$

where $\mathcal{E}_l$ is the relaxed LP of $\mathcal{E}$. Hence we have proof of the first part.

For the second part, using similar arguments as above, we can show that solving Q'' is equivalent to solving an SSR instance. Hence, by Lemma 1, we have that $OPT(Q'') \leq 2 \cdot OPT(Q''_l)$. Hence the proof follows. □

Proof of Theorem 3: Lemma 5 implies that solving Q' (resp. Q'') is equivalent to solving SRS (resp. SSR) problem instance. Let A be the union of the solutions returned by 2-approximation algorithm for SRS problem and SSR-algorithm, used to solve Q' and Q'' respectively. Hence,

$$|A| \leq 2(OPT(Q'_l) + OPT(Q''_l)) \leq 8 \cdot OPT(Q_l) \leq 8 \cdot OPT(Q)$$

Since Q_l consists of n variables where $n = |V|$, solving Q_l takes $O(n^5)$ time [19]. Solving both the SSR and SRS instances takes a total of $O(n \log n)$ time and therefore the total running time of the algorithm is $O(n^5)$.

4 Proof of Theorem 1 and Theorem 2

First we prove the NP-hardness for the MDS problem on unit B_1-VPG graphs. The (h, w)-*grid* is the undirected graph G with vertex set $\{(x, y) \colon x, y \in \mathbb{Z}, 1 \leq x \leq h, 1 \leq y \leq w\}$ and edge set $\{(u, v)(x, y) \colon |u - x| + |v - y| = 1\}$. A graph G is a *grid* graph if G is an induced subgraph of (h, w)-grid for some positive integers h, w.

Proof of Theorem 1: We shall reduce the NP-complete MDS problem on grid graphs [6] to the MDS problem on unit B_1-VPG graphs. It is sufficient to show that for any positive integers h, w the (h, w)-grid has a unit B_1-VPG representation. Let $G = (V, E)$ be a (h, w)-grid and $\epsilon = \frac{1}{hw}$. For each $(x, y) \in V$ consider the unit L-path $\mathrm{L}_{(x,y)}$ such that top endpoint of vertical segment of $\mathrm{L}_{(x,y)}$ is $(x, y - \epsilon(x - 1))$. The set $\mathcal{R} = \{\mathrm{L}_{(x,y)} \colon (x, y) \in V\}$ is a unit B_1-VPG representation of G. This completes the proof.

To prove Theorem 2 we will need Lemma 6. For sake of clarity, first we shall prove Theorem 2 assuming Lemma 6 to be true. The proof of Lemma 6 uses Lemma 1, but it is omitted due to space constraints.

Lemma 6. *Let S_1 and S_2 be sets of orthogonal unit length segments. Let $\mathcal{C}$ be the ILP of the problem of finding a minimum cardinality subset D of S_2 such that every segment in S_1 intersects some segment in D. There is an $O(n^5)$-time algorithm to compute a set D' which gives a solution of $\mathcal{C}$ and $|D'| \leq 18 \cdot OPT(\mathcal{C}_l)$ where $n = |S_1 \cup S_2|$ and $\mathcal{C}_l$ is the relaxed LP of $\mathcal{C}$.*

Completion of Proof of Theorem 2 Assuming Lemma 6: Let $\mathcal{R}$ be a unit B_k-VPG representation of a unit B_k-VPG graph $G = (V, E)$. We shall assume that every vertex of G has a self loop (this does not contradict the intersection model as every rectilinear path intersects itself). For a vertex $v \in V$, let $P(v)$ denote the path in $\mathcal{R}$ that corresponds to v. For a vertex $v \in V$, $N(v)$ and $N[v]$ denote the *open neighbourhood* and *closed neighbourhood* of v, respectively. Throughout this section, we assume that the segments of each path $P \in \mathcal{R}$ are numbered consecutively starting from the leftmost segment by $1, 2, \ldots, t$ where $t(\leq k+1)$ is the number of segments in P.

Define $\phi\colon E \to \mathbb{N} \times \mathbb{N}$ such that for an edge uv, $\phi(uv) = (i, j)$ if and only if the ith segment of $P(u)$ intersects the jth segment of $P(v)$, and for all $1 \leq a < i$ and $1 \leq b < j$, the ath segment of $P(u)$ does not intersect the bth segment of $P(v)$.

For a vertex $u \in V$, let $X_u(i, j) = \{v \in N[u] \colon \phi(uv) = (i, j)\}$. For distinct pairs (i, j) and (i', j') the sets $X_u(i, j)$ and $X_u(i', j')$ are disjoint. Let $\mathcal{K}$ denote the set $\{1, 2, \ldots, k+1\} \times \{1, 2, \ldots, k+1\}$. Based on these we have the following ILP of the MDS problem on G.

$$\begin{array}{ll} \text{minimize} & \sum_{v \in V} x_v \\ \text{subject to} & \sum_{(i,j) \in \mathcal{K}} \sum_{v \in X_u(i,j)} x_v \geq 1, \ \forall u \in V \\ & x_v \in \{0, 1\}, \ \forall v \in V \end{array}$$

Q

First step of our algorithm is to solve the relaxed LP formulation (say Q_l) of Q. Let $\mathbf{Q}_l = \{x_v \colon v \in V\}$ be an optimal solution of Q_l. For each vertex $u \in V$, there is a pair $(i, j) \in \mathcal{K}$ such that $\sum_{v \in X_u(i,j)} x_v \geq \frac{1}{(k+1)^2}$. For each pair $(i, j) \in \mathcal{K}$, define

$$A(i, j) = \left\{ u \in V \colon \sum_{v \in X_u(i,j)} x_v \geq \frac{1}{(k+1)^2} \right\}, B(i, j) = \bigcup_{u \in A(i,j)} X_u(i, j)$$

Based on these, we have the following ILP for each pair $(i, j) \in \mathcal{K}$.

$$\begin{array}{ll} \text{minimize} & \sum_{v \in B(i,j)} x'_v \\ \text{subject to} & \sum_{v \in X_u(i,j)} x'_v \geq 1, \ \forall u \in A(i, j) \\ & x'_v \in \{0, 1\}, \ \forall v \in B(i, j) \end{array}$$

$Q(i, j)$

For each pair $(i, j) \in \mathcal{K}$, let $Q_l(i, j)$ be the relaxed LP of $Q(i, j)$. We have the following

$$OPT(Q) \leq \sum_{(i,j) \in \mathcal{K}} OPT(Q(i, j))$$

For each $x_v \in \mathbf{Q}_l$, define $y_v = \min\{1, x_v(k+1)^2\}$ and define $\mathbf{Y}_l = \{y_v\}_{x_v \in \mathbf{Q}_l}$. Clearly, $\mathbf{Y}_l$ gives a solution to $Q_l(i,j)$ for each $(i,j) \in \mathcal{K}$. Moreover,

$$\sum_{(i,j)\in\mathcal{K}} OPT(Q_l(i,j)) \leq (k+1)^4 \cdot OPT(Q_l)$$

Now we have the following lemma.

Lemma 7. *For each pair $(i,j) \in \mathcal{K}$, there is a solution $D(i,j)$ for $Q(i,j)$ such that $|D(i,j)| \leq 18 \cdot OPT(Q_l(i,j))$.*

Proof. For any $(i,j) \in \mathcal{K}$, solving $Q(i,j)$ is equivalent to finding a minimum cardinality subset D of $B(i,j)$ such that each vertex $u \in A(i,j)$ has a neighbour in $D \cap X_u(i,j)$. Notice that, for each $u \in A(i,j)$ the set $X_u(i,j)$ is non-empty and each $v \in X_u(i,j)$, the ith segment of $P(u)$ intersects the jth segment of $P(v)$. Let $S = \{i^{th}$ segment of $P(u)\colon u \in A(i,j)\}$, $T = \{j^{th}$ segment of $P(v)\colon v \in B(i,j)\}$.

Solving $Q(i,j)$ is equivalent to the problem of finding a minimum cardinality subset D of T such that every segment in S intersect at least one segment in D. Moreover, every segment in $S \cup T$ have unit length. Hence by Lemma 6, we have a solution (say $D(i,j)$) for $Q(i,j)$ such that $|D(i,j)| \leq 18 \cdot OPT(Q(i,j))$. □

For each pair $(i,j) \in \mathcal{K}$, due to Lemma 7, we have a solution $D(i,j)$ of $Q(i,j)$ such that $|D(i,j)| \leq 18 \cdot OPT(Q_l(i,j))$ in polynomial time. Let D be the union of $D(i,j)$'s for all $(i,j) \in \mathcal{K}$. We have that

$$\begin{aligned}
|D| &= \sum_{(i,j)\in\mathcal{K}} |D(i,j)| \\
&\leq \sum_{(i,j)\in\mathcal{K}} 18 \cdot OPT(Q_l(i,j)) \\
&\leq 18 \cdot (k+1)^4 \cdot OPT(Q_l) \leq 18 \cdot (k+1)^4 \cdot OPT(Q)
\end{aligned}$$

This completes the proof of Theorem 2.

5 Conclusion

We initiated the study of approximating MDS of string graphs in terms of their unit B_k-VPG representation. It is unlikely that there is $o(\log k)$-approximaiton algorithm for MDS problem on B_k-VPG graphs. This naturally leads to the following question.

Question 1. Is there an $O(\log k)$-approximation algorithm for the MDS problem on B_k-VPG graphs or unit B_k-VPG graphs?

References

1. Asinowski, A., Cohen, E., Golumbic, M.C., Limouzy, V., Lipshteyn, M., Stern, M.: Vertex intersection graphs of paths on a grid. J. Graph Algorithms Appl. **16**(2), 129–150 (2012)
2. Bandyapadhyay, S., Maheshwari, A., Mehrabi, S., Suri, S.: Approximating dominating set on intersection graphs of rectangles and L-frames. In: MFCS, pp. 37:1–37:15 (2018)
3. de Berg, M., Cheong, O., van Kreveld, M., Overmars, M.: Computational Geometry: Algorithms and Applications. Springer, Heidelberg (2008). https://doi.org/10.1007/978-3-540-77974-2
4. Bousquet, N., Gonçalves, D., Mertzios, G.B., Paul, C., Sau, I., Thomassé, S.: Parameterized domination in circle graphs. In: Golumbic, M.C., Stern, M., Levy, A., Morgenstern, G. (eds.) WG 2012. LNCS, vol. 7551, pp. 308–319. Springer, Heidelberg (2012). https://doi.org/10.1007/978-3-642-34611-8_31
5. Chlebík, M., Chlebíková, J.: Approximation hardness of dominating set problems in bounded degree graphs. Inf. Comput. **206**(11), 1264–1275 (2008)
6. Clark, B.N., Colbourn, C.J., Johnson, D.S.: Unit disk graphs. Discret. Math. **86**(1–3), 165–177 (1990)
7. Colbourn, C.J., Stewart, L.K.: Permutation graphs: connected domination and steiner trees. Discret. Math. **86**(1–3), 179–189 (1990)
8. Damian, M., Pemmaraju, S.V.: APX-hardness of domination problems in circle graphs. Inf. Process. Lett. **97**(6), 231–237 (2006)
9. Damian-Iordache, M., Pemmaraju, S.V.: A (2+ ε)-approximation scheme for minimum domination on circle graphs. J. Algorithms **42**(2), 255–276 (2002)
10. Erlebach, T., van Leeuwen, E.J.: Domination in geometric intersection graphs. In: Laber, E.S., Bornstein, C., Nogueira, L.T., Faria, L. (eds.) LATIN 2008. LNCS, vol. 4957, pp. 747–758. Springer, Heidelberg (2008). https://doi.org/10.1007/978-3-540-78773-0_64
11. Farber, M., Keil, J.M.: Domination in permutation graphs. J. Algorithms **6**(3), 309–321 (1985)
12. Gibson, M., Pirwani, I.A.: Algorithms for dominating set in disk graphs: breaking the log$n$ barrier. In: de Berg, M., Meyer, U. (eds.) ESA 2010. LNCS, vol. 6346, pp. 243–254. Springer, Heidelberg (2010). https://doi.org/10.1007/978-3-642-15775-2_21
13. Golumbic, M.C., Rotem, D., Urrutia, J.: Comparability graphs and intersection graphs. Discrete Math. **43**(1), 37–46 (1983)
14. Gonçalves, D., Isenmann, L., Pennarun, C.: Planar graphs as L-intersection or L-contact graphs. In: SODA, pp. 172–184 (2018)
15. Katz, M.J., Mitchell, J.S.B., Nir, Y.: Orthogonal segment stabbing. Comput. Geom.: Theory Appl. **30**(2), 197–205 (2005)
16. Lahiri, A., Mukherjee, J., Subramanian, C.R.: Maximum independent set on B_1-VPG graphs. In: Lu, Z., Kim, D., Wu, W., Li, W., Du, D.-Z. (eds.) COCOA 2015. LNCS, vol. 9486, pp. 633–646. Springer, Cham (2015). https://doi.org/10.1007/978-3-319-26626-8_46
17. McGuinness, S.: On bounding the chromatic number of L-graphs. Discrete Math. **154**(1–3), 179–187 (1996)
18. Mehrabi, S.: Approximating domination on intersection graphs of paths on a grid. In: Solis-Oba, R., Fleischer, R. (eds.) WAOA 2017. LNCS, vol. 10787, pp. 76–89. Springer, Cham (2018). https://doi.org/10.1007/978-3-319-89441-6_7
19. Tardos, E.: A strongly polynomial algorithm to solve combinatorial linear programs. Oper. Res. **34**(2), 250–256 (1986)

Classified Rank-Maximal Matchings and Popular Matchings – Algorithms and Hardness

Meghana Nasre[1], Prajakta Nimbhorkar[2], and Nada Pulath[1(✉)]

[1] Indian Institute of Technology, Chennai, India
nadapulath1710@gmail.com

[2] UMI ReLaX, Chennai Mathematical Institute, Chennai, India

Abstract. In this paper, we consider the problem of computing an optimal matching in a bipartite graph $G = (A \cup P, E)$ where elements of A specify preferences over their neighbors in P, possibly involving ties, and each vertex can have capacities and classifications. A classification $\mathcal{C}_u$ for a vertex u is a collection of subsets of neighbors of u. Each subset (class) $C \in \mathcal{C}_u$ has an *upper quota* denoting the maximum number of vertices from C that can be matched to u. The goal is to find a matching that is *optimal* amongst all the *feasible matchings*, which are matchings that respect quotas of all the vertices and classes.

We consider two well-studied notions of optimality namely *popularity* and *rank-maximality*. The notion of *rank-maximality* involves finding a matching in G with maximum number of rank-1 edges, subject to that, maximum number of rank-2 edges and so on. We present an $O(|E|^2)$-time algorithm for finding a feasible rank-maximal matching, when each classification is a *laminar* family. We complement this with an NP-hardness result when classes are non-laminar even under strict preference lists, and even when only posts have classifications, and each applicant has a quota of one. We show an analogous dichotomy result for computing a popular matching amongst feasible matchings (if one exists) in a bipartite graph with posts having capacities and classifications and applicants having a quota of one.

To solve the classified rank-maximal and popular matchings problems, we present a framework that involves computing max-flows iteratively in multiple flow networks. Besides giving polynomial-time algorithms for classified rank-maximal and popular matching problems, our framework unifies several algorithms from literature [1,10,12,15].

Keywords: Bipartite graphs · Popularity · Rank-maximality · Matchings under classifications

1 Introduction

Matchings under preferences have been well-studied in literature. The input consists of a bipartite graph $G = (A \cup P, E)$ where A is a set of applicants, P

I. Sau and D. M. Thilikos (Eds.): WG 2019, LNCS 11789, pp. 244–257, 2019.
https://doi.org/10.1007/978-3-030-30786-8_19

is a set of posts, and applicants have preferences over their neighbors in P. The preference list of an applicant can involve ties. An edge (a, p) is said to have rank k, if p is a kth choice of a. The setting where applicants and posts both can be assigned at most one element from the opposite set is called the *one-to-one setting*. This has been generalized to a model where a post p can accommodate more than one applicant, and has a positive quota $q(p)$ denoting the maximum number of applicants it can accommodate. An allocation of applicants to posts in this setting is referred to as a *many-to-one* matching. Additionally, if an applicant a can be allotted more than one post simultaneously, up to a positive quota $q(a)$, it is called the *many-to-many* setting.

In all the above settings the goal is to match applicants to posts *optimally* with respect to the preferences of the applicants. Two well-studied notions of optimality in this setting are rank-maximality [10] and popularity [1,6]. The notion of *signature* defined in the literature [10] is useful to compare two matchings with respect to rank-maximality. The *signature* σ_M of a matching M is an r-tuple $(x_1, \ldots, x_r)$ where r denotes the largest rank used by an applicant to rank any post. For $1 \leq k \leq r$, x_k denotes the number of rank k edges in M.

Definition 1 (Rank-maximal matching). *A matching is* rank-maximal *if it matches the maximum number of applicants to their rank-*1 *posts, subject to that, the maximum number of applicants to their rank-*2 *posts, and so on. That is, M has the maximum signature when comparing the signatures of two matchings lexicographically.*

Popularity is defined in terms of votes of applicants between two matchings M and M'. An applicant a prefers M over M' if and only if a is matched to a higher preferred post in M as compared to that in M', and any applicant prefers to be matched to one of its neighbors over remaining unmatched.

Definition 2 (Popular matching). *A matching M is* more popular than *another matching M' if more applicants prefer M over M' than M' over M. A matching M is popular if there is no matching that is more popular than M.*

Rank-maximal matchings have been studied in the many-to-many setting [15], and popular matchings have been considered in the many-to-one setting [12], polynomial-time algorithms being known in both the cases.

We consider a natural generalization of the many-to-many setting, where each vertex u can additionally specify a *classification* $\mathcal{C}_u$ on its set of neighbors $N(u)$. A classification is a set of subsets of $N(u)$. Each subset $C_u^i \in \mathcal{C}_u$ is called a *class*, and each class has its own quota $0 < q(C_u^i) \leq q(u)$. A matching now requires to respect the quotas of all the classes, and we call such a matching a *feasible* matching. Classifications arise naturally in matching problems. While allotting courses to students, a student does not want to be allotted too many courses on closely related topics. An instructor may not want a course to have too many students from the same department. While assigning doctors to hospitals in the well-studied hospital-residents setting, a hospital has an upper limit on the number of doctors in a specialization. These constraints are readily modeled using classifications.

Figure 1 shows an example instance of the classified matching problem where $A = \{a_1, a_2, a_3\}$ and $P = \{p_1, p_2\}$. The preferences of the applicants, and the classifications and quotas can be read from the figure. Consider a feasible matching $M = \{(a_1, p_2), (a_2, p_1), (a_3, p_2)\}$ which has a signature of $(2, 1)$. The matching $M' = \{(a_1, p_1), (a_2, p_1), (a_3, p_2)\}$ has signature $(3, 0)$ but is infeasible because of the classification $C^1_{p_1}$. We will show that the matching is M is both rank-maximal and popular in this instance.

Preference Lists	Classifications	Quotas
$a_1 : p_1, p_2$	$\mathcal{C}_{p_1} = \{C^1_{p_1} = \{a_1, a_2\}\}$	$q(p_1) = q(p_2) = 2$
$a_2 : (p_1, p_2)$	$\mathcal{C}_{p_2} = \{C^1_{p_2} = \{a_2, a_3\}\}$	$q(C^1_{p_1}) = q(C^1_{p_2}) = 1$
$a_3 : p_2, p_1$		$q(a_i) = 1; \quad i = \{1, 2, 3\}$

Fig. 1. Preferences to be read as: a_1 treats p_1 as rank-1 post and p_2 as rank-2 post. Applicant a_2 has ties in its preference list and treats both p_1 and p_2 as its rank-1 posts. Although $q(p_1) = 2$, the class $C^1_{p_1} \in \mathcal{C}_{p_1}$ implies that in any feasible matching post p_1 can be matched to at most one applicant from $\{a_1, a_2\}$.

We address the problems of finding a rank-maximal matching in the many-to-many setting and a popular matching (if it exists) in the many-to-one setting, both in the presence of classifications. We refer to these problems as CRMM and CPM problems respectively. We show that both the problems are NP-hard for arbitrary classifications, and complement the hardness results by giving polynomial-time algorithms when the classification of each vertex is a *laminar* family. A family $\mathcal{F}$ of subsets of a set S is said to be *laminar* if, for every pair of sets $X, Y \in \mathcal{F}$, either $X \subseteq Y$ or $Y \subseteq X$ or $X \cap Y = \emptyset$. Laminar classifications are natural in settings like student allocation to schools where schools may want at most a certain number of students from a particular region, district, state, country and so on. Laminar classification includes the special case of *partition*, where the classes are disjoint – a natural classification arising in practice.

Classifications have been previously studied for the two-sided preference model [4,7] – called the stable marriage setting. In the stable matching case, existence of a stable matching respecting the classifications can be determined in polynomial-time if the classes specified by each vertex form a *laminar* family [4,7], and otherwise the problem is NP-complete [7]. In our setting, the preferences being only on one side and the optimality criteria being rank-maximality or popularity are very different from the stable matching setting. Yet we show similar results as those of [7] and [4].

Our Contribution: We show the following new results in this paper. Let $G = (A \cup P, E)$ denote an instance of the CRMM problem or the CPM problem.

Theorem 1. *There is an $O(|E|^2)$-time algorithm for the* CRMM *problem when the classification for every vertex is a laminar family.*

We also show the above result for the CPM problem in the many-to-one setting.

Theorem 2. *There is an $O(|A||E|)$-time algorithm for the* CPM *problem when the classification for every post is a laminar family.*

We complement the above results with a matching hardness result:

Theorem 3. *The* CRMM *and* CPM *problems are* NP*-hard when the classes are non-laminar even when all the preferences are strict, and classifications exist on only one side of the bipartition.*

The hardness holds even when the intersection of the classes in a family is at most one, and the preference lists have length at most 2. Our reduction also shows the following hardness result even when there are no ranks on edges.

Theorem 4. *The problem of finding a maximum cardinality matching is* NP*-hard in the presence of non-laminar classifications.*

Related Work: Irving introduced the rank-maximal matchings problem as "greedy matchings" in [9] for the one-to-one case of strict preferences. Irving et al. [10] generalized the same to preference lists with ties allowed and this was further generalized by Paluch [15] for the many-to-many setting. Abraham et al. [1] initiated the study of Popular Matchings problem in the one-to-one setting and subsequently there have been several results [8,12,13] on generalization of this model. In all the above results where the model is without classifications, the algorithms for computing a rank-maximal matching [10,15] and for computing popular matching in [1,8,12,13] have the following template: The algorithms are iterative and make crucial use of the well-known Dulmage-Mendelsohn decomposition w.r.t. maximum matchings in bipartite graphs. The main use of the decomposition theorem in all the literature mentioned above is to identify edges that can not belong to any optimal matching. Such edges are deleted in each iteration, resulting in a *reduced graph*, such that every maximum matching in the reduced graph is an optimal matching in the given instance.

Our Technique and Comparison with Prior Work: Although the high-level idea of our algorithm is similar as in the Irving et al. [10] and Abraham et al. [1], due to the presence of classifications the proof techniques and the details are completely different. The Dulmage-Mendelsohn decomposition [3] crucially used in [10,15] for identifying unnecessary edges relies on the fact that a partial matching computed in an iteration is a *maximum matching* in the graph considered in that iteration. In our setting, a feasible matching need not be a maximum matching in any graph; this poses a difficulty in obtaining a decomposition. We overcome this by exploiting the connection between matchings and flows. We use the fact that in *any* flow network, w.r.t. *any* max-flow the vertices can be decomposed into three disjoint sets and this decomposition is *invariant* of the flow. To the best of our knowledge, the decomposition of the vertices of a flow network w.r.t. *any* max-flow and its invariance w.r.t. a max-flow have not been used in the literature on matchings with preferences. Besides giving polynomial-time algorithms for CRMM and CPM problems, our work also unifies several known results [1,10,12,15].

We emphasize that the techniques used in the capacitated setting, namely by Manlove and Sng [12] in the capacitated house allocation problem and by Paluch [15] in the capacitated rank-maximal matchings do not work in the presence of classifications. We believe that exploiting the natural connection between bipartite matchings and network flows makes our approach elegant and also practical. The problems of capacitated matchings with preferences are not only of theoretical interest but have a wide range of practical applications. We finally note that the CRMM problem can also be solved using min-cost flows with slightly higher time complexity, but that approach involves using exponential weights. Analogously, the CPM problem can be solved using the matroid generalization considered by Kamiyama [11], however, the running time of the corresponding algorithm is $O(m^4)$. In contrast, our algorithms for the CRMM and CPM problems are simple, combinatorial and use only elementary flow computations.

Organization of the Paper: In Sect. 2 we describe our flow network for the laminar CRMM problem and prove properties of the network. In Sect. 3 we present our algorithm and prove its correctness. In Sect. 4 we give the hardness for the non-laminar CRMM problem. In the interest of space and readability we present the detailed algorithmic results for the CPM problem in the full-version [14].

2 Construction of the Flow Network

In this section, we present the construction of our flow-networks used by the polynomial-time algorithms for the CRMM and the CPM problems when the classes of each vertex form a laminar family. Recall that the given instance is a bipartite graph $G = (A \cup P, E)$, along with a preference list for each $a \in A$, and a laminar classification $\mathcal{C}_u$ for each $u \in A \cup P$ for the CRMM problem, and for each $u \in P$ for the CPM problem. We describe the more general case of the CRMM problem here. We construct the flow network H_0 corresponding to the input bipartite graph G with classifications. We apply the following pre-processing for every vertex in G:

For every $u \in A \cup P$ with classification C_u, we add the following classes to $\mathcal{C}_u$.

- C_u^*: We include a class $C_u^* = N(u)$ into $\mathcal{C}_u$ with capacity $q(C_u^*) = q(u)$.
- C_u^w: For every $w \in N(u)$ and $u \in A \cup P$, we add a class C_u^w to $\mathcal{C}_u$ with capacity $q(C_u^w) = 1$.

It is easy to see that this does not change the set of feasible matchings. In the rest of the paper, we refer to this modified instance as our instance G.

Definition 3 (Classification tree). *Let every vertex $u \in A \cup P$ have a laminar family of classes $\mathcal{C}_u$. Then, the classes in $\mathcal{C}_u$ can be represented as a tree called the* *classification tree* $\mathcal{T}_u$ *with* C_u^* *being the root of* $\mathcal{T}_u$. *For two classes* $C_u^1, C_u^2 \in \mathcal{C}_u$, *the class* C_u^1 *is the parent of* C_u^2 *in* $\mathcal{T}_u$ *iff* C_u^1 *is the smallest class in* $\mathcal{C}_u$ *containing* C_u^2. *For every* $w \in N(u)$, *the corresponding singleton class* C_u^w *is a leaf of* $\mathcal{T}_u$.

Throughout the paper, we refer to the vertices V of H_0 as "nodes". The network H_0 has nodes corresponding to every element of $\mathcal{T}_u$ for each $u \in A \cup P$. In addition, there is a source s and a sink t. The edges of H_0 include an edge from s to the root of $\mathcal{T}_a$ for each $a \in A$, and an edge from the root of $\mathcal{T}_p$ to t, for each $p \in P$. Each edge of $\mathcal{T}_a$, for each $a \in A$, is directed from parent to child whereas each edge of $\mathcal{T}_p$, $p \in P$ is directed from child to parent in H_0. Thus,

$$V = \{s, t\} \cup \{C_u^i \mid C_u^i \in \mathcal{T}_u \text{ and } u \in A \cup P\}$$

The set of all edges of H_0 represented by F_0 and their capacities are as follows:

- For every $a \in A$, F_0 contains an edge (s, C_a^*) with capacity $q(C_a^*)$.
- For every $p \in P$, F_0 contains an edge (C_p^*, t) with capacity $q(C_p^*)$.
- For $a \in A$ and edge $(C_a^1, C_a^2) \in \mathcal{T}_a$ such that C_a^1 is the parent of C_a^2, F_0 contains an edge (C_a^1, C_a^2) with capacity $q(C_a^2)$.
- For $p \in P$ and edge $(C_p^1, C_p^2) \in \mathcal{T}_p$ such that C_p^2 is the parent of C_p^1, F_0 contains an edge (C_a^1, C_a^2) with capacity $q(C_a^1)$.

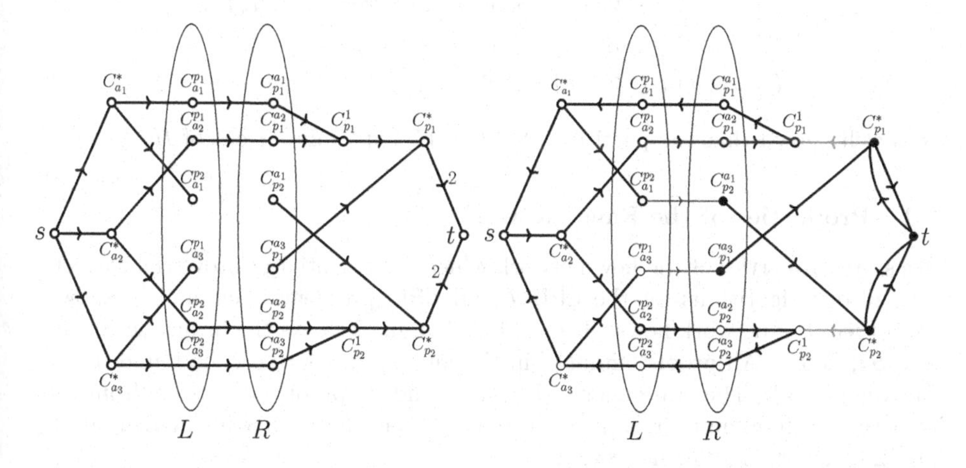

Fig. 2. Flow Network H_1 corresponding to example in Fig. 1. H_0 is the network without any edges between L and R. All edges except the one labeled have unit capacity.

Fig. 3. Black edges and red edges form the network $H_1(f_1)$. Black edges alone form the network H_1'. Black and blue edges together form the network H_2. The white and black nodes represent S_1, T_1 respectively. In this example $U_1 = \emptyset$. (Colro figure online)

We collectively refer to the set of leaves of $\mathcal{T}_a$ for all $a \in A$ as L and similarly, the set of leaves of $\mathcal{T}_p$ for all $p \in P$ as R. Thus

$$L = \{C_a^p \mid a \in A \text{ and } p \in N(a)\}; \ R = \{C_p^a \mid p \in P \text{ and } a \in N(p)\}$$

Figure 2 (ignoring the edges from L to R) shows the flow network H_0 corresponding to the example in Fig. 1. The nodes in L (respectively R) (shown in the two ellipses in the figure) have a unique predecessor (successor) in H_0. Moreover, H_0 can be seen as a disjoint union of two trees, one rooted at s and another at t, the edges of the former being directed from parent to child and those of the latter from child to parent. We call the two trees as *applicant-tree* and *post-tree* respectively. As evident, the graph H_0 admits no path from s to t, hence has a zero max-flow. Our algorithm in Sect. 3 iteratively adds edges to H_0. Let H be any such flow network constructed by our algorithm in some iteration and let f be a max-flow in H.

Decomposition of Vertices: In this section, we present a decomposition of the vertices of the flow network w.r.t. a max-flow f. We prove in Sect. 2.1 that the decomposition is *invariant* of the max-flow. The decomposition of the vertices allows us to delete certain edges in H that ensures that signature of the matching M corresponding to H is preserved in the future iterations. For a flow network H and a max-flow f in H, let $H(f)$ denote the residual network. We define the sets S_f, T_f, U_f as follows. Since f is a max-flow, it is immediate that the sets partition the vertex set V.

$$S_f = \{v \mid v \in V \text{ and } v \text{ is reachable from } s \text{ in } H(f)\}$$
$$T_f = \{v \mid v \in V \text{ and } v \text{ can reach } t \text{ in } H(f)\}$$
$$U_f = \{v \mid v \in V \text{ and } v \notin S_f \cup T_f\}$$

It is well-known that the partition $(S_f, T_f \cup U_f)$ is a min-s-t-cut of H.

2.1 Properties of the Flow Network

We state properties of the flow network which are essential to prove the correctness of our algorithms for the CRMM and CPM problems. Lemma 1 is known from theory of network flows (See e.g. [5]). Lemma 2 shows the invariance of the sets S_f, T_f, U_f. All proofs ommited in the interest of space can be found in the full-version [14]. The statement of Lemma 2 and its proof are implicit from the structure of minimum cuts [16]. We remark that the properties in Lemmas 1, and 2 hold for any flow network H.

Lemma 1. *[5] Let $H = (V, F)$ be any flow network where V is the vertex set and F is the edge set. Let f be a max-flow in H and (X, Y) be any min-s-t-cut of H. Then the following hold:*

- *For any edge $(a, b) \in F$ such that $a \in X, b \in Y$, we have $f(a, b) = c(a, b)$.*
- *For any edge $(b, a) \in F$ such that $a \in X, b \in Y$, we have $f(a, b) = 0$.*

Lemma 2. *The sets S_f, T_f and U_f are invariant of the max-flow f in H.*

The next lemma is specific to our flow network and is useful in proving the rank-maximality of our algorithm. An analogous claim can be proved for the leaf classes in the post tree.

Lemma 3. *Consider a node $C_a^i \in T \cup U$ such that either the parent C of C_a^i in $\mathcal{T}_a$ is in S or $C_a^i = C_a^*$. Then the following hold:*

- *(i) Every leaf C_a^p in the subtree of C_a^i in $\mathcal{T}_a$ belongs to $T \cup U$.*
- *(ii) Every max-flow f must saturate the edge (C, C_a^i).*

Conversely, in the applicant-tree, every leaf node $C_a^p \in T \cup U$ has an ancestor $C_a^i \in T \cup U$ such that the predecessor C of C_a^i (possibly s) is in S and the edge (C, C_a^i) is saturated in every max-flow.

3 Algorithm for Laminar CRMM

This section gives the detailed pseudo-code for our iterative algorithm for computing a laminar CRMM (see Algorithm 1). We begin by constructing the flow network H_0 as described in Sect. 2. For all $e \in F_0$, the max-flow $f_0(e) = 0$ since there is no s-t path in H_0. Start with $G'_0 = G_0 = (A \cup P, \emptyset)$. We partition the edges of G into sets E_k, $1 \leq k \leq r$ where r is the maximum rank on any edge of G and E_k contains the edges of rank k from G. Our algorithm repeatedly constructs the network H_k and maintains the reduced bipartite graph G'_k. In each iteration our algorithm operates as follows: it computes a max-flow f_k in a flow network H_k (Step 5) and computes the partition of the vertices S_k, T_k, U_k w.r.t f_k (Step 6). The algorithm then deletes forward and reverse edges of min-cut $(S_k, T_k \cup U_k)$ (Step 7). This step is crucial to ensure that the signature of the matching corresponding to the flow in the subsequent iterations does not degrade. Finally, the algorithm deletes certain edges of rank higher than k from the given bipartite graph (Step 8) – we prove that these edges cannot belong to any CRMM and hence can be removed. Finally, the output of our algorithm is the R-L edges of the flow network H'_r constructed in the final iteration.

We illustrate these steps on the example in Fig. 1. Add to H_0 edges of the form (C_a^p, C_p^a) for every rank-1 edge in G to obtain the flow network H_1 (see Fig. 2). Let f_1 be a max-flow in H_1 corresponding to the matching $M_1 = \{(a_1, p_1), (a_3, p_2)\}$. That is, for an edge $(a, p) \in M_1$ the unique $s - t$ path containing the edge (C_a^p, C_p^a) in H_1 carries unit flow. Figure 3 (black and red edges) shows the residual network $H_1(f_1)$ along with the partition of the vertices as S_1, T_1, U_1 (the set U_1 is empty in this example). The edges $\{(C_{p_1}^*, C_{p_1}^1), (C_{p_2}^*, C_{p_2}^1)\}$ in $H_1(f_1)$ are of the form (T_1, S_1) and hence are deleted as a reverse edges of the min-s-t cut to obtain H'_1. The flow network H_2 is obtained by adding to H'_1 the edges $\{(C_{a_1}^{p_2}, C_{p_2}^{a_1}), (C_{a_3}^{p_1}, C_{p_1}^{a_3})\}$. We remark that if the reverse edges of the min-s-t cut were not deleted, an augmenting path in H_2 of the form $\rho_1 = \langle s, C_{a_2}^*, C_{a_2}^{p_1}, \ldots, C_{a_1}^*, \ldots, C_{p_2}^*, C_{p_2}^1, C_{a_3}^*, \ldots C_{p_1}^{a_3}, C_{p_1}^*, t\rangle$ can be used to degrade the signature on rank-1 edges. We prove in the subsequent sections that our deletions ensure that the signature is never degraded.

In Lemma 4 and Corollary 1 we state properties of edges deleted by our algorithm.

Lemma 4. *Any edge between C_a^p and C_p^a in $H_k(f_k)$ is of the form S_kS_k, T_kT_k or U_kU_k, irrespective of its direction in $H_k(f_k)$. Hence an edge between L and R is not deleted in Step 7 of Algorithm 1.*

Algorithm 1. Laminar CRMM

1: Construct the flow network $H_0 = (V, F_0)$ as described in Section 2.
2: Let $F'_0 = F_0$ and for each i set $E'_i = E_i$.
3: **for** $k = 1$ to r **do**
4: $H_k = (V, F_k)$ where $F_k = F'_{k-1} \cup \{(C^p_a, C^a_p) \mid (a,p) \in E'_k\}$.
5: Let f_k be a max-flow in H_k. Compute the residual graph $H_k(f_k)$ w.r.t. flow f_k.
6: Compute the sets S_k, T_k and U_k.
7: Delete all edges of the form $(T_k \cup U_k, S_k)$ in $H_k(f_k)$.
8: Delete an edge $(a,p) \in E'_j$ where $j > k$ if $C^p_a \in T_k \cup U_k$ or $C^a_p \in S_k \cup U_k$.
9: Let $H'_k = (V, F'_k)$ be the modified $H_k(f_k)$ and let $G'_k = (A \cup P, \bigcup_{i=1}^k E'_i)$.
10: Let $M_k = \{(a,p) | (C^a_p, C^p_a) \in H'_k\}$.
11: **end for**
12: Return M_r.

Proof. Let $e = (C^p_a, C^a_p)$ be an edge in H_k. Recall that this is the only outgoing edge for C^p_a and only incoming edge for C^a_p in H_k. Also, C^p_a has an incoming edge of capacity 1 from its parent and C^a_p has an outgoing edge with capacity 1 to its parent.

Case 1: Edge e does not carry a flow in f_k. Then C^p_a and C^a_p do not receive any flow. In $H_k(f_k)$, e retains its direction. Thus if C^p_a is in S_k, so is C^a_p. Conversely, if C^a_p is in S_k, then C^p_a has to be in S_k, since C^a_p has no other incoming edge, and hence the path from s to C^a_p must use the edge e. Similarly, C^p_a is in T_k if and only if C^a_p is in T_k. If C^p_a is in U_k, then by the same argument as above, C^a_p can not be in S_k or T_k and hence must be in U_k.

Case 2: Edge e carries a flow of 1 unit in f_k. Then the direction of e is reversed in $H_k(f_k)$, thus (C^a_p, C^p_a) is in $H_k(f_k)$. Similarly, the direction of the edge to C^p_a from its parent and of the edge from C^a_p to its parent is also reversed. Thus, both C^p_a and C^a_p still have only one incoming and one outgoing edge in $H_k(f_k)$. Now, if C^p_a is in S_k, the only path possible from s to C^p_a has to be through C^a_p and hence C^a_p must be in S_k. Conversely, if C^a_p is in S_k, so is C^p_a since $(C^a_p, C^p_a) \in H_k(f_k)$. An analogous argument holds for containment in T_k, and hence in U_k as well. □

Corollary 1. *For every edge (C^p_a, C^a_p) in H_k that carries unit flow in f_k, either one edge on the path from s to C^p_a in H_k or an edge on the path from C^a_p to t in H_k, but not both, is deleted in the k-th iteration of Step 7 of Algorithm 1.*

Proof. By Lemma 4, each edge (C^p_a, C^a_p) has both its end-point in the same set i.e. S, U, or T. If both the end-points are in S, by analogue of Lemma 3, an edge on the path from C^a_p to t is deleted in Step 7 of the algorithm. We argue that no edge on the path from s to C^p_a gets deleted. Let ρ_A be the path from s to C^p_a that carried flow in H_k. Then every edge on the path ρ_A is reversed in $H_k(f_k)$ and because $C^p_a \in S$, every vertex on ρ_A also belongs to S. This implies that no edge on the path ρ_A gets deleted.

If both the end-points are in U or T, by Lemma 3, an edge on the path from s to C^p_a is saturated and hence deleted in Step 7 of the algorithm. An argument similar to above shows that no edge on the path from C_a^p to t gets deleted in this case. □

3.1 Rank-Maximality of the Output

To prove correctness, we consider flow networks $X_i = (V, F_0 \cup \{(C_a^p, C_p^a) \mid (a,p) \in \bigcup_{j \leq i} E_j\})$ and first establish a one-to-one correspondence between matchings in G_i and flows in X_i. With an abuse of notation, we call an edge (C_a^p, C_p^a) in any flow network H a rank k edge if the corresponding edge (a,p) in G has rank k. Also, we refer to directed edges from leaves in the applicant-tree to leaves in the post-tree as *L-R edges* and directed edges from leaves in the post-tree to leaves in the applicant-tree as *R-L edges*. In the following lemma, we establish a correspondence between matchings in G_i and flows in X_i.

Lemma 5. *For every feasible matching M_i in G_i, there is a corresponding feasible flow g_i in X_i and vice versa. Moreover, the edges present in M_i are precisely the L-R edges in X_i that carry one unit flow in g_i and hence appear as R-L edges in the residual network $X_i(g_i)$.*

Define signature of a flow to be signature of the corresponding matching in G.

Definition 4 (Rank-maximal flow). *We call a flow g_i in a network X_i to be* rank-maximal *if the corresponding matching M_i is rank-maximal in G_i.*

Thus g_i is a rank-maximal flow in X_i if it uses the maximum number of rank 1 edges, subject to that, maximum number of rank 2 edges and so on. By flow-decomposition theorem (see e.g. [2]), a flow g_i in X_i can be decomposed into flow on $s-t$ paths, such that each path uses exactly one L-R edge. Thus, based on the ranks of the L-R edges used, g_i can be decomposed into flows $g_i^1, \ldots, g_i^i$ such that, for each j: $1 \leq j \leq i$, g_i^j uses paths only through L-R edges of rank j. Thus $g_i = g_i^1 + \ldots + g_i^i$. We call g_i^j to be the jth component of g_i.

Lemma 6. *Suppose, for each $j \leq i$, the jth component g_i^j of every rank-maximal flow g_i in X_i is a max-flow in H_j. Then the $(i+1)$st component g_{i+1}^{i+1} of any rank-maximal flow g_{i+1} in X_{i+1} is a max-flow in H_{i+1}.*

Proof. The statement clearly holds for $i = 0$, since H_1 is same as X_1. Now assume the statement for all $j \leq i < r$. We will prove it for $i+1$. Moreover, by the definition of rank-maximal flow, $g_{i+1}^1 + \ldots + g_{i+1}^i$ is a rank-maximal flow in X_i, call it g_i.

Let e be an edge with residual capacity c in X_i when the flow g_i is set up in X_i. We show that e has the same residual capacity in $H_i(g_{i+1}^i)$, and hence in H_{i+1}. This clearly holds in $H_1(g_{i+1}^1)$ since H_1 and X_1 are the same networks. Inductively, each g_{i+1}^j is a flow in H_j for $1 \leq j < i$ and hence the same amount of flow is sent through e in X_j as the total flow sent in $H_1, \ldots, H_j$. Hence the residual capacity of e is the same in $X_i(g_{i+1}^i)$ as in $H_i(g_{i+1}^i)$.

Consider a path ρ in X_{i+1} that carries a flow of one unit from g_{i+1}^{i+1}. Let e_ρ be the rank $i+1$ L-R edge on ρ. Moreover ρ_A and ρ_P be the subpaths of ρ from s to the leaf node in applicant-tree and from the leaf node to t in the post-tree.

Every edge e on ρ must be unsaturated by $g_{i+1}^1 + \ldots + g_{i+1}^i$. If this is not the case, then g_{i+1}^{i+1} can not be routed through e without reducing some flow from $g_{i+1}^1 + \ldots + g_{i+1}^i$ and the resulting flow will not be rank-maximal. Since each g_{i+1}^j for $1 \le j \le i$ is a max-flow in H_j, and all the edges on ρ_A and ρ_P are unsaturated in each of the flows, every node on ρ_A is in S and each node on ρ_P is in T in each of the first i iterations of the algorithm. Thus no edge of ρ_A or ρ_P is deleted from H_j in the jth iteration of the algorithm for any $1 \le j \le i$, and also, e_ρ is not deleted in Step 7 in any iteration.

Thus, in the flow-decomposition of g_{i+1}, every path that carries some flow along a rank $i+1$ edge, is also present in H_{i+1}. Moreover, if c such paths pass through an edge e, then as proved above, e has a capacity c in H_{i+1}. Hence g_{i+1}^{i+1} is a valid flow in H_{i+1}. It has to be a max-flow in H_{i+1}, otherwise g_{i+1} will not be a rank-maximal flow in X_{i+1}. □

Lemma 7. *Define Y_i as the set of R-L edges in H_i'. For every i, j, with $j > i$, the number of edges of rank at most i is the same in Y_i and Y_j.*

Let f_i be a max-flow in H_i and $H_i(f_i)$ denote the corresponding residual network. Let Y denote the set of R-L edges in $H_i(f_i)$. Corresponding to the R-L edges in Y, we can set up a flow g_i which is a feasible flow in X_i. To obtain such a flow, we start with every edge having $g_i(e) = 0$. Repeatedly select an unselected edge e from Y. Let ρ_e denote the unique $s - t$ path in X_i containing e. We increase the flow along every edge in ρ_e by one unit. Using arguments similar to Lemma 5 we conclude that g_i is a feasible flow in X_i.

Lemma 8. *For every $1 \le k \le r$, the following hold:*

1. *For every rank-maximal flow $g_k = g_k^1 + \ldots + g_k^k$ of X_k, g_i is a max-flow in H_i for $1 \le i \le k$.*
2. *Conversely, the flow g_k (constructed as above) corresponding to the R-L edges of $H_k(f_k)$ is a rank-maximal flow in X_k.*

Proof. We prove this by induction on k. When $k = 1$, X_1 and H_1 are the same networks. A rank-maximal flow g_1 in X_1 is just a max-flow in X_1 and hence in H_1. Algorithm 1 also computes a max-flow in H_1. Hence both the statements hold for $k = 1$. Assume the statements to be true for each $j \le i$. We prove them for $i+1$. The first statement follows from Lemma 6. We prove the second statement. By induction hypothesis, g_i corresponding to f_i is a rank-maximal flow in X_i, let its signature be $(\sigma_1, \ldots, \sigma_i)$. Let the signature of a rank-maximal flow in X_{i+1} be $(\sigma_1, \ldots, \sigma_{i+1})$. By Lemma 7, the number of R-L edges of rank j in H_{i+1}' and hence in $H_{i+1}(f_{i+1})$ is the same as in H_i', for each $j \le i$. Thus the signature of g_{i+1} in X_{i+1} corresponding to f_{i+1} is $(\sigma_1, \ldots, \sigma_i, \sigma_{i+1}')$ where $\sigma_{i+1}' \le \sigma_{i+1}$. However, by Lemma 6, the $(i+1)$st component of a rank-maximal flow in X_{i+1} is a max-flow in H_{i+1}. Since f_{i+1} is also a max-flow in H_{i+1} it must be of the same value and hence the corresponding flow g_{i+1} of f_{i+1} must have signature $(\sigma_1, \ldots, \sigma_{i+1})$. □

Running Time: The size of our flow network is determined by the total number of classes. Due to the tree structure of T_u, the size of the flow network is equal to the total size of all preference lists which is $O(|E|)$. The maximum matching size in our instance is upper bounded by $|E|$ and the max-flow in our network is also at most $O(|E|)$. By Ford-Fulkerson algorithm, each augmentation takes $O(|E|)$ time and $O(|E|)$ augmentations are sufficient. This gives an upper bound of $O(|E|^2)$ on the running time. Thus we establish Theorem 1.

4 Hardness for Non-laminar Classifications

Here, we consider the CRMM and CPM problems where the classifications are not necessarily laminar. We show that the following decision version of the CRMM problem is NP-hard: Given an instance $G = (A \cup P, E)$ of the CRMM problem and a signature vector $\sigma = (\sigma_1, \ldots, \sigma_r)$, does there exist a feasible matching M in G such that M has a signature ρ such that $\rho \succeq \sigma$? We give a reduction from the monotone 1-in-3 SAT problem to the above decision version of CRMM. Throughout this section, we refer to this decision version as the CRMM problem.

The monotone 1-in-3 SAT problem is a variant of the boolean satisfiability problem where the input is a conjunction of m clauses. Each clause is a disjunction of exactly three variables and no variable appears in negated form. The goal is to decide whether there exists a truth assignment to the variables such that every clause has exactly one true variable and hence two false variables. This problem is known to be NP-hard [17]. Let ϕ be the given instance of the monotone 1-in-3 SAT problem, with n variables $x_1, \ldots, x_n$ and m clauses $C_1, C_2, \ldots, C_m$. We construct an instance $G = (A \cup P, E)$ of the CRMM problem as follows:

Applicants: For each variable x_i in ϕ, there are two applicants a_i, b_i in A. For each occurrence of x_i in clause C_j, there are two applicants a_{ij}, b_{ij}. Thus $A = \{a_i, b_i, a_{ij}, b_{ij} \mid x_i \in \phi, x_i \in C_j\}$ and $|A| = 2n + 6m$.

Posts: For each variable x_i, there are three posts p_i, p_i^t and p_i^f. For each clause C_j, there is a post $\hat{p}_j$. Thus $P = \{p_i, p_i^t, p_i^f \mid x_i \in \phi\} \cup \{\hat{p}_j \mid C_j \in \phi\}$ and $|P| = 3n + m$.

Preferences of Applicants: The applicants have following preferences:

$$a_i : \; p_i, \; p_i^t \qquad\qquad a_{ij} : \; \hat{p}_j, \; p_i^t$$
$$b_i : \; p_i, \; p_i^f \qquad\qquad b_{ij} : \; \hat{p}_j, \; p_i^f$$

Quotas and Classifications of Posts:

1. Let $C_j = x_i \vee x_{i'} \vee x_{i''}$; the corresponding post $\hat{p}_j$, $q(\hat{p}_j) = 3$ and has classes:
   (a) $S_{ij} = \{a_{ij}, b_{ij}\}$ with quota 1 for each $x_i \in C_j$.

   (b) $S_{1j} = \{a_{ij}, a_{i'j}, a_{i''j}\}$ with quota 1.
   (c) $S_{2j} = \{b_{ij}, b_{i'j}, b_{i''j}\}$ with quota 2.
2. Each post p_i^t has quota $k_i =$ the number of occurrences of x_i in ϕ and classes: $S_j^t = \{a_{ij}, a_i\}$ with quota 1, for each j such that $x_i \in C_j$.
3. Each post p_i^f has quota $k_i =$ the number of occurrences of x_i in ϕ and classes: $S_j^f = \{b_{ij}, b_i\}$ with quota 1, for each j such that $x_i \in C_j$.
4. Each post p_i has quota 1 and no classes.

This completes the description of our CRMM instance. Our reduction also works for showing the hardness for CPM problem, since only posts have classifications, and each applicant can be matched to at most one post. We note that the preferences are strict and are of length two; thus the hardness for both the problems applies even under these restrictions. This establishes Theorems 3 and 4.

References

1. Abraham, D.J., Irving, R.W., Kavitha, T., Mehlhorn, K.: Popular matchings. SIAM J. Comput. **37**(4), 1030–1045 (2007)
2. Ahuja, R.K., Magnanti, T.L., Orlin, J.B.: Network Flows: Theory, Algorithms, and Applications. Prentice-Hall Inc., Upper Saddle River (1993)
3. Dulmage, A.L., Mendelsohn, N.S.: Coverings of bipartite graphs. Can. J. Math. **10**, 517–534 (1958)
4. Fleiner, T., Kamiyama, N.: A matroid approach to stable matchings with lower quotas. In: Proceedings of the Twenty-Third Annual ACM-SIAM Symposium on Discrete Algorithms, SODA, pp. 135–142 (2012)
5. Ford, D.R., Fulkerson, D.R.: Flows in Networks. Princeton University Press, Princeton (1962)
6. Gärdenfors, P.: Match making: assignments based on bilateral preferences. Behav. Sci. **20**, 166–173 (1975)
7. Huang, C-C.: Classified stable matching. In: Proceedings of the Twenty-First Annual ACM-SIAM Symposium on Discrete Algorithms, SODA, pp. 1235–1253 (2010)
8. Huang, C.-C., Kavitha, T., Michail, D., Nasre, M.: Bounded unpopularity matchings. Algorithmica **61**(3), 738–757 (2011)
9. Irving, R.W.: Greedy Matchings. Technical report TR-2003-136, University of Glasgow, April 2003
10. Irving, R.W., Kavitha, T., Mehlhorn, K., Michail, D., Paluch, K.: Rank-maximal matchings. ACM Trans. Algorithms **2**(4), 602–610 (2006)
11. Kamiyama, N.: Popular matchings with ties and matroid constraints. SIAM J. Discrete Math. **31**(3), 1801–1819 (2017)
12. Manlove, D.F., Sng, C.T.S.: Popular matchings in the capacitated house allocation problem. In: Azar, Y., Erlebach, T. (eds.) ESA 2006. LNCS, vol. 4168, pp. 492–503. Springer, Heidelberg (2006). https://doi.org/10.1007/11841036_45
13. Mestre, J.: Weighted popular matchings. In: Bugliesi, M., Preneel, B., Sassone, V., Wegener, I. (eds.) ICALP 2006. LNCS, vol. 4051, pp. 715–726. Springer, Heidelberg (2006). https://doi.org/10.1007/11786986_62
14. Nasre, M., Nimbhorkar, P., Pulath, N.: Dichotomy results for classified rank-maximal matchings and popular matchings. CoRR, abs/1805.02851 (2018)

15. Paluch, K.: Capacitated rank-maximal matchings. In: Spirakis, P.G., Serna, M. (eds.) CIAC 2013. LNCS, vol. 7878, pp. 324–335. Springer, Heidelberg (2013). https://doi.org/10.1007/978-3-642-38233-8_27
16. Picard, J.-C., Queyranne, M.: On the structure of all minimum cuts in a network and applications. Math. Program. Study **13**, 8–16 (1980)
17. Schaefer, T.J.: The complexity of satisfiability problems. In: Proceedings of the Tenth Annual ACM Symposium on Theory of Computing, pp. 216–226 (1978)

Maximum Matchings and Minimum Blocking Sets in Θ_6-Graphs

Therese Biedl[1], Ahmad Biniaz[1], Veronika Irvine[1], Kshitij Jain[2], Philipp Kindermann[3(✉)], and Anna Lubiw[1]

[1] David R. Cheriton School of Computer Science, University of Waterloo, Waterloo, Canada
{biedl,virvine,alubiw}@uwaterloo.ca, ahmad.biniaz@gmail.com
[2] Borealis AI, Waterloo, Canada
kshitij.jain.1@uwaterloo.ca
[3] Lehrstuhl für Informatik I, Universität Würzburg, Würzburg, Germany
philipp.kindermann@uni-wuerzburg.de

Abstract. Θ_6-graphs are important geometric graphs that have many applications especially in wireless sensor networks. They are equivalent to Delaunay graphs where empty equilateral triangles take the place of empty circles. We investigate lower bounds on the size of maximum matchings in these graphs. The best known lower bound is $n/3$, where n is the number of vertices of the graph, which comes from half-Θ_6-graphs that are subgraphs of Θ_6-graphs. Babu et al. (2014) conjectured that any Θ_6-graph has a (near-)perfect matching (as is true for standard Delaunay graphs). Although this conjecture remains open, we improve the lower bound to $(3n-8)/7$.

We also relate the size of maximum matchings in Θ_6-graphs to the minimum size of a *blocking set*. Every edge of a Θ_6-graph on point set P corresponds to an empty triangle that contains the endpoints of the edge but no other point of P. A *blocking set* has at least one point in each such triangle. We prove that the size of a maximum matching is at least $\beta(n)/2$ where $\beta(n)$ is the minimum, over all Θ_6-graphs with n vertices, of the minimum size of a blocking set. In the other direction, lower bounds on matchings can be used to prove bounds on β, allowing us to show that $\beta(n) \geq 3n/4 - 2$.

Keywords: Theta-six graphs · Proximity graphs · Maximum matching · Minimum blocking set · Triangular-distance Delaunay graph

1 Introduction

One of the many beautiful properties of Delaunay triangulations is that they always contain a (near-)perfect matching, that is, at most one vertex is unmatched, as proved by Dillencourt [21]. This is one example of a structural property of a so-called *proximity graph*. A proximity graph is determined by a set

I. Sau and D. M. Thilikos (Eds.): WG 2019, LNCS 11789, pp. 258–270, 2019.
https://doi.org/10.1007/978-3-030-30786-8_20

$\mathcal{S}$ of geometric objects in the plane, such as all disks, or all axis-aligned squares. Given such a set $\mathcal{S}$ and a finite point set P, we construct a proximity graph with vertex set P and with an edge (p, q) if there is an object from $\mathcal{S}$ that contains p and q and no other point of P. When $\mathcal{S}$ consists of all disks, then we get the Delaunay triangulation. Proximity graphs are often defined in a more general way, with constraints on how the objects may touch points p and q, but this narrow definition suffices for our purposes.

Various structural properties have been proved for different classes of proximity graphs. Another example is that the L_∞-Delaunay graph, which is a proximity graph defined in terms of the set $\mathcal{S}$ of all axis-aligned squares, has the even stronger property of always having a Hamiltonian path [2].

Our paper is about structural properties of Θ_6-graphs, which are the proximity graphs determined by equilateral triangles with a horizontal edge. More precisely, for any finite point set P, define $G^{\triangle}(P)$ to be the proximity graph of P with respect to *upward* equilateral triangles $\triangle$, define $G^{\triangledown}(P)$ to be the proximity graph of P with respect to *downward* equilateral triangles $\triangledown$, and define $G^{\davidsstar}(P)$, the Θ_6-graph of P, to be their union. In particular, $G^{\davidsstar}(P)$ has an edge between points p and q if and only if there is an equilateral triangle with a horizontal side that contains p and q and no other point of P. Such a triangle can be shrunk to an *empty triangle* that has one of p or q at a corner, the other point on its boundary, and no points of P in its interior.

The graphs $G^{\triangle}(P)$ and $G^{\triangledown}(P)$ are *triangular-distance* (or "TD") Delaunay graphs, first introduced by Chew [18]. Clarkson [19] and Keil [24] first introduced Θ_6-graphs(via a different definition), and the equivalence with the above definition was proved by Bonichon et al. [14]. See Sect. 1.1 for more information.

We explore two conjectures about Θ_6-graphs.

Conjecture 1 (Babu et al. [8]*).* Every Θ_6-graph has a (near-)perfect matching.

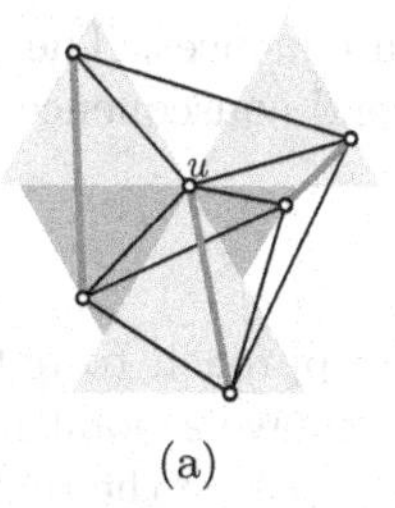

(a)

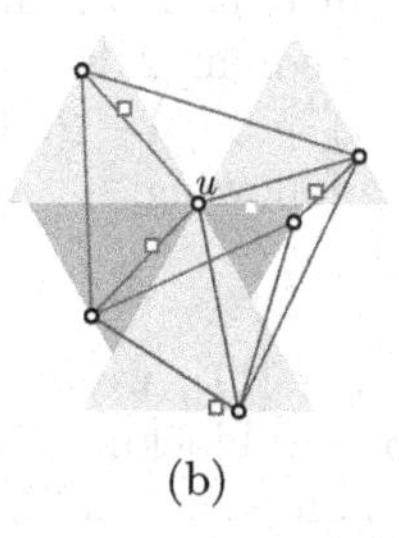

(b)

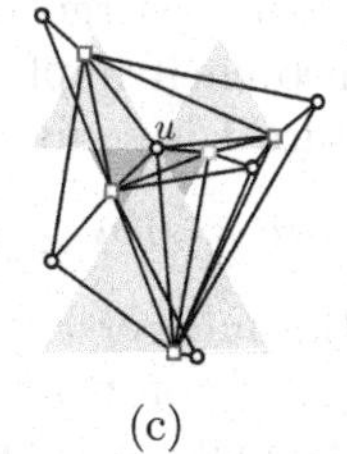

(c)

Fig. 1. A Θ_6-graph on $n = 6$ points with a perfect matching and a blocking set of size 5. (a) A perfect matching. Empty triangles corresponding to edges of u are highlighted. (b) A blocking set B of size $n - 1$. Edges have the same color as their blocking point. (c) $G^{\davidsstar}(P \cup B)$. For every edge, one endpoint is in B. (Color figure online)

See Fig. 1a for an example. The best known bound is that every Θ_6-graph on n points has a matching of size at least $n/3$ minus a small constant—in fact, this bound holds for any planar graph with minimum degree 3 [27], hence for any triangulation and in particular for each of $G^{\triangle}$ and $G^{\triangledown}$ (modulo the small additive constant)—see Babu et al. [8] for the exact bound of $\lceil (n-1)/3 \rceil$. Our main result is an improvement of this lower bound:

Theorem 1. *Every Θ_6-graph on n points has a matching of size* $(3n-8)/7$.

We prove Theorem 1 in Sect. 2 using the same technique that has been used for matchings in planar proximity graphs, namely the Tutte-Berge theorem, which relates the size of a maximum matching in a graph to the number of components of odd cardinality after removing some vertices. In our case, this approach is more complicated because Θ_6-graphs are not planar.

Our second main result relates the size of matchings to the size of *blocking* or *stabbing* sets of proximity graphs, which were introduced by Aronov et al. [5] for purposes unrelated to matchings. For a proximity graph $G(P)$ defined in terms of a set of objects $\mathcal{S}$, we say that a set B of points *blocks* $G(P)$ if B has a point in the interior of any object from $\mathcal{S}$ that contains exactly two points of P, i.e., the set B destroys all the edges of $G(P)$, or equivalently, $G(P \cup B)$ has no edges between vertices in P; see Fig. 1b–c.

For a set of points P, let $\beta(P)$ be the minimum size of a blocking set of $G^{\varhexstar}(P)$. Let $\beta(n)$ be the minimum, over all point sets P of size n, of $\beta(P)$. It is known that $\beta(n) \geq \lceil (n-1)/2 \rceil$ since that is a lower bound for blocking all $G^{\triangle}$-graphs of n points [12]. Let $\mu(n)$ be the minimum, over all point sets P of size n, of the size of a maximum matching in $G^{\varhexstar}(P)$. Conjecture 1 can hence be restated as $\mu(n) \geq \lceil (n-1)/2 \rceil$. We relate the parameters μ and β as follows.

Theorem 2. *(a) For any point set P of n points in the plane, $G^{\varhexstar}(P)$ has a matching of size $\beta(n)/2$, i.e., $\mu(n) \geq \beta(n)/2$. (b) On the other hand, if $\mu(n) \geq cn + d$ for some constants c, d, then $\beta(n) \geq (cn+d)/(1-c)$.*

The two statements in the theorem are proved in Sect. 3. The idea of using bounds on blocking sets to obtain bounds on matchings is new, and is proved via the Tutte-Berge theorem. Theorem 2 has two consequences. The first is that Theorem 1 implies that $\beta(n) \geq 3n/4 - 2$. The second consequence is that Conjecture 1 is equivalent to the following:

Conjecture 2. $\beta(n) \geq n - 1$.

In the full version of the paper [11], we explore an approach to obtaining lower bounds on $\beta(n)$. For B to be a blocking set, it must have a point in every empty triangle of P that defines an edge in $G^{\varhexstar}(P)$. Let $\alpha(n)$ be the maximum number of pairwise internally-disjoint empty triangles of any point set of size n. Clearly, $\beta(P) \geq \alpha(P)$ and $\beta(n) \geq \alpha(n)$. Conjecture 1 would be proved if we could show that $\alpha(n) \geq n - 1$. However, we give an example of a point set P of size n with $\alpha(P) \leq 3n/4$, which shows that $\alpha(n) \leq 3n/4$.

We also explore a previously-studied variant where the empty triangles must be completely disjoint. If D is such a set, then every empty triangle in D corresponds to an edge in $G^{\varhexstar}(P)$, and these edges share no endpoint because the triangles are disjoint. Then D corresponds to a *strong matching* in $G^{\varhexstar}(P)$. Strong matchings were introduced by Ábrego et al. [1,2] for the case where the empty objects are line segments, rectangles, disks, or squares. They who showed that Delaunay and L_∞-Delaunay need not have strong (near-)perfect matchings (for disks and squares, respectively). See the following subsection for further background. Biniaz et al. [12] proved that for any point set of size n, $G^{\triangle}(P)$ has a strong matching of at least $\lceil (n-1)/9 \rceil$ edges and $G^{✡}(P)$ has a strong matching of at least $\lceil (n-1)/4 \rceil$ edges. We prove an upper bound by giving an example where the maximum strong matching in $G^{✡}(P)$ has $2n/5$ edges.

In the full version of the paper [11], we prove some additional bounds on the number of edges, maximum vertex degree, and maximum independent set of Θ_6-graphs.

1.1 Background

Θ_6-graphs and TD-Delaunay Graphs. The Θ_6-graph on a set P of points in the plane, as originally defined by Clarkson [19] and Keil [24], is a geometric graph with vertex set P and edges constructed as follows. For every point $p \in P$, place 6 rays emanating from p at angles that are multiples of $\pi/3$ radians from the positive x-axis. These rays partition the plane into 6 cones with apex p, which we label $C_1, \dots, C_6$ in counterclockwise order starting from the positive x-axis; see Fig. 2a. Add an edge from p to the *closest* point in each cone C_i, where the distance between the apex p and a point q in C_i is measured by the Euclidean distance from p to the projection of q on the bisector of C_i as depicted in Fig. 2a. It is straight-forward to show that this definition of Θ_6-graphs is equivalent to the definition of $G^{✡}(P)$. The edges of $G^{\triangle}(P)$ come from the odd cones, and the edges of $G^{\triangledown}(P)$ come from the even cones, so the TD-Delaunay graphs $G^{\triangle}(P)$ and $G^{\triangledown}$ are known as "half-Θ_6" graphs.

TD-Delaunay graphs are called TD-Delaunay "triangulations". In fact, they might fall short of being triangulations. As discussed by Drysdale [22] and Chew [18] (see also [7]), they are plane graphs that consist of a "support hull" which need not be convex, and a complete triangulation of the interior (an explicit proof can be found in [8]). This anomaly is often remedied by surrounding the point set with a large bounding triangle. We will use a similar approach.

The more general Θ_k-graphs, which are defined in terms of k cones, have some properties that are relevant in a number of application areas. In particular, they are *sparse*—$\Theta_k(P)$ has at most $k|P|$ edges [26]—and they are *spanners*—the ratio (known as the *spanning ratio*) of the length of the shortest path between any two vertices in Θ_k, $k \geq 4$, to the Euclidean distance between the vertices is at most a constant [15,17,18,24]. Because of these properties, Θ_k-graphs have applications in many areas including wireless networking [4,16], motion planning [19], real-time animation [23], and approximating complete Euclidean graphs [18,25].

Among Θ_k-graphs, Θ_6 has some nice properties that make it suitable for communications in wireless sensor networks. In particular, $k = 6$ is the smallest integer for which: (i) Θ_k has spanning ratio 2 [14,15,17]; (ii) the so-called $\Theta\Theta_k$-graph, which is a subgraph of Θ_k where each vertex has only one incoming edge per cone, is a spanner [20]; and (iii) half-Θ_k-graphs admit a deterministic local competitive routing strategy [16].

Convex Distance Delaunay Graphs. For a set $\mathcal{S}$ of homothets of a convex polygon, the corresponding proximity graphs are the *convex distance Delaunay graphs.* This concept has been thoroughly studied, see, e.g., [7,22]. Some of the helper lemmas we need for half-Θ_6-graphs come from more general results that hold for all convex distance Delaunay graphs.

Blocking Sets in Proximity Graphs. Blocking or "stabbing" sets were introduced by Aronov et al. [5] as a more flexible way to represent graphs via proximity. The idea was explored further by Aichholzer et al. [3] who showed that $3n/2$ points are sufficient and at least $n-1$ points are necessary to block any Delaunay triangulation with n vertices. Biniaz et al. [12] showed that at least $\lceil (n-1)/2 \rceil$ points are necessary to block any $G^{\triangle}$-graph with n vertices. This bound is tight for $G^{\triangle}$-graphs and provides a lower bound on $\beta(n)$. For results on blocking [higher order] Gabriel graphs, see [6,13].

Strong Matchings in Proximity Graphs. The idea of *strong matchings* in proximity graphs—i.e., pairwise disjoint objects from $\mathcal{S}$ each with two points of P on the boundary and no points in the interior—was introduced by Ábrego et al. [1,2] for line segments, rectangles, disks, and squares. They show that strong (near-)perfect matchings always exist in the first two cases, but not always for disks (Delaunay graphs) and squares (L_∞-Delaunay graphs). In fact, they prove upper bounds of $36n/73$ and $5n/11$, respectively, on the size of a strong matching. They also give lower bounds of $\lceil (n-1)/8 \rceil$ and $\lceil n/5 \rceil$, respectively. The lower bound for squares was improved to $\lceil (n-1)/4 \rceil$ by Biniaz et al. [12] who also proved lower bounds of $\lceil (n-1)/9 \rceil$ for $G^{\triangle}$ and $\lceil (n-1)/4 \rceil$ for $G^{\text{✡}}$.

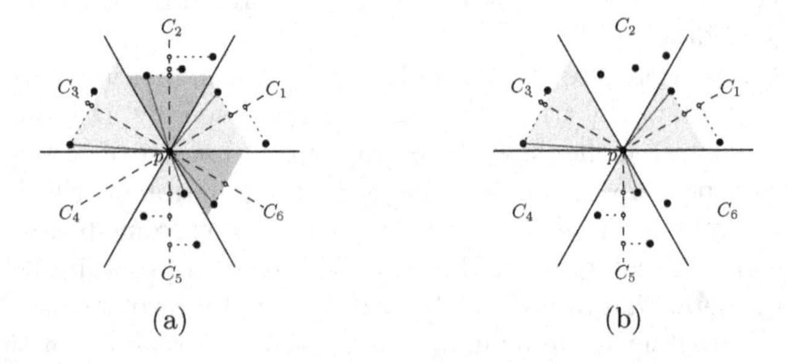

Fig. 2. The construction of (a) the Θ_6-graph, and (b) the odd half-Θ_6-graph.

1.2 Preliminaries

We assume that points are in general position and that no line passing through two points of P makes an angle of 0°, 60° or 120° with the horizontal.

Notation. For two points p and q in the plane, we denote by $\triangle(p,q)$ (resp., by $\triangledown(p,q)$) the smallest upward (resp., downward) equilateral triangle that has p and q on its boundary. We say that a triangle is *empty* if it has no points of P

in its interior. With these definitions, the Θ_6-graph has an edge between p and q if and only if $\triangle(p,q)$ is empty or $\bigtriangledown(p,q)$ is empty, in which case we say that the edge (p,q) is *introduced* by $\triangle(p,q)$ or by $\bigtriangledown(p,q)$. Let P be a set of points. We use the following notation:

$$\mu(P) = \text{maximum number of edges in a matching of } G^{✡}(P)$$
$$\beta(P) = \text{minimum size of a set of points that block all empty triangles of } P$$
$$\alpha(P) = \text{maximum number of pairwise internally-disjoint empty triangles}$$
$$\mu^*(P) = \text{maximum number of edges in a strong matching of } G^{✡}(P)$$

Furthermore, we define $\mu(n), \beta(n), \alpha(n), \mu^*(n)$ to be the minimum of the corresponding parameter over all sets of n points.

Properties of Θ_6-Graphs. We need the following two properties of Θ_6-graphs:

Lemma 1 (Babu et al. [8]). *Let P be a set of points in the plane, and let p and q be any two points in P. There is a path between p and q in $G^{\triangle}(P)$ that lies entirely in $\triangle(p,q)$. Moreover, the triangles that introduce the edges of this path also lie entirely in $\triangle(p,q)$. Analogous statements hold for $G^{\bigtriangledown}(P)$ and $\bigtriangledown(p,q)$.*

We remark that this lemma holds more generally for any convex-distance Delaunay graph, as does the second property we need. It generalizes the fact that the (standard) Delaunay triangulation contains the minimum spanning tree with respect to Euclidean distances. We state the result for the special case of equilateral triangles. For any two points p and q in the plane, define the weight function $w^{\triangle}(p,q)$ to be the area of the smallest $\triangle$-triangle containing p and q.

Lemma 2 (Aurenhammer and Paulini [7]). *The minimum spanning tree of points P with respect to the weight function $w^{\triangle}(p,q)$ is contained in $G^{\triangle}(P)$.*

A consequence of Lemma 2 (as noted by Aurenhammer and Paulini in their more general setting) is that the minimum spanning tree of points P with respect to the weight function $w^{\triangle}(p,q)$ is contained in both $G^{\triangle}(P)$ and $G^{\bigtriangledown}(P)$, because $w^{\triangle}(p,q) = w^{\bigtriangledown}(p,q)$. In particular, this means that the intersection of $G^{\triangle}(P)$ and $G^{\bigtriangledown}(P)$ is connected, as was proved with a different method by Babu et al. [8].

The Tutte-Berge Matching Theorem. Let G be a graph and let S be an arbitrary subset of vertices of G. Removing S splits G into a number, $\text{comp}(G \setminus S)$, of connected components. Let $\text{odd}(G \setminus S)$ denote the number of odd components of $G \setminus S$, i.e., components with an odd number of vertices. In 1947, Tutte [28] characterized graphs that have a perfect matching as exactly those graphs that have at most $|S|$ odd components for any subset S. In 1957, Berge [10] extended this result to a formula (today known as the Tutte-Berge formula) for the size of maximum matchings in graphs. The following is an alternate way of stating this formula in terms of the number of unmatched vertices.

Theorem 3 (Tutte-Berge Formula; Berge [10]). *The number of unmatched vertices of a maximum matching in G is equal to the maximum over subsets $S \subseteq V$ of $\text{odd}(G \setminus S) - |S|$.*

To obtain a lower bound on the size of a maximum matching it suffices, by Theorem 3, to find an upper bound on $\mathrm{odd}(G \setminus S) - |S|$ that holds for any S. We will use this approach in our proofs of Theorems 1 and 2. In fact, as in Dillencourt's proof [21] that Delaunay graphs have perfect matchings we will find an upper bound on $\mathrm{comp}(G \setminus S) - |S|$ that holds for any S, i.e., we establish a bound on the *toughness* of the graph [9].

2 Bounding the Size of a Matching

In this section, we prove Theorem 1. Let P be a set of n points in the plane and let $G^{\davidsstar}(P)$ be the Θ_6-graph on P. We will prove that $G^{\davidsstar}(P)$ contains a matching of size at least $(3n-8)/7$. As implied by Theorem 3, in order to prove a lower bound on the size of maximum matching in $G^{\davidsstar}(P)$, it suffices to prove an upper bound on $\mathrm{odd}(G^{\davidsstar}(P) \setminus S) - |S|$ that holds for any subset S of P. Since it is hard to argue about odd components, we will in fact prove an upper bound on $\mathrm{comp}(G^{\davidsstar}(P) \setminus S) - |S|$. Such a bound applies to $\mathrm{odd}(G^{\davidsstar}(P) \setminus S) - |S|$ because $\mathrm{odd}(G^{\davidsstar}(P) \setminus S) \leq \mathrm{comp}(G^{\davidsstar}(P) \setminus S)$.

Our proof will depend on an analysis of the faces of $G^{\triangle}(P) \setminus S$ and $G^{\nabla}(P) \setminus S$ for which we need some preliminary results. Consider a planar graph G with a fixed planar embedding. Such an embedding divides the plane into connected regions, called faces. For every face f of G, we define its *degree* as the number of triangles in a triangulation of f plus 2; see Fig. 3 for some examples. Let $\mathcal{F}_d(G)$ denote the set of faces of G with degree d. An easy counting argument shows that if $|V| \geq 3$, then $\sum_{d \geq 3}(d-2)|\mathcal{F}_d(G)| = 2|V| - 4$, since a face of degree d gives rise to $d-2$ faces in a triangulation of G, which has $2|V|-4$ faces by Euler's polyhedra formula.

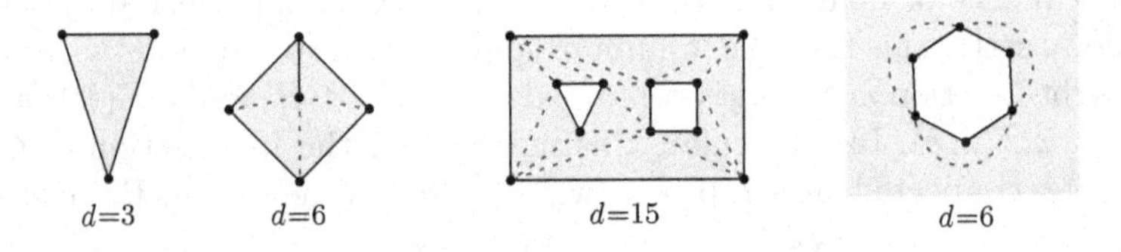

Fig. 3. The notion of degree of a face.

We will utilize the following lemma that Dillencourt used in his proof that every Delaunay triangulation contains a (near-)perfect matching. Let $G[S]$ denote the subgraph of G that is induced by a subset S of its vertices.

Lemma 3 (Dillencourt [21], Lemma 3.4). *Let G be a triangulated planar graph and let S be a subset of vertices of G. Then every face of $G[S]$ contains at most one component of $G \setminus S$.*

We aim to apply this result to $G^{\triangle}(P)$ and $G^{\triangledown}(P)$. As noted in Sect. 1.1, the interior faces of $G^{\triangle}(P)$ and $G^{\triangledown}(P)$ are triangles, but their outer faces need not be the convex hull of P. For this reason, we add a set $A = \{a_1, \ldots, a_6\}$ of *surrounding points* as follows. Find the smallest $\triangle$-triangle $T^{\triangle}$ and $\triangledown$-triangle $T^{\triangledown}$ containing all points of P. Let $\mathcal{R}(P)$ be the region $T^{\triangle} \cup T^{\triangledown}$. Observe that all of the empty triangles that introduce edges of $G^{\text{✡}}(P)$ lie in $\mathcal{R}(P)$, so adding points outside $\mathcal{R}(P)$ does not remove any edge from the graph. We now place points $a_1, \ldots, a_6$ near the corners of $T^{\triangle}$ and $T^{\triangledown}$ while maintaining the general position of the point set (see Fig. 4a): at each corner, place a point in the cone opposite to the cone that contains the triangle, and name the points in such a way that every point of P has a_i in cone C_i.

Now fix a set S for which we want to bound $\text{comp}(G^{\text{✡}}(P) \setminus S) - |S|$, and define $S_A = S \cup A$. Pick an arbitrary representative point from every connected component of $G^{\text{✡}}(P) \setminus S$, and let Q be the set of these points, so $|Q| = \text{comp}(G^{\text{✡}}(P) \setminus S)$.

Define $G_A^{\triangle} = G^{\triangle}(P \cup A)$ and consider its subgraph $G_A^{\triangle}[S_A]$ induced by S_A. By construction, the outer face of both $G_A^{\triangle}$ and $G_A^{\triangle}[S_A]$ is the hexagon formed by A; we add three graph edges (not segments) to triangulate the outer face, so that $G_A^{\triangle}$ is triangulated. Note that none of the points of P (and in particular therefore no points of Q) are inside the four newly introduced triangular faces.

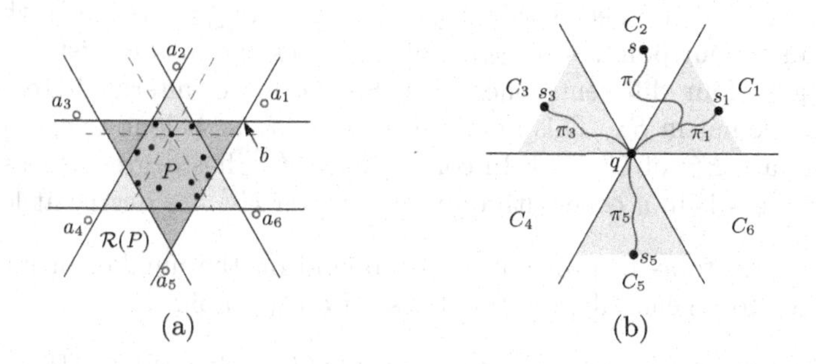

Fig. 4. (a) Augmentation of P: the shaded region is $\mathcal{R}(P)$, and $A = \{a_1, \ldots, a_6\}$. (b) Illustration for the proof of Lemma 4.

Let $f_d^{\triangle}$ be the number of faces of degree d in $G_A^{\triangle}[S_A]$ that contain some point of Q. We define $f_{4+}^{\triangle} = \sum_{d \geq 4} f_d^{\triangle}$. Since all faces of $G_A^{\triangle}$ are now triangles, Lemma 3 applies and every face of $G_A^{\triangle}[S_A]$ contains at most one component, hence at most one point of Q. Therefor,

$$|Q| = f_3^{\triangle} + f_{4+}^{\triangle} \qquad \text{and similarly} \qquad |Q| = f_3^{\triangledown} + f_{4+}^{\triangledown}, \tag{1}$$

where $f_d^{\triangledown}$ is defined in a symmetric manner on graph $G_A^{\triangledown}[S_A]$.

Let $\mathcal{F}_d$ be the set of faces of degree d in $G_A^{\triangle}[S_A]$ and observe that, since no point of Q appears in the four triangles outside the hexagon of A, we have $f_3^{\triangle} \leq |\mathcal{F}_3| - 4$. As a consequence,

$$f_3^{\triangle} + 2f_{4+}^{\triangle} \leq \sum_{d\geq 3}(d-2)f_d^{\triangle} \leq \sum_{d\geq 3}(d-2)|\mathcal{F}_d| - 4$$
$$\leq 2|V(G_A^{\triangle}[S_A])| - 4 - 4 = 2|S| + 2|A| - 8 = 2|S| + 4 \qquad (2)$$

and similarly $f_3^{\triangledown} + 2f_{4+}^{\triangledown} \leq 2|S| + 4$.

The crucial insight for getting an improved matching bound is that no component can reside inside a face of degree 3 in both $G^{\triangle}$ and $G^{\triangledown}$.

Lemma 4. *We have* $f_3^{\triangle} \leq f_{4+}^{\triangledown}$ *and* $f_3^{\triangledown} \leq f_{4+}^{\triangle}$.

Proof. Consider any point $q \in Q$, hence $q \notin S_A$. Let $F^{\triangle}$ and $F^{\triangledown}$ be the faces of $G_A^{\triangle}[S_A]$ and $G_A^{\triangledown}[S_A]$ that contain q, respectively. It suffices to show that one of $F^{\triangle}$ and $F^{\triangledown}$ has degree at least 4.

By Lemma 2, the minimum-weight spanning tree T of $P \cup A$ belongs to both $G_A^{\triangle}$ and $G_A^{\triangledown}$. Find a path π in T that connects q to some point $s \in S_A$ such that no vertex of π except s belongs to S_A.

Assume first that s is in a cone with even index. Let s_1, s_3, s_5 be the points of S_A that are closest to q in cones C_1, C_3, C_5, respectively; since $A \subseteq S_A$, such points s_i exist. Refer to Fig. 4b. By Lemma 1, for every $i \in \{1, 3, 5\}$, there exists a path π_i between q and s_i in $G^{\triangle}$ that lies fully in $\triangle(q, s_i)$. By our choices of s_i, no vertex of π_i except s_i is in S_A.

So we have four (not necessarily disjoint) paths π, π_1, π_3, π_5 in $G^{\triangle}$ that begin at q and end at four points s, s_1, s_3, s_5 of S_A. These points are distinct because they belong to four different cones of q. Furthermore, intermediate points of these paths are not in S_A. This implies that s, s_1, s_3, s_5 belong to the boundary of the same face $F^{\triangle}$ of $G^{\triangle}[S_A]$. In consequence, $F^{\triangle}$ has degree at least 4.

Similarly, if s is in a cone with odd index, then $F^{\triangledown}$ has degree at least 4. □

Now we have tools to prove an upper bound on the number of unmatched vertices and, more generally, the toughness of a Θ_6-graph.

Lemma 5. *For any* $S \subseteq P$, *we have* $\text{comp}(G^{\text{✡}}(P) \setminus S) - |S| \leq (|P| + 16)/7$.

Proof. Recall that we fixed a set Q of points in $P \backslash S$ with $|Q| = \text{comp}(G^{\text{✡}}(P) \backslash S)$. So $n = |P| \geq |S| + |Q|$, or equivalently $n - |Q| - |S| \geq 0$. Combining this with the above inequalities, we get

$$\begin{aligned}
7\left(\text{comp}(G^{\text{✡}}(P) \setminus S) - |S|\right) &\leq 7|Q| - 7|S| + (n - |Q| - |S|) \\
&= n + 3|Q| + 3|Q| - 8|S| \\
\text{(by (1))} \qquad &= n + 3\left(f_3^{\triangle} + f_{4+}^{\triangle}\right) + 3\left(f_3^{\triangledown} + f_{4+}^{\triangledown}\right) - 8|S| \\
\text{(by Lemma 4)} \qquad &\leq n + 2f_3^{\triangle} + 4f_{4+}^{\triangle} + 2f_3^{\triangledown} + 4f_{4+}^{\triangledown} - 8|S| \\
\text{(by (2))} \qquad &\leq n + (4|S| + 8) + (4|S| + 8) - 8|S| \\
&= n + 16.
\end{aligned}$$

□

Therefore, $\text{odd}(G^{\text{✡}}(P) \setminus S) - |S| \leq \text{comp}(G^{\text{✡}}(P) \setminus S) - |S| \leq (n+16)/7$. In consequence of the Tutte-Berge formula, therefore any maximum matching M of $G^{\text{✡}}(P)$ has at most $(n+16)/7$ unmatched vertices, hence at least $(6n-16)/7$ matched vertices and $|M| \geq (3n-8)/7$. This completes the proof of Theorem 1.

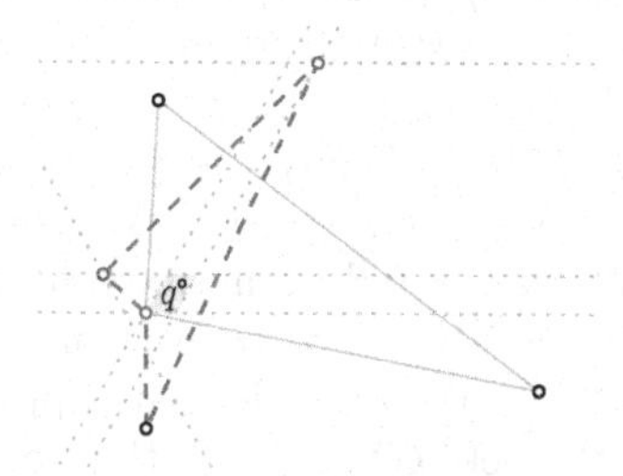

Fig. 5. A set P of seven points and a subset S (the six larger points). The graph $G^{\text{✡}}(P) \setminus S$ contains a singleton-component q which lies in a face of degree 3 in $G^{\triangledown}[S]$ (green solid edges) and a face of degree 4 in $G^{\triangle}[S]$ (blue dashed edges). (Color figure online)

Remark. If we knew $f_3^{\triangle} \leq f_{5+}^{\triangledown}$ and $f_3^{\triangledown} \leq f_{5+}^{\triangle}$ (where $f_{5+}^{\triangle} = \sum_{d \geq 5} f_d^{\triangle}$ etc.), then a similar analysis would show $\text{odd}(G^{\text{✡}}(P) \setminus S) - |S| \leq 4$, which would imply Conjecture 1 except for a small constant term. However, Fig. 5 shows an example where a point $q \notin S$ lies in a face of degree 3 in $G_A^{\triangledown}$ and a face of degree 4 in $G_A^{\triangle}$, so our proof-approach cannot be used to prove such a claim.

3 Relationship Between Blocking Sets and Matchings

In this section, we prove Theorem 2—that a lower bound on the blocking size function $\beta(n)$ implies a lower bound on the size $\mu(n)$ of a maximum matching, and vice versa.

Lemma 6. *For any $n \geq 1$, we have $\beta(n+1) \leq \beta(n) + 1$.*

Proof. Consider a set P with n points such that $\beta(n) = \beta(P)$. Let $T^{\triangledown}$ be a downward equilateral triangle that strictly encloses all points of P. Let b be the rightmost point of $T^{\triangledown}$. Then P lies in cone C_4 of b. Let a_1 be a point strictly inside cone C_1 of b; see also Fig. 4a. Every upward or downward equilateral triangle between a_1 and any point of P contains the point b. Set $P' = P \cup \{a_1\}$, and observe that we can block $G^{\text{✡}}(P')$ by using a minimum blocking set B of $G^{\text{✡}}(P)$ and adding b to it. Since $|B| = \beta(P) = \beta(n)$, we have $\beta(P') \leq \beta(n) + 1$, and $\beta(n+1)$ cannot be larger than that.

Since $\beta(1) = 0$, this lemma also shows that $\beta(n) \leq n - 1$, or in other words, that the '$n-1$' in Conjecture 2 is tight.

Theorem 2 (a). *For any set* P *of* n *points in the plane,* $G^{\hexstar}(P)$ *has a matching of size* $\beta(n)/2$.

Proof. Consider the Θ_6-graph $G^{\hexstar}(P)$ on a set P of n points in the plane. We again use the Tutte-Berge formula (Theorem 3) to prove that $G^{\hexstar}(P)$ contains a matching of size at least $\beta(n)/2$. Fix an arbitrary set $S \subseteq P$ and consider the connected components of $G^{\hexstar}(P) \setminus S$. As in the proof of Theorem 1, fix one representative point in each component, and let Q be the set of these points.

Consider the Θ_6-graph $G^{\hexstar}(Q)$ of only the points in Q, and let (q_1, q_2) be an edge in it; say it is introduced by $\triangle(q_1, q_2)$. By Lemma 1, there is a path π between q_1 and q_2 in $G^{\triangle}(P)$ that is fully contained in $\triangle(q_1, q_2)$. Since q_1 and q_2 are in different components of $G^{\hexstar}(P) \setminus S$, at least one point of π belongs to S.

Thus, for any edge in $G^{\hexstar}(Q)$, the triangle that supports that edge contains a point in S. Put differently, S blocks $G^{\hexstar}(Q)$, and thus $|S| \geq \beta(|Q|)$. Furthermore, $\beta(n) \leq \beta(|Q|) + n - |Q|$ by Lemma 6 since $|Q| \leq n$. Combining this with Theorem 3, it follows that the size of maximum matching in $G^{\hexstar}(P)$ is at least

$$\frac{n - (|Q| - |S|)}{2} \geq \frac{n - (|Q| - \beta(|Q|))}{2} \geq \frac{n - (n - \beta(n))}{2} = \frac{\beta(n)}{2}.$$

□

In particular, if $\beta(n) \geq n - 1$, then $\mu(n) \geq \beta(n)/2 \geq (n-1)/2$, so by integrality $\mu(n) \geq \lceil (n-1)/2 \rceil$. In other words, Conjecture 2 implies Conjecture 1.

We now turn to the other half of Theorem 2. Note that Aichholzer et al. [3] proved a similar result (for $c = d = 1/2$ and Delaunay graphs), and our proof is a modification of theirs. (In fact, the proof applies to any proximity graphs.)

Theorem 2 (b). *Assume that we know that* $\mu(n) \geq cn + d$ *for some constants* c, d. *Then* $\beta(n) \geq (cn + d)/(1 - c)$.

Proof. Let P be a set of n points such that $\beta(P) = \beta(n) = b$, and let B be a minimum blocking set of $G^{\hexstar}(P)$. Let M be a matching of size at least $\mu(b+n) \geq cb + cn + d$ in $G^{\hexstar}(P \cup B)$. Since P is an independent set in $G^{\hexstar}(P \cup B)$, it contains at most one endpoint of each edge in M, as well as some unmatched points, so

$$n = |P| \leq |M| + (n + b - 2|M|) \leq n + b - (cb + cn + d)$$

Solving for b gives $\beta(n) = b \geq (cn + d)/(1 - c)$.

In particular, if Conjecture 1 holds, then $\mu(n) \geq (n-1)/2$. Hence, $c = 1/2$ and $d = -1/2$, therefore $\beta(n) \geq 2(n-1)/2 = n - 1$ and Conjecture 2 holds. So Conjecture 1 implies Conjecture 2. As a second consequence, we know that $(3n-8)/7$ is a valid lower bound on $\mu(n)$ by Theorem 1, therefore (with $c = 3/7$) we have $\beta(n) \geq 7/4 \cdot (3n - 8)/7 = 3n/4 - 2$.

4 Conclusions, Additional Properties, Open Problems

We have improved the lower bound on the size of a matching in any Θ_6-graph on n points to $(3n-8)/7$. A main open problem is to prove the conjecture that any Θ_6-graph has a (near-)perfect matching.

We have shown that this conjecture is equivalent to proving that every Θ_6-graph on n points requires at least $n-1$ points to block all its edges. More generally, we proved a relationship between the minimum size of maximum matchings and the minimum size of blocking sets so that any improvement in the lower bound for one of these parameters will also improve the other.

We can also give additional bounds on several parameters of Θ_6-graphs. In particular, we can show that $\alpha(n) \leq 3n/4$ and $\mu^* \leq 2n/5$. Further, we can prove that $\hat{\beta}(n) \geq (5n-6)/4$, where $\hat{\beta}(n)$ is the maximum number of points that may be needed to block any Θ_6-graph on n points. Finally, we can show that any Θ_6-graph on $n \geq 3$ points has at most $5n-12$ edges and minimum degree at most 9, while there are Θ_6-graphs with $5n-17$ edges and Θ_6-graphs with minimum degree 7. The proofs are given in the full version of the paper [11].

Acknowledgements. This work was done by a University of Waterloo problem solving group. We thank the other participants, Alexi Turcotte and Anurag Murty Naredla, for inspiring discussions, and the anonymous reviewers for helpful comments.

References

1. Ábrego, B.M., et al.: Matching points with circles and squares. In: Akiyama, J., Kano, M., Tan, X. (eds.) JCDCG 2004. LNCS, vol. 3742, pp. 1–15. Springer, Heidelberg (2005). https://doi.org/10.1007/11589440_1
2. Ábrego, B.M., et al.: Matching points with squares. Discrete Comput. Geom. **41**(1), 77–95 (2009)
3. Aichholzer, O., et al.: Blocking Delaunay triangulations. Comput. Geom.: Theory Appl. **46**(2), 154–159 (2013)
4. Alzoubi, K.M., Li, X., Wang, Y., Wan, P., Frieder, O.: Geometric spanners for wireless ad hoc networks. IEEE Trans. Parallel Distrib. Syst. **14**(4), 408–421 (2003)
5. Aronov, B., Dulieu, M., Hurtado, F.: Witness (Delaunay) graphs. Comput. Geom.: Theory Appl. **44**(6–7), 329–344 (2011)
6. Aronov, B., Dulieu, M., Hurtado, F.: Witness Gabriel graphs. Comput. Geom.: Theory Appl. **46**(7), 894–908 (2013)
7. Aurenhammer, F., Paulini, G.: On shape Delaunay tessellations. Inf. Process. Lett. **114**(10), 535–541 (2014)
8. Babu, J., Biniaz, A., Maheshwari, A., Smid, M.H.M.: Fixed-orientation equilateral triangle matching of point sets. Theor. Comput. Sci. **555**, 55–70 (2014). Also in WALCOM 2013
9. Bauer, D., Broersma, H., Schmeichel, E.: Toughness in graphs a survey. Graphs Comb. **22**(1), 1–35 (2006)
10. Berge, C.: Sur le couplage maximum d'un graphe. Comptes Rendus de l'Académie des Sciences, Paris **247**, 258–259 (1958)

11. Biedl, T., Biniaz, A., Irvine, V., Jain, K., Kindermann, P., Lubiw, A.: Maximum matchings and minimum blocking sets in θ_6-graphs. Arxiv report (2019). https://arxiv.org/abs/1901.01476
12. Biniaz, A., Maheshwari, A., Smid, M.H.M.: Higher-order triangular-distance Delaunay graphs: graph-theoretical properties. Comput. Geom.: Theory Appl. **48**(9), 646–660 (2015). Also in CALDAM 2015
13. Biniaz, A., Maheshwari, A., Smid, M.H.M.: Matchings in higher-order Gabriel graphs. Theor. Comput. Sci. **596**, 67–78 (2015)
14. Bonichon, N., Gavoille, C., Hanusse, N., Ilcinkas, D.: Connections between theta-graphs, delaunay triangulations, and orthogonal surfaces. In: Thilikos, D.M. (ed.) WG 2010. LNCS, vol. 6410, pp. 266–278. Springer, Heidelberg (2010). https://doi.org/10.1007/978-3-642-16926-7_25
15. Bose, P., De Carufel, J.L., Hill, D., Smid, M.H.M.: On the spanning and routing ratio of theta-four. In: Chan, T.M. (ed.) Proceedings of the Thirtieth Annual ACM-SIAM Symposium on Discrete Algorithms (SODA), pp. 2361–2370. SIAM (2019)
16. Bose, P., Fagerberg, R., Van Renssen, A., Verdonschot, S.: Competitive routing in the half-θ_6-graph. In: Rabani, Y. (ed.) Proceedings of the 23rd Annual ACM-SIAM Symposium on Discrete Algorithms (SODA), pp. 1319–1328. SIAM (2012)
17. Bose, P., Morin, P., van Renssen, A., Verdonschot, S.: The θ_5-graph is a spanner. Comput. Geom. **48**(2), 108–119 (2015). Also in WG 2013
18. Chew, P.: There are planar graphs almost as good as the complete graph. J. Comput. Syst. Sci. **39**(2), 205–219 (1989)
19. Clarkson, K.L.: Approximation algorithms for shortest path motion planning. In: Aho, A.V. (ed.) Proceedings of the 19th Annual ACM Symposium on Theory of Computing (STOC), pp. 56–65. ACM (1987)
20. Damian, M., Iacono, J., Winslow, A.: Spanning properties of Theta-Theta-6. arXiv:1808.04744 (2018)
21. Dillencourt, M.B.: Toughness and Delaunay triangulations. Discrete Comput. Geom. **5**, 575–601 (1990)
22. Drysdale III, R.L.S.: A practical algorithm for computing the Delaunay triangulation for convex distance functions. In: Proceedings of the 1st Annual ACM-SIAM Symposium on Discrete Algorithms (SODA), pp. 159–168 (1990)
23. Fischer, M., Lukovszki, T., Ziegler, M.: Geometric searching in walkthrough animations with weak spanners in real time. In: Bilardi, G., Italiano, G.F., Pietracaprina, A., Pucci, G. (eds.) ESA 1998. LNCS, vol. 1461, pp. 163–174. Springer, Heidelberg (1998). https://doi.org/10.1007/3-540-68530-8_14
24. Keil, J.M.: Approximating the complete Euclidean graph. In: Karlsson, R., Lingas, A. (eds.) SWAT 1988. LNCS, vol. 318, pp. 208–213. Springer, Heidelberg (1988). https://doi.org/10.1007/3-540-19487-8_23
25. Keil, J.M., Gutwin, C.A.: Classes of graphs which approximate the complete Euclidean graph. Discrete Comput. Geom. **7**, 13–28 (1992)
26. Morin, P., Verdonschot, S.: On the average number of edges in Theta graphs. Online J. Anal. Comb., page to appear (2014). Also in ANALCO 2014
27. Nishizeki, T., Baybars, I.: Lower bounds on the cardinality of the maximum matchings of planar graphs. Discrete Math. **28**(3), 255–267 (1979)
28. Tutte, W.T.: The factorization of linear graphs. J. Lond. Math. Soc. **22**, 107–111 (1947)

A Polynomial-Time Algorithm for the Independent Set Problem in $\{P_{10}, C_4, C_6\}$-Free Graphs

Edin Husić[1](✉) and Martin Milanič[2,3]

[1] LSE, Houghton Street, London WC2A 2AE, UK
e.husic@lse.ac.uk
[2] IAM, University of Primorska, Muzejski trg 2, 6000 Koper, Slovenia
martin.milanic@upr.si
[3] FAMNIT, University of Primorska, Glagoljaška 8, 6000 Koper, Slovenia

Abstract. We consider the independent set problem, a classical NP-hard optimization problem that remains hard even under substantial restrictions on the input graphs. The complexity status of the problem is unknown for the classes of P_k-free graphs for all $k \geq 7$ and for the class of even-hole-free graphs, that is, graphs not containing any even induced cycles. Using the technique of augmenting graphs we show that the independent set problem is solvable in polynomial time in the class of even-hole-free graphs not containing an induced path on 10 vertices. Our result is developed in the context of the more general class of $\{P_{10}, C_4, C_6\}$-free graphs.

Keywords: Independent set · Augmenting graph · Polynomial-time algorithm

1 Introduction

Given a (finite, simple, undirected) graph $G = (V, E)$, a set $I \subseteq V$ is an *independent set* if no two vertices in I are adjacent. An independent set is said to be *maximal* if it is not contained in any other independent set, and *maximum* if it is of maximum possible size. In the Maximum Independent Set problem we are given a graph G as input and the goal is to find an independent set in G of maximum cardinality. This NP-hard problem is one of the central problems in theoretical computer science and combinatorial optimization. The problem is known to remain NP-hard even under substantial restrictions on the input graphs, for instance for triangle-free graphs [20], planar graphs of maximum

---

E. Husić did most of his work on the paper while he was a student at the University of Primorska and École normale supérieure de Lyon. Several ideas for the proofs were developed in his master thesis [14].

M. Milanič—The work is supported in part by the Slovenian Research Agency (I0-0035, research program P1-0285 and research projects J1-9110, N1-0102).

I. Sau and D. M. Thilikos (Eds.): WG 2019, LNCS 11789, pp. 271–284, 2019.
https://doi.org/10.1007/978-3-030-30786-8_21

degree at most three [9], and for H-free graphs whenever not every component of H is a subdivision of a path or the claw [2].

Among the most prominent examples of polynomial-time solvable cases, let us mention the classes of P_5-free graphs [16], P_6-free graphs [12], claw-free graphs [18,22], their generalizations fork-free graphs [3] and ℓclaw-free graphs [5], and perfect graphs [11]. While MAXIMUM INDEPENDENT SET is known to admit subexponential-time algorithms in any class of P_k-free graphs [6], the determination of the exact complexity status of the problem in the classes of graphs excluding a fixed path as induced subgraph is a notorious open problem. Another related open question is whether MAXIMUM INDEPENDENT SET admits a polynomial-time algorithm in the class of even-hole-free graphs, where a graph is said to be *even-hole-free* if all its induced cycles are odd (see, e.g., the survey by Vušković [24]). It is worth pointing out that even-hole-free graphs (without clique cutsets) have unbounded clique-width [1] and hence we cannot solve the problem in this class by using Tarjan's decomposition [23] and the metatheorem of Courcelle et al. [7].

Using the technique of augmenting graphs we show that MAXIMUM INDEPENDENT SET is solvable in polynomial time in the class of even-hole-free graphs not containing an induced path on 10 vertices. Our result is developed in the context of the more general class of $\{P_{10}, C_4, C_6\}$-free graphs. These result generalize the fact that MAXIMUM INDEPENDENT SET is solvable in polynomial time in the class of $\{P_8$, even-hole$\}$-free graphs, which follows from polynomial-time solvability of the problem in the more general class of $\{P_8$, banner$\}$-free graphs, as shown by Gerber et al. [10].

Structure of the Paper. In Sect. 2 we present the necessary definitions and the method of augmenting graphs. A characterization of minimal augmenting graphs in the class of even-hole-free graphs is given in Sect. 3. We present a polynomial-time algorithm for the independent set problem in $\{P_9, C_4, C_6\}$-free graphs in Sect. 4. Section 5 is devoted to the main result. Due to space limitations, several proofs are omitted.

2 Preliminaries

Let $G = (V, E)$ be a graph. As usual, we write $V(G)$ for V, $E(G)$ for E, and $uv \in E$ for an edge $\{u, v\} \in E$. For a non-empty subset $W \subseteq V$, the induced subgraph $G[W]$ is defined as the graph $H = (W, E \cap \binom{W}{2})$, where $\binom{W}{2}$ is the set of all unordered pairs in W. Given two graphs G and H, we say that G is *H-free* if it does not contain an induced subgraph isomorphic to H. Let $\mathcal{F}$ be a family of graphs. A graph G is $\mathcal{F}$-free if it is H-free for every $H \in \mathcal{F}$. We denote by $\text{Free}(\mathcal{F})$ the class of $\mathcal{F}$-free graphs. The graph $G[V \backslash W]$ is denoted as $G \backslash W$. For $W = \{v\}$, we will write $G \backslash v$ for simplicity. A *hole* in a graph is a chordless cycle of length at least four. A hole is *even* (resp. *odd*) if it contains an even (resp. odd) number of vertices. For $n \geq 3$, the cycle with n vertices is denoted as C_n. For $k \geq 2$, a path on k vertices (denoted P_k) is obtained by deleting a vertex from C_{k+1}. The *neighborhood* of a vertex v is the set of vertices $N_G(v) = \{u \in V \mid uv \in E\}$. We define $N_G[v] = N_G(v) \cup \{v\}$. When the graph is clear from the context, we may omit the subscript G. The (closed and open) neighborhoods of a set $W \subseteq V$ are defined as $N_G[W] = \bigcup_{v \in W} N_G[v]$ and $N_G(W) = N_G[W] \setminus W$, respectively. Let $S_1, S_2 \subseteq V(G)$. We say that S_1 *dominates* S_2 if every vertex in S_2 is adjacent to a vertex in S_1. Assume additionally that $S_1 \cap S_2 = \emptyset$. We say that S_1 and S_2 are *complete* to each other if every vertex of S_1 is adjacent to every vertex S_2 and we say that S_1 and S_2 are *anticomplete* to each other if there is no edge with one endpoint in S_1 and the other in S_2. We will denote by n the number of vertices of the graph under consideration.

Augmenting Graphs

Given a graph $G = (V, E)$, its *line graph* $L(G)$ is the graph with vertex set $E(G)$, with two distinct vertices of $L(G)$ adjacent if and only if they share a vertex as edges in G. If G is isomorphic to the line graph of H, we say that H is a *root graph* of G. Every independent set in $L(H)$ corresponds to a matching in H. Hence, in the class of line graphs MAXIMUM INDEPENDENT SET is equivalent to the maximum matching problem in the root graph. Since we can find in linear time a root graph of a given line graph [15,21] and the maximum matching is solvable in polynomial time for any graph G [8], it follows that MAXIMUM INDEPENDENT SET is polynomial-time solvable in the class of line graphs.

Generalizing the notion of augmenting paths for the maximum matching problem to the notion of augmenting graphs for MAXIMUM INDEPENDENT SET led to several new polynomial-time algorithms for the problem in particular classes of graphs, starting with the result for the class of claw-free graphs, obtained in 1980 independently by Minty [18] and Sbihi [22]. The key definition is as follows.

Definition 1. *Given a graph G and an independent set $I \subseteq V(G)$, an induced bipartite subgraph $H = (W, B; E)$ in G is* I-augmenting *if the following holds:*

- $W \subseteq I$,
- $B \subseteq V(G) \setminus I$,
- $|W| < |B|$ *and*
- $N_G(B) \cap I \subseteq W$.

The following theorem, which can be found, e.g., in Mosca [19], suggests a possible application of augmenting graphs.

Theorem 1. *An independent set I in a graph G is maximum if and only if there is no I-augmenting graph.*

The theorem leads to the following approach for finding a maximum independent set. Start with an independent set I, then find an I-augmenting graph $(W, B; E)$ if one exists, replace I with $(I \setminus W) \cup B$ and iterate. Since the size of an independent set is at most $n = |V(G)|$, we will repeat the step at most

$n-1$ times and obtain a maximum independent set. This approach leads to efficient algorithms for the problem for particular classes of graphs, see [4,10,13,17]. Moreover, it suffices to consider only minimal augmenting graphs. A definition and a characterization follow.

Definition 2. *An augmenting graph H for an independent set I in a graph G is* minimally I-augmenting *if no proper induced subgraph of H is I-augmenting.*

Lemma 1 (Lozin and Milanič [17]). *Given a graph G and an independent set $I \subseteq V(G)$, an I-augmenting graph $H = (W, B; E)$ is minimally I-augmenting if and only if the following conditions hold:*

(i) $|W| = |B| - 1$,
(ii) for every non-empty subset $A \subseteq W$, we have $|A| < |N_H(A)|$, and
(iii) H is connected.

Note that conditions *(i)*–*(iii)* are independent of G and I. This leads to the following.

Definition 3. *A* minimal augmenting graph *is a connected bipartite graph $H = (W, B; E)$ such that $|W| = |B| - 1$ and for every non-empty subset $A \subseteq W$, we have $|A| < |N_H(A)|$.*

3 Minimal Augmenting Even-Hole-Free Graphs

In order to characterize the minimal augmenting graphs in the class of even-hole-free graphs, the following definition will be helpful.

Definition 4. *We say that a tree T is a* black-white tree *if its vertex set can be partitioned into two sets B and W such that both sets are independent and every vertex in W has degree exactly 2. We say that vertices in W are* white *and vertices in B are* black.

Observation 2. *A graph T is a black-white tree if and only if it can be obtained by subdividing each edge of a tree exactly once.*

Lemma 2. *A tree is a minimal augmenting graph if and only if it is a black-white tree.*

Proof. Let $H = (W, B; E)$ be a tree. Suppose first that H is a minimal augmenting graph. We will show that every vertex in part W is of degree 2. Since $|N_H(A)| > |A|$ for all non-empty subsets $A \subseteq W$, there exists no white vertex of degree 1. Moreover, since $|B| = |W| + 1$, we have $|B| + |W| = 2|W| + 1$. Since H is a tree, the latter implies that H has exactly $2|W|$ edges. Suppose that there exists a vertex $w \in W$ of degree 3 or more. Since every vertex in W has degree at least 2 and W is a part of bipartition (W, B) of H, we infer that H has more than $2|W|$ edges; a contradiction.

Suppose now that H is a black-white tree. Consider the graph H' with vertex set B in which two vertices are adjacent if and only if they have a common neighbor in H. Since every white vertex in H has degree exactly 2 and H is acyclic, there is a bijective correspondence between edges of H' and vertices in W. Thus, $|W| = |E(H')| = |V(H')| - 1 = |B| - 1$, where the second equality holds since H' is a tree. We still need to show that $|N_H(A)| > |A|$ for all non-empty subsets $A \subseteq W$. Such a set A can be identified with a set of edges of H', say A'. Then, $N_H(A)$ corresponds to the set of endpoints of edges in A'. Since the subgraph of H' induced by $N_H(A)$ is acyclic and contains all edges in A, we have $|A| \leq |N_H(A)| - 1$, as claimed. Thus, H is a minimal augmenting graph. □

Note that every even-hole-free graph that is bipartite is acyclic. Since minimal augmenting graphs are connected and bipartite, it follows that every even-hole-free minimal augmenting graphs is a tree. In particular, by Lemma 2 we have the following corollary.

Corollary 1. *An even-hole-free graph H is a minimal augmenting graph if and only if it is a black-white tree.*

We will refer to a black-white tree also as a (minimal) augmenting tree.

Let us define some particular black-white trees for later use. For positive integers r and s, we define a black-white rooted tree $T_{r,s}$ as follows. The root of the tree $T_{r,s}$ is white and has exactly two children, both black. The two children of the root have exactly r and s white children, respectively. Recall that each non-root white vertex is of degree two and hence has a unique child. There are no further vertices in $T_{r,s}$. Furthermore, for a non-negative integer s we define a black-white tree T_s as a rooted tree having a black root vertex with exactly s white children, each having a unique child. There are no further vertices in T_s. See Fig. 1 for concrete examples. Note that the trees $T_{1,1}$ and T_2 are isomorphic to paths P_7 and P_5, respectively.

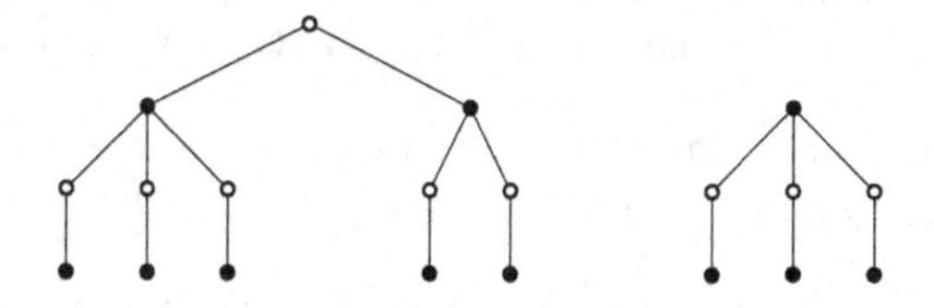

Fig. 1. $T_{3,2}$ (left) and T_3 (right).

The following lemma is a consequence of Corollary 1.

Lemma 3. *Let H be $\{P_9$, even-hole$\}$-free graph. Then, H is a minimal augmenting graph if and only if H is either isomorphic to a $T_{r,s}$ for some positive integers r and s, or to a T_s for some non-negative integer s.*

Proof. By Corollary 1, it suffices to prove that trees of the form $T_{r,s}$ or T_s are the only P_9-free black-white trees. Let H be a P_9-free black-white tree. By Observation 2, there exists a tree T such that H can be obtained from T by subdividing each edge exactly once. Since H is P_9-free, T is P_5-free. If T contains an induced P_4, then there exist positive integers r and s such that T is isomorphic to the graph obtained from the disjoint union of $K_{1,r}$ and $K_{1,s}$ by connecting the two centers with an edge. In this case H is isomorphic to $T_{r,s}$. On the other hand, if T is P_4-free, then T is isomorphic to some $K_{1,s}$ for some non-negative integer s, which implies that H is isomorphic to a T_s for some non-negative integer s.

In the remaining two sections we apply the method of augmenting graphs to obtain a polynomial-time algorithm for MAXIMUM INDEPENDENT SET in the class of $\{P_{10}, C_4, C_6\}$-free graphs. We start by developing a polynomial-time algorithm for the case of $\{P_9, C_4, C_6\}$-free graphs.

4 MAXIMUM INDEPENDENT SET in $\{P_9, C_4, C_6\}$-Free Graphs

Suppose that we want to check, given a graph G and an independent set I in G, if G contains an I-augmenting graph. Trivially, we can assume that I is a maximal independent set, since every vertex $w \in V(G)\backslash I$ with no neighbor in I induces an I-augmenting graph. Moreover, we can check in polynomial time if there exists an I-augmenting graph isomorphic to P_3. Thus, we may assume the following without loss of generality.

Assumption 1. *Set I is a maximal independent set in G that does not admit an augmenting P_3.*

Note that the trees T_0 and T_1 are isomorphic to K_1 and P_3, respectively. Thus, if a graph G admits an I-augmenting T_s, then $s \geq 2$.

For a vertex $w \in I$ we define $K(w) = \{v \in V(G) : N_G(v) \cap I = \{w\}\}$.

Observation 3. *If I is a maximal independent set in a graph G that does not admit any I-augmenting P_3, then for every $w \in I$, the set $K(w)$ is a clique in G.*

Our goal is to show that we can efficiently check if a maximal independent set I in a given graph G (that satisfies some extra assumptions) admits an augmenting graph. We try to build an augmenting graph starting from some prescribed set of vertices. Note that, once we add a vertex $b \in V(G)\backslash I$ to our augmenting graph, then, by the definition of an augmenting graph, all the neighbors of b that are in the independent set I must also be included in the augmenting graph.

We start with two general lemmas establishing sufficient conditions for checking if a maximal independent set I in a given graph G without an I-augmenting P_3 admits an augmenting graph of the form T_s or $T_{r,s}$.

Lemma 4. *Let $\mathcal{F}$ be a set of graphs and $k \geq 3$ such that the* MAXIMUM INDEPENDENT SET *problem can be solved in polynomial time in the class of $(\{P_{k-1}\} \cup \mathcal{F})$-free graphs. Let G be an $(\{P_k\} \cup \mathcal{F})$-free graph and let I be a maximal independent set in G such that G has no I-augmenting P_3. Then it can be tested in polynomial time whether G has an I-augmenting graph isomorphic to T_s for some integer $s \geq 2$.*

Proof. For every vertex $x \in V(G) \backslash I$ with $|N_G(x) \cap I| \geq 2$, we show how to check in polynomial time whether G has an I-augmenting T_s with root x, where $s = |N_G(x) \cap I|$. We will try to build such a tree T. By the definition of an augmenting graph, all vertices in $N_G(x) \cap I$ are in T, so add them to T. Note that since $s \geq 2$, vertex x does not belong to $K(w)$ for any $w \in N_G(x) \cap I$. Let $S = \left(\bigcup_{w \in N_G(x) \cap I} K(w)\right) \backslash N_G(x)$. By Observation 3, set S is a union of s cliques. Therefore, there exists an augmenting T_s with root x if and only if there exists an independent set I^* in $G[S]$ such that $|I^*| = s$. Moreover, any such set I^* is necessarily a maximum independent set in $G[S]$. It thus suffices to show that the graph $G[S]$ is $(\{P_{k-1}\} \cup \mathcal{F})$-free, since the assumption of the lemma will then imply that a maximum independent set in $G[S]$ can be computed in polynomial time. Since G is $\mathcal{F}$-free, so is $G[S]$. Suppose that $G[S]$ contains an induced P_{k-1} with vertices $p_1, \ldots, p_{k-1}$ along the path. Let $w \in N_G(x) \cap I$ such that $p_{k-1} \in K(w)$. Then $\{p_1, \ldots, p_{k-1}, w\}$ induces a P_k in G unless $p_{k-2} \in K(w)$. Assume $p_{k-2} \in K(w)$. Then the set $\{p_1, \ldots, p_{k-2}, w, x\}$ induces a P_k in G; a contradiction. □

Lemma 5. *Let $\mathcal{F}$ be a set of graphs and $k \geq 4$ such that the* MAXIMUM INDEPENDENT SET *problem can be solved in polynomial time in the class of $(\{P_{k-3}\} \cup \mathcal{F})$-free graphs. Let G be a $(\{P_k\} \cup \mathcal{F})$-free graph and let I be a maximal independent set in G such that G has no I-augmenting P_3. Then it can be tested in polynomial time whether G has an I-augmenting graph isomorphic to $T_{r,s}$ for some positive integers r, s.*

Proof. For every vertex $v \in I$, we show how to test in polynomial time if there exists an I-augmenting $T_{r,s}$ with root v. We consider all pairs of non-adjacent vertices $b', b'' \in N_G(v)$ and check if there exists an I-augmenting tree T of type $T_{r,s}$ with root at v and containing vertices v, b', b''. Since b' and b'' are black vertices, any I-augmenting tree T containing b' and b'' also contains all vertices in $N_G(b') \cap I$ and $N_G(b'') \cap I$. If $N_G(b') \cap N_G(b'') \cap I \neq \{v\}$ then such an augmenting tree T does not exist. For the rest of the proof, suppose that $N_G(b') \cap N_G(b'') = \{v\}$. Denote $q = |N_G(\{b', b''\}) \cap (I \backslash \{v\})|$. Let $Q = \left(\bigcup_{w \in N_G(\{b',b''\}) \cap (I \backslash \{v\})} K(w)\right) \backslash N_G(\{b', b''\})$. By Observation 3, set Q is a union of q cliques. Therefore, there exists an augmenting $T_{r,s}$ with root v and containing $\{b', b''\}$ if and only if there exists an independent set I^* in $G[Q]$ such that $|I^*| = q$. Moreover, any such set I^* is necessarily a maximum independent set in $G[Q]$. It thus suffices to show that the graph $G[Q]$ is $(\{P_{k-3}\} \cup \mathcal{F})$-free, since the assumption of the lemma will then imply that a maximum independent set in $G[Q]$ can be computed in polynomial time.

Since G is $\mathcal{F}$-free, so is $G[Q]$. For the sake of a contradiction, suppose that there exists an induced P_{k-3} in $G[Q]$. Denote its vertices as $p_1, \ldots, p_{k-3}$ along the path. Denote by w a vertex in $N_G(\{b', b''\}) \cap (I \backslash \{v\})$ such that $p_{k-3} \in K(w) \backslash N_G(\{b', b''\})$. We may assume without loss of generality that $wb'' \in E$. Then $wb' \notin E$. Then the set $\{p_1, \ldots, p_{k-3}, w, b'', v\}$ induces a P_k in G unless $p_{k-4} \in K(w)$. Assume $p_{k-4} \in K(w)$. Then $\{p_1, \ldots, p_{k-4}, w, b'', v, b'\}$ induces a P_k in G; a contradiction. □

Gerber et al. showed in [10] that Maximum Independent Set is solvable in polynomial time in the class of $\{P_8$, banner$\}$-free graphs. The banner is a graph containing an induced C_4, therefore every $\{P_8, C_4\}$-free graph is also $\{P_8$, banner$\}$-free. It follows that Maximum Independent Set is solvable in polynomial time in the class of $\{P_8$, banner$\}$-free graphs. Applying Lemmas 4 and 5 with $\mathcal{F} = \{C_4\}$ and $k = 9$ we thus obtain the following.

Corollary 2. *Let G be a $\{P_9, C_4\}$-free graph and let I be a maximal independent set in G such that G has no I-augmenting P_3. Then it can be tested in polynomial time whether G has an I-augmenting graph isomorphic to T_s for some integer $s \geq 2$ or to $T_{r,s}$ for some positive integers r, s.*

Corollary 2 leads to the following theorem.

Theorem 4. Maximum Independent Set *is polynomial-time solvable in the class of $\{P_9$, even-hole$\}$-free graphs.*

Proof. By Lemma 3, the minimal augmenting graphs in the class of $\{P_9$, even-hole$\}$-free are of type T_s or $T_{r,s}$. By Assumption 1 and Corollary 2, we conclude that Maximum Independent Set is polynomial-time solvable in the class of $\{P_9$, even-hole$\}$-free graphs. □

Observe that the $\{P_9$, even-hole$\}$-free graphs are exactly the $\{P_9, C_4, C_6, C_8\}$-free graphs. We strengthen the result of Theorem 4 to the class of $\{P_9, C_4, C_6\}$-free graphs. We do this by showing that there are only finitely many minimal augmenting $\{P_9, C_4, C_6\}$-free graphs that are not T_s or $T_{r,s}$. This follows from Lemma 3 and the following observation.

Lemma 6. *Every minimal augmenting $\{P_9, C_4, C_6\}$-free graph containing an induced C_8 has at most 15 vertices.*

Assumption 1, Corollary 2, and Lemma 6 lead to the following theorem.

Theorem 5. Maximum Independent Set *is polynomial-time solvable in the class of $\{P_9, C_4, C_6\}$-free graphs.*

5 Maximum Independent Set in $\{P_{10}, C_4, C_6\}$-Free Graphs

In this section we generalize the previous result on $\{P_9, C_4, C_6\}$-free graphs to the class of $\{P_{10}, C_4, C_6\}$-free graphs. As before, we first classify the minimal augmenting graphs and then present an algorithm for finding them. We consider two cases: augmenting trees and augmenting graphs containing a C_8 or a C_{10}.

Augmenting Trees. Next to the already mentioned trees T_s and $T_{r,s}$, a minimal augmenting tree in the class of $\{P_9, C_4, C_6\}$-free graphs can be a black-white tree containing an induced P_9. These are exactly black-white trees with the root in the middle vertex of a P_9 and every vertex being at distance at most 4 from the root. An example of such a tree is depicted in Fig. 2.

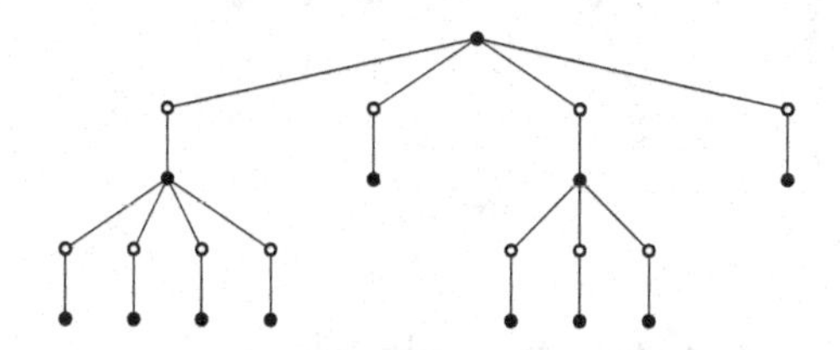

Fig. 2. A minimal augmenting tree containing a P_9.

We already know that MAXIMUM INDEPENDENT SET is solvable in polynomial time in the class of $\{P_9, C_4, C_6\}$-free graphs. Therefore, by Lemmas 4 and 5 we have a polynomial-time algorithm for checking whether a maximal independent set I in a $\{P_{10}, C_4, C_6\}$-free graph G admits a minimal augmenting graph isomorphic to a T_s or a $T_{r,s}$. Hence, we need to show how to test whether such a maximal independent set I admits a minimal augmenting tree containing a P_9.

Lemma 7. *Let G be a $\{P_{10}, C_4, C_6\}$-free graph and I a maximal independent set in G. It can be tested in polynomial time whether I admits a minimal augmenting tree $H = (W, B; E)$ containing an induced P_9.*

Augmenting Graphs Containing a C_8 or a C_{10}. Let us show that a minimal augmenting graph in the class of $\{P_{10}, C_4, C_6\}$-free graphs cannot contain an induced C_{10}.

Lemma 8. *Let $H = (W, B; E)$ be a minimal $\{P_{10}, C_4, C_6\}$-free augmenting graph. Then H is C_{10}-free.*

Proof. We prove the lemma by contradiction. Let C be an induced C_{10} in H. Denote the vertices of C by $v_1, \ldots, v_{10}$ so that consecutive vertices are adjacent (indices modulo 10). Since H is a connected graph with an odd number of vertices, it contains a vertex adjacent to C that is not in C. Let v_i be a vertex in C adjacent to some vertex $x \in V(H)\backslash V(C)$. Since G is $\{C_4, C_6\}$-free, we infer that $N_H(x) \cap V(C) = \{v_i\}$. But now, $(V(C)\backslash\{v_{i+1}\}) \cup \{x\}$ (indices modulo 10) induces a P_{10} in G; a contradiction. $\square$

By Lemma 8, we know that in the class of $\{P_{10}, C_4, C_6\}$-free graphs a minimal augmenting graph H is either a black-white tree or contains a C_8. It turns out that a minimal augmenting graph containing a C_8 is either one of two specific graphs or belongs to a particular family of graphs.

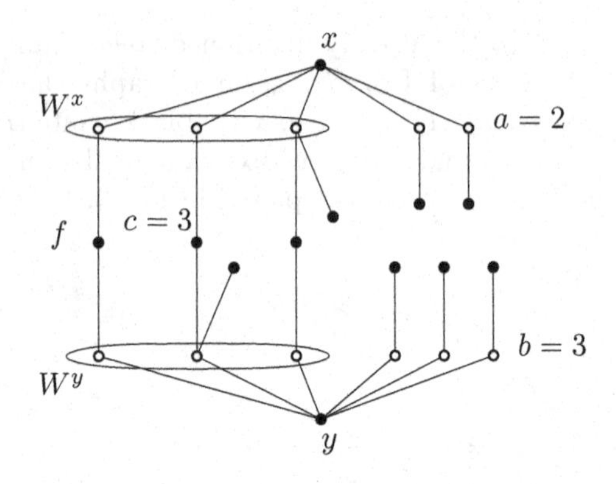

Fig. 3. A $(2, 3, 3)$-crab.

Definition 5. *For non-negative integers a, b, c with $c \geq 2$, an (a, b, c)*-crab *is a bipartite graph $H = (W, B; E)$ that can be built from two disjoint trees of type T_r, say T and T', as follows (see Fig. 3):*

- *T is isomorphic to T_{a+c} and T' is isomorphic to T_{b+c}.*
- *Exactly c leaves of T are identified with c leaves of T'. We call these* inner leaves. *Non-identified leaves of T and T' are* leaves *of H.*
- *We name the root of T as x and the root of T' as y. We say that x and y are the* root *vertices of H. We denote by W^x the set of white vertices of T whose children are identified with some leaves of T'. Similarly for W^y.*
- *A set of $c - 1$ inner leaves (out of c) is chosen and for each such vertex ℓ, exactly one of its two neighbors in $W^x \cup W^y$ is adjacent to a unique vertex $\ell_p \notin V(T) \cup V(T')$. Moreover ℓ_p has no other neighbors in H. We say that ℓ_p is the* additional leaf *of ℓ. The unique vertex $f \in N_H(W^x) \cap N_H(W^y)$ that does not have an additional leaf will be called the* flat *vertex of H.*

Lemma 9. *There exist two graphs F_1 and F_2 such that the following holds. Let G be a $\{P_{10}, C_4, C_6\}$-free graph and I a maximal independent set in G that admits neither an augmenting P_3 nor an augmenting P_5. Let $H = (W, B; E)$ be a minimally augmenting graph for I that contains a C_8. Then H is either F_1, F_2, or an (a, b, c)-crab.*

Since F_1 and F_2 are two fixed graphs, an augmenting graph isomorphic to F_1 or F_2 can be trivially detected in polynomial time. Thus, Lemma 9 and the next lemma imply that we can efficiently detect the existence of an augmenting graph that contains a C_8.

Lemma 10. *Let G be a $\{P_{10}, C_4, C_6\}$-free graph and I a maximal independent set in G such that G has no I-augmenting P_3. Then we can test in polynomial time if I admits an augmenting (a, b, c)-crab.*

Proof. By the definition of an (a, b, c)-crab it holds that $c \geq 2$. It follows that every (a, b, c)-crab contains an induced C_8. We will check if there exists an induced cycle C on 8 vertices in G that can be extended to an induced (a,b,c)-crab. For the rest of the proof, we fix an induced cycle C on 8 vertices in G and show how one can check if C can be extended to an (a,b,c)-crab. By the definition of an augmenting graph it must hold $|V(C) \cap I| = |C \backslash I| = 4$. Denote the vertices of C by $v_1, \ldots, v_8$ so that consecutive vertices are adjacent (indices modulo 8) and $\{v_1, v_3, v_5, v_7\} \subseteq I$.

More precisely, we will show how to check if C can be extended to an (a,b,c)-crab with two particular opposite black vertices of C set as roots of the crab; furthermore, we may also guess the flat vertex, as well as the additional leaf. By symmetry, it suffices to consider the case when C can be extended to an (a,b,c)-crab H with roots v_2 and v_6, flat vertex v_8, and an additional leaf s_3 that has unique neighbor v_3 in C. See Fig. 4.

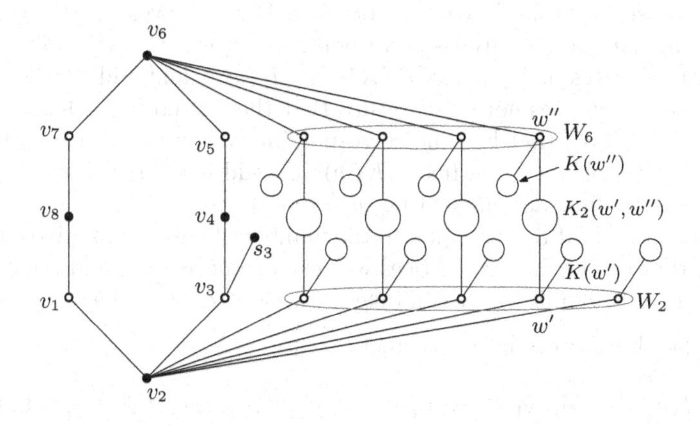

Fig. 4. Testing for an augmenting (a,b,c)-crab.

In the rest of the proof, whenever we say (a,b,c)-crab we mean an (a,b,c)-crab with the properties described in the previous sentence. Therefore, it suffices to consider only the non-neighbors of $\{v_1, v_3, v_4, v_5, v_7, v_8, s_3\}$ (except v_2, v_6). Define $W_6 = N_G(v_6) \cap (I \backslash \{v_5, v_7\})$ and $W_2 = N_G(v_2) \cap (I \backslash \{v_1, v_3\})$. By the definition of an augmenting graph, $W_2 \cup W_6 \subseteq V(H)$. Moreover, by the definition of an (a,b,c)-crab, the set of all white vertices that should be in H is exactly the set $\{v_1, v_3, v_5, v_7\} \cup W_2 \cup W_6$. So we only need to check whether we can find a set a black vertices to complete our augmenting graph. We only consider vertices in $V(G) \backslash I$ non-adjacent to v_2 and v_6 that are contained in some $K(w)$ for $w \in W_2 \cup W_6$ (we will call these *type 1* vertices) or vertices with exactly one neighbor in W_2 and exactly one neighbor in W_6 (we will call these *type 2* vertices).

Let x be a type 2 vertex and denote by w', w'' its neighbors in W_2 and W_6, respectively. Suppose that there exists a type 2 vertex y such that its neighbors in W_2 and W_6 are w' and w''_y ($w''_y \neq w''$), respectively. If x is non-adjacent to y, then the set $\{y, w', x, w'', v_6, w''_y\}$ induces a C_6. It follows that $xy \in E(G)$ and thus $\{y, x, w'', v_6, v_7, v_8, v_1, v_2, v_3, s_3\}$ induces a P_{10}; a contradiction. Hence for each $w' \in W_2$ having a type 2 neighbor we can define its unique *corresponding* vertex $w'' \in W_6$. Moreover, we define $K_2(w') = K_2(w'') = K_2(w', w'')$ as the set of all type 2 neighbors of w' and w''. Since G is C_4-free, every such set is a clique.

Suppose that $w' \in W_2$ and $w'' \in W_6$ are a pair of corresponding vertices. Then $K_2(w', w'') \neq \emptyset$. Let u and v be two non-adjacent vertices in $K_2(w', w'') \cup K(w') \cup K(w'')$. Since G does not admit any I-augmenting P_3, sets $K(w')$ and $K(w'')$ are cliques by Observation 3. Thus, u and v must belong to different sets among $K(w')$, $K(w'')$, and $K_2(w', w'')$. If $u \in K_2(w', w'')$ and, say, $v \in K(w'')$, then we could add u as an inner leaf and v as the additional leaf of u to H. If on the other hand, $u \in K(w')$ and $v \in K(w'')$ (or vice versa), then we could add both u and v as leaves to H.

The above suggests the following procedure. For every vertex $w' \in W_2$ having a type 2 neighbor, we identify its corresponding vertex $w'' \in W_6$, take a pair of non-adjacent vertices in $K_2(w', w'') \cup K(w') \cup K(w'')$, and add the two vertices to H. If such a pair does not exist, return that there is no (a, b, c)-crab. For the vertices $w \in W_2 \cup W_6$ that have no corresponding vertex, i.e., have no neighbor of type 2, pick an arbitrary vertex in $K(w)$ and add it to H. If some $K(w) = \emptyset$, then I does not admit an augmenting (a, b, c)-crab.

One can check that $c-2$ is equal to the number of added non-adjacent vertex pairs such that one of the two vertices was from some $K_2(w')$ and the other one from some $K(w'')$. Similarly, we can take $a = |W_2|-(c-2)$ and $b = |W_6|-(c-2)$.

Claim. The above procedure is correct.

Proof of Claim: For all type 1 vertices $w \in W_2 \cup W_6$, we set $K_2(w) = \emptyset$. Observe that for the correctness of the above procedure it suffices to prove the following statement: For every two distinct vertices $w', w'' \in W_2 \cup W_6$ such that either

1. at least one of w', w'' is of type 1, or
2. they are both of type 2 but they are not corresponding vertices,

sets $K(w') \cup K_2(w')$ and $K(w'') \cup K_2(w'')$ are anticomplete to each other.

To prove the above statement, suppose for a contradiction that there exist two distinct vertices $w', w'' \in W_2 \cup W_6$ such that either at least one of w', w'' is of type 1, or they are both of type 2 but they are not corresponding vertices, a vertex $s \in K(w') \cup K_2(w')$, and a vertex $t \in K(w'') \cup K_2(w'')$ such that $st \in E(G)$. The assumptions on w' and w'' imply that s and t have no common neighbors in $W_2 \cup W_6$. Now, if $w' \in W_2$, then $\{t, s, w', v_2, v_3, v_4, v_5, v_6, v_7, v_8\}$ induces a P_{10}, while if $w' \in W_6$, then $\{t, s, w', v_6, v_7, v_8, v_1, v_2, v_3, v_4\}$ induces a P_{10}. In either case, we reach a contradiction. ▲

Clearly, the procedure runs in polynomial time. □

We obtain the following theorem.

Theorem 6. Maximum Independent Set *is solvable in polynomial time in the class of* $\{P_{10}, C_4, C_6\}$*-free graphs.*

Corollary 3. Maximum Independent Set *is solvable in polynomial time in the class of* $\{P_{10},$ even-hole$\}$*-free graphs.*

Acknowledgment. The authors are grateful to the anonymous reviewers for many helpful remarks.

References

1. Adler, I., Le, N.K., Müller, H., Radovanović, M., Trotignon, N., Vušković, K.: On rank-width of (diamond, even hole)-free graphs. Discrete Math. Theor. Comput. Sci. **19**(1), 12 (2017). Paper No. 24
2. Alekseev, V.E.: The effect of local constraints on the complexity of determination of the graph independence number. In: Combinatorial-Algebraic Methods in Applied Mathematics, pp. 3–13 (1982)
3. Alekseev, V.E.: Polynomial algorithm for finding the largest independent sets in graphs without forks. Discrete Appl. Math. **135**(1–3), 3–16 (2004). [mr1760726], Russian translations. II
4. Alekseev, V.E., Lozin, V.V.: Augmenting graphs for independent sets. Discrete Appl. Math. **145**(1), 3–10 (2004)
5. Brandstädt, A., Mosca, R.: Maximum weight independent set for ℓclaw-free graphs in polynomial time. Discrete Appl. Math. **237**, 57–64 (2018)
6. Brause, C.: A subexponential-time algorithm for the maximum independent set problem in P_t-free graphs. Discrete Appl. Math. **231**, 113–118 (2017)
7. Courcelle, B., Makowsky, J.A., Rotics, U.: Linear time solvable optimization problems on graphs of bounded clique-width. Theory Comput. Syst. **33**(2), 125–150 (2000)
8. Edmonds, J.: Paths, trees, and flowers. Can. J. Math. **17**(3), 449–467 (1965)
9. Garey, M.R., Johnson, D.S.: The rectilinear Steiner tree problem is NP-complete. SIAM J. Appl. Math. **32**(4), 826–834 (1977)
10. Gerber, M.U., Hertz, A., Lozin, V.V.: Stable sets in two subclasses of banner-free graphs. Discrete Appl. Math. **132**(1–3), 121–136 (2003)
11. Grötschel, M., Lovász, L., Schrijver, A.: Polynomial algorithms for perfect graphs. North-Holland Math. Stud. **88**, 325–356 (1984)
12. Grzesik, A., Klimošová, T., Pilipczuk, M., Pilipczuk, M.: Polynomial-time algorithm for maximum weight independent set on P_6-free graphs. In: Proceedings of the Thirtieth Annual ACM-SIAM Symposium on Discrete Algorithms, pp. 1257–1271. SIAM (2019)
13. Hertz, A., Lozin, V.V.: The maximum independent set problem and augmenting graphs. In: Avis, D., Hertz, A., Marcotte, O. (eds.) Graph Theory and Combinatorial Optimization. GERAD 25th Anniversary Series, vol. 8, pp. 69–99. Springer, New York (2005). https://doi.org/10.1007/0-387-25592-3_4
14. Husić, E.: The maximum independent set problem and equistable graphs. Master's thesis, University of Primorska (2017)
15. Lehot, P.G.H.: An optimal algorithm to detect a line graph and output its root graph. J. ACM **21**(4), 569–575 (1974)
16. Lokshtanov, D., Vatshelle, M., Villanger, Y.: Independent set in P_5-free graphs in polynomial time. In: Proceedings of the Twenty-Fifth Annual ACM-SIAM Symposium on Discrete Algorithms, pp. 570–581. ACM, New York (2014)

17. Lozin, V.V., Milanič, M.: On finding augmenting graphs. Discrete Appl. Math. **156**(13), 2517–2529 (2008)
18. Minty, G.J.: On maximal independent sets of vertices in claw-free graphs. J. Comb. Theory Ser. B **28**(3), 284–304 (1980)
19. Mosca, R.: Polynomial algorithms for the maximum stable set problem on particular classes of P_5-free graphs. Inf. Process. Lett. **61**(3), 137–143 (1997)
20. Poljak, S.: A note on stable sets and colorings of graphs. Comment. Math. Univ. Carol. **15**(2), 307–309 (1974)
21. Roussopoulos, N.: A max $\{m, n\}$ algorithm for determining the graph H from its line graph C. Inf. Process. Lett. **2**(4), 108–112 (1973)
22. Sbihi, N.: Algorithme de recherche d'un stable de cardinalité maximum dans un graphe sans étoile. Discrete Math. **29**(1), 53–76 (1980)
23. Tarjan, R.E.: Decomposition by clique separators. Discrete Math. **55**(2), 221–232 (1985)
24. Vušković, K.: Even-hole-free graphs: a survey. Appl. Anal. Discrete Math. **4**(2), 219–240 (2010)

Independent Set Reconfiguration Parameterized by Modular-Width

Rémy Belmonte[1], Tesshu Hanaka[2], Michael Lampis[3], Hirotaka Ono[4], and Yota Otachi[5(✉)]

[1] The University of Electro-Communications, Chofu, Tokyo 182-8585, Japan
remy.belmonte@uec.ac.jp
[2] Chuo University, Bunkyo-ku, Tokyo 112-8551, Japan
hanaka.91t@g.chuo-u.ac.jp
[3] Université Paris-Dauphine, PSL University, CNRS, LAMSADE, 75016 Paris, France
michail.lampis@dauphine.fr
[4] Nagoya University, Nagoya 464-8601, Japan
ono@nagoya-u.jp
[5] Kumamoto University, Kumamoto 860-8555, Japan
otachi@cs.kumamoto-u.ac.jp

Abstract. Independent Set Reconfiguration is one of the most well-studied problems in the setting of combinatorial reconfiguration. It is known that the problem is PSPACE-complete even for graphs of bounded bandwidth. This fact rules out the tractability of parameterizations by most well-studied structural parameters as most of them generalize bandwidth. In this paper, we study the parameterization by modular-width, which is not comparable with bandwidth. We show that the problem parameterized by modular-width is fixed-parameter tractable under all previously studied rules TAR, TJ, and TS. The result under TAR resolves an open problem posed by Bonsma [WG 2014, JGT 2016].

Keywords: Reconfiguration · Independent set · Modular-width

1 Introduction

In a reconfiguration problem, we are given an instance of a search problem together with two feasible solutions. The algorithmic task there is to decide whether one solution can be transformed to the other by a sequence of prescribed local modifications while maintaining the feasibility of intermediate states. Recently, reconfiguration versions of many search problems have been studied (see [14,23]).

Partially supported by JSPS and MAEDI under the Japan-France Integrated Action Program (SAKURA) Project GRAPA 38593YJ, and by JSPS/MEXT KAKENHI Grant Numbers JP24106004, JP17H01698, JP18K11157, JP18K11168, JP18K11169, JP18H04091, JP18H06469.

I. Sau and D. M. Thilikos (Eds.): WG 2019, LNCS 11789, pp. 285–297, 2019.
https://doi.org/10.1007/978-3-030-30786-8_22

Independent Set Reconfiguration is one of the most well-studied reconfiguration problems. In this problem, we are given a graph and two independent sets. Our goal is to find a sequence of independent sets that represents a step-by-step modification from one of the given independent sets to the other. There are three local modification rules studied in the literature: Token Addition and Removal (TAR) [3,19,22], Token Jumping (TJ) [4,5,16–18], and Token Sliding (TS) [1,2,8,10,13,15,21]. Under TAR, given a threshold k, we can remove or add any vertices as long as the resultant independent set has size at least k. (When we want to specify the threshold k, we call the rule TAR(k).) TJ allows to swap one vertex in the current independent set with another vertex not dominated by the current independent set. TS is a restricted version of TJ that additionally asks the swapped vertices to be adjacent.

It is known that Independent Set Reconfiguration is PSPACE-complete under all three rules for general graphs [16], for perfect graphs [19], and for planar graphs of maximum degree 3 [13] (see [4]). For claw-free graphs, the problem is solvable in polynomial time under all three rules [4]. For even-hole-free graphs (graphs without induced cycles of even length), the problem is known to be polynomial-time solvable under TAR and TJ [19], while it is PSPACE-complete under TS even for split graphs [1]. Under TS, forests [8] and interval graphs [2] form maximal known subclasses of even-hole-free graphs for which Independent Set Reconfiguration is polynomial-time solvable. For bipartite graphs, the problem is PSPACE-complete under TS and, somewhat surprisingly, it is NP-complete under TAR and TJ [21].

Independent Set Reconfiguration is studied also in the setting of parameterized computation. (See the recent textbook [7] for basic concepts in parameterized complexity.) It is known that there is a constant b such that the problem is PSPACE-complete under all three rules even for graphs of bandwidth at most b [25]. Since bandwidth is an upper bound of well-studied structural parameters such as pathwidth, treewidth, and clique-width, this result rules out FPT (and even XP) algorithms with these parameters. Given this situation, Bonsma [3] asked whether Independent Set Reconfiguration parameterized by modular-width is tractable under TAR and TJ. The main result of this paper is to answer this question by presenting an FPT algorithm for Independent Set Reconfiguration under TAR and TJ parameterized by modular-width. We also show that under TS the problem allows a much simpler FPT algorithm.

Our results in this paper can be summarized as follows:[1]

Theorem 1.1. *Under all three rules* TAR*,* TJ*, and* TS*,* Independent Set Reconfiguration *parameterized by modular-width* mw *can be solved in time* $O^*(2^{\mathrm{mw}})$.

In Sect. 3, we give our main result for TAR (Theorem 3.9), which implies the result for TJ (Corollary 3.10). The FPT algorithm under TS is given in Sect. 4 (Theorem 4.7).

The proofs marked with ★'s are omitted due to the space limitation.

[1] The $O^*(\cdot)$ notation suppresses factors polynomial in the input size.

2 Preliminaries

Let $G = (V, E)$ be a graph. For a set of vertices $S \subseteq V$, we denote by $G[S]$ the subgraph induced by S. For a vertex set $S \subseteq V$, we denote by $G - S$ the graph $G[V \backslash S]$. For a vertex $u \in V$, we write $G - u$ instead of $G - \{u\}$. For $u, v \in S$, we denote $S \cup \{u\}$ by $S + u$ and $S \backslash \{v\}$ by $S - v$, respectively. We use $\alpha(G)$ to denote the size of a maximum independent set of G. For two sets S, R we use $S \vartriangle R$ to denote their symmetric difference, that is, the set $(S \backslash R) \cup (R \backslash S)$. For an integer k we use $[k]$ to denote the set $\{1, \ldots, k\}$. For a vertex $v \in V$, its *(open) neighborhood* is denoted by $N(v)$. The *open neighborhood* of a set $S \subseteq V$ of vertices is defined as $N(S) = \bigcup_{v \in S} N(v) \backslash S$. A *component* of G is a maximal vertex set $S \subseteq V$ such that G contains a path between each pair of vertices in S.

In the rest of this section, we are going to give definitions of the terms used in the following formalization of the main problem:

Problem: Independent Set Reconfiguration under $\mathsf{TAR}(k)$
Input: A graph G, an integer k, and independent sets S and S' of G.
Parameter: The modular-width of the input graph $\text{mw}(G)$.
Question: Does $S \leftrightsquigarrow_k S'$ hold?

2.1 TAR(k) Rule

Let S and S' be independent sets in a graph G and k an integer. Then we write $S \overset{G}{\leftrightarrow}_k S'$ if $|S \vartriangle S'| \leq 1$ and $\min\{|S|, |S'|\} \geq k$. If G is clear from the context we simply write $S \leftrightarrow_k S'$. Here $S \leftrightarrow_k S'$ means that S and S' can be reconfigured to each other in one step under the $\mathsf{TAR}(k)$ rule, which stands for "Token Addition and Removal", under the condition that no independent set contains fewer than k vertices (tokens). We write $S \overset{G}{\leftrightsquigarrow}_k S'$, or simply $S \leftrightsquigarrow_k S'$ if G is clear, if there exists $\ell \geq 0$ and a sequence of independent sets $S_0, \ldots, S_\ell$ with $S_0 = S$, $S_\ell = S'$ and for all $i \in [\ell]$ we have $S_{i-1} \leftrightarrow_k S_i$. If $S \leftrightsquigarrow_k S'$ we say that S' is *reachable* from S under the $\mathsf{TAR}(k)$ rule.

We recall the following basic facts.

Observation 2.1. *For all integers k the relation defined by $\leftrightsquigarrow_k$ is an equivalence relation on independent sets of size at least k. For any graph G, integer k, and independent sets S, R, if $S \leftrightsquigarrow_k R$, then $S \leftrightsquigarrow_{k-1} R$. For any graph G and independent sets S, R we have $S \leftrightsquigarrow_0 R$.*

2.2 TJ and TS Rules

Under the TJ rule, one step is formed by a removal of a vertex and an addition of a vertex. As this rule does not change the size of the independent set, we assume that the given initial and target independent sets are of the same size. In other words, two independent sets S and S' with $|S| = |S'|$ can be reconfigured to each other in one step under the TJ rule if $|S \vartriangle S'| = 2$. It is known that the TJ reachability can be seen as a special case of TAR reachability as follows.

Proposition 2.2 ([19]). *Let S and R be independent sets of G with $|S| = |R|$. Then, R is reachable from S under* TJ *if and only if $S \leftrightsquigarrow_{|S|-1} R$.*

One step under the TS rule is a TJ step with the additional constraint that the removed and added vertices have to be adjacent. Intuitively, one step in a TS sequence "slides" a token along an edge. We postpone the introduction of notation for TS until Sect. 4 to avoid any confusions.

2.3 Modular-Width

In a graph $G = (V, E)$ a module is a set of vertices $M \subseteq V$ with the property that for all $u, v \in M$ and $w \in V \backslash M$, if $\{u, w\} \in E$, then $\{v, w\} \in E$. In other words, a module is a set of vertices that have the same neighbors outside the module. A graph $G = (V, E)$ has *modular-width* at most k if it satisfies at least one of the following conditions (i) $|V| \leq k$, or (ii) there exists a partition of V into at most k sets $V_1, V_2, \ldots, V_s$, such that $G[V_i]$ has modular-width at most k and V_i is a module in G, for all $i \in [s]$. We will use $\mathrm{mw}(G)$ to denote the minimum k for which G has modular-width at most k. We recall that there is a polynomial-time algorithm which, given a graph $G = (V, E)$ produces a non-trivial partition of V into at most $\mathrm{mw}(G)$ modules [6,12,24] and that deleting vertices from G can only decrease the modular-width. We also recall that Maximum Independent Set is solvable in time $O^*(2^{\mathrm{mw}})$. Indeed, a faster algorithm with running time $O^*(1.7347^{\mathrm{mw}})$ is known [9].

A graph has neighborhood diversity at most k if its vertex set can be partitioned into k modules, such that each module induces either a clique or an independent set. We use $\mathrm{nd}(G)$ to denote the minimum neighborhood diversity of G, and recall that $\mathrm{nd}(G)$ can be computed in polynomial time [20] and that $\mathrm{nd}(G) \geq \mathrm{mw}(G)$ for all graphs G [11].

It can be seen that the modular-width of a graph is not smaller than its clique-width. On the other hand, we can see that treewidth, pathwidth, and bandwidth are not comparable to modular-width. To see this, observe that the complete graph of n vertices has treewidth $n-1$ and modular-width 2, and that the path of n vertices has treewidth 1 and modular-width n for $n \geq 4$. Our positive result and the hardness result by Wrochna [25] together give Fig. 1 that depicts a map of structural graph parameters with a separation of the complexity of Independent Set Reconfiguration.

3 FPT Algorithm for Modular-Width Under TAR

In this section we present an FPT algorithm for the TAR(k)-reachability problem parameterized by modular-width. The main technical ingredient of our algorithm is a sub-routine which solves a related problem: given a graph G, an independent set S, and an integer k, what is the largest size of an independent set reachable from S under TAR(k)? This sub-routine relies on dynamic programming: we present (in Lemma 3.4) an algorithm which answers this "maximum extensibility" question, if we are given tables with answers for the same question for all

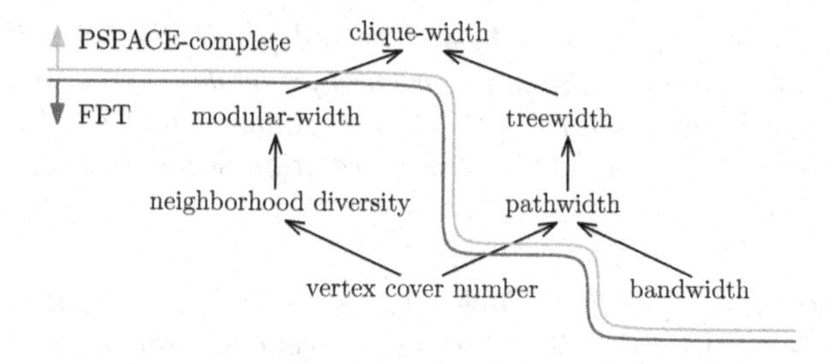

Fig. 1. The complexity of INDEPENDENT SET RECONFIGURATION under TAR, TJ, and TS parameterized by structural graph parameters. "$X \to Y$" implies that there is a function f such that $X(G) \geq f(Y(G))$ for every graph G.

the modules in a non-trivial partition of the input graph. This results in an algorithm (Theorem 3.5) that solves this problem on graphs of small modular-width, which we then put to use in Sect. 3.2 to solve the reconfiguration problem.

3.1 Computing a Largest Reachable Set

In this section we present an FPT algorithm (parameterized by modular-width) which computes the following value:

Definition 3.1. Given a graph G, an independent set S, and an integer k, we define $\lambda(G, S, k)$ as the largest size of the independent sets S' such that $S \leftrightsquigarrow_k S'$.

In particular, we will present a *constructive* algorithm which, given G, S, k will return an independent set S' such that $|S'| = \lambda(G, S, k)$, as well as a reconfiguration sequence proving that $S \leftrightsquigarrow_k S'$.

We begin by tackling an easier case: the case when the parameter is the neighborhood diversity.

Lemma 3.2 (★). *There is an algorithm which, given a graph G, an independent set S, and an integer k, returns an independent set S', with $|S'| = \lambda(G, S, k)$, and a reconfiguration sequence proving that $S \leftrightsquigarrow_k S'$, in time $O^*(2^{\text{nd}(G)})$.*

Before presenting the main algorithm of this section, let us also make a useful observation: once we are able to reach a configuration that contains a sufficiently large number of vertices from a module, we can safely delete a vertex from the module (bringing us closer to the case where Lemma 3.2 will apply).

Lemma 3.3 (★). *Let G be a graph, S be an independent set of G, k an integer, and M a module of G. Suppose there exists an independent set $A \subset M$ such that $(S \cap M) \subseteq A$ and $|A| = \alpha(G[M])$. Then, for all $u \in M \backslash A$ we have $\lambda(G, S, k) = \lambda(G - u, S, k)$.*

We are now ready to present our main dynamic programming procedure.

Lemma 3.4 (★). *Suppose we are given the following input:*

1. A graph $G = (V, E)$, an integer k, and an independent set S with $|S| \geq k$.
2. A partition of V into $r \leq \mathrm{mw}(G)$ non-empty modules, $V_1, \ldots, V_r$.
3. For each $i \in [r]$, for each $j \in [|S \cap V_i|]$ an independent set $R_{i,j}$, such that $|R_{i,j}| = \lambda(G[V_i], S \cap V_i, j)$, and a transformation sequence proving that $(S \cap V_i) \overset{G[V_i]}{\leftrightsquigarrow_j} R_{i,j}$.

Then, there exists an algorithm which returns an independent set R of G, such that $|R| = \lambda(G, S, k)$, and a transformation sequence proving that $S \leftrightsquigarrow_k R$, running in time $O^*(2^{\mathrm{mw}(G)})$.

We thus arrive to the main theorem of this section.

Theorem 3.5. *There exists an algorithm which, given a graph G, an independent set S, and an integer k, runs in time $O^*(2^{\mathrm{mw}(G)})$ and outputs an independent set S' such that $|S'| = \lambda(G, S, k)$ and a $\mathsf{TAR}(k)$ transformation $S \leftrightsquigarrow_k S'$.*

Proof. We perform dynamic programming using Lemma 3.4. More precisely, our goal is, given G and S, to produce for each value of $j \in [|S|]$ an independent set R_j such that $S \leftrightsquigarrow_j R_j$ and $|R_j| = \lambda(G, S, j)$. Clearly, if we can solve this more general problem in time $O^*(2^{\mathrm{mw}(G)})$ we are done.

Our algorithm works as follows: first, it computes a modular decomposition of G of minimum width, which can be done in time at most $O(n^2)$ [12]. If $|V(G)| \leq \mathrm{mw}(G)$ then the problem can be solved in $O^*(2^{\mathrm{mw}(G)})$ by brute force (enumerating all independent sets of G), or even by Lemma 3.2. We therefore assume that G has a non-trivial partition into $r \leq \mathrm{mw}(G)$ modules $V_1, \ldots, V_r$. We call our algorithm recursively for each $G[V_i]$, and obtain for each $i \in [r]$ and $j \in [|S \cap V_i|]$ a set $R_{i,j}$ such that $|R_{i,j}| = \lambda(G[V_i], S \cap V_i, j)$ and a transformation $(S \cap V_i) \overset{G[V_i]}{\leftrightsquigarrow_j} R_{i,j}$. We use this input to invoke the algorithm of Lemma 3.4 for each value of $j \in [|S|]$. This allows us to produce the sets R_j and the corresponding transformations.

Suppose that $\beta \geq 2$ is a constant such that the algorithm of Lemma 3.4 runs in time at most $O(2^{\mathrm{mw}(G)} n^{\beta})$. Our algorithm runs in time at most $O(2^{\mathrm{mw}(G)} n^{\beta+2})$. This can be seen by considering the tree representing a modular decomposition of G. In each node of the tree (that represents a module of G) our algorithm makes at most n calls to the algorithm of Lemma 3.4. Since the modular decomposition has at most $O(n)$ nodes, the running time bound follows. □

3.2 Reachability

In this section we will apply the algorithm of Theorem 3.5 to obtain an FPT algorithm for the $\mathsf{TAR}(k)$ reconfiguration problem parameterized by modular-width. The main ideas we will need are that (i) using the algorithm of Theorem 3.5 we can decide if it is possible to arrive at a configuration where a module is empty of tokens (Lemma 3.6) (ii) if a module is empty in both the initial and target configurations, we can replace it by an independent set (Lemma 3.7) and (iii) the reconfiguration problem is easy on graphs with small neighborhood diversity (Lemma 3.8). Putting these ideas together we obtain an algorithm which can always identify an irrelevant vertex which we can delete if the input graph is connected. If the graph is disconnected, we can use ideas similar to those of [3] to reduce the problem to appropriate sub-instances in each component.

Lemma 3.6. *There is an algorithm which, given a graph G, an independent set S, a module M of G, and an integer k, runs in time $O^*(2^{\text{mw}(G)})$ and either returns a set S' with $S' \cap M = \emptyset$ and $S \leftrightsquigarrow_k S'$ or correctly concludes that no such set exists.*

Proof. We assume that $S \cap M \neq \emptyset$ (otherwise we simply return S).

Let H be the graph obtained by deleting from G all vertices of $V \setminus M$ that have a neighbor in M. We invoke the algorithm of Theorem 3.5 to compute a set R in H such that $S \overset{H}{\leftrightsquigarrow}_k R$ and $|R| = \lambda(H, S, k)$. If $|R \setminus M| \geq k$ then we return as solution the set $R \setminus M$, and as transformation the transformation sequence returned by the algorithm, to which we append moves that delete all vertices of $R \cap M$. If $|R \setminus M| < k$ we answer that no such set exists.

Let us now argue for correctness. If the algorithm returns a set $S' := R \setminus M$, it also returns a $\mathsf{TAR}(k)$ transformation from S to S' in H; this is also a transformation in G, and since $S' \cap M = \emptyset$, the solution is correct.

Suppose then that the algorithm returns that no solution exists, but for the sake of contradiction there exists a T with $S \overset{G}{\leftrightsquigarrow}_k T$ and $T \cap M = \emptyset$. Among all such sets T select the one at minimum reconfiguration distance from S and let $S_0 = S, S_1, \ldots, S_\ell = T$ be a shortest reconfiguration sequence. We claim that this is also a valid reconfiguration sequence in H. Indeed, for all $j \in [\ell - 1]$, the set S_j contains a vertex from M (otherwise we would have a shorter sequence), therefore may not contain any deleted vertex. As a result, if a solution T exists, then $S \overset{H}{\leftrightsquigarrow}_k T$. Let A be a maximum independent set of $G[M]$. We observe that (i) $|T \setminus M| \geq k$ since T is reachable with $\mathsf{TAR}(k)$ moves and $T \cap M = \emptyset$ (ii) $T \overset{H}{\leftrightsquigarrow}_k (T \cup A)$. However, this gives a contradiction, because we now have $S \overset{H}{\leftrightsquigarrow}_k (T \cup A)$ and this set is strictly larger than the set returned by the algorithm of Theorem 3.5 when computing $\lambda(H, S, k)$. □

Lemma 3.7. *Let G be a graph, k an integer, M a module of G, and S, T two independent sets of G such that $S \cap M = T \cap M = \emptyset$. Let A be a maximum independent set of $G[M]$. Then, for all $u \in M \setminus A$ we have $S \overset{G}{\leftrightsquigarrow}_k T$ if and only if $S \overset{G-u}{\leftrightsquigarrow}_k T$.*

Proof. The proof is similar to that of Lemma 3.3. Specifically, since $u \notin S$ and $u \notin T$, it is easy to see that $S \overset{G-u}{\leftrightsquigarrow}_k T$ implies $S \overset{G}{\leftrightsquigarrow}_k T$. Suppose then that $S \overset{G}{\leftrightsquigarrow}_k T$ and we have a sequence $S_0 = S, S_1, \ldots, S_\ell = T$. We construct a sequence $S'_0 = S, S'_1, \ldots, S'_\ell$ such that for all $i \in [\ell]$ we have $|S_i| = |S'_i|$, $S_i \setminus M = S'_i \setminus M$, and $S'_i \cap M \subseteq A$. This can be done inductively: for S'_0 the desired properties hold; and for all $i \in [\ell]$ we can prove that if the properties hold for S'_{i-1}, then we can construct S_i' in the same way as in the proof of Lemma 3.3 (namely, we perform the same moves as S_i outside of M, and pick an arbitrary vertex of A when S_i adds a vertex of M). □

Lemma 3.8. *There is an algorithm which, given a graph G, an integer k, and two independent sets S, T, decides if $S \leftrightsquigarrow_k T$ in time $O^*(2^{\mathrm{nd}(G)})$.*

Proof. The proof is similar to that of Lemma 3.2, but we need to carefully handle some corner cases. We are given a partition of G into $r \leq \mathrm{nd}(G)$ sets $V_1, \ldots, V_r$, such that each V_i induces a clique or an independent set. Suppose V_i induces a clique. We use the algorithm of Lemma 3.6 with input (G,S,V_i,k) and with input (G,T,V_i,k) to decide if it is possible to empty V_i of tokens. If the algorithm gives different answers we immediately reject, since there is a configuration that is reachable from S but not from T. If the algorithm returns S', T' with $S' \cap V_i = T' \cap V_i = \emptyset$, then the problem reduces to deciding if $S' \leftrightsquigarrow_k T'$. However, by Lemma 3.7 we can delete all the vertices of V_i except one and this does not change the answer. Finally, if the algorithm responds that V_i cannot be empty in any configuration reachable from S or T then, if $S \cap V_i \neq T \cap V_i$ we immediately reject, while if $S \cap V_i = T \cap V_i$ we delete from the input V_i and all its neighbors and solve the reconfiguration problem in the instance $(G[V \setminus N[V_i]], k-1, S \setminus V_i, T \setminus V_i)$.

After this preprocessing all sets V_i are independent. We now construct an auxiliary graph G' as in Lemma 3.2, namely, our graph has a vertex for every independent set S of G with $|S| \geq k$ such that for all $i \in [r]$ either $S \cap V_i = \emptyset$ or $V_i \subseteq S$. Again, we have an edge between S_1, S_2 if $S_1 \triangle S_2 = V_i$ for some $i \in [r]$. We can assume without loss of generality that S, T are represented in this graph (if there exists V_i such that $0 < |S \cap V_i| < |V_i|$ we add to S all remaining vertices of V_i). Now, $S \leftrightsquigarrow T$ if and only if S is reachable from T in G', and this can be checked in time linear in the size of G'. □

Theorem 3.9 (TAR). *There is an algorithm which, given a graph G, an integer k, and two independent sets S, T, decides if $S \leftrightsquigarrow_k T$ in time $O^*(2^{\mathrm{mw}(G)})$.*

Proof. Our algorithm considers two cases: if G is connected we will attempt to simplify G in a way that eventually produces either a graph with small neighborhood diversity or a disconnected graph; if G is disconnected we will recursively solve an appropriate subproblem in each component.

First, suppose that G is connected. We compute a modular decomposition of G which gives us a partition of V into $r \leq \mathrm{mw}(G)$ modules $V_1, \ldots, V_r$. We may assume that $r \geq 2$ since otherwise G has at most $\mathrm{mw}(G)$ vertices and the claimed running time is trivial in that case. If for all $i \in [r]$ we have that $G[V_i]$ is an independent set, then $\mathrm{nd}(G) \leq r$ and we invoke the algorithm of Lemma 3.8. Suppose then that for some $i \in [r]$, $G[V_i]$ contains at least one edge. We invoke the algorithm of Lemma 3.6 on input (G, S, V_i, k) and on input (G, T, V_i, k). If the answers returned are different, we decide that S is not reachable from T in G, because from one set we can reach a configuration that contains no vertex of V_i and from the other we cannot.

If the algorithm of Lemma 3.6 returned to us two sets S', T' with $S' \cap V_i = T' \cap V_i = \emptyset$ then by transitivity we know $S \leftrightsquigarrow T$ if and only if $S' \leftrightsquigarrow T'$. We compute a maximum independent set A of $G[V_i]$ and delete from our graph a vertex $u \in V_i \setminus A$. Such a vertex exists, since $G[V_i]$ is not an independent set. By Lemma 3.7 deleting u does not affect whether $S' \leftrightsquigarrow T'$, so we call our algorithm with input $(G - u, k, S', T')$, and return its response.

On the other hand, if the algorithm of Lemma 3.6 concluded that no set reachable from either S or T has empty intersection with V_i, we find a vertex $u \in V \setminus V_i$ that has a neighbor in V_i and delete it, that is, we call our algorithm with input $(G - u, k, S, T)$. Such a vertex u exists because G is connected. This recursive call is correct because any configuration reachable from S or T contains some vertex of V_i, which is a neighbor of u, so no reachable configuration uses u.

We note that if G is connected, all the cases described above will make a single recursive call on an input that has strictly fewer vertices.

Suppose now that G is not connected and there are s connected components $C_1, C_2, \ldots, C_s$. We will assume that $|S| = \lambda(G, S, k)$ and $|T| = \lambda(G, T, k)$. This is without loss of generality, since we can invoke the algorithm of Theorem 3.5 and in case $|S| < \lambda(G, S, k)$ replace S with the set S' returned by the algorithm while keeping an equivalent instance (similarly for T).

As a result, we can assume that $|S| = |T|$, otherwise the answer is trivially no. More strongly, if there exists a component C_i such that $|S \cap C_i| \neq |T \cap C_i|$ we answer no. To see that this is correct, we argue that for all S' such that $S \leftrightsquigarrow_k S'$ we have $|S' \cap C_i| \leq |S \cap C_i|$. Indeed, suppose there exists S' such that for some $i \in [s]$ we have $|S' \cap C_i| > |S \cap C_i|$ and $S \leftrightsquigarrow S'$. Among such configurations S' select one that is at minimum reconfiguration distance from S and let $S_0 = S, S_1, \ldots, S_\ell = S'$ be a shortest reconfiguration from S to S'. Then for all $j \in [\ell]$ we have $|S \setminus C_i| \geq |S_j \setminus C_i|$ (otherwise we would have an S' that is at shorter reconfiguration distance from S). This means that the sequence $S_0 \cap C_i, S_1 \cap C_i, \ldots, S_\ell \cap C_i$ is a $\mathsf{TAR}(k - |S \setminus C_i|)$ transformation of $S \cap C_i$ to $S' \cap C_i$ in $G[C_i]$. But this transformation proves that the set $(S \setminus C_i) \cup (S' \cap C_i)$ is $\mathsf{TAR}(k)$ reachable from S in G, and since this set is larger than S we have a contradiction.

For each $i \in [s]$ we now consider the reconfiguration instance given by the following input: $(G[C_i], k - |S \setminus C_i|, S \cap C_i, T \cap C_i)$. We call our algorithm recursively for each such instance. If the answer is yes for all these instances we reply that S is reachable from T, otherwise we reply that the sets are not reachable.

To argue for correctness we use induction on the depth of the recursion. Suppose that the algorithm correctly concludes that the answer to all sub-instances is yes. Then, there does indeed exist a transformation $S \leftrightsquigarrow T$ as follows: starting from S, for each $i \in [s]$ we keep $S \setminus C_i$ constant and perform in $G[C_i]$ the transformation $(S \cap C_i) \overset{G[C_i]}{\leftrightsquigarrow}_{k - |S \setminus C_i|} (T \cap C_i)$. At each step this gives a configuration where S and T agree in more components. Furthermore, since $|S \cap C_i| = |T \cap C_i|$ for all $i \in [s]$, this is a valid $\mathsf{TAR}(k)$ reconfiguration.

Suppose now that the answer is no for the instance $(G[C_i], k - |S \backslash C_i|, S \cap C_i, T \cap C_i)$. Suppose also, for the sake of contradiction, that there exists a $\mathsf{TAR}(k)$ reconfiguration $S_0 = S, S_1, \dots, S_\ell = T$. As argued above, any configuration S' reachable from S has $|S' \cap C_{i'}| \le |S \cap C_{i'}|$ for all $i' \in [s]$. This means that $|S \backslash C_i| \ge |S_j \backslash C_i|$ for all $j \in [\ell]$. Hence, the sequence $S_0 \cap C_i, S_1 \cap C_i, \dots, S_\ell \cap C_i$ gives a valid $\mathsf{TAR}(k - |S \backslash C_i|)$ reconfiguration in $G[C_i]$, which is a contradiction.

Finally, it is not hard to see that the algorithm runs in time $O^*(2^{\mathrm{mw}(G)})$, because in the case of disconnected graphs we make a single recursive call for each component. □

Theorem 3.9 and Proposition 2.2 give an FPT algorithm with the same running time for TJ.

Corollary 3.10 (TJ). *There is an algorithm which, given a graph G and two independent sets S, T, decides the* TJ *reachability between S and T in time $O^*(2^{\mathrm{mw}(G)})$.*

4 FPT Algorithm for Modular-Width Under TS

We now present an FPT algorithm deciding the TS-reachability parameterized by modular-width. The problem under TS is much easier than the one under TAR since we can reduce the problem to a number of constant-size instances that can be considered separately. To see this, we first observe that the components can be considered separately. We then further observe that we only need to solve the case where each maximal nontrivial module contains at most one vertex of the current independent set. Finally, we show that the reachability problem on the reduced case thus far is equivalent to a generalized reachability problem on a graph of order at most $\mathrm{mw}(G)$, where G is the original graph.

Let S and S' be independent sets of G with $|S| = |S'|$. Recall that S and S' can be reached by one step under TS if $|S \triangle S'| = 2$ and the two vertices in $S \triangle S'$ are adjacent. We denote this relation by $S \overset{G}{\leftrightarrow} S'$, or simply by $S \leftrightarrow S'$ if G is clear from the context. We write $S \overset{G}{\leftrightsquigarrow} S'$ (or simply $S \leftrightsquigarrow S'$) if there exists $\ell \ge 0$ and a sequence of independent sets $S_0, \dots, S_\ell$ with $S_0 = S$, $S_\ell = S'$ and for all $i \in [\ell]$ we have $S_{i-1} \leftrightarrow S_i$. If $S \leftrightsquigarrow S'$ we say that S' is *reachable* from S under the TS rule. Observe that the relation defined by $\leftrightsquigarrow$ is an equivalence relation on independent sets.

The first easy observation is that the TS rule cannot move a token to a different component since a TS step is always along an edge (and thus within a component). This is formalized as follows.

Observation 4.1. *Let G be a graph, S, S' independent sets of G, and $C_1, \dots, C_c$ the components of G. Then, $S \overset{G}{\leftrightsquigarrow} S'$ if and only if $(S \cap V(C_i)) \overset{G[V(C_i)]}{\leftrightsquigarrow} (S' \cap V(C_i))$ for all $i \in [c]$.*

The next lemma, which is still an easy one, is a key tool in our algorithm.

Lemma 4.2 (★). *Let G be a graph, M a module of G, and S an independent set in G such that $|S \cap M| \geq 2$. Then, for every independent set S' in G, $S \overset{G}{\leftrightsquigarrow} S'$ if and only if $S' \cap N(M) = \emptyset$ and $S \overset{G-N(M)}{\leftrightsquigarrow} S'$.*

Lemma 4.2 implies that S with $|S \cap M| \geq 2$ and S' with $|S' \cap M| \leq 1$ are not reachable to each other. This fact in the following form will be useful later.

Corollary 4.3. *Let G be a graph, M a module of G, and S an independent set in G such that $|S \cap M| \leq 1$. Then, for every independent set S' in G such that $S \leftrightsquigarrow S'$, it holds that $|S' \cap M| \leq 1$.*

We now show that a module sharing at most one vertex with both initial and target independent sets can be replaced with a single vertex, under an assumption that we may solve a slightly generalized reachability problem (which is still trivial on a graph of constant size).

Lemma 4.4 (★). *Let G be a graph, M a module of G with $|M| \geq 2$, and S, S' independent sets of G with $|S| = |S'|$. If $|M \cap (S \cup S')| \leq 1$, then $S \overset{G}{\leftrightsquigarrow} S'$ if and only if $S \overset{G-v}{\leftrightsquigarrow} S'$ for every $v \in M \setminus (S \cup S')$.*

Lemma 4.5 (★). *Let G be a graph, M a module of G, and S, S' independent sets of G with $|S| = |S'|$. If $M \cap S = \{u\}$, $M \cap S' = \{v\}$, $u \neq v$, and u and v are in the same component of $G[M]$, then $S \overset{G}{\leftrightsquigarrow} S'$ if and only if $S \overset{G-v}{\leftrightsquigarrow} S' - v + u$.*

Lemma 4.6 (★). *Let G be a graph, M a module of G, and S, S' independent sets of G with $|S| = |S'|$. If $M \cap S = \{u\}$, $M \cap S' = \{v\}$, $u \neq v$, and u and v are in different components of $G[M]$, then $S \overset{G}{\leftrightsquigarrow} S'$ if and only if $S \overset{G-v}{\leftrightsquigarrow} S' - v + u$ and there is an independent set T in $G - v$ such that $T \cap M = \emptyset$ and $S \overset{G-v}{\leftrightsquigarrow} T$.*

Now we are ready to present our algorithm for the TS-reachability problem.

Theorem 4.7 (TS, ★). *There is an algorithm which, given a graph G and two independent sets S, T, decides if $S \leftrightsquigarrow T$ in time $O^*(2^{\mathrm{mw}(G)})$.*

References

1. Belmonte, R., Kim, E.J., Lampis, M., Mitsou, V., Otachi, Y., Sikora, F.: Token sliding on split graphs. In: STACS. LIPIcs, vol. 126, pp. 13:1–13:17. Schloss Dagstuhl - Leibniz-Zentrum fuer Informatik (2019)
2. Bonamy, M., Bousquet, N.: Token sliding on chordal graphs. In: Bodlaender, H.L., Woeginger, G.J. (eds.) WG 2017. LNCS, vol. 10520, pp. 127–139. Springer, Cham (2017). https://doi.org/10.1007/978-3-319-68705-6_10
3. Bonsma, P.S.: Independent set reconfiguration in cographs and their generalizations. J. Graph Theory **83**(2), 164–195 (2016)
4. Bonsma, P., Kamiński, M., Wrochna, M.: Reconfiguring independent sets in claw-free graphs. In: Ravi, R., Gørtz, I.L. (eds.) SWAT 2014. LNCS, vol. 8503, pp. 86–97. Springer, Cham (2014). https://doi.org/10.1007/978-3-319-08404-6_8

5. Bousquet, N., Mary, A., Parreau, A.: Token jumping in minor-closed classes. In: Klasing, R., Zeitoun, M. (eds.) FCT 2017. LNCS, vol. 10472, pp. 136–149. Springer, Heidelberg (2017). https://doi.org/10.1007/978-3-662-55751-8_12
6. Cournier, A., Habib, M.: A new linear algorithm for modular decomposition. In: Tison, S. (ed.) CAAP 1994. LNCS, vol. 787, pp. 68–84. Springer, Heidelberg (1994). https://doi.org/10.1007/BFb0017474
7. Cygan, M., et al.: Parameterized Algorithms. Springer, Cham (2015). https://doi.org/10.1007/978-3-319-21275-3
8. Demaine, E.D., et al.: Linear-time algorithm for sliding tokens on trees. Theor. Comput. Sci. **600**, 132–142 (2015)
9. Fomin, F.V., Liedloff, M., Montealegre, P., Todinca, I.: Algorithms parameterized by vertex cover and modular width, through potential maximal cliques. Algorithmica **80**(4), 1146–1169 (2018)
10. Fox-Epstein, E., Hoang, D.A., Otachi, Y., Uehara, R.: Sliding token on bipartite permutation graphs. In: Elbassioni, K., Makino, K. (eds.) ISAAC 2015. LNCS, vol. 9472, pp. 237–247. Springer, Heidelberg (2015). https://doi.org/10.1007/978-3-662-48971-0_21
11. Gajarský, J., Lampis, M., Ordyniak, S.: Parameterized algorithms for modular-width. In: Gutin, G., Szeider, S. (eds.) IPEC 2013. LNCS, vol. 8246, pp. 163–176. Springer, Cham (2013). https://doi.org/10.1007/978-3-319-03898-8_15
12. Habib, M., Paul, C.: A survey of the algorithmic aspects of modular decomposition. Comput. Sci. Rev. **4**(1), 41–59 (2010)
13. Hearn, R.A., Demaine, E.D.: PSPACE-completeness of sliding-block puzzles and other problems through the nondeterministic constraint logic model of computation. Theor. Comput. Sci. **343**(1–2), 72–96 (2005)
14. van den Heuvel, J.: The complexity of change. In: Blackburn, S.R., Gerke, S., Wildon, M. (eds.) Surveys in Combinatorics 2013. London Mathematical Society Lecture Note Series, vol. 409, pp. 127–160. Cambridge University Press, Cambridge (2013)
15. Hoang, D.A., Uehara, R.: Sliding tokens on a cactus. In: ISAAC. LIPIcs, vol. 64, pp. 37:1–37:26. Schloss Dagstuhl - Leibniz-Zentrum fuer Informatik (2016)
16. Ito, T., et al.: On the complexity of reconfiguration problems. Theor. Comput. Sci. **412**(12–14), 1054–1065 (2011)
17. Ito, T., Kamiński, M., Ono, H., Suzuki, A., Uehara, R., Yamanaka, K.: On the parameterized complexity for token jumping on graphs. In: Gopal, T.V., Agrawal, M., Li, A., Cooper, S.B. (eds.) TAMC 2014. LNCS, vol. 8402, pp. 341–351. Springer, Cham (2014). https://doi.org/10.1007/978-3-319-06089-7_24
18. Ito, T., Kamiński, M., Ono, H.: Fixed-parameter tractability of token jumping on planar graphs. In: Ahn, H.-K., Shin, C.-S. (eds.) ISAAC 2014. LNCS, vol. 8889, pp. 208–219. Springer, Cham (2014). https://doi.org/10.1007/978-3-319-13075-0_17
19. Kamiński, M., Medvedev, P., Milanič, M.: Complexity of independent set reconfigurability problems. Theor. Comput. Sci. **439**, 9–15 (2012)
20. Lampis, M.: Algorithmic meta-theorems for restrictions of treewidth. Algorithmica **64**(1), 19–37 (2012)
21. Lokshtanov, D., Mouawad, A.E.: The complexity of independent set reconfiguration on bipartite graphs. In: SODA 2018, pp. 185–195 (2018)
22. Mouawad, A.E., Nishimura, N., Raman, V., Simjour, N., Suzuki, A.: On the parameterized complexity of reconfiguration problems. Algorithmica **78**(1), 274–297 (2017)
23. Nishimura, N.: Introduction to reconfiguration. Algorithms **11**(4), 52 (2018)

24. Tedder, M., Corneil, D., Habib, M., Paul, C.: Simpler linear-time modular decomposition via recursive factorizing permutations. In: Aceto, L., Damgård, I., Goldberg, L.A., Halldórsson, M.M., Ingólfsdóttir, A., Walukiewicz, I. (eds.) ICALP 2008. LNCS, vol. 5125, pp. 634–645. Springer, Heidelberg (2008). https://doi.org/10.1007/978-3-540-70575-8_52
25. Wrochna, M.: Reconfiguration in bounded bandwidth and tree-depth. J. Comput. Syst. Sci. **93**, 1–10 (2018)

Counting Independent Sets in Graphs with Bounded Bipartite Pathwidth

Martin Dyer[1], Catherine Greenhill[2], and Haiko Müller[1](✉)

[1] School of Computing, University of Leeds, Leeds LS2 9JT, UK
{M.E.Dyer,H.Muller}@leeds.ac.uk
[2] School of Mathematics and Statistics, UNSW, Sydney 2052, Australia
C.Greenhill@unsw.edu.au

Abstract. The Glauber dynamics can efficiently sample independent sets almost uniformly at random in polynomial time for graphs in a certain class. The class is determined by boundedness of a new graph parameter called bipartite pathwidth. This result, which we prove for the more general hardcore distribution with fugacity λ, can be viewed as a strong generalisation of Jerrum and Sinclair's work on approximately counting matchings. The class of graphs with bounded bipartite pathwidth includes line graphs and claw-free graphs, which generalise line graphs. We consider two further generalisations of claw-free graphs and prove that these classes have bounded bipartite pathwidth.

Keywords: Markov chain Monte Carlo algorithm · Fully polynomial-time randomized approximation scheme · Independent set · Pathwidth

1 Introduction

We will show that we can approximate the number of independent sets in graphs for which all bipartite induced subgraphs are well structured, in a sense that we will define precisely. Our approach is to generalise the Markov chain analysis of Jerrum and Sinclair [19] for the corresponding problem of counting matchings. Their canonical path argument relied on the fact that the symmetric difference of two matchings of a given graph G is a bipartite subgraph of G consisting of a disjoint union of paths and even-length cycles. We introduce a new graph parameter, which we call bipartite pathwidth, to enable us to give the strongest generalisation of the approach of [19].

1.1 Independent Set Problems

For a given graph G, let $\mathcal{I}(G)$ be the set of all independent sets in G. The *independence number* $\alpha(G) = \max\{|I| : I \in \mathcal{I}(G)\}$ is the size of the largest

M. Dyer and H. Müller—Research supported by EPSRC grant EP/S016562/1.
C. Greenhill—Research supported by Australian Research Council grant DP19010097.

I. Sau and D. M. Thilikos (Eds.): WG 2019, LNCS 11789, pp. 298–310, 2019.
https://doi.org/10.1007/978-3-030-30786-8_23

independent set in G. The problem of finding $\alpha(G)$ is NP-hard in general, even in various restricted cases, such as degree-bounded graphs. However, polynomial time algorithms have been constructed for finding a maximum independent set, for various graph classes. The most important case has been *matchings*, which are independent sets in the *line graph* $L(G)$ of G. This has been generalised to larger classes of graphs, for example *claw-free* graphs [24], which include line graphs [4], and *fork-free* graphs [1], which include claw-free graphs.

Counting independent sets in graphs is known to be #P-complete in general [26], and in various restricted cases [15,30]. Exact counting is known only for some restricted graph classes. Even approximate counting is NP-hard in general, and is unlikely to be in polynomial time for bipartite graphs [11].

For some classes of graphs, for example line graphs, approximate counting is known to be possible [19,20]. The most successful Markov chain approach relies on a close correspondence between approximate counting and sampling uniformly at random [21]. It was applied to degree-bounded graphs in [23] and [12]. In his PhD thesis [22], Matthews used a Markov chain for sampling independent sets in *claw-free* graphs. His chain, and its analysis, generalises that of [19].

Several other approaches to approximate counting have been successfully applied to the independent set problem. Weitz [31] used the *correlation decay* approach on degree-bounded graphs, resulting in an FPTAS for counting independent sets in graphs with degree at most 5. Sly [29] gave a matching NP-hardness result. The correlation decay method was also applied to matchings in [3], and was extended to complex values of λ in [16]. Recently, Efthymiou et al. [14] proved that the Markov chain approach can (almost) produce the best results obtainable by other methods.

The *independence polynomial* $P_G(\lambda)$ of a graph G is defined in (1) below. The *Taylor series* approach of Barvinok [2] was used by Patel and Regts [25] to give a FPTAS for $P_G(\lambda)$ in degree-bounded claw-free graphs. The success of the method depends on the location of the roots of the independence polynomial. Chudnovsky and Seymour [7] proved that all these roots are real, and hence they are all negative. Then the algorithm of [25] is valid for all complex λ which are not real and negative. In this extended abstract (for proofs see [13]), we return to the Markov chain approach.

1.2 Preliminaries

We write $[m] = \{1, 2, \ldots, m\}$ for any positive integer m, and let $A \oplus B$ denote the symmetric difference of sets A, B. For graph theoretic definitions not given here, see [10]. Throughout this paper, all graphs are simple and undirected. $G[S]$ denotes the subgraph of G induced by the set S and $N(v)$ denotes the neighbourhood of vertex v. Given a graph $G = (V, E)$, let $\mathcal{I}_k(G)$ be the set of independent sets of G of size k. The *independence polynomial* of G is the *partition function*

$$P_G(\lambda) = \sum_{I \in \mathcal{I}(G)} \lambda^{|I|} = \sum_{k=0}^{\alpha(G)} N_k \lambda^k, \tag{1}$$

where $N_k = |\mathcal{I}_k(G)|$ for $k = 0, \ldots, \alpha$. Here $\lambda \in \mathbb{C}$ is called the *fugacity*. We consider only real λ and assume $\lambda \geq 1/n$ to avoid trivialities. We have $N_0 = 1$, $N_1 = n$ and $N_k \leq \binom{n}{k}$ for $k = 2, \ldots, n$. Thus it follows that for any $\lambda \geq 0$,

$$1 + n\lambda \leq P_G(\lambda) \leq \sum_{k=0}^{\alpha(G)} \binom{n}{k} \lambda^k \leq (1 + \lambda)^n. \tag{2}$$

Note also that $P_G(0) = 1$ and $P_G(1) = |\mathcal{I}(G)|$.

An *almost uniform sampler* for a probability distribution π on a state Ω is a randomised algorithm which takes as input a real number $\delta > 0$ and outputs a sample from a distribution μ such that the *total variation distance* $\frac{1}{2}\sum_{x\in\Omega} |\mu(x) - \pi(x)|$ is at most δ. The sampler is a *fully polynomial almost uniform sampler (FPAUS)* if its running time is polynomial in the input size n and $\log(1/\delta)$. The word "uniform" here is historical, as it was first used in the case where π is the uniform distribution. We use it in a more general setting.

If $w : \Omega \to \mathbb{R}$ is a *weight function*, then the *Gibbs distribution* π satisfies $\pi(x) = w(x)/W$ for all $x \in \Omega$, where $W = \sum_{x\in\Omega} w(x)$. If $w(x) = 1$ for all $x \in \Omega$ then π is uniform. For independent sets with $w(I) = \lambda^{|I|}$, we have

$$\pi(I) = \lambda^{|I|}/P_G(\lambda), \tag{3}$$

and is often called the hardcore distribution. Jerrum, Valiant and Vazirani [21] showed that approximating W is equivalent to the existence of an FPAUS for π, provided the problem is *self-reducible*. Counting independent sets in a graph is a self-reducible problem. (2) can be tightened to

$$P_G(\lambda) \leq \sum_{k=0}^{\alpha} \binom{n}{k} \lambda^k \leq \sum_{k=0}^{\alpha} \frac{(n\lambda)^k}{k!} \leq (n\lambda)^\alpha \sum_{k=0}^{\alpha} \frac{1}{k!} \leq e(n\lambda)^\alpha. \tag{4}$$

2 Markov Chains

2.1 Mixing Time

For general information on Markov chains and approximate counting see [17,18].

Consider a Markov chain on state space Ω with stationary distribution π and transition matrix $\mathbf{P}$. Let p_n be the distribution of the chain after n steps. We will assume that p_0 is the distribution which assigns probability 1 to a fixed initial state $x \in \Omega$. The *mixing time* of the Markov chain, from initial state $x \in \Omega$, is $\tau_x(\varepsilon) = \min\{n : d_{\mathrm{TV}}(p_n, \pi) \leq \varepsilon\}$, where $d_{\mathrm{TV}}(p_n, \pi)$ is the total variation distance between p_n and π. In the case of the Glauber dynamics for independent sets, the stationary distribution π satisfies (3), and in particular $\pi(\varnothing)^{-1} = P_G(\lambda)$. We will always use $\varnothing$ as our starting state.

Let $\beta_{\max} = \max\{\beta_1, |\beta_{|\Omega|-1}|\}$, where β_1 is the second-largest eigenvalue and $\beta_{|\Omega|-1}$ is the smallest eigenvalue of $\mathbf{P}$. From [9, Proposition 3] follows $\tau_x(\varepsilon) \leq (1-\beta_{\max})^{-1}\left(\ln(\pi(x)^{-1}) + \ln(1/\varepsilon)\right)$, see also [28, Proposition 1(i)]. Hence for $\lambda \geq 1/n$,

$$\tau_{\varnothing}(\varepsilon) \leq (1-\beta_{\max})^{-1}\left(\alpha(G)\ln(n\lambda) + 1 + \ln(1/\varepsilon)\right), \tag{5}$$

using (4). We can easily prove that $(1+\beta_{|\Omega|-1})^{-1}$ is bounded above by $\min\{\lambda, n\}$, see (9). It is more difficult to bound the *relaxation time* $(1-\beta_1)^{-1}$.

2.2 Canonical Paths Method

To bound the mixing time of our Markov chain we will apply the *canonical paths* method of Jerrum and Sinclair [19]. This may be summarised as follows. Let the problem size be n (in our setting, n is the number of vertices in the graph G, $\Omega = \mathcal{I}(G)$ and hence $|\Omega| \leq 2^n$). For each pair of states $X, Y \in \Omega$ we define a path γ_{XY} from X to Y, namely $X = Z_0 \to Z_2 \to \cdots \to Z_\ell = Y$ such that successive pairs along the path are given by a transition of the Markov chain. Write $\ell_{XY} = \ell$ for the length of the path γ_{XY}, and let $\ell_{\max} = \max_{X,Y} \ell_{XY}$. We require $\ell_{\max}$ to be at most polynomial in n. This is usually easy to achieve, but the set of paths $\{\gamma_{XY}\}$ must also satisfy the following property.

For any transition (Z, Z') of the chain there must exist an *encoding* W, such that, given (Z, Z') and W, there are at most ν distinct possibilities for X and Y such that $(Z, Z') \in \gamma_{XY}$. That is, each transition of the chain can lie on at most $\nu\,|\Omega^*|$ canonical paths, where Ω^* is some set which contains all possible encodings. We usually require ν to be polynomial in n. It is common to refer to the additional information provided by ν as "guesses", and we will do so here. In our situation, all encodings will be independent sets, so we may assume that $\Omega^* = \Omega$. The *congestion* ϱ of the chosen set of paths is given by

$$\varrho = \max_{(Z,Z')}\left\{\frac{1}{\pi(Z)\mathbf{P}(Z,Z')}\sum_{X,Y:\gamma_{XY}\ni(Z,Z')}\pi(X)\pi(Y)\right\}, \tag{6}$$

where the maximum is taken over all pairs (Z, Z') with $\mathbf{P}(Z, Z') > 0$ and $Z' \neq Z$ (that is, over all transitions of the chain), and the sum is over all paths containing the transition (Z, Z'). A bound on the relaxation time $(1-\beta_1)^{-1}$ will follow from a bound on congestion, using Sinclair's result [28, Cor. 6]:

$$(1-\beta_1)^{-1} \leq \ell_{\max}\,\varrho. \tag{7}$$

2.3 Glauber Dynamics

The Markov chain we employ will be the *Glauber dynamics*. In fact, we will consider a weighted version of this chain, for a given value of the fugacity (also called activity) $\lambda > 0$. Define $\pi(Z) = \lambda^{|Z|}/P_G(\lambda)$ for all $Z \in \mathcal{I}(G)$, where $P_G(\lambda)$ is the independence polynomial defined in (1). A transition from $Z \in \mathcal{I}(G)$ to $Z' \in \mathcal{I}(G)$ will be as follows. Choose a vertex v of G uniformly at random.

- If $v \in Z$ then $Z' \leftarrow Z \setminus \{v\}$ with probability $1/(1+\lambda)$.
- If $v \notin Z$ and $Z \cup \{v\} \in \mathcal{I}(G)$ then $Z' \leftarrow Z \cup \{v\}$ with probability $\lambda/(1+\lambda)$.
- Otherwise $Z' \leftarrow Z$.

This Markov chain is irreducible and aperiodic, and satisfies the detailed balance equations $\pi(Z)\,\mathbf{P}(Z,Z') = \pi(Z')\,\mathbf{P}(Z',Z)$ for all $Z, Z' \in \mathcal{I}(G)$. Therefore, the Gibbs distribution π is the stationary distribution of the chain. If Z' is obtained from Z by deleting a vertex v then

$$\mathbf{P}(Z,Z') = \frac{1}{n(1+\lambda)} \quad \text{and} \quad \mathbf{P}(Z',Z) = \frac{\lambda}{n(1+\lambda)}. \tag{8}$$

The unweighted version is given by setting $\lambda = 1$, and has uniform stationary distribution. Since the analysis for general λ is hardly any more complicated than that for $\lambda = 1$, we will work with the weighted case.

It follows from the transition procedure that $\mathbf{P}(Z,Z) \geq \min\{1,\lambda\}/(1+\lambda)$ for all states $Z \in \mathcal{I}(G)$. That is, every state has a self-loop probability of at least this value. Using a result of Diaconis and Saloff-Coste [8, p. 702], we conclude that the smallest eigenvalue $\beta_{|\mathcal{I}(G)|-1}$ of $\mathbf{P}$ satisfies

$$(1+\beta_{|\mathcal{I}(G)|-1})^{-1} \leq \frac{1+\lambda}{2\min\{1,\lambda\}} \leq \min\{\lambda, n\} \tag{9}$$

for $\lambda \geq 1/n$. This bound will be dominated by our bound on the relaxation time. We will always use the initial state $Z_0 = \varnothing$, since $\varnothing \in \mathcal{I}(G)$ for any graph G.

In order to bound the relaxation time $(1-\beta_1)^{-1}$ we will use the canonical path method. A key observation is that for any $X, Y \in \mathcal{I}(G)$, the induced subgraph $G[X \oplus Y]$ of G is bipartite. This can easily be seen by colouring vertices in $X \setminus Y$ black and vertices in $Y \setminus X$ white, and observing that no edge in G can connect vertices of the same colour. To exploit this observation, we introduce the *bipartite pathwidth* of a graph in Sect. 3. In Sect. 4 we show how to use the bipartite pathwidth to construct canonical paths for independent sets, and analyse the congestion of this set of paths to prove our main result, Theorem 1.

3 Pathwidth and Bipartite Pathwidth

The *pathwidth* of a graph was defined by Robertson and Seymour [27], and has proved a very useful notion in graph theory [6,10]. A *path decomposition* of a graph $G = (V, E)$ is a sequence $\mathcal{B} = (B_1, B_2, \ldots, B_r)$ of subsets of V such that

1. for every $v \in V$ there is some $i \in [r]$ such that $v \in B_i$,
2. for every $e \in E$ there is some $i \in [r]$ such that $e \subseteq B_i$, and
3. for every $v \in V$ the set $\{i \in [r] : v \in B_i\}$ forms an interval in $[r]$.

The *width* and *length* of this path decomposition $\mathcal{B}$ are $w(\mathcal{B}) = \max\{|B_i| : i \in [r]\} - 1$ and $\ell(\mathcal{B}) = r$ and the *pathwidth* $\mathrm{pw}(G)$ of a given graph G is $\mathrm{pw}(G) = \min_{\mathcal{B}} w(\mathcal{B})$, where the minimum taken over all path decompositions

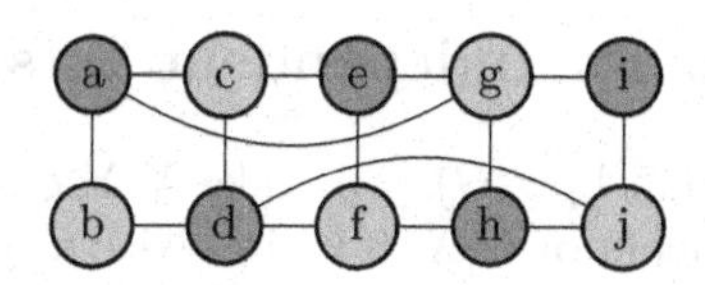

Fig. 1. A bipartite graph

$\mathcal{B}$ of G. Condition 3 is equivalent to $B_i \cap B_k \subseteq B_j$ for all i, j and k with $1 \le i \le j \le k \le r$. If we refer to a bag with index $i \notin [r]$ then by default $B_i = \varnothing$.

The graph in Fig. 1 has a path decomposition with the following bags:

$$B_1 = \{\mathrm{a,b,d,g}\}\ B_2 = \{\mathrm{a,c,d,g}\}\ B_3 = \{\mathrm{c,d,g,e}\}\ B_4 = \{\mathrm{d,e,f,g}\}$$
$$B_5 = \{\mathrm{d,f,g,j}\}\ B_6 = \{\mathrm{f,g,h,j}\}\ B_7 = \{\mathrm{g,h,i,j}\}$$

This path decomposition has length 7 and width 3. If P is a path, C is a cycle, K_n is a complete graph and $K_{a,b}$ is a complete bipartite graph then

$$\mathrm{pw}(P) = 1, \quad \mathrm{pw}(C) = 2, \quad \mathrm{pw}(K_n) = n - 1, \quad \mathrm{pw}(K_{a,b}) = \min\{a, b\}. \tag{10}$$

The following result will be useful for bounding the pathwidth. The first statement is [5, Lemma 11], while the second appears in [27, Eq. (1.5)].

Lemma 1. *For every subgraph H of G, $\mathrm{pw}(H) \le \mathrm{pw}(G)$ holds. If $W \subseteq V(G)$ then $\mathrm{pw}(G) \le \mathrm{pw}(G - W) + |W|$.*

The *bipartite pathwidth* $\mathrm{bpw}(G)$ of a graph G is the maximum pathwidth of an induced subgraph of G that is bipartite. For any integer $p \ge 2$, let $\mathcal{C}_p$ be the class of graphs G with $\mathrm{bpw}(G) \le p$. By Lemma 1 $\mathcal{C}_p$ is a hereditary class.

Clearly $\mathrm{bpw}(G) \le \mathrm{pw}(G)$, but the bipartite pathwidth of G may be much smaller than its pathwidth. A more general example is the class of *unit interval graphs.* These may have cliques of arbitrary size, and hence arbitrary pathwidth. However they are claw-free, so their induced bipartite subgraphs are linear forests, and hence they have bipartite pathwidth at most 1 from Eq. 10. The even more general *interval graphs* do not contain a tripod (depicted in Sect. 5.3), so their bipartite subgraphs are forests of caterpillars, and hence they have bipartite pathwidth at most 2.

Lemma 2. *Let p be a positive integer.*

(i) *Every graph with at most $2p + 1$ vertices belongs to $\mathcal{C}_p$.*
(ii) *No element of $\mathcal{C}_p$ can contain $K_{p+1,p+1}$ as an induced subgraph.*

A fixed linear order $<$ on the vertex set V of a graph G, extends to subsets of V as follows: if $A, B \subseteq V$ then $A < B$ if and only if (a) $|A| < |B|$; or (b) $|A| = |B|$ and the smallest element of $A \oplus B$ belongs to A. Next, given two path decompositions $\mathcal{A} = (A_j)_{j=1}^r$ and $\mathcal{B} = (B_j)_{j=1}^s$ of G, we say that $\mathcal{A} < \mathcal{B}$ if and only if (a) $r < s$; or (b) $r = s$ and $A_j < B_j$, where $j = \min\{i : A_i \ne B_i\}$.

4 Canonical Paths for Independent Sets

Suppose that $G \in \mathcal{C}_p$, so that $\text{bpw}(G) \leq p$. Take $X, Y \in \mathcal{I}(G)$ and let $H_1, \ldots, H_t$ be the connected components of $G[X \oplus Y]$ in lexicographical order. The graph $G[X \oplus Y]$ is bipartite, so every component $H_1, \ldots, H_t$ is connected and bipartite. We will define a canonical path γ_{XY} from X to Y by processing the components $H_1, \ldots, H_t$ in order. Let H_a be the component of $G[X \oplus Y]$ which we are currently processing, and suppose that after processing $H_1, \ldots, H_{a-1}$ we have a partial canonical path $X = Z_0, \ldots, Z_N$. If $a = 0$ then $Z_N = Z_0 = X$. The encoding W_N for Z_N is defined by

$$Z_N \oplus W_N = X \oplus Y \quad \text{and} \quad Z_N \cap W_N = X \cap Y. \tag{11}$$

In particular, when $a = 0$ we have $W_0 = Y$. We remark that (11) will not hold during the processing of a component, but always holds immediately after the processing of a component is complete. Because we process components one-by-one, in order, and due to the definition of the encoding W_N, we have

$$Z_N \cap H_s = Y \cap H_s \text{ for } s = 1, \ldots, a-1 \text{ (processed)}, \tag{12}$$

$$Z_N \cap H_s = X \cap H_s \text{ for } s = a, \ldots, t \text{ (not processed)}, \tag{13}$$

$$W_N \cap H_s = X \cap H_s \text{ for } s = 1, \ldots, a-1 \text{ (processed)}, \tag{14}$$

$$W_N \cap H_s = Y \cap H_s \text{ for } s = a, \ldots, t \text{ (not processed)}. \tag{15}$$

We now describe how to extend this partial canonical path by processing the component H_a. Let $h = |H_a|$. We will define a sequence $Z_N, Z_{N+1}, \ldots, Z_{N+h}$ of independent sets, and a corresponding sequence $W_N, W_{N+1}, \ldots, W_{N+h}$ of encodings, such that $Z_\ell \oplus W_\ell \subseteq X \oplus Y$ and $Z_\ell \cap W_\ell = X \cap Y$ for $j = N, \ldots, N+h$. Define the set of "remembered vertices" $R_\ell = (X \oplus Y) \backslash (Z_\ell \oplus W_\ell)$ for $\ell = N, \ldots, N+h$. By definition, the triple $(Z, W, R) = (Z_\ell, W_\ell, R_\ell)$ satisfies

$$(Z \oplus W) \cap R = \varnothing \quad \text{and} \quad (Z \oplus W) \cup R = X \oplus Y. \tag{16}$$

This immediately implies that $|Z_\ell| + |W_\ell| + |R_\ell| = |X| + |Y|$ for $\ell = N, \ldots, N+h$.

Let $\mathcal{B} = (B_1, \ldots, B_r)$ be the lexicographically-least path decomposition of H_a. Here we use the ordering on path decompositions defined at the end of Sect. 3. Since $G \in \mathcal{C}_p$, the maximum bag size in $\mathcal{B}$ is $d \leq p+1$.

We process H_a by processing the bags $B_1, \ldots, B_r$ in order. Initially $R_N = \varnothing$, by (11). If bag B_i is currently being processed and the current independent set is Z and the current encoding is W, then

$$(X \cap (B_1 \cup \cdots \cup B_{i-1})) \backslash B_i = (W \cap (B_1 \cup \cdots \cup B_{i-1})) \backslash B_i, \tag{17}$$

$$(Y \cap (B_1 \cup \cdots \cup B_{i-1})) \backslash B_i = (Z \cap (B_1 \cup \cdots \cup B_{i-1})) \backslash B_i, \tag{18}$$

$$(X \cap (B_{i+1} \cup \cdots \cup B_r)) \backslash B_i = (Z \cap (B_{i+1} \cup \cdots \cup B_r)) \backslash B_i, \tag{19}$$

$$(Y \cap (B_{i+1} \cup \cdots \cup B_r)) \backslash B_i = (W \cap (B_{i+1} \cup \cdots \cup B_r)) \backslash B_i. \tag{20}$$

Let Z_ℓ, W_ℓ, R_ℓ denote the current independent set, encoding and set of remembered vertices, immediately after the processing of bag B_{i-1}. We will write

$R_\ell = R_\ell^+ \cup R_\ell^-$ where vertices in R_ℓ^+ are added to R_ℓ during the preprocessing phase (and must eventually be inserted into the current independent set), and vertices in R_ℓ^- are added to R_ℓ due to a deletion step (and will go into the encoding during the postprocessing phase). When $i = 0$ we have $\ell = N$ and in particular, $R_N = R_N^+ = R_N^- = \varnothing$.

Preprocessing: *We "forget" the vertices of* $B_i \cap B_{i+1} \cap W_\ell$ *and add them to* R_ℓ^+*. This does not change the independent set or add to the canonical path.*
$R_\ell^+ \leftarrow R_\ell^+ \cup (B_i \cap B_{i+1} \cap W_\ell)$; $W_\ell \leftarrow W_\ell \backslash (B_i \cap B_{i+1})$;

Deletion steps: **for** each $u \in B_i \cap Z_\ell$, in lexicographical order, **do**
$Z_{\ell+1} \leftarrow Z_\ell \backslash \{u\}$;
if $u \notin B_{i+1}$ **then** $W_{\ell+1} \leftarrow W_\ell \cup \{u\}$; $R_{\ell+1}^- \leftarrow R_\ell^-$
else $W_{\ell+1} \leftarrow W_\ell$; $R_{\ell+1}^- \leftarrow R_\ell^- \cup \{u\}$;
$\ell \leftarrow \ell + 1$;

Insertion steps: **for** each $u \in \big(B_i \cap (W_\ell \cup R_\ell^+)\big) \backslash B_{i+1}$, in lexicogr. order, **do**
$Z_{\ell+1} \leftarrow Z_\ell \cup \{u\}$;
if $u \in W_\ell$ **then** $W_{\ell+1} \leftarrow W_\ell \backslash \{u\}$; $R_{\ell+1}^+ \leftarrow R_\ell^+$;
else $W_{\ell+1} \leftarrow W_\ell$; $R_{\ell+1}^+ \leftarrow R_\ell^+ \cup \{u\}$;
$\ell \leftarrow \ell + 1$;

Postprocessing: *All elements of* $R_{\ell+1}^-$ *which do not belong to* B_{i+1} *can now be safely added to* W_ℓ*. This does not change the current independent set or add to the canonical path.*
$W_\ell \leftarrow W_\ell \cup (R_\ell^- \backslash B_{i+1})$; $R_\ell^- \leftarrow R_\ell^- \cap B_{i+1}$;

By construction, vertices added to R_ℓ^+ are removed from W_ℓ, so the "otherwise" case for insertion is precisely $u \in R_\ell^+$.

Observe that both Z_ℓ and W_ℓ are independent sets at every step. This is true initially (when $\ell = N$) and remains true. The preprocessing phases removes all vertices of $B_i \cap B_{i+1}$ from W_ℓ, which makes room for other vertices to be inserted into the encoding later. A deletion step shrinks the current independent set and adds the removed vertex into W_ℓ or R_ℓ^-. A deleted vertex is only added to R_ℓ^- if it belongs to $B_i \cap B_{i+1}$, and so might have a neighbour in W_ℓ. In the insertion steps we add vertices from $\big(B_i \cap (W_\ell \cup R_\ell^+)\big) \backslash B_{i+1}$ to Z_ℓ, now that we have made room. Here B_i is the last bag which contains the vertex being inserted into the independent set, so any neighbour of this vertex in X has already been deleted from the current independent set. This phase can only shrink the encoding W_ℓ. Also observe that (16) holds for $(Z, W, R) = (Z_\ell, W_\ell, R_\ell)$ at every point. Finally, by construction we have $R_\ell \subseteq B_i$ at all times. Table 1 illustrates this construction for the graph in Fig. 1.

Each step of the canonical path alters the current independent set Z_i by exactly one element of $X \oplus Y$. Every vertex of $X \backslash Y$ is removed from the current independent set at some point, and is never re-inserted, while every vertex of $Y \backslash X$ is inserted into the current independent set once, and is never removed. Vertices outside $X \oplus Y$ are never altered and belong to all or none of the independent sets in the canonical path. Therefore $\ell_{\max} \leq 2\alpha(G)$.

Table 1. The steps of the canonical path, processing each bag in order.

B_i	preprocessing after 3rd step	after 1st step	after 2nd step postprocessing
B_1	d–b–a–g	d–b–$\bar{a}$–g	$\bar{d}$–b–$\bar{a}$–g
	$\bar{d}$–b–$\bar{a}$–g		
B_2	$\bar{d}$–c–$\bar{a}$–g		$\bar{d}$–c–a–g
B_3	$\bar{d}$–c–e–g	$\bar{d}$–c–$\bar{e}$–g	$\bar{d}$–c–$\bar{e}$–g
B_4	$\bar{d}$–f–$\bar{e}$–g		$\bar{d}$–f–e–g
B_5	f–$\bar{d}$–j g		f–d–j g
B_6	f–h–j g	f–$\bar{h}$–j g	f–$\bar{h}$–j g
B_7	$\bar{h}$–j–i–g	$\bar{h}$–j–i–g	$\bar{h}$–j–i–g
	$\bar{h}$–j–i–g		h–j–i–g

Lemma 3. *At any transition (Z, Z') which occurs during the processing of bag B_i, the set R of remembered vertices satisfies $R \subseteq B_i$, with $|R| \leq p$ unless $Z \cap B_i = W \cap B_i = \varnothing$. In this case $R = B_i$, which gives $|R| = p + 1$, and $Z' = Z \cup \{u\}$ for some $u \in B_i$.*

Lemma 4. *Given a transition (Z, Z'), the encoding W of Z and the set R of remembered vertices, we can uniquely reconstruct (X, Y) with $(Z, Z') \in \gamma_{XY}$.*

Theorem 1. *Let $G \in \mathcal{C}_p$ be a graph with n vertices and let $\lambda \geq 1/n$, where $p \geq 2$ is an integer. Then the Glauber dynamics with fugacity λ on $\mathcal{I}(G)$ (and initial state $\varnothing$) has mixing time*

$$\tau_{\varnothing}(\varepsilon) \leq 2e\alpha(G)\, n^{p+1}\, \lambda^p \Big(1 + \max(\lambda, 1/\lambda)\Big)\Big(\alpha(G)\, \ln(n\lambda) + 1 + \ln(1/\varepsilon)\Big).$$

When p is constant, this upper bound is polynomial in n and $\max(\lambda, 1/\lambda)$.

5 Recognisable Subclasses of $\mathcal{C}_p$

Theorem 1 shows that the Glauber dynamics for independent sets is rapidly mixing for any graph G in the class $\mathcal{C}_p$, where p is a fixed positive integer. However, the complexity of recognising membership in the class $\mathcal{C}_p$ is unknown. Therefore, we consider here three classes of graphs determined by small excluded subgraphs. These classes have polynomial time recognition algorithms. Note that we must always exclude large complete bipartite subgraphs. The three classes are nested. We will obtain better bounds for pathwidth in the smaller classes, and hence better mixing time bounds in Theorem 1.

5.1 Claw-Free Graphs

Claw-free graphs exclude the $K_{1,3}$, the *claw*. Claw-free graphs form an important superclass of *line graphs* [4], and independent sets in line graphs are *matchings*.

Lemma 5. *Let G be a claw-free graph with independent sets $X, Y \in \mathcal{I}(G)$. Then $G[X \oplus Y]$ is a disjoint union of paths and cycles.*

Lemma 6. *Claw-free graphs are a proper subclass of $\mathcal{C}_2$.*

5.2 Graphs with No Fork or Complete Bipartite Subgraph

Fork-free graphs exclude the following induced subgraph, the fork:

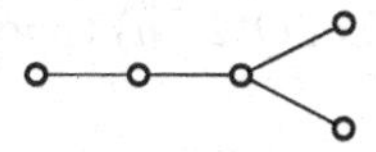

Two vertices u and v are *false twins* if $N(u) = N(v)$. In Fig. 2, vertices to which false twins can be added are indicated by red colour. Hence each graph containing a red vertex represents an infinite family of augmented graphs.

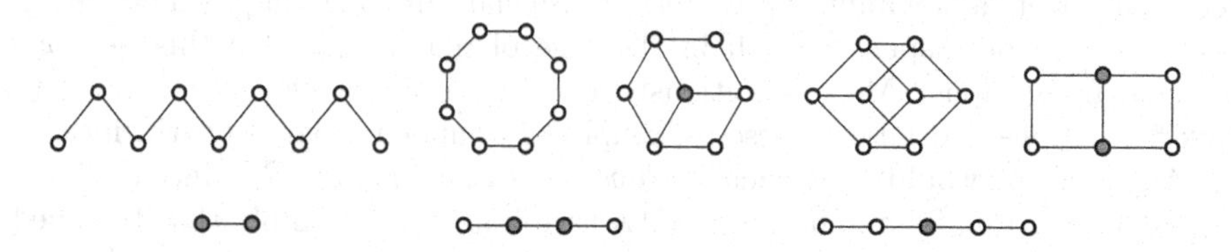

Fig. 2. The path P_9, the cycle C_8, the augmented bipartite wheel BW_3^*, the cube Q_3, an augmented domino, followed by augmented paths P_2^*, P_4^* and P_5^*. (Color figure online)

Lemma 7. *A bipartite graph is fork-free if and only if every connected component is a path, a cycle of even length, a BW_3^*, a cube Q_3, or can be obtained from a complete bipartite graph by removing at most two edges that form a matching.*

Lemma 8. *For all integers $d \geq 1$ the fork-free graphs without induced $K_{d+1,d+1}$ have bipartite pathwidth at most $\max(4, d+2)$.*

5.3 Graphs Free of Armchairs, Stirrers and Tripods

The graphs depicted below are called *armchair*, *stirrer* and *tripod.* A *fast graph* is a graph that contains none of these three as an induced subgraph.

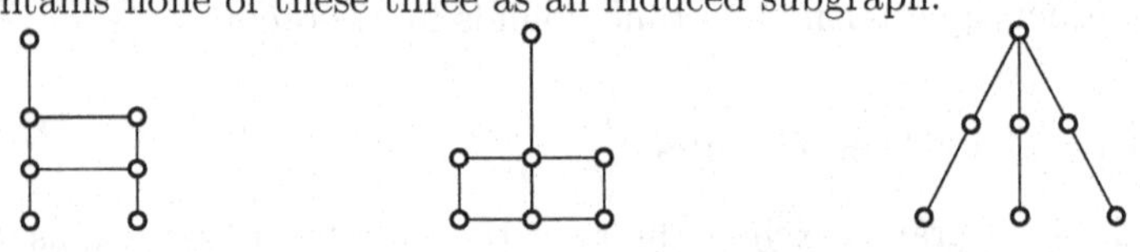

Theorem 2. *For every integer $d \geq 1$, a fast bipartite graph that does not contain $K_{d+1,d+1}$ as an induced subgraph has pathwidth at most $4d - 1$.*

6 Conclusions and Further Work

It is clearly NP-hard in general to determine the bipartite pathwidth of a graph, since it is NP-complete to determine the pathwidth of a bipartite graph. However, we need only determine whether $\text{bpw}(G) \leq d$ for some constant d. The complexity of this question is less clear. Bodlaender [5] has shown that the question $\text{pw}(G) \leq d$, can be answered in $O(2^{d^2} n)$ time. However, this implies nothing about $\text{bpw}(G)$.

In the case of claw-free graphs we can prove stronger sampling results using log-concavity. How far does log-concavity extends in this setting? Does it hold for fork-free graphs? Does some generalisation of log-concavity hold for graphs of bounded bipartite pathwidth? Where log-concavity holds, it allows us to approximate the number of independent sets of a given size. However, there is still the requirement of "amenability" [19]. Jerrum, Sinclair and Vigoda [20] have shown that this can be dispensed with in the case of matchings. Can this be done for claw-free graphs? More ambitiously, can the result of [20] be extended to fork-free graphs and larger classes of graphs of bounded bipartite pathwidth?

An extension would be to consider *bipartite treewidth*, $\text{btw}(G)$. Since $\text{tw}(G) = O(\text{pw}(G) \log n)$ [6, Thm. 66], our results here immediately imply that bounded bipartite treewidth implies *quasipolynomial* mixing time for the Glauber dynamics. Can this be improved to polynomial time?

Finally, can approaches to approximate counting be employed for the independent set problem in particular graph classes? Patel and Regts [25] have used the Taylor expansion approach for claw-free graphs. Could this be extended?

References

1. Alekseev, V.E.: Polynomial algorithm for finding the largest independent sets in graphs without forks. Discrete Appl. Math. **135**, 3–16 (2004)
2. Barvinok, A.: Computing the partition function of a polynomial on the Boolean cube. In: Loebl, M., Nešetřil, J., Thomas, R. (eds.) A Journey Through Discrete Mathematics, pp. 135–164. Springer, Cham (2017). https://doi.org/10.1007/978-3-319-44479-6_7

3. Bayati, M., Gamarnik, D., Katz, D., Nair, C., Tetali, P.: Simple deterministic approximation algorithms for counting matchings. In: Proceedings of the STOC, pp. 122–127 (2007)
4. Beineke, L.: Characterizations of derived graphs. J. Comb. Theory **9**, 129–135 (1970)
5. Bodlaender, H.L.: A linear time algorithm for finding tree-decompositions of small treewidth. SIAM J. Comput. **25**, 1305–1317 (1996)
6. Bodlander, H.L.: A partial k-arboretum of graphs with bounded treewidth. Theor. Comput. Sci. **209**, 1–45 (1998)
7. Chudnovsky, M., Seymour, P.: The roots of the independence polynomial of a clawfree graph. J. Comb. Theory (Ser. B) **97**, 350–357 (2007)
8. Diaconis, P., Saloff-Coste, L.: Comparison theorems for reversible Markov chains. Ann. Appl. Probab. **3**, 696–730 (1993)
9. Diaconis, P., Stroock, D.: Geometric bounds for eigenvalues of Markov chains. Ann. Appl. Probab. **1**, 36–61 (1991)
10. Diestel, R.: Graph Theory. Graduate Texts in Mathematics, vol. 173. Springer, Heidelberg (2017). https://doi.org/10.1007/978-3-662-53622-3
11. Dyer, M., Goldberg, L.A., Greenhill, C., Jerrum, M.: On the relative complexity of approximate counting problems. Algorithmica **38**, 471–500 (2003)
12. Dyer, M., Greenhill, C.: On Markov chains for independent sets. J. Algorithms **35**, 17–49 (2000)
13. Dyer, M., Greenhill, C., Müller, H.: Counting independent sets in graphs with bounded bipartite pathwidth. Preprint: arXiv:1812.03195 (2018)
14. Efthymiou, C., Hayes, T., Stefankovic, D., Vigoda, E., Yin, Y.: Convergence of MCMC and loopy BP in the tree uniqueness region for the hard-core model. In: Proceedings of the FOCS 2016, pp. 704–713. IEEE (2016)
15. Greenhill, C.: The complexity of counting colourings and independent sets in sparse graphs and hypergraphs. Comput. Complex. **9**, 52–72 (2000)
16. Harvey, N.J.A., Srivastava, P., Vondrák, J.: Computing the independence polynomial: from the tree threshold down to the roots. In: Proceedings of the SODA 2018, pp. 1557–1576 (2018)
17. Jerrum, M.: Mathematical foundations of the Markov chain Monte Carlo method. In: Habib, M., McDiarmid, C., Ramirez-Alfonsin, J., Reed, B. (eds.) Probabilistic Methods for Algorithmic Discrete Mathematics. AC, vol. 16, pp. 116–165. Springer, Heidelberg (1998). https://doi.org/10.1007/978-3-662-12788-9_4
18. Jerrum, M.: Counting, Sampling and Integrating: Algorithms and Complexity. Lectures in Mathematics - ETH Zürich. Birkhäuser, Basel (2003)
19. Jerrum, M., Sinclair, A.: Approximating the permanent. SIAM J. Comput. **18**, 1149–1178 (1989)
20. Jerrum, M., Sinclair, A., Vigoda, E.: A polynomial-time approximation algorithm for the permanent of a matrix with non-negative entries. J. ACM **51**, 671–697 (2004)
21. Jerrum, M.R., Valiant, L.G., Vazirani, V.V.: Random generation of combinatorial structures from a uniform distribution. Theoret. Comput. Sci. **43**, 169–188 (1986)
22. Matthews, J.: Markov chains for sampling matchings, Ph.D. thesis, University of Edinburgh (2008)
23. Luby, M., Vigoda, E.: Approximately counting up to four. In: Proceedings of the STOC 1995, pp. 150–159. ACM (1995)
24. Minty, G.J.: On maximal independent sets of vertices in claw-free graphs. J. Comb. Theory Ser. B **28**, 284–304 (1980)

25. Patel, V., Regts, G.: Deterministic polynomial-time approximation algorithms for partition functions and graph polynomials. SIAM J. Comput. **46**, 1893–1919 (2017)
26. Provan, J.S., Ball, M.O.: The complexity of counting cuts and of computing the probability that a graph is connected. SIAM J. Comput. **12**, 777–788 (1983)
27. Robertson, N., Seymour, P.D.: Graph minors I: excluding a forest. J. Comb. Theory Ser. B **35**, 39–61 (1983)
28. Sinclair, A.: Improved bounds for mixing rates of Markov chains and multicommodity flow. Comb. Probab. Comput. **1**, 351–370 (1992)
29. Sly, A.: Computational transition at the uniqueness threshold. In: Proceedings of the FOCS 2010, pp. 287–296. IEEE (2010)
30. Vadhan, S.P.: The complexity of counting in sparse, regular, and planar graphs. SIAM J. Comput. **31**, 398–427 (2001)
31. Weitz, D.: Counting independent sets up to the tree threshold. In: Proceedings of the STOC 2006, pp. 140–149. ACM (2006)

Intersection Graphs of Non-crossing Paths

Steven Chaplick(✉)

Lehrstuhl für Informatik I, Universität Würzburg, Würzburg, Germany
steven.chaplick@uni-wuerzburg.de

Abstract. We study graph classes modeled by families of non-crossing (NC) connected sets. Two classic graph classes in this context are disk graphs and proper interval graphs. We focus on the cases when the sets are paths and the host is a tree. Forbidden induced subgraph characterizations and linear time certifying recognition algorithms are given for intersection graphs of NC paths of a tree (and related subclasses). For intersection graphs of NC paths of a tree, the dominating set problem is shown to be solvable in linear time. Also, each such graph is shown to have a Hamiltonian cycle if and only if it is 2-connected, and to have a Hamiltonian path if and only if its block-cutpoint tree is a path.

Keywords: Clique trees · Non-crossing models · Domination · Hamiltonicity

1 Introduction

Intersection models of graphs are ubiquitous in graph theory and covered in many graph theory textbooks, see, e.g., [20,29]. Generally, for a given graph G, a collection $\mathcal{S}$ of sets, $\{S_v\}_{v \in V(G)}$, is an *intersection model* of G when $S_u \cap S_v \neq \emptyset$ if and only if $uv \in E(G)$. Similarly, we say that G is the intersection graph of $\mathcal{S}$. One quickly sees that all graphs have intersection models (e.g., by choosing, for every $v \in V(G)$, S_v as the edges incident to v). Thus, one often considers restrictions either on the *host* set (i.e., the domain from which the elements of the S_v's can be chosen), collection $\mathcal{S}$, and/or on the individual sets S_v.

In this paper we consider classes of intersection graphs where the sets are taken from a topological space, *(path) connected*, and pairwise *non-crossing*. A set S is *(path) connected* when any two of its points can be connected by a *curve* within the set (note: a *curve* is a homeomorphic image of a closed interval). Notice that, when the topological space is a graph, connectedness is precisely the usual connectedness of a graph and curves are precisely paths. Two connected sets S_1, S_2 are said to be *non-crossing* when both $S_1 \backslash S_2$ and $S_2 \backslash S_1$ are connected. Our focus will be on intersection graphs of non-crossing paths.

The full version of this article with the appendix referred to herein is on arxiv.org [3].
S. Chaplick—Research supported by DFG grant WO 758/11-1.

I. Sau and D. M. Thilikos (Eds.): WG 2019, LNCS 11789, pp. 311–324, 2019.
https://doi.org/10.1007/978-3-030-30786-8_24

The most general case of intersection graphs of non-crossing sets which have been studied are those of non-crossing connected (NC-C) sets in the plane [23]. These were considered together with another non-crossing class, the intersection graphs of disks in the plane or simply *disk* graphs. The recognition of both NC-C graphs and disk graphs is NP-hard [23]. More recently [22], disk graph recognition was shown to complete for the *existential theory of the reals* ($\exists\mathbb{R}$)[1].

One of the simplest cases of connected sets one can consider are those which reside in $\mathbb{R}$, i.e., the intervals of $\mathbb{R}$. The corresponding intersection graphs are precisely the well studied *interval graphs*. Moreover, imposing the non-crossing property on these intervals leads to the *proper* interval graphs[2]. It has often been considered how to generalize proper interval graphs to more complicated hosts, but simple attempts to do so involving the property that the sets are *proper* are often uninteresting. For example, the intersection graphs of proper paths in trees or proper subtrees of a tree are easily seen as the same as their non-proper versions. We will see that the non-crossing property leads to natural new classes which generalize proper interval graphs.

We formalize the setting as follows. For graph classes $\mathcal{S}$ and $\mathcal{H}$, a graph G is an $\mathcal{S}$-$\mathcal{H}$ graph when each $v \in V(G)$ has an $S_v \in \mathcal{S}$ such that:

- the graph $H = \bigcup_{v \in V(G)} S_v$ is in $\mathcal{H}$, and
- uv is an edge of G if and only if $S_u \cap S_v \neq \emptyset$.

Additionally, we say that $(\{S_v\}_{v \in V(G)}, H)$ is an $\mathcal{S}$-$\mathcal{H}$ *model* of G where H is the *host* and each S_v is a *guest*. We further state that G is a *non-crossing-$\mathcal{S}$-$\mathcal{H}$* (NC-$\mathcal{S}$-$\mathcal{H}$) graph when the sets S_v are pairwise non-crossing. In this context the proper interval graphs are the NC-path-path graphs.

Many classes of $\mathcal{S}$-$\mathcal{H}$ graphs have been studied in the literature; see, e.g., [29]. Some of these are described in the table below together with the complexity of their recognition problems and whether a *forbidden induced subgraph characterization (FISC)* is known. The table utilizes the following terminology. A *directed tree (d.tree)* is a tree in which every edge uv has been assigned one direction. A *rooted tree (r.tree)* is a directed tree where there is exactly one source node. A survey of path-tree graph classes is given in [30].

	Graph class	Guest	Host	Recognition	FISC?
1	Interval	path	path	$O(n+m)$ [8]	yes [26]
2	Rooted path tree (RPT)	path	r.tree	$O(n+m)$ [11]	open
3	Directed path tree (DPT)	path	d.tree	$O(nm)$ [5]	yes [33]
4	Path tree (PT)	path	tree	$O(nm)$ [36]	yes [27]
5	Chordal	tree	tree	$O(n+m)$ [35]	by definition[a]

[a] A graph is *chordal* when it has no induced cycles of length four or more.

[1] Note: all $\exists\mathbb{R}$-hard problems are NP-hard, see [28] for an introduction to $\exists\mathbb{R}$.
[2] Usually defined as having no interval strictly contained within any other.

Results and Outline. We study the non-crossing classes corresponding to graph classes 1–4 given in the table. Section 2 provides background and notation concerning intersection models. In Sect. 3 we provide forbidden induced subgraph characterizations for the non-crossing classes corresponding to 1–4 and certifying linear time recognition algorithms for them. Interestingly, this implies that one can test whether a chordal graph contains a claw in linear time. Then, for NC-path-tree graphs, in Sect. 4, we solve the minimum dominating set (MDS) problem in linear time by showing that there is an independent set which is also an MDS and using a known algorithm [12]. In Sect. 5, we show that 2-connectedness implies that each plane drawing of the NC-path-tree model leads to a distinct Hamiltonian cycle (HC) which can also be found in linear time, and a similar necessary condition implies the presence of Hamiltonian path (HP). For the MDS, HC and HP problems, we use the special structure of NC-path-tree models obtained in Sect. 3.1. Notably, the MDS problem is NP-complete on PT graphs [2], and *split* graphs[3] [9], but it is polynomial time solvable on RPT graphs [2]. Also, the HC and HP problems are NP-complete on *strongly chordal* split graphs [31], and DPT graphs [32], but easily solved on proper interval graphs [1]. We conclude with avenues for further research.

2 Preliminaries

Notation. Unless explicitly stated otherwise, all the graphs we discuss in this work are connected, undirected, simple, and loopless. For a graph G with a vertex v, we use $N_G(v)$ to denote the *neighborhood* of v, and $N_G[v]$ to denote the *closed neighborhood* of v, i.e., $N_G[v] = N_G(v) \cup \{v\}$. The subscript G will be omitted when it is clear. For a subset S of $V(G)$, we use $G[S]$ to denote the subgraph of G induced by S. For a set of graphs $\mathcal{F}$, we say that a graph G is $\mathcal{F}$-free when G does not contain any $F \in \mathcal{F}$ as an induced subgraph.

For graph classes $\mathcal{S}$ and $\mathcal{H}$, and an $\mathcal{S}$-$\mathcal{H}$ model $(\{S_v\}_{v \in V(G)}, H)$ of a graph G, we use the following notation. We refer to elements of $V(G)$ as *vertices* and use symbols u and v to refer them whereas we call elements of $V(H)$ *nodes* and use x, y, and z to refer to them. For a node x of H we use G_x to denote the set of vertices v in G where S_v contains x. Observe that every set G_x induces a clique in G. Note that Sect. 3.1 defines the terms *terminal*, *junction*, and *mixed* that are also used in later sections of the paper.

Several special graphs are named and depicted in Fig. 1 along with models of them. We will refer to these throughout this paper.

Twin-Free Graphs. For a graph G, two vertices x and y are called *twins* when they have the same closed neighborhood, i.e., $N[x] = N[y]$. Note that, for the MDS problem, it is an easy exercise to show that it suffices to consider twin-free graphs. Also, as the vertex set of a graph can be easily partitioned into its

[3] A graph is a split graph when its vertices can be partitioned into a clique and an independent set. It is easy to see that split graphs are chordal.

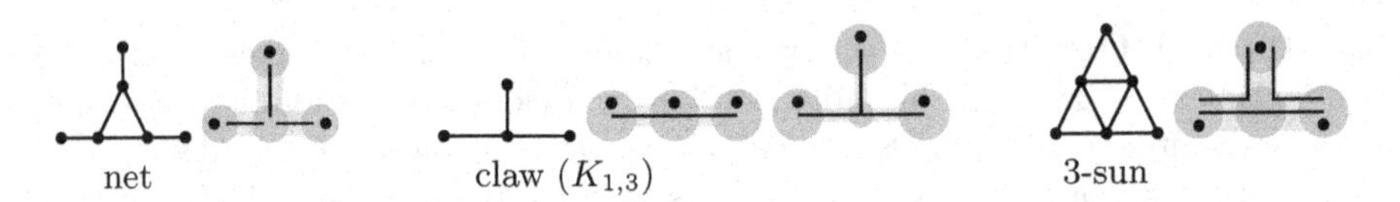

Fig. 1. Some small graphs and tree models of them. In the models the nodes of the host graph are given as darkly shaded circles and its edges are lightly shaded corridors connecting them. Each subset S_v is depicted by a tree (or single point) overlaid on the drawing of the host graph.

equivalence classes of twins in linear time, one can distill the relevant twin-free induced subgraph of G in linear time.

Chordality and Clique Trees. This area is deeply studied and while there are many interesting results related to our work, we only pick out a few concepts and results which are useful in this paper. The starting point is that the chordal graphs are well-known to be the tree-tree graphs [16].

For a chordal graph G, a *clique tree* T of G has the maximal cliques of G as its vertices, and for every vertex v of G, the set K_v of maximal cliques which contain v induces a subtree of T. In other words, it is a tree-tree model of G whose nodes are in bijection with the maximal cliques of G. Clique trees are very useful when discussing models where the host graph is a tree. When a graph has a tree-tree [16], path-tree [18], path-d.tree [30], path-r.tree [17], or path-path [14] model, then it also has one that is a clique tree. Such results are summarized in [29].

We establish similar clique tree results for the corresponding NC graphs when the guests are paths. However, note that when the guests are trees, we cannot rely on clique trees. For example, the claw is an NC-tree-tree graph, but it does not have an NC-tree-tree model that is a clique tree (see Fig. 1). We discuss this further in the conclusions.

An important property of clique trees for our linear time algorithms is the following. For a chordal graph G, $\sum_{v \in V(G)} |K_v| \in O(n+m)$ [20]. This implies that the total size of a clique tree T is $O(n+m)$. So, any algorithm that is linear in the size of T is also linear in the size of G. Additionally, one can produce a clique tree of a chordal graph in linear time [15].

3 Non-crossing Paths in Trees: Structure and Recognition

This section concerns classes of intersection graphs of non-crossing paths in trees; namely, NC-path-tree, NC-path-d.tree, NC-path-r.tree, and NC-path-path. We first note that the *claw* ($K_{1,3}$) is not an NC-path graph regardless of the host.

Observation 1. *If G is an NC-path graph, then G is claw-free.*

Proof. Suppose G contains a claw with central vertex u and pendant vertices a, b, c. Let $\mathcal{P}$ be a path-$\mathcal{H}$ model of G where $\mathcal{P} = \{P_v\}_{v \in V(G)}$. Clearly, $P_a \cap P_u$,

$P_b \cap P_u$ and $P_c \cap P_u$ are disjoint. As such, at most two of them include an endpoint of P_u. Thus, for some $d \in \{a,b,c\}$, $P_u \backslash P_d$ is disconnected. ∎

This section proceeds as follows. The NC-path-tree graphs are shown to be the claw-free chordal graphs and the structure of NC-path-tree models is described. From this structure, we then show that NC-path-d.tree = NC-path-r.tree = (claw, 3-sun)-free chordal. This provides, as a nearly direct consequence, the classic result that proper interval graphs are precisely the (claw, 3-sun, net)-free chordal graphs [34]. We conclude with linear time certifying recognition algorithms for NC-path-tree and NC-path-r.tree graphs.

3.1 The Structure of NC-Path-Tree Models

In this subsection we explore the structure of NC-path-tree models and prove our FISCs along the way. We first take a slight detour to claw-free chordal graphs and prove the FISC of NC-path-tree graphs. In doing so we obtain the first insight into NC-path-tree models. Namely, that it suffices to consider clique trees and that the clique trees of these graphs are unique (see Theorem 2). We then take a closer examination of these clique NC-path-tree models and carefully describe the nodes they contain – this will be used repeatedly in the rest of the paper.

Theorem 2. *A graph G is claw-free chordal iff it is an NC-path-tree graph. Moreover, G has a unique clique tree and it is an NC-path-tree model.*

Proof. $\Leftarrow$ Observation 1 and chordal graphs being tree-tree graphs imply this.
$\Rightarrow$ Let T be a clique tree of a claw-free chordal graph G. We first show that every subtree T_v must be a path, and then we show that these paths are non-crossing. These two claims prove the characterization. The uniqueness of the clique tree of every claw-free chordal graph has been shown previously [24].

Claim 1. For every $v \in V(G)$, T_v is a path.

Suppose T_v is not a path. Then T_v contains some claw x_0, x_1, x_2, x_3 with center x_0. However, since G_{x_j} is a maximal clique (for each $j \in \{0,1,2,3\}$), for each $i \in \{1,2,3\}$, there is $v_i \in G_{x_i} \backslash G_{x_0}$. Thus v, v_1, v_2, v_3 induces a claw in G. △

Claim 2. The set $\{T_v : v \in V(G)\}$ is non-crossing.

Suppose that T_u intersects T_v but does not include either end of T_v. Let x_1 and x_2 be the endpoints of T_v. Now there must be $v_1 \in G_{x_1} \backslash N_G(u)$ and $v_2 \in G_{x_2} \backslash N_G(u)$. That is, v, u, v_1, v_2 induces a claw in G. △ ∎

We now study the structure of the clique NC-path-tree model $(\{P_v\}_{v \in V(G)}, T)$ of a graph G. We introduce some terminology. A node x of T is called a *terminal* when it is a leaf of every path which contains it, i.e., x is not an internal node of any P_v. For example, the leaves of T are terminals. Similarly, a node x of T is a *junction* when it is an internal node of every path which contains it, i.e., x is not a leaf of any P_v. A node of T which is neither a terminal nor a junction is called *mixed*. We now present the main lemma describing T in these terms and an observation connecting these terms with certain induced subgraphs of G.

Lemma 1. *For an NC-path-tree graph G, let $(\{P_v\}_{v \in V(G)}, T)$ be its clique NC-path-tree model. A node x of T must satisfy the following properties:*

1. *If x is mixed, then x has degree two.*
2. *If x is a junction, then (i) x has degree 3 and (ii) x's neighbors are terminals.*
3. *If x has degree four or more, then x is a terminal.*

Proof. **1.** Suppose that x has degree at least 3, is a leaf of P_v, and is an internal node of P_u. Further, let y be the unique neighbor of x in P_v. We see that P_u includes y (otherwise, P_v and P_u cross). Let y' be the neighbor of x in $P_u \backslash P_v$ and let y'' be a neighbor of x which is not in P_u. Since G is connected, there exists $u' \in G_x \cap G_{y''}$. Furthermore, x is not a leaf of $P_{u'}$ (otherwise, P_u crosses $P_{u'}$). Thus, similarly to P_u, y belongs to $P_{u'}$. Now, since G_x and G_y are maximal cliques, there is $u'' \in G_x \backslash G_y$. Thus for $P_{u''}$ to neither cross P_u nor $P_{u'}$ it must include both y' and y''. However, this means $P_{u''}$ and P_v cross. △

2. Suppose that x is a junction and let $y_1, \ldots, y_k$ be the neighbors of x. Since x is a junction, for every $v \in G_x$, P_v contains exactly two y_i's. Thus, if $k = 2$, then $G_x \subseteq G_{y_1}$ – contradicting T being a clique tree. Now suppose $k \geq 3$ and consider $v \in G_x$ where (w.l.o.g.) P_v contains y_1 and y_2. Since G is connected, there must be $v' \in G_x \cap G_{y_3}$. Furthermore, (w.l.o.g.) $P_{v'}$ contains y_1 (otherwise, P_v and $P_{v'}$ cross). Now, since G_x and G_{y_1} are maximal cliques, there is $v'' \in G_x \backslash G_{y_1}$. Notice that $P_{v''}$ must contain y_2 and y_3 in order for $P_{v''}$ to cross neither P_v nor $P_{v'}$. Finally consider any $u \in G_x \backslash \{v, v', v''\}$. Notice that, in order for P_u to not cross any of P_v, $P_{v'}$, or $P_{v''}$, it must contain at least two of y_1, y_2, y_3. In particular, if $k \geq 4$, then $G_x \cap G_{y_4} = \emptyset$ – contradicting G being connected. Thus, $k = 3$ (establishing (i)).

Now, suppose that y_1 is not a terminal. By 1. and 2.(i), y_1 is either a junction with degree 3 or mixed with degree 2.

Case 1: *y_1 is a junction with neighbors x, z_1, z_2.* Notice that each of P_v and $P_{v'}$ must contain exactly one of z_1 or z_2. Moreover, w.l.o.g. they both must contain z_1 otherwise they will cross. However, since y_1 is a junction, we have vertices w, w', w'' such that $P_w \supseteq \{x, y_1, z_1\}$, $P_{w'} \supseteq \{x, y_1, z_2\}$ and $P_{w''} \supseteq \{z_1, y_1, z_2\}$. Moreover, both P_w and $P_{w'}$ must contain either y_2 or y_3. Regardless of this choice, we end up with a crossing between either $P_{w'}$ and P_v or $P_{w'}$ and $P_{v'}$. Thus, junctions cannot be neighbors.

Case 2: *y_1 has degree 2 and is mixed.* Let z be the neighbor of y_1 other than x and let w be a vertex of G where y_1 is not a leaf of P_w, i.e., w.l.o.g. $P_w \supseteq \{z, y_1, x, y_2\}$. Notice that, $P_{v'}$ must also contain z otherwise $P_{v'}$ and P_w would cross. Similarly, since $P_{v'}$ now contains z, P_v must also contain z otherwise P_v and $P_{v'}$ would cross. However, now a vertex $u \in G_{y_1} \backslash G_z$ must have $P_u = \{y_1\}$ but then P_u crosses P_w. Thus, no neighbor of a junction is mixed. △

3. This follows immediately from 1. and 2.(i). △ ■

Observation 3. *For an NC-path-tree graph G, let $(\{P_v : v \in V(G)\}, T)$ be its clique NC-path-tree model. Let x be a node of T of degree at least three.*

1. *If x is a junction, then G contains a 3-sun. Also, if G is twin-free, $|G_x| = 3$.*
2. *If x is a terminal, then G contains a net.*

Proof. **1.** As in the proof of Lemma 1.2.(i) a junction x in T has three neighbors y_1, y_2, y_3 and vertices $v, v', v'' \in G_x$ such that $P_v \supseteq \{y_1, x, y_2\}$, $P_{v'} \supseteq \{y_1, x, y_3\}$ and $P_{v''} \supseteq \{y_2, x, y_3\}$. Additionally, since x, y_1, y_2, y_3 are maximal cliques, there are vertices $u_1, u_2, u_3 \in V(G)$ such that $u_i \in G_x \backslash G_{y_i}$ for each $i \in \{1, 2, 3\}$. Moreover, all of these vertices are distinct due to their paths being incomparable. Thus, by considering the 3-sun and its clique tree model given in Fig. 1, it is now easy to see that $G[v, v', v'', u_1, u_2, u_3]$ is a 3-sun. Furthermore, since y_1, y_2, y_3 are terminals, the paths $P_v, P_{v'}, P_{v''}$ are the only distinct paths which are possible for vertices in G_x. In other words, every vertex in $G_x \backslash \{v, v', v''\}$ is a twin of one of v, v', or v''. △

2. Let y_1, y_2, y_3 be distinct neighbors of x. Since G is connected and x, y_1, y_2, y_3 are maximal cliques, we have $v_i \in G_x \cap G_{y_i}$ and $u_i \in G_{y_i} \backslash G_x$ for each $i \in \{1, 2, 3\}$. The v_i's are distinct since x is a terminal, and the u_i's are distinct since their paths are disjoint. Thus, by considering the net and its clique tree model given in Fig. 1, it is easy to see that $G[v_1, v_2, v_3, u_1, u_2, u_3]$ is a net. △ ■

3.2 Restricted Host Trees

Here we relate and characterize the classes of NC-path-d.tree, NC-path-r.tree, and NC-path-path graphs as stated in the next two theorems. The proofs are in the appendix and follow from Theorem 2, Lemma 1, and Observation 3.

Theorem 4. *A graph G is (claw, 3-sun)-free chordal if and only if it is NC-path-r.tree. Moreover, a graph has an NC-path-**d**.tree model if and only if it has a clique NC-path-**r**.tree.*

Theorem 5. *A graph G is (claw, 3-sun, net)-free chordal if and only if it is NC-path-path, i.e., proper interval.*

3.3 Recognition Algorithms

From our characterizations, there are straightforward polynomial-time certifying algorithms for the classes of NC-path-tree and NC-path-r.tree graphs. Specifically, since these classes are characterized as chordal graphs with an additional finite set of forbidden induced subgraphs, we can apply a linear time certifying algorithm for chordal graphs [35], and then apply brute-force search for our additional forbidden induced subgraphs. If no forbidden induced subgraph is found, we can simply construct the unique clique tree of the given graph and it will be an NC-path-tree (or NC-path-r.tree) model as needed to positively certify membership in our classes. However, we can do this more carefully and obtain linear time certifying algorithms as in the next theorem.

Theorem 6. *The classes NC-path-tree and NC-path-r.tree (= NC-path-d.tree) have linear-time certifying algorithms.*

Proof. Recall that the size $\sum_{v \in V(G)} |K_v|$ of a clique tree is $O(n+m)$ (we use this implicitly throughout the following). First, we run a linear-time certifying algorithm for chordal graphs, e.g., [35]. Then, we construct a clique tree T in linear-time [15]. We then annotate the clique tree to mark, for each vertex, for each maximal clique K in K_v, if K is a leaf or an internal node of the model of v. If some vertex v uses $\geq$3 cliques as leaves, we produce a claw as in Claim 1 of the proof of Theorem 2. If there is a mixed node x of degree $\geq$3, then we proceed as in the proof of Lemma 1.1. This provides us a pair of paths which cross in linear time. Then, proceeding as in Claim 2 in the proof of Theorem 2, we identify a claw. Now all of the nodes of degree $\geq$3 are either terminals or junctions, and we mark them as such. So, if there is a junction x with degree $\geq$4, we proceed as in Lemma 1.2.(i) to identify a pair of paths which cross and as before to find a corresponding claw. Furthermore, if a junction x neighbors a non-terminal y, we proceed as in Lemma 1.2.(i) to identify a pair of paths which cross and (again) a corresponding claw.

Now, no crossing between two paths can involve a node of degree ≥ 3. So, it remains just to ensure no crossings occur between such nodes. In particular, since the neighbors of all junctions are terminals, such a crossing must occur on a path connecting two terminals (where all nodes in between are mixed). Let $x_1, \ldots, x_k$ be such a path. Clearly, this path of cliques represents an interval graph. Moreover, we will find a pair of crossing paths on it precisely when this interval graph is not a proper interval graph. Conveniently, this problem is known to be solvable in linear time [10]. However, to obtain linear time in total (when processing all such paths) we need to be a bit careful. Namely, rather than simply checking whether each $G[\bigcup_{i=1}^{k} G_{x_i}]$ is a proper interval graph, for each such path we create the following auxiliary graph G'. The vertex set of G' is $\{u_1, u_k\} \cup \bigcup_{i=2}^{k-1} G_{x_i}$. In G', for each $i \in \{2, \ldots, k-1\}$, G_{x_i} is a clique. Also, u_1 is adjacent to $G_{x_1} \cap G_{x_2}$ and u_k is adjacent to $G_{x_{k-1}} \cap G_{x_k}$. In this way, the size of G' can easily be seen as linear in the size of $G[\bigcup_{i=2}^{k-1} G_{x_i}]$. Moreover, since we only consider paths connecting terminals, each vertex and edge of G is contained in at most one G'. Finally, observe that G' is interval and is a proper interval graph if and only if each $G[\bigcup_{i=2}^{k-1} G_{x_i}]$ is as well. Thus, running the certifying algorithm for proper interval graphs on G' will provide a claw when G' is not a proper interval graph, and such a claw is easily mapped back to a claw in G.

This completes the case of NC-path-tree graphs. For NC-path-r.tree graphs, we additionally look for junctions and proceed as in Observation 3.1. ■

4 Minimum Dominating Set

Recall that a *dominating set* in a graph G is a subset D of $V(G)$ such that every vertex is either in D or adjacent to a vertex in D. The MDS problem is NP-complete on PT graphs [2], and split graphs [9], and line graphs of planar graphs [37] (which are of course claw-free). Interestingly, the *minimum independent dominating set* (MIDS) problem can be solved on chordal graphs in linear time [12]. We will show that, for NC-path-tree graphs, the size of an MIDS is the same as the size of an MDS. Thus, by using [12], we can solve the MDS problem on NC-path-tree graphs in linear time. We assume graphs are twin-free here.

Theorem 7. *For any NC-path-tree graph G, there is an independent dominating set that is also a minimum dominating set. Moreover, such an independent dominating set can be found in linear time.*

Proof. Let $(\{P_v\}_{v \in V(G)}, T)$ be the clique NC-path-tree model of G. We root T at a leaf r and call the result $\overrightarrow{T}$. For each node x of $\overrightarrow{T}$, let $p(x)$ denote the parent of x. Now, if there is an MDS D of G where each node x has $|D \cap G_x| \leq 1$, then D is an independent set. For an MDS D, let $\overrightarrow{T}(D)$ be the subtree of $\overrightarrow{T}$ that contains the root and consists strictly of nodes with $|D \cap G_x| \leq 1$ (if $|D \cap G_r| \geq 2$, set $\overrightarrow{T} := \emptyset$). Let D be an MDS of G where $\overrightarrow{T}(D)$ is *maximal* ($\overrightarrow{T}(D)$ not strictly contained in $\overrightarrow{T}(D')$ for any other MDS D') and secondly, for each node x of $\overrightarrow{T} \backslash \overrightarrow{T}(D)$ where $p(x) \in V(\overrightarrow{T}(D))$, $|D \cap G_x|$ is minimized.

Suppose that there is a node in $\overrightarrow{T} \backslash \overrightarrow{T}(D)$, and let x be a node of $\overrightarrow{T} \backslash \overrightarrow{T}(D)$ whose parent is in $\overrightarrow{T}(D)$ (if $\overrightarrow{T}(D)$ is empty, we set $x = r$). By our choice of x, $|G_x \cap D| \geq 2$, and $|D \cap G_{p(x)}| \leq 1$. We consider the three cases regarding x, namely, x being mixed, a terminal, or a junction.

Case 1: *x is mixed.* Note that x has exactly one child. Let z be the closest descendant of x that is a terminal, and $P_{x,z} = (x = x_1, \ldots, x_k = z)$ be the (x, z)-path in $\overrightarrow{T}$. Note that, by Lemma 1, each x_i $(2 \leq i < k)$ is mixed and has degree two. For each vertex u of $D \cap G_x$, P_u contains a prefix $P^*(u)$ of this path. Let u be a vertex of $D \cap G_x$ where $|P^*(u)|$ is maximum. Since $|G_{p(x)} \cap D| \leq 1$, u cannot belong to $G_{p(x)}$ as otherwise replacing D by $\{u\} \cup (D \backslash (D \cap G_x)$ would result in a smaller dominating set. Similarly, $D \cap (G_x \backslash G_{p(x)}) = \{u\}$ as otherwise replacing D by $\{u\} \cup (D \cap G_{p(x)}) \cup (D \backslash (D \cap G_x))$ would result in a smaller dominating set. Thus, $D \cap G_x = \{u, v\}$ where v is a vertex of $G_{p(x)}$. Now, let u' be any vertex of $G_{x_1} \backslash G_x$. In order for the path $P_{u'}$ of u' to not cross P_u, $P_{u'}$ must contain $P^*(u) \backslash \{x_1\}$. Now, replacing u by u' in D results in an MDS D' where $\overrightarrow{T}(D')$ strictly contains $\overrightarrow{T}(D)$, contradicting the choice of D.

Case 2: *x is a junction.* Let y_1 and y_2 be the two children of x. Recall that, as G is twin-free, by Lemma 1, G_x contains exactly three vertices $v_1, v_2, v_{1,2}$ and these vertices have the paths $(p(x), x, y_1), (p(x), x, y_2)$, and (y_1, x, y_2) respectively. Since $|D \cap G_{p(x)}| \leq 1$, D does not contain both of v_1 and v_2. So, since $|D \cap G_x| \geq 2$, w.l.o.g., suppose that D contains v_1 and $v_{1,2}$. However, now, as y_2 is a terminal, replacing $v_{1,2}$ by any vertex of $G_{y_2} \backslash G_x$ results in a new MDS D' in which $\overrightarrow{T}(D')$ is strictly larger than $\overrightarrow{T}(D)$ as $T(D) \cup \{x\} \subseteq T(D')$.

Case 3: *x is a terminal.* Note that, since $|D \cap G_x| \geq 2$, D cannot contain a vertex v where $P_v = (x)$ as this would contradict the minimality of D. So, since x is a terminal, for every vertex $v \in D \cap G_x$, P_v must contain exactly one neighbor of x. Let $y_1, \ldots, y_t$ be the children of x where $|D \cap G_x \cap G_{y_i}| \geq 1$.

Suppose some y_i is a junction. Let z_1 and z_2 be the children of y_i, and let $u_1, u_2, u_{1,2}$ be the vertices of G_{y_i} where $P_{u_1} = (x, y_i, z_1)$, $P_{u_2} = (x, y_i, z_2)$,

$P_{u_{1,2}} = (z_1, y_i, z_2)$ respectively. Suppose, w.l.o.g., that u_1 is in D. Now, either there is $v \in D \cap (G_x \backslash G_{y_i})$, or u_2 is also in D. In both cases replacing, u_1 with $u_{1,2}$ leads to a contradiction in our choice of D (either due to the maximality of $\overrightarrow{T}(D)$ or due to the second condition). Thus, no y_i is a junction.

So, y_1 is not a junction. We observe that $|D \cap G_x \cap G_{y_1}| = 1$ as follows. Suppose we have $u, u' \in D \cap G_x \cap G_{y_1}$. Then, since x is a terminal and y_1 is not a junction, w.l.o.g., P_u contains $P_{u'}$, i.e., we must have $u = u'$. Note that, for each $u^* \in G_{y_1} \backslash G_x$, in order for P_{u^*} to not cross P_u, P_{u^*} must extend as least as far down $\overrightarrow{T}$ as P_u. Now, since $|D \cap G_x| \geq 2$ there is a vertex v which is either in $D \cap G_x \cap G_{p(x)}$ or in $D \cap G_x \cap G_{y_2}$. Thus, due to the presence of v, by replacing u with u^* we obtain a new MDS that contradicts our choice of D. ■

5 Hamiltonian Cycles and Paths

As mentioned before the HC and HP problems are NP-complete on DPT graphs and split graphs. They are also NP-complete on line graphs of bipartite graphs, i.e., (claw, diamond, odd-hole)-free graphs [25]. In contrast, we show that, like proper interval graphs [1], 2-connectivity suffices for Hamiltonicity in NC-path-tree graphs, but additionally, every tracing of a clique NC-path-tree model provides a distinct HC of its graph. We similarly characterize the presence of an HP.

Theorem 8. *An NC-path-tree graph G has a Hamiltonian cycle if and only if it is 2-connected and has at least three vertices. Also, for each plane layout of G's clique NC-path-tree model T, we obtain a distinct a Hamiltonian cycle of G.*

Proof. We build on the fact that 2-connected proper interval graphs are not only Hamiltonian but have an HC with quite special structure, established in [1], and described as follows. Consider a proper interval graph G. Let $x_1, \ldots, x_k$ be the maximal cliques G ordered according to the clique NC-path-path model of G. Further, let u_1 be a vertex of $G_{x_1} \backslash G_{x_2}$ and let u_k be a vertex of $G_{x_k} \backslash G_{x_{k-1}}$. When G is 2-connected there are internally disjoint (u_1, u_k)-paths P_1 and P_2 such that every vertex of G belongs to either P_1 or P_2. In essence, we will see (through an auxiliary multigraph Q constructed below) that such paths also occur in 2-connected NC-path-tree graphs by considering the proper interval graphs occurring between terminals.

Now consider a 2-connected NC-path-tree graph G and its clique NC-path-tree model T. Recall that, as we noted when designing our certifying algorithm for NC-path-tree graphs, for a path $x_1, \ldots, x_k$ in T where x_1 and x_k are terminals and each inner node is mixed, the graph $G[\bigcup_{i=1}^{k} G_{x_i}]$ is a proper interval graph. Moreover, since G is 2-connected, each such subgraph is also 2-connected. Additionally, the graph G' created from $G[\bigcup_{i=1}^{k} G_{x_i}]$ as before is also 2-connected. However, there is one special case where we use a slightly different auxiliary graph (otherwise we simply use the G' defined before). When $k = 2$, the graph G' is the clique $G_{x_1} \cap G_{x_k}$ together with new vertices u_1 and u_k where $N(u_1) = N(u_k) = G_{x_1} \cap G_{x_k}$. Now, it is easy to see that our graphs G' are

2-connected and proper interval, and since u_1 and u_k are not adjacent, we have two non-empty disjoint paths that both start with a vertex of $G_{x_1} \cap G_{x_2}$, and end with a vertex of $G_{x_{k-1}} \cap G_{x_k}$.

We now consider the case when a neighbor y of x is a junction before completing our construction of the HC. Let the other two neighbors of the junction y be x' and x''. Due to the fact that x, x', x'' are all terminals, the vertices of G_y form three equivalence classes A, A', A'' of twins, where each vertex in A is represented by the path x, y, x', each vertex in A' is represented by the path x', y, x'', and each vertex in A'' is represented by the path x'', y, x. Namely, using A, A', A'' we can "traverse" from x to x', from x' to x'', and from x'' back to x.

Based on the above observations, we can now build our HCs. To do this we will trace the outline of T by using the paths guaranteed by the above arguments. This trace can be described by a multigraph Q formed on the terminals of T where each Eulerian tour of Q will correspond to a distinct HC of G. Namely, for each terminal x, and each neighbor y of x:

- if y is a terminal, then in Q, x and y are connected by two edges (representing the two paths present in the corresponding G').
- if y is a mixed node and z is the terminal so that y occurs on the (x, z)-path in T, then, in Q, x and z are connected by two edges (representing the two paths present in the corresponding G').
- if y is a junction and x' and x'' are its two other neighbors, then in Q, we have the edges xx' and xx''.
- finally, if G_x contains vertices that do not belong to any other $G_{x'}$ (e.g., when x is a leaf of T), we also add a self-loop to x and map to this self-loop the vertices of $G_x \backslash (\bigcup_{x' \in N(x)} G_{x'})$.

We note the following properties of Q to complete the proof. The edges of Q partition the vertices of G and each edge xy corresponds to a path in G where the first vertex belongs G_x and the last vertex belongs to G_y. Furthermore, Q is Eulerian, each Eulerian cycle C provides an HC, and C describes a plane layout of T, i.e., a cyclic order of the edges around each node of T so that C traces the outline of this plane layout of T. Note that, each such plane layout will often arise from multiple Eulerian cycles in Q, but no two distinct layouts arise from the same cycle. ■

The *block-cutpoint* tree $BC(G)$ of a graph G contains a node for each cut-vertex of G, a node for each maximal 2-connected subgraph (*block*) of G, and its edge set is $\{cB : c \text{ is a cut-vertex, and } B \text{ is a block of } G \text{ containing } c\}$. It is well-known that $BC(G)$ can be computed in linear time [21], and is indeed always a tree. Clearly, if G has an HP, $BC(G)$ is a path. We show that this is sufficient to have an HP in NC-path-tree graphs. The main idea is to observe where the cut-vertices occur in the model and then reuse our Eulerian structure Q from the previous proof (see the appendix for the proof).

Theorem 9. *An NC-path-tree graph G contains a Hamiltonian path if and only if its block-cutpoint tree is a path.*

6 Concluding Remarks

A natural next step would be to study the NC-tree-tree graphs. But, it is not safe to simply work with clique trees in this case as the claw requires the use of a non-clique tree model. We conjecture that the NC-tree-tree graphs can be characterized as chordal graphs avoiding finite set of forbidden induced subgraphs.

Other host domains have been considered in the literature. Notice that similar to proper interval graphs being NC-path-path graphs, the proper circular arc graphs are precisely the NC-path-cycle graphs. A simple host graph class which generalizes both trees and cycles is that of *cacti*. A *cactus* is a connected graph in which every 2-connected component is a single vertex, a single edge or a chordless cycle. The intersection graphs of subtrees of a cactus were studied by Gavril [19]. So, one might consider the NC-path/tree/cactus-cactus graphs.

Finally, an alternative view of host domains has been considered quite recently through the notion of *H-graphs* [4,6,7,13], i.e., for a fixed graph H, a graph G is an H-graph when it is an intersection graph of connected subgraphs of a subdivision of H. Here, interval graphs are the K_2-graphs and circular arc graphs are the K_3-graphs. While there is a natural notion of proper H-graphs [4] (which indeed restrict H-graphs for every H), the more restrictive non-crossing H-graphs might have a nicer structure and lead to easier (and faster) algorithms.

References

1. Bertossi, A.A.: Finding Hamiltonian circuits in proper interval graphs. Inf. Process. Lett. **17**(2), 97–101 (1983). https://doi.org/10.1016/0020-0190(83)90078-9
2. Booth, K.S., Johnson, J.H.: Dominating sets in chordal graphs. SIAM J. Comput. **11**(1), 191–199 (1982). https://doi.org/10.1137/0211015
3. Chaplick, S.: Intersection graphs of non-crossing paths. CoRR abs/1907.00272 (2019)
4. Chaplick, S., Fomin, F.V., Golovach, P.A., Knop, D., Zeman, P.: Kernelization of graph hamiltonicity: proper H-graphs. In: Friggstad, Z., Sack, J.-R., Salavatipour, M.R. (eds.) WADS 2019. LNCS, vol. 11646, pp. 296–310. Springer, Cham (2019). https://doi.org/10.1007/978-3-030-24766-9_22
5. Chaplick, S., Gutierrez, M., Lévêque, B., Tondato, S.B.: From path graphs to directed path graphs. In: Thilikos, D.M. (ed.) WG 2010. LNCS, vol. 6410, pp. 256–265. Springer, Heidelberg (2010). https://doi.org/10.1007/978-3-642-16926-7_24
6. Chaplick, S., Töpfer, M., Voborník, J., Zeman, P.: On H-topological intersection graphs. In: Bodlaender, H.L., Woeginger, G.J. (eds.) WG 2017. LNCS, vol. 10520, pp. 167–179. Springer, Cham (2017). https://doi.org/10.1007/978-3-319-68705-6_13
7. Chaplick, S., Zeman, P.: Combinatorial problems on H-graphs. In: EUROCOMB, vol. 61, pp. 223–229 (2017). https://doi.org/10.1016/j.endm.2017.06.042
8. Corneil, D.G., Olariu, S., Stewart, L.: The LBFS structure and recognition of interval graphs. SIAM J. Discrete Math. **23**(4), 1905–1953 (2009). https://doi.org/10.1137/S0895480100373455
9. Corneil, D.G., Perl, Y.: Clustering and domination in perfect graphs. Discrete Appl. Math. **9**(1), 27–39 (1984). https://doi.org/10.1016/0166-218X(84)90088-X

10. Deng, X., Hell, P., Huang, J.: Linear-time representation algorithms for proper circular-arc graphs and proper interval graphs. SIAM J. Comput. **25**(2), 390–403 (1996). https://doi.org/10.1137/S0097539792269095
11. Dietz, P.F.: Intersection graph algorithms. Ph.D. thesis, Cornell University (1984)
12. Farber, M.: Independent domination in chordal graphs. Oper. Res. Lett. **1**(4), 134–138 (1982). https://doi.org/10.1016/0167-6377(82)90015-3
13. Fomin, F.V., Golovach, P.A., Raymond, J.: On the tractability of optimization problems on H-graphs. In: ESA. LIPIcs, vol. 112, pp. 30:1–30:14. Schloss Dagstuhl - Leibniz-Zentrum fuer Informatik (2018). https://doi.org/10.4230/LIPIcs.ESA.2018.30
14. Fulkerson, D., Gross, O.: Incidence matrices and interval graphs. Pac. J. Math. **15**, 835–855 (1965)
15. Galinier, P., Habib, M., Paul, C.: Chordal graphs and their clique graphs. In: Nagl, M. (ed.) WG 1995. LNCS, vol. 1017, pp. 358–371. Springer, Heidelberg (1995). https://doi.org/10.1007/3-540-60618-1_88
16. Gavril, F.: The intersection graphs of subtrees of trees are exactly the chordal graphs. J. Comb. Theory Ser. B **16**, 47–56 (1974)
17. Gavril, F.: A recognition algorithm for the intersection graphs of directed paths in directed trees. Discrete Math. **13**(3), 237–249 (1975). https://doi.org/10.1016/0012-365X(75)90021-7
18. Gavril, F.: A recognition algorithm for the intersection of graphs of paths in trees. Discrete Math. **23**, 211–227 (1978)
19. Gavril, F.: Intersection graphs of helly families of subtrees. Discrete Appl. Math. **66**, 45–56 (1996)
20. Golumbic, M.: Algorithmic Graph Theory and Perfect Graphs. Annals of Discrete Mathematics. Elsevier, Amsterdam (2004)
21. Hopcroft, J., Tarjan, R.: Algorithm 447: efficient algorithms for graph manipulation. Commun. ACM **16**(6), 372–378 (1973). https://doi.org/10.1145/362248.362272
22. Kang, R.J., Müller, T.: Sphere and dot product representations of graphs. Discrete Comput. Geom. **47**(3), 548–568 (2012)
23. Kratochvíl, J.: Intersection graphs of noncrossing arc-connected sets in the plane. In: North, S. (ed.) GD 1996. LNCS, vol. 1190, pp. 257–270. Springer, Heidelberg (1997). https://doi.org/10.1007/3-540-62495-3_53
24. Kumar, P., Madhavan, C.: Clique tree generalization and new subclasses of chordal graphs. Discrete Appl. Math. **117**(1–3), 109–131 (2002). https://doi.org/10.1016/S0166-218X(00)00336-X
25. Lai, T., Wei, S.: The edge hamiltonian path problem is NP-complete for bipartite graphs. Inf. Process. Lett. **46**(1), 21–26 (1993). https://doi.org/10.1016/0020-0190(93)90191-B
26. Lekkerkerker, C., Boland, J.: Representation of a finite graph by a set of intervals on a real line. Fundam. Math. **51**(1), 45–64 (1962)
27. Lèvêque, B., Maffray, F., Preissmann, M.: Characterizing path graphs by forbidden induced subgraphs. J. Graph Theory **62**, 4 (2009)
28. Matousek, J.: Intersection graphs of segments and $\exists\mathbb{R}$. CoRR abs/1406.2636 (2014). http://arxiv.org/abs/1406.2636
29. McKee, T., McMorris, F.: Intersection Graph Theory. SIAM (1999)
30. Monma, C.L., Wei, V.K.: Intersection graphs of paths in a tree. J. Comb. Theory Ser. B **41**(2), 141–181 (1986)
31. Müller, H.: Hamiltonian circuits in chordal bipartite graphs. Discrete Math. **156**(1–3), 291–298 (1996). https://doi.org/10.1016/0012-365X(95)00057-4

32. Narasimhan, G.: A note on the hamiltonian circuit problem on directed path graphs. Inf. Process. Lett. **32**(4), 167–170 (1989). https://doi.org/10.1016/0020-0190(89)90038-0
33. Panda, B.S.: The forbidden subgraph characterization of directed vertex graphs. Discrete Math. **196**(1–3), 239–256 (1999). https://doi.org/10.1016/S0012-365X(98)00127-7
34. Roberts, F.S.: On nontransitive indifference. J. Math. Psychol. **7**(2), 243–258 (1970). https://doi.org/10.1016/0022-2496(70)90047-7
35. Rose, D.J., Tarjan, R.E., Lueker, G.S.: Algorithmic aspects of vertex elimination on graphs. SIAM J. Comput. **5**(2), 266–283 (1976). https://doi.org/10.1137/0205021
36. Schäffer, A.A.: A faster algorithm to recognize undirected path graphs. Discrete Appl. Math. **43**, 261–295 (1993). https://doi.org/10.1016/0166-218X(93)90116-6
37. Yannakakis, M., Gavril, F.: Edge dominating sets in graphs. SIAM J. Appl. Math. **38**(3), 364–372 (1980)

Reconfiguring Hamiltonian Cycles in L-Shaped Grid Graphs

Rahnuma Islam Nishat(✉) and Sue Whitesides

Department of Computer Science,
University of Victoria, Victoria, Canada
{rnishat,sue}@uvic.ca

Abstract. Given a pair of 1-complex Hamiltonian cycles C and C' in an L-shaped grid graph G, we show that one is reachable from the other under two operations, *flip* and *transpose*, while remaining in the family of 1-complex Hamiltonian cycles throughout the reconfiguration. Operations *flip* and *transpose* are local in G. We give a reconfiguration algorithm that uses $O(|G|)$ operations.

Keywords: Hamilton cycle · Reconfiguration · Grid graph · Algorithm

1 Introduction

An *L-shaped grid graph* G is a finite, embedded, vertex-induced subgraph of the 2D integer grid, determined by an L-shaped orthogonal polygon drawn on the grid together with the grid vertices in its closure. The L-shaped polygon has six edges serving as the six *boundaries* of G and exactly one *reflex corner* vertex as shown in Fig. 1(a). In this paper we study reconfiguration of 1-*complex Hamiltonian cycles* in G, where a cycle is 1-complex if it connects each vertex of G to a boundary with a turn-free subpath. See Fig. 1(b).

Many kinds of *reconfiguration problems* have been proposed (e.g., [5,8,11, 16]), where one structure is to be transformed to another by applying a given set of operations. Hamiltonian cycle problems on grid graphs have been studied both combinatorially (e.g., [9,10,15]) as well as with regard to computational complexity (e.g., [1,7,17]). We initiated the study of reconfiguration problems for Hamiltonian cycles in grid graphs in a previous paper [13], where we studied rectangular grid graphs. Here we take the essential next step towards more general grid graphs by dealing with a single reflex corner.

Takaoka [16] has recently shown that the problem of deciding whether there is a sequence of "switch" operations between two given Hamiltonian cycles is PSPACE-complete for chordal bipartite graphs, strongly chordal split graphs, and bipartite graphs with maximum degree 6. In contrast to our work, the graph

R. I. Nishat—Travel supported by a grant from Faculty of Graduate Studies, University of Victoria.

S. Whitesides—Research supported in part by NSERC.

I. Sau and D. M. Thilikos (Eds.): WG 2019, LNCS 11789, pp. 325–337, 2019.
https://doi.org/10.1007/978-3-030-30786-8_25

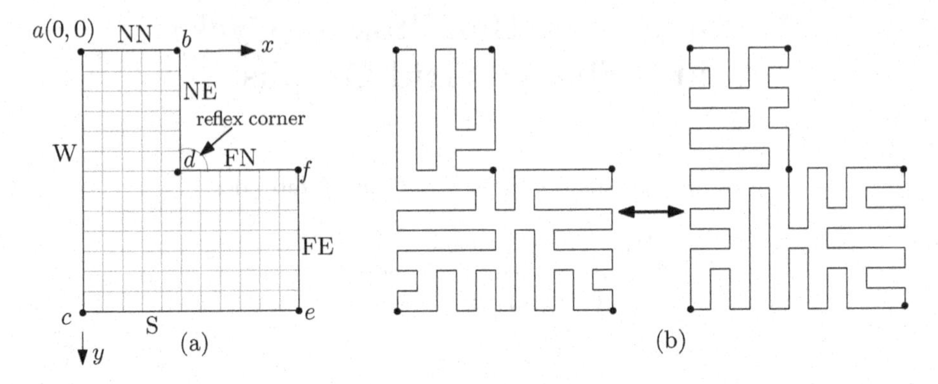

Fig. 1. (a) An L-shaped grid graph G with labeled corners and boundary edges. (b) A reconfiguration problem on G: can the first 1-*complex Hamiltonian cycle* be reconfigured to the second by *flips* and *transposes*?

classes he considered are not necessarily grid graphs or embedded. In [13], we defined two simple local operations *flip* and *transpose* for Hamiltonian cycles on embedded grid graphs and showed that any two given 1-complex Hamiltonian cycles in a rectangular grid graph are connected by a sequence of *flip* and *transpose* operations.

Grid graphs are used in path planning problems for robots and machine tools that must vacuum, explore, mill, or print material in a region that can be overlaid by a grid graph (e.g., [6,12]). A Hamiltonian path or cycle problem arises when each vertex should be visited only once. The combinatorial enumeration of Hamiltonian paths and cycles in grid graphs has found application in polymer science (e.g., [3,14]). We are particularly interested in 1-*complex* Hamiltonian cycles in grid graphs because they can be used to reduce turn costs and travel time in milling problems and traveling salesman tours; also they may improve accuracy of robot navigation (e.g., [2,4,18]).

This paper presents a reachability result for a problem posed in [13]. Namely, we show that there is a sequence of $O(|G|)$ flip and transpose operations between any two 1-complex Hamiltonian cycles in G, via *canonical* forms we define in Sect. 2. The rest of the paper is organized as follows. Section 2 defines terminology and presents a key lemma. Sections 3 and 4 give algorithms to transform a 1-complex Hamiltonian cycle to a canonical form. Section 5 handles reconfiguration between any two canonical forms, thus yielding the main result of the paper. Section 6 concludes with some open problems.

2 Preliminaries

Let G be an L-shaped grid graph with *reflex corner* d, embedded on the integer grid as shown in Fig. 1(a), so that vertex a is at the origin $(0, 0)$ of a coordinate system with positive y downward. A *rectangular grid graph* G' is defined by a rectangle drawn on the 2D integer grid together with the grid vertices in its closure. G' is a *rectangular subgrid* of G if the rectangle defining G' belongs to the closure of the L-shaped polygon defining G. Similarly, an L-shaped grid graph G'' is an *L-shaped subgrid* of G if the polygon defining G'' belongs to the closure of the L-shaped polygon for G.

For a vertex v of G, $x(v)$ and $y(v)$ denote its x– and y–coordinates. *Row* j is the set of vertices of G with y-coordinate j, and *Column* i is the set of vertices of G with x-coordinate i. A *horizontal track* t_j^h is the rectangle determined by the pair of Rows j and $j+1$; a *vertical track* t_i^v is the rectangle determined by the pair of Columns i and $i+1$. We call Columns 0 and $x(e)$ the *west (W)* and *far-east (FE) boundaries*; Rows 0 and $y(e)$ are the *near-north (NN)* and *south (S) boundaries.* The vertices on Column $x(d)$ from b to d (inclusive) comprise the *near-east (NE) boundary* and the vertices on Row $y(d)$ from d to f (inclusive) comprise the *far-north (FN) boundary.* The words "near" and "far" evoke the closeness of a boundary to the origin. See Fig. 1(a). The definitions of *flip* and *transpose* are recalled from [13] in Fig. 2.

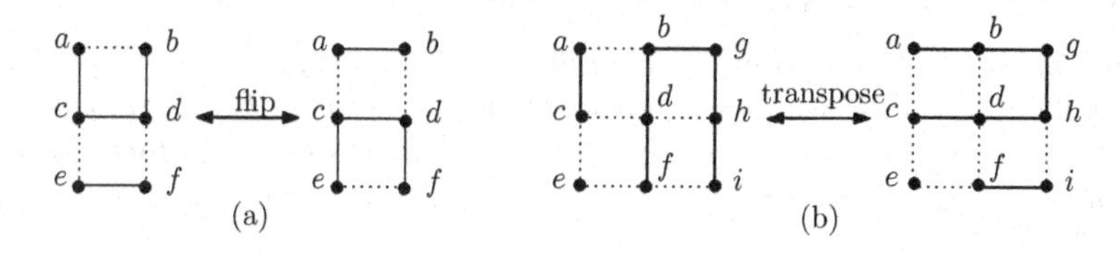

Fig. 2. On a Hamiltonian cycle C in a grid graph: (a) a *flip* interchanges path a, c, d, b with edge (a, b), and interchanges edge (e, f) with path e, c, d, f; (b) a *transpose* interchanges path f, d, b, g, h, i with edge (f, i), and interchanges edge (a, c) with path a, b, g, h, d, c.

A *cookie* c of a 1-complex Hamiltonian cycle C of G is a path in C that begins and ends on boundaries of G, has no intermediate points on boundaries, and has exactly two bends. Observe that each internal vertex of G lies on some cookie of C, and also that the endpoints of a cookie must be adjacent grid points on the same boundary of G. Thus we have six *types* of cookies: NN, FN, S, NE, FE and W. We say cookies are *from* one of the four axis-aligned *directions*: W cookies from west, NE and FE from east, NN and FN from north, and S from south. The *size* of a cookie is the distance it extends along its track tr.

A *zip* operation $Z = zip_{tp}(tr, sz)$ in a Hamiltonian cycle C of G is a sequence of zero or more transpose operations followed by zero or more flip operations in track tr such that the size of the desired cookie of type tp in track tr is sz after the zip. We define the *zone* of Z to be the closed $1 \times sz$ rectangle that contains the desired cookie (see Fig. 3). For the next definition, consider the two closed line segments s and s' that are $\frac{1}{2}$-unit translates of the sz-length sides of the zone, such that s and s' are outside the zone and each has one endpoint on the tp boundary. We call them the two *sidelines* of the zone.

Definition 1 ***(Zippability).*** *The zone of* $Z = zip_{tp}(tr, sz)$ *is* zippable *if (i) at least one of its sidelines does not intersect any cookie of* C*, (ii) any cookie perpendicular to* tr *that intersects the zone covers exactly 4 internal vertices of* G *in the zone, and (iii) any cookie of* C *of type* tp *in* tr *has size* $\leq sz$.

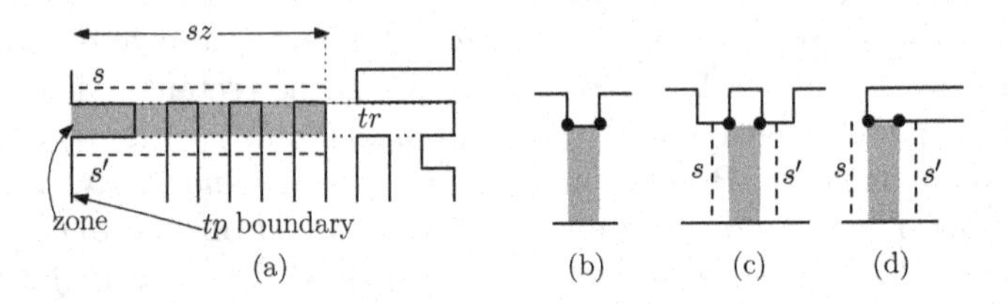

Fig. 3. Here, zones are shown in gray. Zones in (a)–(b) are zippable, but in (c)–(d) are not zippable. The complete cycle C is not shown. Sidelines are shown dashed.

The following lemma justifies the terminology.

Lemma 1. *Let* C *be a* 1*-complex Hamiltonian cycle in an L-shaped grid graph* G *and let* $Z = zip_{tp}(tr, sz)$ *be a zip operation. If the zone of* Z *is zippable, then* C *can be transformed with at most* sz *flips and transposes to another* 1*-complex Hamiltonian cycle* C' *such that* C' *has a cookie of type* tp *and size* sz *in* tr.

It is easy to see that performing a flip or transpose operation preserves Hamiltonicity, but the resultant Hamiltonian cycle need not be 1-complex. However, in this paper, all flips and transposes are performed on 1-complex Hamiltonian cycles in the context of zip operations with zippable tracks. From Lemma 1 we note the following.

Remark 1. *Performing a zip operation in a zippable zone on a* 1*-complex Hamiltonian cycle results in another* 1*-complex Hamiltonian cycle.*

Definition 2. *A* ***canonical Hamiltonian cycle*** $\mathbb{C}$ *in an L-shaped grid graph is a* 1*-complex Hamiltonian cycle that has at most two sets of cookies, where within each set, all are of the same type and size. (See Fig. 8 of Sect. 5).*

3 1-Complex Cycles to Canonical Forms: Special Cases

From now on, "cycle" C means a 1-complex Hamiltonian cycle in an L-shaped grid graph unless otherwise stated. In this section, we consider some special cases defined by *forbidding* certain cookie types in cycle C. We give algorithms that reconfigure these special cases of the input cycle to a canonical form. These algorithms are used in Sect. 4 to handle the general case. The algorithms in this section are sweep algorithms whose details depend on parities. All sweeps throughout the paper are inclusive of their end positions and are designed to ensure that the zippability conditions hold. Hence, intermediate cycles remain 1-complex Hamiltonian. The algorithms share the same proof of correctness, given in Sect. 3.4.

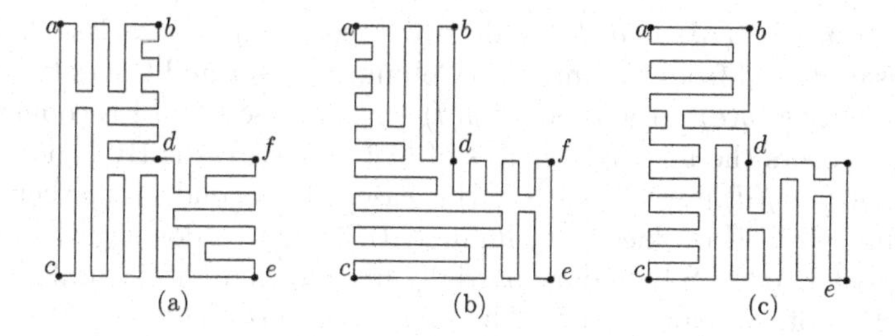

Fig. 4. Special cases of 1-complex cycles with no cookies of type (a) W, (b) NE or FE, (c) NN or FE, respectively. See Sects. 3.1, 3.2 and 3.3, respectively.

3.1 1-Complex Cycles Without W Cookies

Let C be a cycle no W cookies (see Fig. 4(a)). We give an algorithm that we call XW to reconfigure C to a canonical Hamiltonian cycle $\mathbb{C}$ of G. We consider four cases (see Fig. 5) based on the parities of $x(d)$ and $x(e)$:

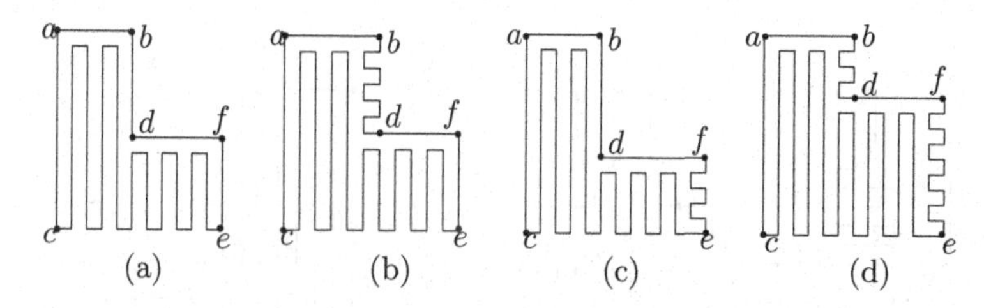

Fig. 5. The intermediate Hamiltonian cycles after the first sweep in Cases 1–4, respectively. For Case 1, the intermediate cycle is already the final canonical cycle.

1. *$x(e)$ and $x(d)$ are both odd.* We do one sweep across G from Column 1 to Column $x(e)-1$, filling alternate vertical tracks with two sets of S cookies of sizes $y(e)-1$ and $y(e)-y(d)-1$, respectively.
2. *$x(e)$ is odd and $x(d)$ is even.* We do two sweeps. The first is eastward from Column 1 to Column $x(e)-1$, leaving two sets of S cookies, one set with size $y(e)-1$ and the other with size $y(e)-y(d)-1$, and also one set of NE cookies of unit size covering the internal vertices of Column $x(d)-1$ from Row 1 through Row $y(d)$. Therefore, $y(d)$ must be even in this case. Otherwise, if $y(d)$ were odd then the 1-complex Hamltonian cycle produced by the zip operations done so far would leave some internal vertex in Column $x(d)-1$ between Rows 1 and $y(d)$ uncovered, contradicting Remark 1 and Hamiltonicity of C. The second sweep then expands the unit size NE cookies toward the W boundary, namely a downward sweep from Row 1 to Row $y(d)$ fills alternate tracks with NE cookies of size $x(d)-1$, which shortens the S cookies to size $y(e)-y(d)-1$. The final cycle has one set of NE cookies of size $x(d)-1$ and one set of S cookies of size $y(e)-y(d)-1$.

3. *$x(e)$ is even and $x(d)$ is odd.* We do two sweeps, similar to Case 2. The first sweep is eastward from Column 1 to Column $x(e)-1$ and leaves two sets of S cookies, of sizes $y(e)-1$ and $y(e)-y(d)-1$, and one set of FE cookies of unit size that covers the internal vertices of Column $x(e)-1$. By logic similar to Case 2, $y(e)-y(d)$ must be even in this case. The second sweep then expands these unit size FE cookies to Column $x(d)$, namely a downward sweep from Row $y(d)+1$ to the S boundary partially fills alternate horizontal tracks with FE cookies of size $x(e)-x(d)$. The final cycle has one set of FE cookies of size $x(e)-x(d)$ and one set of S cookies of size $y(e)-1$.
4. *$x(e)$ and $x(d)$ are both even.* We do two sweeps. The first sweep is eastward from Column 1 to Column $x(e)-1$ and leaves two sets of S cookies, of sizes $y(e)-1$ and $y(e)-y(d)-1$, and also one set of NE cookies of unit size that covers the vertices of Column $x(d)-1$ from Row 1 through Row $y(d)$, and also one set of unit size FE cookies that covers the internal vertices of Column $x(e)-1$. Thus, reasoning as above, $y(d)$ and $y(e)$ must be even. Figure 6 shows the first few steps of the eastward sweep. The second sweep is downward from Row 1 to Row $y(e)-1$ and expands the unit size NE and FE cookies westward to Column 1.

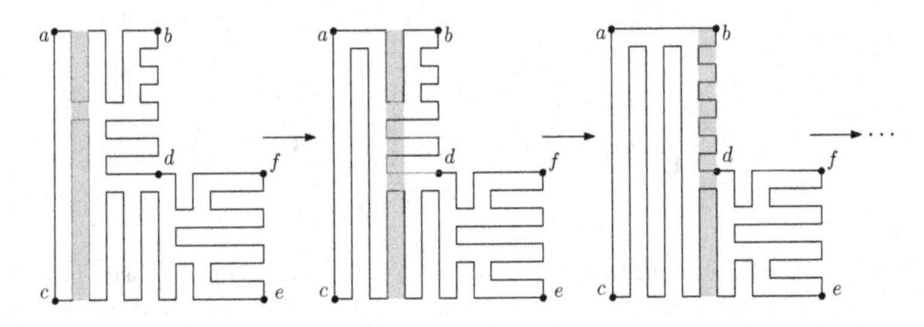

Fig. 6. First few steps of the eastward sweep in Case 4, in which $x(e)$ and $x(d)$ are found to be even in the course of the sweep. The sweeping track is highlighted in pink. (Color figure online)

3.2 1-Complex Cycles Without Any NE or FE Cookies

Let C be a 1-complex Hamiltonian cycle of G without any NE or FE cookies (see Fig. 4(b)). We give an algorithm we call XNEFE to reconfigure C to a canonical Hamiltonian cycle. We consider four cases based on the parities of $x(d)$ and $x(e)$:

1. *$x(e)$ and $x(d)$ are both odd.* We do one sweep westward from Column $x(e)-1$ to Column 1 that fills alternate vertical tracks with two sets of S cookies of sizes $y(e)-y(d)-1$ and $y(e)-1$, respectively.
2. *$x(e)$ and $x(d)$ are both even.* We do two sweeps. The first sweep is westward from Column $x(e)-1$ to Column 1 and leaves two sets of S cookies, of sizes $y(e)-y(d)-1$ and $y(e)-1$, and also one set of W cookies of unit size that covers the internal vertices of Column 1. In this case, $y(e)$ must be found to

be even as a consequence of Remark 1 and the Hamiltonicity of C. The second sweep expands the unit size W cookies eastward to Column $x(d)-1$, namely a downward sweep from Row 1 to Row $y(e)-1$ partially fills alternate horizontal tracks with a set of W cookies of size $x(d)-1$. The final cycle has one set of W cookies of size $x(d)-1$ and one set of S cookies of size $y(e)-y(d)-1$.

3. $x(e)$ *is even and* $x(d)$ *is odd.* We do two sweeps. The first is westward from Column $x(e)-1$ to Column 1 and leaves one set of S cookies of size $y(e)-y(d)-1$ and one set of NN cookies of size $y(d)$, and also one set of W cookies of unit size that covers the vertices of Column 1 from Row $y(d)+1$ to Row $y(e)-1$. In this case, reasoning as above, $y(e)-y(d)$ must be found to be even. The second sweep expands the unit size W cookies eastward to Column $x(e)-1$: a downward sweep from Row $y(d)+1$ to Row $y(e)-1$ partially fills alternate horizontal tracks with W cookies of size $x(e)-1$. The final cycle has one set of NN cookies of size $y(d)$ and one set of W cookies of size $x(e)-1$.
4. $x(e)$ *is odd and* $x(d)$ *is even.* We do two sweeps. The first is westward from Column $x(e)-1$ to Column 1 and leaves one set of S cookies of size $y(e)-y(d)-1$ and one set of NN cookies of size $y(d)$, and also one set of W cookies of unit size. The W cookies cover the vertices of Column 1, but this time the vertices from Row 1 to Row $y(d)$ are covered by W cookies. Reasoning as in the previous cases, $y(d)$ must be found to be even. The second sweep expands the unit size W cookies eastward to Column $x(d)-1$: an upward sweep from Row $y(d)$ to Row 1 fills alternate tracks with W cookies of size $x(d)-1$. The final cycle has one set of S cookies of size $y(e)-y(d)-1$ and one set of W cookies of size $x(d)-1$.

3.3 1-Complex Cycles Without Any NN or FE Cookies

Let C be a 1-complex Hamiltonian cycle of G without any NN or FE cookies (see Fig. 4(c)). We give an algorithm called XNNFE that sweeps downward from Row 1 either to Row $y(d)$ (if $y(d)$ is odd) or to Row $y(d)+1$ (if $y(d)$ is even), and fills alternate horizontal tracks with W cookies of size $x(d)-1$. This removes any initial NE cookies in C. Let C' be the 1-complex Hamiltonian cycle after this sweep. Since C' does not have any NN, NE or FE cookies, we can then call Algorithm XNEFE on C' as a procedure.

3.4 Proof of Correctness

The following theorem establishes the correctness of the Algorithms XW, XNEFE and XNNFE.

Theorem 1. *Let C be a 1-complex Hamiltonian cycle in an L-shaped grid graph G. Algorithms XW, XNEFE and XNNFE compute canonical Hamiltonian cycles of G using $O(|G|)$ flips and transposes such that the forbidden cookie types do not appear in the canonical cycles. Also the intermediate cycle computed after each operation remains 1-complex Hamiltonian.*

Proof Sketch. Our algorithms consist of a sequence of zips. By Lemma 1 we have a 1-complex Hamiltonian cycle during and after each zip operation. It is easy to see that the forbidden cookie types do not appear at any step of the algorithms, and by Remark 1 the cycle remains 1-complex Hamiltonian after each operation. Now we check zippability of the sweeps. In the first sweep in all three algorithms, the zone of the first zip in any loop is next to a boundary of forbidden cookie type, and hence is zippable. The zones of the remaining zip operations that follow in the same loop all have the same size, so we can advance the sideline by 2 and it still does not intersect any cookies. Thus, zippability holds for all the zips of the first sweep. The second sweep (if it is carried out) is orthogonal to the first sweep and expands cookies of unit size. The zone of the first zip in any loop of the sweep would have a sideline either next to the NN boundary of G or between Rows $y(d)$ and $y(d)+1$, so the zippability conditions hold. The zips that follow in the same loop of the sweep have zones that are zippable. Since each of the algorithms performs at most two zips in any track (horizontal or vertical) of G, a total of $O(|G|)$ flips and transposes is done. □

4 1-Complex Cycles to Canonical Forms: General Case

Let C be a 1-complex Hamiltonian cycle in an L-shaped grid graph G. Now we consider the general case when C has cookies from all four directions (east, west, north, and south); otherwise (possibly after repositioning G to make e the origin and a the bottom right corner) C falls into a special case of Sect. 3. The problem we want to solve algorithmically is as follows.

Problem $\Pi(C)$: *Given a 1-complex Hamiltonian cycle C in a rectangular or L-shaped grid graph G, reconfigure C to a canonical Hamiltonian cycle $\mathbb{C}$ of G.*

Our algorithm for solving $\Pi(C)$ runs in three steps. First it creates certain *subproblems* of $\Pi(C)$. Then it solves the subproblems using the algorithms from Sect. 3 and [13]. Finally, it merges the solutions to the subproblems to obtain $\mathbb{C}$. Below, a subscript on Π indicates the shape of C and G, e.g., $\Pi_L(C)$, $\Pi_R(C)$.

Creating the Subproblems. To create the particular subproblems of $\Pi(C)$ required by our algorithm, we use properties of C given in the next lemma.

Lemma 2 (Splitting track). *Let C be a 1-complex Hamiltonian cycle in an L-shaped grid graph G, where C has cookies from all four directions. Then there is a track tr of G whose interior intersects no cookies of C and such that tr has cookies to each side. (We call tr a* splitting track *of C.)*

Proof Sketch. We sweep a vertical line i eastward from Column 1 to Column $x(e)-1$ until one of the following two events occurs.

1. Sweepline i intersects cookies from both east and west directions. Then it is easy to show that there is a pair of cookies from east and west such that if one of them is in track t^h_j then the other is in t^h_{j+2}, where $1 \leq j \leq y(e)-4$. Then the horizontal track t^h_{j+1} is a splitting track.

2. Sweepline i does not intersect any W cookies. If t^v_{i-1} does not contain any cookies, then it is a splitting track. Otherwise, t^v_{i-1}contains at least one cookie from the north or south, say without loss of generality, the south. Let the size of the S cookie be k. Then track $t^h_{y(e)-k-1}$ is a splitting track. □

We now identify the subproblems we need. Let t^v_i be a splitting track of C. We remove the two cycle edges (u', u'') and (v', v'') of C at the two ends of t^v_i on boundaries of G, to partition C into two disjoint paths $P'_{u'v'}$ and $P''_{u''v''}$. Let the *subgrids* of G that contain P' and P'' be G' and G'', where G' is to the left of t^v_i and G'' is to the right. Now add a copy of Column i from u' to v' (inclusive) one unit to the right of G' to create the augmented vertex-induced grid graph $(G')^+$. Join u' to v' by a path through the new vertices to obtain a Hamiltonian cycle C' of $(G')^+$. Note that if $i \geq x(d)$ then $(G')^+$ is L-shaped; otherwise it is rectangular. Similarly obtain a Hamiltonian cycle C'' of $(G'')^+$, the vertex-induced grid graph created from G'' by adding a column to its left. One of $(G')^+$ and $(G'')^+$ is rectangular and the other is L-shaped. The columns added to create $(G')^+$ and $(G'')^+$ are their *augmented boundaries.*

Definition 3. *The subproblems $\Pi(C')$ and $\Pi(C'')$ are the* subproblems *of $\Pi(C)$ we need. One of them has the form $\Pi_L(C_L)$ and the other has the form $\Pi_R(C_R)$.*

We now explain how the subproblems $\Pi(C')$ and $\Pi(C'')$ will be used to solve $\Pi(C)$. Clearly C_L and C_R have no cookies from the augmented boundaries. Suppose we can solve $\Pi_L(C_L)$ and $\Pi_R(C_R)$, obtaining canonical cycles $\mathbb{C}_L$ and $\mathbb{C}_R$, such that $\mathbb{C}_L$ and $\mathbb{C}_R$ have no cookies from the augmented boundaries. Then we can merge $\mathbb{C}_L$ and $\mathbb{C}_R$ by removing the paths along the augmented boundaries and adding back edges (u', u'') and (v', v'') to obtain a new 1-complex Hamiltonian cycle C_1 of G. (We will later show that C_1 can easily be reconfigured to a canonical form, as C_1 fails to have certain cookie types.)

In addition to the goal of creating subproblems whose solutions are easy to merge, we would also like to achieve a second goal, namely to create subproblems that can be solved with the algorithms of Sect. 3 and [13]. To achieve these two goals, we choose a splitting track that is either the leftmost vertical splitting track (if it exists) or the southernmost horizontal splitting track. As will be seen, this enables us to achieve the two goals above. The next algorithm generates all the subproblems needed to solve $\Pi(C)$.

Algorithm SPLIT: Find the leftmost vertical splitting track if it exists (e.g., Fig. 7(a) and (b)); otherwise, find the southernmost horizontal splitting track, creating two subproblems $\Pi_R(C_R)$ and $\Pi_L(C_L)$. If C_L has cookies from all four directions, find a splitting track for C_L and replace $\Pi_L(C_L)$ with two subproblems $\Pi_R(C'_R)$ and $\Pi_L(C'_L)$.

Theorem 2. *Let C be a 1-complex cycle of an L-shaped grid graph G such that C has cookies from all four directions. Then Algorithm* SPLIT *creates from $\Pi(C)$ in $O(|G|)$ time either the subproblems $\Pi_R(C_R)$ and $\Pi_L(C_L)$, or the subproblems*

$\Pi_R(C_R)$, $\Pi_L(C'_L)$ and $\Pi_R(C'_R)$; here each of C_R and C'_R has at most three types of cookies, and each of C_L and C'_L falls into one of the cases of Sect. 3.

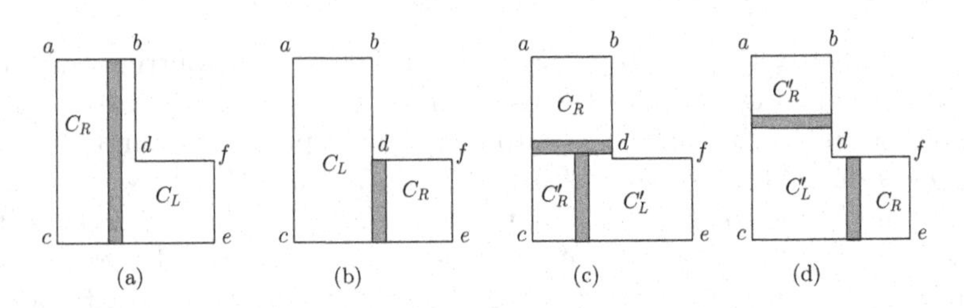

Fig. 7. (a)–(b) $\Pi(C)$ splits into $\Pi_L(C_L)$ and $\Pi_R(C_R)$. (c) The two splitting tracks of C and C_L intersect, and (d) the splitting tracks do not intersect.

Proof Sketch. It is easy to see that SPLIT generates either (i) $\Pi_L(C_L)$ and $\Pi_R(C_R)$, or (ii) $\Pi_R(C_R)$, $\Pi_L(C'_L)$ and $\Pi_R(C'_R)$. In (i), C_L has no cookies from a direction perpendicular to its splitting track, and hence falls into a special case from Sect. 3. In (ii), it can be shown that the splitting tracks of C and C_L must be perpendicular to each other, as otherwise the splitting track choice is contradicted. If the two splitting tracks intersect, then NN and W cookies do not appear in C'_L (see Fig. 7(c)); otherwise the splitting tracks do not intersect and NN and FE cookies do not appear in C'_L (see Fig. 7(d)). In either case C'_L falls into a special case. It can be checked that each of C_R and C'_R has at most three types of cookies since their augmented boundaries do not have cookies. □

Solving the Subproblems. To solve $\Pi_R(C_R)$ and $\Pi_R(C'_R)$, we apply the algorithm for rectangular 1-complex cycles with no east cookies from [13] to get canonical forms $\mathbb{C}_R$ and $\mathbb{C}'_R$, where each of them has either a set of west cookies or a set of north cookies. Since C_R has exactly one augmented boundary, we position C_R such that the augmented boundary is the east boundary, as there are no cookies from that boundary. However, C'_R can have two perpendicular augmented boundaries (see Fig. 7(c)). In that case, we position C'_R such that the augmented boundaries are its east and south boundaries, so that the $\mathbb{C}'_R$ produced by the algorithm [13] does not have cookies from the augmented boundaries. Since C_L falls into one of the special cases of Sect. 3 (possibly after repositioning G), we apply the appropriate algorithm to get a canonical form $\mathbb{C}_L$ that does not have cookies from the augmented boundaries.

Merging the Solutions to the Subproblems. We merge the solutions $\mathbb{C}_L$ of $\Pi_L(C_L)$ and $\mathbb{C}_R$ of $\Pi_R(C_R)$, output by the algorithms, as follows. We remove the augmented boundaries from $\mathbb{C}_L$ and $\mathbb{C}_R$ to get two Hamiltonian paths in the original subgrids of G on both sides of the splitting track. We join the paths using the boundary edges at the ends of the splitting track to obtain a 1-complex Hamiltonian cycle C_1 of G. C_1 has at most three sets of cookies as $\mathbb{C}_L$ and $\mathbb{C}_R$

have at most two sets and one set of cookies, respectively. Thus C_1 cannot have cookies from all four directions and must fall into a special case of Sect. 3. We then apply the appropriate algorithm on C_1 to obtain a canonical Hamiltonian cycle $\mathbb{C}$ of G. If $\Pi(C)$ was split into three subproblems $\Pi_R(C_R)$, $\Pi_L(C'_L)$ and $\Pi_R(C'_R)$, we first merge canonical forms $\mathbb{C}'_L$ and $\mathbb{C}'_R$ as above to obtain $\mathbb{C}_L$. Then we merge $\mathbb{C}_L$ and $\mathbb{C}_R$ to obtain $\mathbb{C}$ of G.

Theorem 3. *Let C be a 1-complex Hamiltonian cycle of an L-shaped grid graph G. Then $\Pi_L(C)$ can be solved using $O(|G|)$ flips and transposes.*

Proof. By Theorem 2, SPLIT creates at most three subproblems, with $O(|G|)$ flips and transposes. The subproblems are solved by algorithms that apply $O(|G|)$ flips and transposes in total as seen above. It is easy to see that the merging of the solutions to the subproblems uses $O(|G|)$ flips and transposes. □

5 Reconfiguration Between Any Pair of 1-Complex Cycles

Problem $\Pi(C_1, C_2)$:*Given any pair of 1-complex Hamiltonian cycles C_1 and C_2 of a rectangular or L-shaped grid graph G,reconfigure C_1 to C_2 using flips and transposes.*

To solve $\Pi_L(C_1, C_2)$, we first solve $\Pi_L(C_1)$ and $\Pi_L(C_2)$ yielding $\mathbb{C}_1$ and $\mathbb{C}_2$. Now we give an algorithm for $\Pi_L(\mathbb{C}_1, \mathbb{C}_2)$, a special case of $\Pi_L(C_1, C_2)$.

Let G be an L-shaped grid graph. By definition, any canonical cycle $\mathbb{C}$ of G has at most two sets S_1 and S_2 of cookies, where the cookies of each set are of the same type and size. If there are exactly two sets of cookies then there is a unique track *tr* between the rectangular regions covered by S_1 and S_2. Note that *tr* must be a splitting track of $\mathbb{C}$. We call *tr* the *canonical splitting track* of $\mathbb{C}$. It is easy to see that *tr* is either $t^v_{x(d)-1}$ (Fig. 8(a)) or $t^h_{y(d)}$ (Fig. 8(b)–(c)).

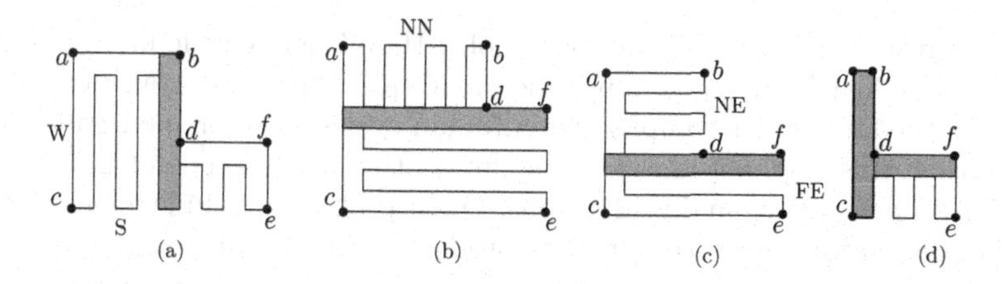

Fig. 8. Some canonical Hamiltonian cycles, canonical splitting tracks shown grey.

Let $\mathbb{C}_1$ and $\mathbb{C}_2$ have the same canonical splitting track $tr = t^v_{x(d)-1}$. We remove the edges of $\mathbb{C}_1$ in *tr* that are on boundaries of G (i.e., NN, NE and S boundaries) to partition $\mathbb{C}_1$ into two disjoint paths, P' from $v_{x(d)-1,0}$ to $v_{x(d)-1,y(e)}$ to the left of *tr* and P'' from d to $v_{x(d),y(e)}$ to the right of *tr*. Let G'

and G'' be the rectangular subgrids of G that contain P' and P''. We generate the augmented vertex-induced grid graphs $(G')^+$ and $(G'')^+$ by adding a copy of Column $x(d)-1$ one unit to the right of G' and a copy of Column $x(d)$ from d to $v_{x(d),y(e)}$ (inclusive) to the left of G'', respectively. Note that both $(G')^+$ and $(G'')^+$ are rectangular. We then join the two endpoints of P' through the new vertices of $(G')^+$ and the endpoints of P'' through the new vertices of $(G'')^+$ to obtain canonical Hamiltonian cycles $\mathbb{C}_1'$ of $(G')^+$ and $\mathbb{C}_1''$ of $(G'')^+$. In a similar way we obtain canonical Hamiltonian cycles $\mathbb{C}_2'$ and $\mathbb{C}_2''$ from $\mathbb{C}_2$ such that $\mathbb{C}_1'$ covers the same rectangular subgrid of G as $\mathbb{C}_2'$ covers, and such that $\mathbb{C}_1''$ covers the same rectangular subgrid of G as $\mathbb{C}_2''$ covers. We say that $\Pi_R(\mathbb{C}_1', \mathbb{C}_2')$ and $\Pi_R(\mathbb{C}_1'', \mathbb{C}_2'')$ are the two *subproblems* of $\Pi_L(\mathbb{C}_1, \mathbb{C}_2)$.

Theorem 4. *Let $\mathbb{C}_1$ and $\mathbb{C}_2$ be canonical Hamiltonian cycles of an L-shaped grid graph G. Then $\Pi_L(\mathbb{C}_1, \mathbb{C}_2)$ can be solved with $O(|G|)$ flips and transposes.*

Proof Sketch. We consider two cases: $\mathbb{C}_1$ and $\mathbb{C}_2$ have the same canonical splitting track, or not. In the second case, as a subgoal, we reconfigure $\mathbb{C}_1$ to some canonical Hamiltonian cycle $\mathbb{C}_3$ of G such that $\mathbb{C}_3$ and $\mathbb{C}_2$ have the same canonical splitting track. We then solve $\Pi_L(\mathbb{C}_3, \mathbb{C}_2)$. □

Theorem 5 Main Result. *Let C_1 and C_2 be any two 1-complex Hamiltonian cycles of an L-shaped grid graph G. Then $\Pi_L(C_1, C_2)$ can be solved with $O(|G|)$ flips and transposes such that the cycle remains 1-complex Hamiltonian after each operation.*

Proof By Theorem 3, C_1 and C_2 can be reconfigured to canonical cycles $\mathbb{C}_1$ and $\mathbb{C}_2$ of G using $O(|G|)$ flips and transposes in total. By Theorem 4, $\Pi_L(\mathbb{C}_1, \mathbb{C}_2)$ can be solved also using $O(|G|)$ flips and transposes. By Remark 1, the cycle remains 1-complex Hamiltonian after each operation. □

6 Conclusion

Our main result is Theorem 5: any 1-complex Hamiltonian cycle in an L-shaped grid graph is reachable from any other in $O(|G|)$ flips and transposes while staying in the family of 1-complex Hamiltonian cycles. As L-shaped grids have a reflex corner, this may provide a key step for continuing the study of Hamiltonian cycles in general orthogonal grid graphs. Open problems include generalization to higher dimensions and to other grids, study of Hamiltonian paths, exploration of fixed parameter approaches based on the *bend complexity* $k \geq 1$ of the cycle as a parameter [13] (here, we studied $k = 1$), and determination of the complexity of the decision problem for the existence of k-complex Hamiltonian cycles in various grid graphs.

References

1. Afrati, F.: The hamilton circuit problem on grids. RAIRO - Theor. Inf. Appl. - Informatique Theorique et Applications **28**(6), 567–582 (1994)
2. Arkin, E.M., Bender, M.A., Demaine, E.D., Fekete, S.P., Mitchell, J.S.B., Sethia, S.: Optimal covering tours with turn costs. SIAM J. Comput. **35**(3), 531–566 (2005). https://doi.org/10.1137/S0097539703434267
3. des Cloizeaux, J., Jannik, G.: Polymers in Solution: Their Modelling and Structure. Clarendon Press, Oxford (1987)
4. Fellows, M., et al.: Milling a graph with turn costs: a parameterized complexity perspective. In: Thilikos, D.M. (ed.) WG 2010. LNCS, vol. 6410, pp. 123–134. Springer, Heidelberg (2010). https://doi.org/10.1007/978-3-642-16926-7_13
5. Gopalan, P., Kolaitis, P.G., Maneva, E., Papadimitriou, C.H.: The connectivity of boolean satisfiability: computational and structural dichotomies. SIAM J. Comput. **38**(6), 2330–2355 (2009)
6. Gorbenko, A., Popov, V., Sheka, A.: Localization on discrete grid graphs. In: He, X., Hua, E., Lin, Y., Liu, X. (eds.) Computer, Informatics, Cybernetics and Applications: CICA 2011. LNEE, vol. 107, pp. 971–978. Springer, Dordrecht (2012). https://doi.org/10.1007/978-94-007-1839-5_105
7. Itai, A., Papadimitriou, C.H., Szwarcfiter, J.L.: Hamilton paths in grid graphs. SIAM J. Comput. **11**(4), 676–686 (1982)
8. Ito, T., et al.: On the complexity of reconfiguration problems. Theor. Comput. Sci. **412**(12), 1054–1065 (2011)
9. Jacobsen, J.L.: Exact enumeration of hamiltonian circuits, walks and chains in two and three dimensions. J. Phys. A: Math. Gen. **40**, 14667–14678 (2007)
10. Keshavarz-Kohjerdi, F., Bagheri, A.: Hamiltonian paths in l-shaped grid graphs. Theor. Comput. Sci. **621**, 37–56 (2016)
11. Mizuta, H., Ito, T., Zhou, X.: Reconfiguration of steiner trees in an unweighted graph. IEICE Trans. Fund. Electr. **E100.A**(7), 1532–1540 (2017)
12. Muller, P., Hascoet, J.Y., Mognol, P.: Toolpaths for additive manufacturing of functionally graded materials (FGM) parts. Rapid Prototyp. J. **20**(6), 511–522 (2014)
13. Nishat, R.I., Whitesides, S.: Bend complexity and hamiltonian cycles in grid graphs. In: Cao, Y., Chen, J. (eds.) COCOON 2017. LNCS, vol. 10392, pp. 445–456. Springer, Cham (2017). https://doi.org/10.1007/978-3-319-62389-4_37
14. Bodroža Pantić, O., Pantić, B., Pantić, I., Bodroža Solarov, M.: Enumeration of hamiltonian cycles in some grid graphs. MATCH - Commun. Math. Comput. Chem. **70**, 181–204 (2013)
15. Pettersson, V.: Enumerating hamiltonian cycles. Electr. J. Comb. 21(4) (2014). P4.7
16. Takaoka, A.: Complexity of hamiltonian cycle reconfiguration. Algorithms **11**(9), 140(15p) (2018)
17. Umans, C., Lenhart, W.: Hamiltonian cycles in solid grid graphs. In: 38th Annual Symposium on Foundations of Computer Science, FOCS 1997, pp. 496–505 (1997)
18. Winter, S.: Modeling costs of turns in route planning. Geoinformatica **6**(4), 345–361 (2002)

Color Refinement, Homomorphisms, and Hypergraphs

Jan Böker(✉)

RWTH Aachen University, Aachen, Germany
boeker@informatik.rwth-aachen.de

Abstract. Recent results show that the structural similarity of graphs can be characterized by counting homomorphisms to them: the Tree Theorem states that the well-known color-refinement algorithm does not distinguish two graphs G and H if and only if, for every tree T, the number of homomorphisms $\mathsf{Hom}(T, G)$ from T to G is equal to the corresponding number $\mathsf{Hom}(T, H)$ from T to H (Dell, Grohe, Rattan 2018). We show how this approach transfers to hypergraphs by introducing a generalization of color refinement. We prove that it does not distinguish two hypergraphs G and H if and only if, for every connected Berge-acyclic hypergraph B, we have $\mathsf{Hom}(B, G) = \mathsf{Hom}(B, H)$. To this end, we show how homomorphisms of hypergraphs and of a colored variant of their incidence graphs are related to each other. This reduces the above statement to one about vertex-colored graphs.

Keywords: Graph isomorphism · Color refinement · Hypergraph homomorphism numbers

1 Introduction

A result by Lovász [8] states that a graph can be characterized up to isomorphism by counting homomorphisms from all graphs to it, i.e., two graphs G and H are isomorphic if and only if, for every graph F, the number of homomorphisms $\mathsf{Hom}(F, G)$ from F to G is equal to the number of homomorphisms $\mathsf{Hom}(F, H)$ from F to H. Equivalently, using the notion of the *homomorphism vector* $\mathsf{HOM}(G) := (\mathsf{Hom}(F, G))_{F \in \mathcal{G}}$ of G, where $\mathcal{G}$ denotes the class of all graphs, we have that two graphs G and H are isomorphic if and only if their homomorphism vectors $\mathsf{HOM}(G)$ and $\mathsf{HOM}(H)$ are equal. However, the problem of computing the entries of a homomorphism vector is #P-complete as it generalizes some well-known counting problems [10, Section 5.1]. Hence, Dell, Grohe, and Rattan [6] considered restrictions $\mathsf{HOM}_{\mathcal{F}}(G) := (\mathsf{Hom}(F, G))_{F \in \mathcal{F}}$ of homomorphism vectors to classes of graphs $\mathcal{F}$ for which these entries can be computed efficiently: under some complexity-theoretic assumption, for a recursively enumerable class of graphs $\mathcal{F}$, counting homomorphisms from the graphs in $\mathcal{F}$ is possible in polynomial time if and only if $\mathcal{F}$ has bounded treewidth [5]. This yields some surprisingly clean results, e.g., for the class $\mathcal{T}$ of all trees, the *Tree*

I. Sau and D. M. Thilikos (Eds.): WG 2019, LNCS 11789, pp. 338–350, 2019.
https://doi.org/10.1007/978-3-030-30786-8_26

Theorem states that the homomorphism vectors $\mathsf{HOM}_{\mathcal{T}}(G)$ and $\mathsf{HOM}_{\mathcal{T}}(H)$ of two graphs G and H are equal if and only if G and H are not distinguished by *color refinement*, a well-known heuristic algorithm for distinguishing non-isomorphic graphs (e.g. [7]).

"Graph matching" is a term used in machine learning for the problem of measuring the similarity of graphs (e.g., [3]), where it has its applications in pattern recognition. However, there is no universally agreed-upon notion of similarity, and a popular notion, the graph edit distance, describing the cost of transforming one graph into another by adding and deleting vertices and edges, is not only hard to compute but also does not reflect the structural similarity of two graphs very well [10, Section 1.5.1]. Restricted homomorphism vectors offer an alternative way of comparing the structural similarity of graphs since, after suitably scaling them, they can be compared using standard vector norms. As demonstrated in [6], one can also define an inner product on these homomorphism vectors, which yields a mapping that is known as a *graph kernel* in machine learning (e.g., [13]). Graph kernels can be used to perform classification on graphs, and to this end, should capture the similarity of graphs well while still being efficiently computable. Similarly to homomorphism vectors, state-of-the-art graph kernels are usually based on counting certain patterns in graphs, e.g., walks or subtrees.

The original observation by Lovász [8], stating that a graph can be characterized up to isomorphism by counting homomorphisms from all graphs, dates back to the 1960s and has led to the theory of graph limits in the recent past [10]. Only very recently, the importance of homomorphism counts for many graph-related counting problems has been recognized [4]: for example, subgraph counts are just linear combinations of homomorphism counts. Even more recent is the approach of characterizing the structural similarity of graphs by counting homomorphisms from restricted classes of graphs [6], which shows that well-known characterizations, e.g., the color-refinement algorithm, can also be stated in terms of homomorphism counts.

1.1 Overview

Color refinement is a simple and efficient but incomplete algorithm for distinguishing non-isomorphic graphs. The algorithm iteratively computes a coloring of the vertices of a graph, and we say that color refinement *distinguishes* two graphs if it computes different color patterns for them. The Tree Theorem [6] states that color refinement can be characterized by counting homomorphisms from trees, i.e., for all graphs G and H, we have $\mathsf{HOM}_{\mathcal{T}}(G) = \mathsf{HOM}_{\mathcal{T}}(H)$ if and only if color refinement does not distinguish G and H. By making use of the initial coloring, color refinement can easily be adapted to vertex-colored graphs. This enables a straight-forward generalization of the Tree Theorem by counting (color-respecting) homomorphisms from vertex-colored trees to vertex-colored graphs. Formally, if we let $\mathcal{CT}$ denote the class of all vertex-colored trees, then for all vertex-colored graphs G and H, we have $\mathsf{HOM}_{\mathcal{CT}}(G) = \mathsf{HOM}_{\mathcal{CT}}(H)$ if and only if color refinement does not distinguish G and H. We refer to this generalization as the *Colored Tree Theorem.*

A possible (although rather conservative) generalization of the notion of a tree to hypergraphs is that of a *connected Berge-acyclic* hypergraph. A hypergraph is called connected and Berge-acyclic if its incidence graph is connected and acyclic, respectively. Similarly to the case of (vertex-colored) graphs, we obtain a surprisingly clean answer when counting homomorphisms from hypergraphs in the class $\mathcal{BA}$ of connected Berge-acyclic hypergraphs.

Theorem 1. *For all hypergraphs G and H, the following are equivalent:*

(1) $\mathsf{HOM}_{\mathcal{BA}}(G) = \mathsf{HOM}_{\mathcal{BA}}(H)$.
(2) Color refinement does not distinguish G and H.

Of course, color refinement in the usual sense is only defined on (vertex-colored) graphs, which is why we propose a generalization of it to hypergraphs in Sect. 2.1; Theorem 1 refers to this generalization. As this generalization turns out to be equivalent to the usual color-refinement algorithm applied to a colored variant of a hypergraph's incidence graph, we are able to "reduce" Theorem 1 to the Colored Tree Theorem instead of adapting the proof of [6]. Here, the interesting (and laborious) part is to show how homomorphisms between hypergraphs are related to homomorphisms between their colored incidence graphs and how counts of these can be obtained from each other. This leads to the notion of an *incidence homomorphism* between hypergraphs in Sect. 2.2, which is used to prove Theorem 1 in Sect. 2.3. Our approach does not only directly generalize to hypergraphs that possibly have parallel edges, but is also simplified by doing so; we nevertheless obtain the corresponding statement about simple hypergraphs as a corollary in Sect. 2.4.

With Theorem 1, one might wonder how the Tree Theorem generalizes to directed graphs. An obvious candidate for a class of directed graphs to count homomorphisms from is the class of connected directed acyclic graphs (DAGs) since a connected DAG can be seen as the directed concept corresponding to a tree. Surprisingly, counting homomorphisms from DAGs is already too expressive and characterizes an arbitrary directed graph up to isomorphism. This result is already implicit in the second homomorphism-related work of Lovász [9], which is concerned with the *cancellation law* among finite relational structures, and we briefly revisit it in Sect. 3.

1.2 Preliminaries

$\mathbb{N}$ denotes the set of non-negative integers. For $n \in \mathbb{N}$, we let $[n] := \{1, \ldots, n\}$. A multiset is denoted using the notation $\{\!\{ 0, 1, 1 \}\!\}$. All relational structures that we consider are finite, and we use standard graph-theoretic terminology and notation without explicitly introducing it, e.g., for any graph-like structure G, the sets of its vertices and edges are denoted by $V(G)$ and $E(G)$, respectively. Unless explicitly specified otherwise, the terms *graph* and *directed graph* refer to simple graphs and simple directed graphs, respectively, while for the sake of brevity, the term *hypergraph* is used for hypergraphs that may have parallel edges. Formally, a hypergraph is a tuple $G = (V, E, f)$ where V is a set of vertices, E a set of edges, and $f\colon E \to 2^V \setminus \{\varnothing\}$ the incidence function assigning a non-empty set of vertices to every edge, where we usually write f_G to denote f. If f is injective, i.e., if G does not have parallel edges, then we call G a *simple hypergraph.* The incidence graph of a hypergraph G is the bipartite graph $I(G)$ with $V(I(G)) := V(G) \dot\cup E(G)$ and $E(I(G)) := \{ ve \mid v \in f_G(e) \text{ for } e \in E(G) \}$.

We work with *infinite matrices*, which are functions $A\colon I \times J \to \mathbb{R}$ where I and J are countable and locally finite posets. The product $A \cdot B\colon I \times J \to \mathbb{R}$ of two infinite matrices $A\colon I \times K \to \mathbb{R}$ and $B\colon K \times J \to \mathbb{R}$ is defined via $(A \cdot B)_{ij} := \sum_{k\in K} A_{ik} \cdot B_{kj}$ for all $i \in I$, $j \in J$ as long as these sums are finite; otherwise, we leave it undefined, which means that this product is not associative, and we follow the convention that this operator is right-associative to reduce the amount of needed parentheses. An infinite matrix A is called lower triangular and upper triangular if we have $A_{ij} = 0$ for all i, j with $j \not\leq i$ and $A_{ij} = 0$ for all i, j with $i \not\leq j$, respectively. As in the finite case, forward substitution yields that lower and upper triangular infinite matrices with non-zero diagonal entries have left inverses [6] that again are lower and upper triangular, respectively. For simplicity, we usually refer to infinite matrices just as matrices.

Since we allow hypergraphs to have parallel edges, a function on vertices is not sufficient to specify a homomorphism between hypergraphs: a homomorphism from a hypergraph F to a hypergraph G is a pair (h_V, h_E) of mappings $h_V\colon V(F) \to V(G)$ and $h_E\colon E(F) \to E(G)$ such that we have $h_V(f_F(e)) = f_G(h_E(e))$ for every $e \in E(F)$, and $\mathsf{Hom}(F, G)$ denotes the number of homomorphisms from F to G. Note that, if F and G are simple hypergraphs, then the mapping h_E on edges is already uniquely determined by h_V. For a hypergraph G, its homomorphism vector is denoted by $\mathsf{HOM}(G)$, and the restriction of $\mathsf{HOM}(G)$ to a class of hypergraphs $\mathcal{F}$ is denoted by $\mathsf{HOM}_{\mathcal{F}}(G)$. For every isomorphism class of hypergraphs, we fix a representative and call it the *isomorphism type* of the hypergraphs in the class. We view Hom as an infinite matrix indexed by the isomorphism types, which are sorted by the sums of their numbers of vertices and edges, where ties are resolved arbitrarily. Then, for a hypergraph G, its homomorphism vector $\mathsf{HOM}(G)$ can be viewed as a column of Hom. We use similar notation for other types of mappings without explicitly introducing it.

Since, to count homomorphisms from a non-connected graph, one can count homomorphisms from its components instead, we usually restrict ourselves to homomorphism counts from connected graphs. The same holds for directed graphs and hypergraphs. Aut is the diagonal matrix whose diagonal entry $\mathsf{Aut}(G, G)$, which we usually denote just by $\mathsf{Aut}(G)$, contains the number of automorphisms of the connected hypergraph G.

2 Hypergraphs

We only outline the main ideas of our proofs in this section. The actual proofs can be found in the preprint of this paper [1].

2.1 Hypergraph Color Refinement

Color refinement colors the vertices of a graph G by setting $C_0^G(v) := 1$ for every $v \in V(G)$ and $C_{i+1}^G(v) := \{\!\{ C_i^G(u) \mid u \in N_G(v) \}\!\}$ for every $v \in V(G)$ and every $i \geq 0$. In a hypergraph, the adjacency of a vertex v is not fully determined by the set of its neighbors, i.e., the set of vertices that share an edge with v, as this does not state *how* v is connected to them. To capture also this information, we rather look at the edges v is incident to: a coloring of the vertices of a hypergraph induces a coloring of its edges. For a hypergraph G, we define $HC_0^G(v) := 1$ for every $v \in V(G)$ and

$$HC_{i+1}^G(v) := \{\!\{ \{\!\{ HC_i^G(u) \mid u \in f_G(e) \}\!\} \mid e \in E \text{ with } v \in f_G(e) \}\!\}$$

for every $v \in V(G)$ and every $i \geq 0$. Color refinement distinguishes two hypergraphs G and H if there is an $i \geq 0$ such that the colorings are *unbalanced*, i.e., that we have $\{\!\{ HC_i^G(v) \mid v \in V(G) \}\!\} \neq \{\!\{ HC_i^H(v) \mid v \in V(H) \}\!\}$.

Thus, two vertices of the same color get different colors in a refinement round if they have a different number of incident edges of an induced color c. Note that such an induced color of an edge is a multiset since distinct vertices of the same edge may have the same color. It is not hard to see that, when interpreting a graph as a hypergraph, the two definitions are equivalent: an inductive argument yields that excluding the color of $v \in V(G)$ itself from the color $\{\!\{ HC_i^G(u) \mid u \in f_G(e) \}\!\}$ induced on the edge $e \in E$ with $v \in f_G(e)$ does not make a difference. Then, the only difference is that each color of a neighbor is placed into its own multiset in the more general definition.

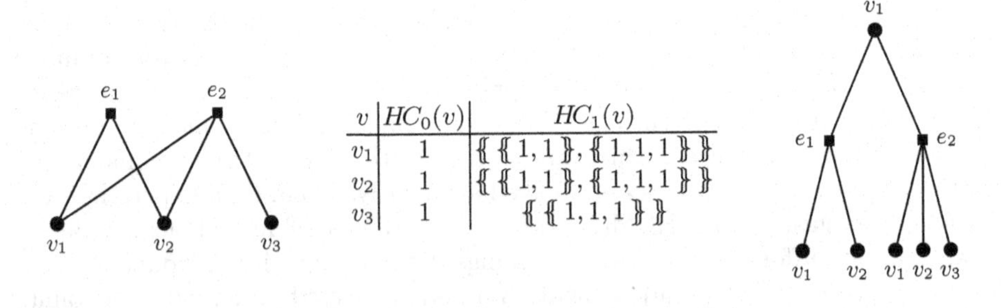

v	$HC_0(v)$	$HC_1(v)$
v_1	1	$\{\!\{ \{\!\{1,1\}\!\}, \{\!\{1,1,1\}\!\} \}\!\}$
v_2	1	$\{\!\{ \{\!\{1,1\}\!\}, \{\!\{1,1,1\}\!\} \}\!\}$
v_3	1	$\{\!\{ \{\!\{1,1,1\}\!\} \}\!\}$

Fig. 1. Color refinement on a hypergraph and the length-one "walk-hypergraph" from v_1

Figure 1 shows an example of color refinement on a hypergraph, which is represented by its incidence graph, where the vertices and edges are depicted as circles and squares, respectively; this distinction is not made in the incidence graph itself. To justify our notion of color refinement, we observe its relation to color refinement on the incidence graph of a hypergraph, which also colors its edges: In a first step, every edge gets assigned the colors of its incident vertices. In a second step, every vertex gets assigned the colors of its incident edges. Hence, a single step of color refinement on a hypergraph corresponds to two steps of color refinement on its incidence graph.

Another way to see this is by considering the tree unfoldings implicitly constructed by color refinement; the color of a vertex v after i refinement rounds can be interpreted as the isomorphism type of the tree obtained by taking all length-i walks from v simultaneously. The proof of the Tree Theorem utilizes this by "redirecting" tree homomorphisms to these tree unfoldings. Analogously, we can think of the colors of hypergraph color refinement as Berge-acyclic hypergraphs, cf. Fig. 1, which can also be thought of as the tree unfoldings obtained by twice the number of steps of color refinement on the incidence graph of the hypergraph.

However, to formally obtain an equivalence between the two notions, we have to deal with the fact that the additional colors of the edges present in color refinement on an incidence graph may obscure unbalanced vertex partitions, which may happen since an incidence graph does not indicate whether one of its vertices is actually a vertex or an edge of the hypergraph, i.e., vertices of the one hypergraph may be confused with edges of the other. To avoid this, we differentiate these right from the beginning by defining the *colored incidence graph* $I_c(G)$ of a hypergraph G, which is the vertex-colored graph obtained by taking the incidence graph $I(G)$ and coloring the elements of $V(G)$ and $E(G)$ with two different colors, say 1 for $V(G)$ and 2 for $E(G)$. In general, for color refinement on a vertex-colored graph, one has to include a vertex's old color in the new one in every refinement round to guarantee that we indeed obtain a refinement. However, a simple inductive argument yields that this is not necessary for colored incidence graphs.

Lemma 1. *For all hypergraphs G and H, the following are equivalent:*

(1) Color refinement does not distinguish G and H.
(2) Color refinement does not distinguish $I_c(G)$ and $I_c(H)$.

2.2 Incidence Homomorphisms

Recall that a hypergraph is connected and Berge-acyclic if and only if its incidence graph is a tree. With the Colored Tree Theorem, Lemma 1 already yields that two hypergraphs G and H are not distinguished by color refinement if and only if $\mathsf{HOM}_{\mathcal{CT}}(I_c(G)) = \mathsf{HOM}_{\mathcal{CT}}(I_c(H))$, i.e., we already have a characterization of color refinement by counting homomorphisms from vertex-colored trees to the hypergraphs' incidence graphs. This motivates a "reduction" to prove Theorem 1, i.e., instead of adapting the proof of the Tree Theorem by defining an unfolding of a hypergraph into a Berge-acyclic one, we relate homomorphisms between colored incidence graphs back to homomorphisms between hypergraphs.

To this end, we first re-formulate $\mathsf{HOM}_{\mathcal{CT}}(I_c(G)) = \mathsf{HOM}_{\mathcal{CT}}(I_c(H))$ in hypergraph terms. Observe that, at this point, it is convenient that we consider hypergraphs with parallel edges because, when interpreting a colored tree as an incidence graph of a hypergraph, it may very well have parallel edges, or more precisely, parallel loops. Thus, when taking the step from vertex-colored trees to hypergraphs, the only noteworthy special case is the colored tree corresponding to an empty edge, which does not have a corresponding hypergraph as empty edges are disallowed by definition.

However, just interpreting vertex-colored trees as hypergraphs does not suffice as, for hypergraphs G and H, the homomorphisms between the colored incidence graphs $I_c(G)$ and $I_c(H)$ do not necessarily correspond to homomorphisms between G and H. While every homomorphism (h_V, h_E) from G to H gives us a corresponding homomorphism $h_V \cup h_E$ from $I_c(G)$ to $I_c(H)$, the converse does not hold: a homomorphism from $I_c(G)$ to $I_c(H)$ does not have to map the vertices of an edge of G to a *full* edge of H but only to a subset of such an edge, cf. Fig. 2. To capture this behavior in terms of hypergraphs, for hypergraphs G and H, we call a pair (h_V, h_E) of mappings $h_V \colon V(G) \to V(H)$ and $h_E \colon E(G) \to E(H)$ satisfying $h_V(f_G(e)) \subseteq f_H(h_E(e))$ for every $e \in E(G)$ an *incidence homomorphism* from G to H. That is, the equality in the definition of a homomorphism is relaxed to an inclusion, which also means that every homomorphism *is* an incidence homomorphism.

Observe that we have a one-to-one correspondence between the incidence homomorphisms from G to H and the homomorphisms from $I_c(G)$ to $I_c(H)$. In particular, if we let $\mathsf{InHom}(G, H)$ denote the number of incidence homomorphisms from G to H, we have $\mathsf{InHom}(G, H) = \mathsf{Hom}(I_c(G), I_c(H))$. This lets us express the requirement $\mathsf{HOM}_{\mathcal{CT}}(I_c(G)) = \mathsf{HOM}_{\mathcal{CT}}(I_c(H))$ in terms of connected Berge-acyclic hypergraphs, where a simple interpolation argument takes care of the colored tree corresponding to an empty edge.

Lemma 2. *For all hypergraphs G and H, the following are equivalent:*

(1) $\mathsf{InHOM}_{\mathcal{BA}}(G) = \mathsf{InHOM}_{\mathcal{BA}}(H)$.
(2) Color refinement does not distinguish G and H.

2.3 Homomorphisms from Berge-Acyclic Hypergraphs

With Lemma 2, it remains to show that counting incidence homomorphisms from $\mathcal{BA}$ is equivalent to counting homomorphisms from $\mathcal{BA}$. To this end, we call an incidence homomorphism (h_V, h_E) from a hypergraph G to a hypergraph H *locally injective*, *locally surjective*, and *locally bijective* if, for every $e \in E(G)$, the restriction $h_V|_{f_G(e)} \colon f_G(e) \to f_H(h_E(e))$ of h_V to the vertices of e is injective, surjective, and bijective, respectively. Note that this definition only concerns the mapping h_V and not h_E, i.e., h_E may still map multiple edges to the same edge as long as the restriction of h_V to each of these edges is injective. For a connected hypergraph G and a hypergraph H, we denote the number of locally injective incidence homomorphisms by $\mathsf{LoInjInHom}(G, H)$ and, since an incidence homomorphism is locally surjective if and only if it is a homomorphism, the number of locally bijective incidence homomorphisms by $\mathsf{LoInjHom}(G, H)$.

The main work is spread across three lemmas: Together, Lemmas 3 and 5 "balance" incidence homomorphisms to locally bijective incidence homomorphisms by first relating incidence homomorphisms to locally injective incidence homomorphisms and then, from there on, to locally bijective incidence homomorphisms. Analogously to Lemmas 3, 4 relates homomorphisms to locally injective homomorphisms or, in other words, locally bijective incidence homomorphisms.

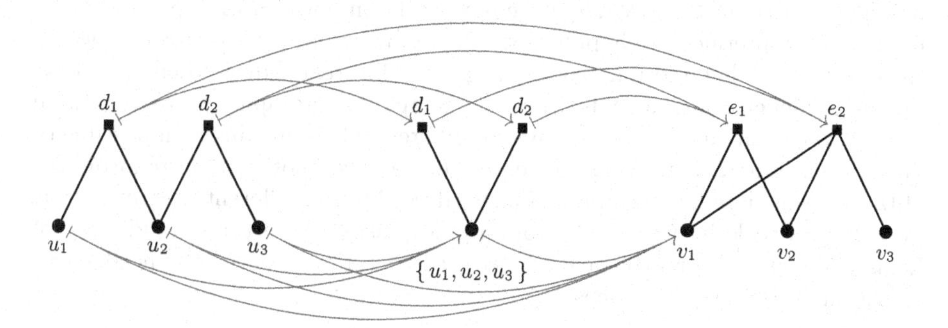

Fig. 2. Decomposition of an incidence homomorphism into a locally merging homomorphism and a locally injective incidence homomorphism

While our goal is to relate incidence homomorphisms to locally surjective incidence homomorphisms, we are forced to take the detour that is local injectivity due to the way we prove Lemma 5: We fill up edges that are mapped non-surjectively by adding leaves, i.e., vertices that are part of exactly one edge. Without this injectivity, which we achieve by merging vertices within an edge that are mapped to the same vertex, these added leaves may be mapped to the same vertex again causing us to overcount endlessly. With local injectivity, also achieving local surjectivity is possible as a locally bijective incidence homomorphism has to map an edge to an edge of exactly the same size. Thus, if we use leaves to fill up an edge to the size of the target edges, we do not overcount as we do not count incidence homomorphisms where adding fewer leaves would have sufficed. Note that, in our setting, it is crucial that we only consider such a local form of injectivity; we have to make sure the Berge-acyclicity is preserved when merging vertices.

To relate incidence homomorphisms to locally injective incidence homomorphisms, we define *locally merging* homomorphisms, which only allow vertices to be mapped to the same vertex if they are part of the same edge. To this end, we first define the relation $\equiv_{h_V} \subseteq V(G) \times V(G)$ for an incidence homomorphism (h_V, h_E) between two hypergraphs G and H by letting $u \equiv_{h_V} v$ if there is a walk $v_0, e_1, \ldots, v_k$ from u to v in G with $h_V(v_{i-1}) = h_V(v_i)$ for every $i \in [k]$. Clearly, $\equiv_{h_V}$ is an equivalence relation, and for all $u, v \in V(G)$, we have that $u \equiv_{h_V} v$ implies $h_V(u) = h_V(v)$. We call a homomorphism (h_V, h_E) between hypergraphs G and H locally merging if

(1) $h_V(u) = h_V(v)$ if and only if $u \equiv_{h_V} v$ for all $u, v \in V(G)$,
(2) h_V is surjective, and
(3) h_E is bijective,

and, for connected hypergraphs G and H, we let $\mathsf{LoMeHom}(G, H)$ be the number of such homomorphisms from G to H.

By decomposing incidence homomorphisms into locally merging homomorphisms and locally injective incidence homomorphisms as in Fig. 2, we obtain Lemma 3. The crucial argument is the fact that the intermediate hypergraph is uniquely determined by (h_V, h_E), i.e., every decomposition of (h_V, h_E) has to use the same intermediate hypergraph. Note that, by merging vertices to obtain the intermediate hypergraph, parallel loops may be created even when decomposing an incidence homomorphism between simple hypergraphs. Moreover, these parallel loops may have to be mapped to different edges, making it impossible to merge them into a single loop. Since, for such a decomposition, automorphisms of the intermediate hypergraph can be used to obtain a different decomposition, we have to divide by the number of automorphisms. Note that the identity of Lemma 3 is stated for arbitrary connected hypergraphs; once it is needed, we restrict it to Berge-acyclic ones.

Lemma 3. *We have* $\mathsf{InHom} = \mathsf{LoMeHom} \cdot \mathsf{Aut}^{-1} \cdot \mathsf{LoInjInHom}$. *The matrix* $\mathsf{LoMeHom}$ *is invertible and lower triangular.*

For the special case of homomorphisms, i.e., locally surjective incidence homomorphisms, the proof of Lemma 3 also directly yields Lemma 4.

Lemma 4. *We have* $\mathsf{Hom} = \mathsf{LoMeHom} \cdot \mathsf{Aut}^{-1} \cdot \mathsf{LoInjHom}$.

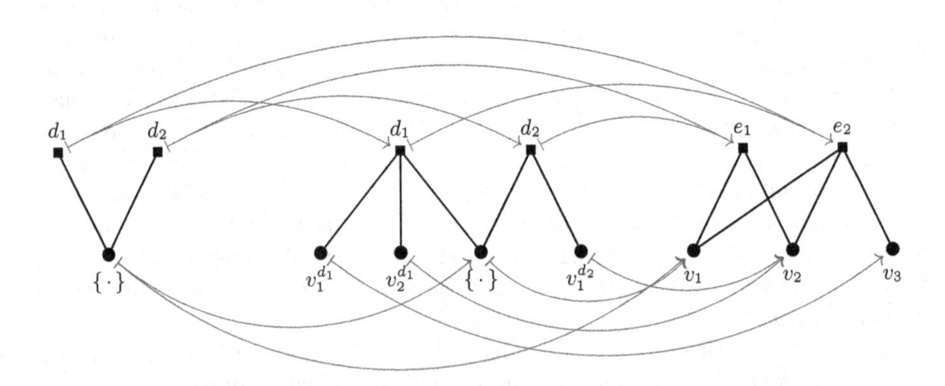

Fig. 3. Decomposition of a locally injective incidence homomorphism into a leaf-adding incidence homomorphism and a locally injective homomorphism

To prove Lemma 5, we define *leaf-adding* incidence homomorphisms, which are embeddings of a hypergraph into another one that has no additional vertices or edges with the exception of leaves. For this, we need the notion of a *strong* incidence homomorphism between hypergraphs G and H, which is an incidence homomorphism (h_V, h_E) from G to H that additionally satisfies the inclusion $h_V^{-1}(f_H(h_E(e)) \cap im(h_V)) \subseteq f_G(e)$ for every $e \in E(G)$; it is actually not hard to see that this is equivalent to requiring that the corresponding homomorphism between the colored incidence graphs $I_c(G)$ and $I_c(H)$ is a strong homomorphism. We call an incidence homomorphism (h_V, h_E) between hypergraphs G and H leaf-adding if

(1) (h_V, h_E) is a strong incidence homomorphism,
(2) h_V is injective,
(3) h_E is bijective, and
(4) the vertices $V(H) \setminus im(h_V)$ are leaves of H,

and, for connected hypergraphs G and H, we let $\mathsf{LeafAddInHom}(G, H)$ be the number of leaf-adding incidence homomorphisms from G to H. Similarly to the proof of Lemma 3, the proof of Lemma 5 decomposes locally injective incidence homomorphisms into leaf-adding incidence homomorphisms and locally injective homomorphisms as in Fig. 3. Again, this identity is proven for arbitrary connected hypergraphs, and we restrict it to Berge-acyclic ones once it is needed.

Lemma 5. *We have* $\mathsf{LoInjInHom} = \mathsf{LeafAddInHom} \cdot \mathsf{Aut}^{-1} \cdot \mathsf{LoInjHom}$. *The matrix* $\mathsf{LeafAddInHom}$ *is invertible and upper triangular.*

We have all we need to prove that counting incidence homomorphisms from $\mathcal{BA}$ is equivalent to counting homomorphisms from $\mathcal{BA}$. Combining Lemmas 3 and 5 yields $\mathsf{InHom} = \mathsf{LoMeHom} \cdot \mathsf{Aut}^{-1} \cdot \mathsf{LeafAddInHom} \cdot \mathsf{Aut}^{-1} \cdot \mathsf{LoInjHom}$, and Lemma 4 states that we have $\mathsf{Hom} = \mathsf{LoMeHom} \cdot \mathsf{Aut}^{-1} \cdot \mathsf{LoInjHom}$. Even with the invertibility of $\mathsf{LoMeHom}$ and $\mathsf{LeafAddInHom}$, the proof of Lemma 6 is not trivial as the inverse of the upper triangular matrix $\mathsf{LeafAddInHom}$ is still an upper triangular matrix, and hence, left multiplication with it may be undefined. This, however, can be avoided by considering finite submatrices as in [6]. This proof finishes our "reduction" and, hence, the proof of Theorem 1 as it follows immediately from Lemmas 2 and 6.

Lemma 6. *For all hypergraphs G and H, the following are equivalent:*

(1) $\mathsf{InHOM}_{\mathcal{BA}}(G) = \mathsf{InHOM}_{\mathcal{BA}}(H)$.
(2) $\mathsf{HOM}_{\mathcal{BA}}(G) = \mathsf{HOM}_{\mathcal{BA}}(H)$.

2.4 Simple Hypergraphs

For a restriction of Theorem 1 to simple hypergraphs, consider a homomorphism (h_V, h_E) from a hypergraph G to a simple hypergraph H. If $e, e' \in E(G)$ are parallel edges of G, i.e., $f_G(e) = f_G(e')$, then we have $f_H(h_E(e)) = h_V(f_G(e)) = h_V(f_G(e')) = f_H(h_E(e'))$, which implies $h_E(e) = h_E(e')$ since H does not have parallel edges. That is, parallel edges of G have to be mapped to the same edge of H since a homomorphism's mapping on edges is determined by its mapping on vertices up to parallel edges. Hence, if we consider the simple hypergraph G' obtained by merging parallel edges of G, then there is a one-to-one correspondence between the homomorphisms from G to H and these from G' to H, and in particular, we have $\mathsf{Hom}(G, H) = \mathsf{Hom}(G', H)$. Thus, for a simple hypergraph, it suffices to count homomorphisms from simple hypergraphs, and we obtain Corollary 1, where $\mathcal{SBA}$ denotes the class of all connected Berge-acyclic simple hypergraphs.

Corollary 1. *For all simple hypergraphs G and H, the following are equivalent:*

(1) $\mathsf{HOM}_{\mathcal{SBA}}(G) = \mathsf{HOM}_{\mathcal{SBA}}(H)$.
(2) Color refinement does not distinguish G and H.

For incidence homomorphisms, however, the situation is not as clear as these may map parallel edges to non-parallel ones. However, with an interpolation argument, it is possible to prove that such a restriction can be made.

Lemma 7. *For all simple hypergraphs G and H, the following are equivalent:*

(1) $\mathsf{InHOM}_{\mathcal{SBA}}(G) = \mathsf{InHOM}_{\mathcal{SBA}}(H)$.
(2) Color refinement does not distinguish G and H.

3 Directed Graphs

To prove that counting homomorphisms from DAGs suffices to characterize arbitrary directed graphs up to isomorphism, one could proceed in a similar fashion to [6], i.e., by defining an unfolding of a directed graph into a DAG and then proving the equivalence of counting homomorphisms and unfolding numbers. This way, one obtains a characterization that is more intuitive than that of homomorphism counts, and one could show that an isomorphism between the directed graphs can be extracted from an isomorphism between appropriate unfoldings. However, as the class of DAGs can also be defined in terms of homomorphism numbers, a proof using the algebraic properties of homomorphism counts turns out to be much simpler.

Lovász's second homomorphism-related work [9] concerns the cancellation law among finite relational structures. For the case of graphs, this asks whether a graph K cancels out from the *tensor products* $G \otimes K \cong H \otimes K$, i.e., whether it satisfies the implication $G \otimes K \cong H \otimes K \implies G \cong H$ for all graphs G and H. Note that the tensor product $G \otimes H$ of two graphs G and H is the graph with vertex set $V(G) \times V(H)$ that has an edge between (u, u') and (v, v') if and only if $uv \in E(G)$ and $u'v' \in E(H)$. Lovász gives the answer that this implication holds if and only if K is not bipartite. Moreover, from his work on the general case of finite relational structures, it follows that the transitive tournament $\overrightarrow{K_n}$ on n vertices satisfies the cancellation law for directed graphs as long as $n \geq 3$.

To see how the cancellation law is related to homomorphism counts, observe that the class of DAGs can be defined as the class of all directed graphs that have a homomorphism into a transitive tournament. Formally, if we let $\mathcal{A}$ denote the class of DAGs and define $\mathcal{A}_n := \{ G \mid \mathsf{Hom}(G, \overrightarrow{K_n}) > 0 \}$ for every $n \in \mathbb{N}$, then we have $\mathcal{A} = \cup_{n \in \mathbb{N}} \mathcal{A}_n$. Then, using the facts that two directed graphs G

and H are isomorphic if and only if we have $\mathsf{Hom}(F, G) = \mathsf{Hom}(F, H)$ for every directed graph F [8] and that $\mathsf{Hom}(F, G \otimes H) = \mathsf{Hom}(F, G) \cdot \mathsf{Hom}(F, H)$ holds for all directed graphs F, G, and H [9], we get that

$$
\begin{aligned}
& G \otimes \overrightarrow{K_n} \cong H \otimes \overrightarrow{K_n} \\
\iff & \forall F.\ \mathsf{Hom}(F, G \otimes \overrightarrow{K_n}) = \mathsf{Hom}(F, H \otimes \overrightarrow{K_n}) \\
\iff & \forall F.\ \mathsf{Hom}(F, G) \cdot \mathsf{Hom}(F, \overrightarrow{K_n}) = \mathsf{Hom}(F, H) \cdot \mathsf{Hom}(F, \overrightarrow{K_n}) \\
\iff & \mathsf{HOM}_{\mathcal{A}_n}(G) = \mathsf{HOM}_{\mathcal{A}_n}(H)
\end{aligned}
$$

holds for all directed graphs G and H and every $n \in \mathbb{N}$. That is, tensor products with $\overrightarrow{K_n}$ are directly related to counting homomorphisms from $\mathcal{A}_n$.

With the work of Lovász [9], this yields that two directed graphs G and H are isomorphic if and only if, for every DAG D, we have $\mathsf{Hom}(D, G) = \mathsf{Hom}(D, H)$. More precisely, we obtain the even stronger statement that it suffices to count homomorphisms from the DAGs in $\mathcal{A}_3$, i.e., from DAGs where the longest directed walk has length two. For the case of undirected graphs, an analogous argument with the complete graph on three vertices K_3, which is not bipartite, yields that arbitrary graphs can be characterized up to isomorphism by counting homomorphisms from all three-colorable graphs.

4 Conclusion

We have proven a generalization of the Tree Theorem for hypergraphs. To this end, we have introduced a generalization of the color refinement algorithm for hypergraphs, which has lead to the notion of an incidence homomorphism. By showing how incidence homomorphisms are related to homomorphisms, we have "reduced" the case of hypergraphs to the case of vertex-colored graphs. For the case of directed graphs, we have revisited a result of Lovász, which shows that the class of DAGs is already too expressive to obtain an analogue of the Tree Theorem.

The central open question posed by our generalization of the Tree Theorem is whether it can further be generalized; the Tree Theorem can be generalized to the k-dimensional Weisfeiler-Leman algorithm (k-WL), a generalization of color refinement that colors k-tuples instead of single vertices, and the class of all graphs of treewidth at most k. More precisely, two graphs G and H are not distinguished by k-WL if and only if $\mathsf{Hom}(F, G) = \mathsf{Hom}(F, H)$ for every graph F of treewidth at most k [6]. An attempt to generalize our result could be to consider k-WL on the colored incidence graphs of hypergraphs as proposed in a recent preprint by Brooksbank et al. [2], in which case, however, our reduction does not generalize as we cannot restrict the identities of Lemmas 3 and 4 to hypergraphs whose incidence graphs have treewidth at most k; merging vertices of a graph may increase its treewidth even when the merged vertices are part of the same neighborhood. A way of interpreting this is that the treewidth of the incidence graph of a hypergraph G is not a meaningful notion for G since it mixes up the vertices and edges of G.

Besides the generalization to k-WL, there are other natural questions: Is there a linear-algebraic characterization of hypergraph color refinement like there is for graphs [11,12]? Do homomorphisms from hypergraph analogues of paths and cycles yield characterizations similar to the ones for graphs [6]?

References

1. Böker, J.: Color refinement, homomorphisms, and hypergraphs. CoRR abs/1903.12432 (2019, preprint). http://arxiv.org/abs/1903.12432
2. Brooksbank, P.A., Grochow, J.A., Li, Y., Qiao, Y., Wilson, J.B.: Incorporating weisfeiler-leman into algorithms for group isomorphism. CoRR abs/1905.02518 (2019). http://arxiv.org/abs/1905.02518
3. Conte, D., Foggia, P., Sansone, C., Vento, M.: Thirty years of graph matching in pattern recognition. Int. J. Pattern Recogn. Artif. Intell. **18**(3), 265–298 (2004)
4. Curticapean, R., Dell, H., Marx, D.: Homomorphisms are a good basis for counting small subgraphs. In: Proceedings of the 49th Annual ACM SIGACT Symposium on Theory of Computing, STOC 2017, pp. 210–223. ACM (2017)
5. Dalmau, V., Jonsson, P.: The complexity of counting homomorphisms seen from the other side. Theor. Comput. Sci. **329**(1), 315–323 (2004). https://doi.org/10.1016/j.tcs.2004.08.008. http://www.sciencedirect.com/science/article/pii/S0304397504005560
6. Dell, H., Grohe, M., Rattan, G.: Lovász meets Weisfeiler and Leman. In: Chatzigiannakis, I., Kaklamanis, C., Marx, D., Sannella, D. (eds.) 45th International Colloquium on Automata, Languages, and Programming (ICALP 2018). Leibniz International Proceedings in Informatics (LIPIcs), vol. 107, pp. 40:1–40:14. Schloss Dagstuhl-Leibniz-Zentrum fuer Informatik (2018)
7. Grohe, M., Kersting, K., Mladenov, M., Schweitzer, P.: Color refinement and its applications. In: An Introduction to Lifted Probabilistic Inference. Cambridge University Press (to appear). https://www.lics.rwth-aachen.de/global/show_document.asp?id=aaaaaaaaabbtcqu
8. Lovász, L.: Operations with structures. Acta Math. Hung. **18**(3–4), 321–328 (1967)
9. Lovász, L.: On the cancellation law among finite relational structures. Periodica Math. Hung. **1**(2), 145–156 (1971)
10. Lovász, L.: Large Networks and Graph Limits. American Mathematical Society (2012)
11. Tinhofer, G.: Graph isomorphism and theorems of birkhoff type. Computing **36**(4), 285–300 (1986). https://doi.org/10.1007/BF02240204
12. Tinhofer, G.: A note on compact graphs. Discrete Appl. Math. **30**(2), 253–264 (1991). https://doi.org/10.1016/0166-218X(91)90049-3. http://www.sciencedirect.com/science/article/pii/0166218X91900493
13. Vishwanathan, S.V.N., Schraudolph, N.N., Kondor, R., Borgwardt, K.M.: Graph kernels. J. Mach. Learn. Res. **11**, 1201–1242 (2010)

3-Colorable Planar Graphs Have an Intersection Segment Representation Using 3 Slopes

Daniel Gonçalves(✉)

LIRMM, Université de Montpellier & CNRS, Montpellier, France
goncalves@lirmm.fr

Abstract. In his PhD Thesis E.R. Scheinerman conjectured that planar graphs are intersection graphs of segments in the plane. This conjecture was proved with two different approaches. In the case of 3-colorable planar graphs E.R. Scheinerman conjectured that it is possible to restrict the set of slopes used by the segments to only 3 slopes. Here we prove this conjecture by using an approach introduced by S. Felsner to deal with contact representations of planar graphs with homothetic triangles.

Keywords: Planar graphs · Segment intersections

1 Introduction

In this paper, we consider intersection representations for planar graphs. A *segment representation* of a graph G maps every vertex $v \in V(G)$ to a segment $\mathbf{v}$ of the plane so that two segments $\mathbf{u}$ and $\mathbf{v}$ intersect if and only if $uv \in E(G)$. Although this graph family is simply defined, it is not easy to manipulate. Actually, even if this class of graphs is small (there are less than $2^{O(n \log n)}$ such graphs with n vertices [12]) a segment representation may be long to encode (in the representations of some of these graphs the endpoints of the segments need at least $2^{\sqrt{n}}$ bits to be coded [10]). There are also interesting open problems concerning this class of graphs. For example, we know that deciding whether a graph G admits a segment representation is NP-hard, indeed it is $\exists\mathbb{R}$-complete [9] but it is still open whether this problem belongs to NP or not. Here we focus on segment representations for planar graphs.

In his PhD Thesis, Scheinerman [13] conjectured that every planar graph has a segment representation. This conjecture attracted a lot of attention. de Fraysseix and Ossona de Mendez [5] proved it for a large family of planar graphs, the planar graphs having a 4-coloring in which every induced cycle of length 4 uses at most 3 colors. In particular, this implies the conjecture for 3-colorable

This research is partially supported by the ANR GATO, under contract ANR-16-CE40-0009.

I. Sau and D. M. Thilikos (Eds.): WG 2019, LNCS 11789, pp. 351–363, 2019.
https://doi.org/10.1007/978-3-030-30786-8_27

planar graphs. Then Chalopin and the author finally proved this conjecture [2]. Recently, a much simpler proof was provided by the author, Isenmann, and Pennarun [7]. Here we focus on segment representations of planar graphs with further restrictions.

In his PhD Thesis, Scheinerman [13] proved that every outerplanar graph has a segment representation where only 3 slopes are used, and where parallel segments do not intersect. Let us call such a representation a *3-slopes segment representation.* This result led Scheinerman conjecture [14] (see also [5]) that such representation exists for every 3-colorable planar graph. Later, several groups proved a related result on bipartite planar graphs [3,6,8]. They proved that every bipartite planar graph has a 2-slopes segment representation, with the extra property that segments do not cross each other. Let us call such a representation a *2-slopes contact segment representation.* More recently de Castro *et al.* [1] considered a particular class of 3-colorable planar graphs. They proved that every triangle-free planar graph has a 3-slopes contact segment representation. Such a contact segment representation cannot be asked for any 3-colorable planar graph. Indeed, up to isomorphism, the octahedron has only one 3-slopes contact segment representation depicted in Fig. 1, and one can easily check that this representation does not extend to the (3-colorable) graph obtained after gluing a copy of an octahedron in each of its faces. However, we will use 3-slopes contact segment representations in the proof of our main result.

Theorem 1. *Every 3-colored planar graph has a 3-slopes segment representation such that parallel segments correspond to the color classes.*

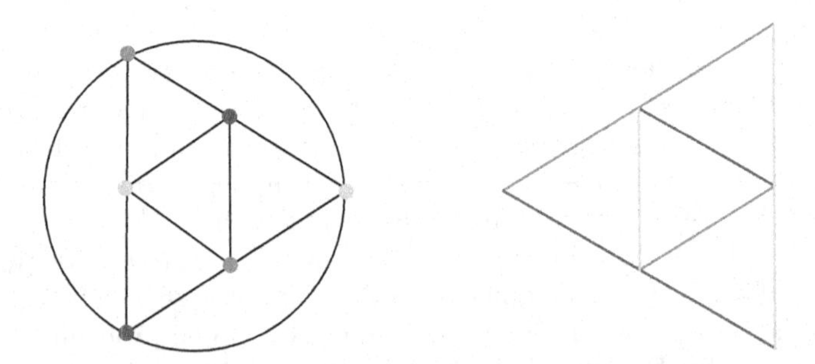

Fig. 1. The octahedron and a 3-slopes contact representation. It is unique, up to vertex automorphism, up to scaling, and once the slopes are set. (Color figure online)

As every 3-colored planar graph is the induced subgraph of some 3-colored triangulation we only consider the case of triangulations in the following. In Sect. 2 we review some basic definitions. Section 3 is devoted to the so-called *triangular contact schemes.* It is shown that every 3-colorable triangulation admits such a scheme. Then, those schemes are used in Sect. 4 to build 3-slopes segment representations. Finally, we conclude with some remarks on 4-slopes segment representations.

2 Terminology

A *triangulation* is a plane graph where every face has size three. Unless stated otherwise, in this paper triangulations are simple, that means without loops nor multiple edges. A triangulation T, simple or not, is *Eulerian* if every vertex has even degree. It is folklore that these triangulations are the 3-colorable triangulations. Actually these triangulations are uniquely 3-colorable (up to color permutation). Hence their vertex set $V(T)$ is canonically partitioned into three independent sets A, B and C. In the following we will denote the vertices of these sets respectively a_i with $0 \leq i < |A|$, b_j with $0 \leq j < |B|$, and c_k with $0 \leq k < |C|$. In such a triangulation T any face is incident to one vertex a_i, one vertex b_j, and one vertex c_k, and these vertices appear in this order either clockwisely or counterclockwisely. In the following, the vertices of the outerface are always denoted a_0, b_0 and c_0, and they appear clockwisely in this order around T. As the orders of two adjacent faces are opposite, the dual graph of T is bipartite. Given an Eulerian triangulation T with face set $F(T)$, let us denote by $F_1(T)$ and $F_2(T)$ (or simply F_1 and F_2 if it is clear from the context) the face sets partitioning $F(T)$, such that no two adjacent faces belong to the same set, and such that $F_2(T)$ contains the outer-face. Note that by construction for any face $f \in F_1(T)$ (resp. $f \in F_2(T)$) its vertices a_i, b_j and c_k appear in clockwise (resp. counterclockwise) order around f. Note that the vertices a_0, b_0 and c_0 appear in clockwise order around T, but in counterclockwise order w.r.t. the outer face. Let $n = |V(T)|$. As T is a triangulation, by Euler's formula it has $2n - 4$ faces. Hence, as T's dual is bipartite and 3-regular, $|F_1(T)| = |F_2(T)| = n - 2$.

In the following we build 3-slopes segment representations. The 3 slopes used are expected to be distinct, but apart from that the exact 3 slopes considered do not matter. Indeed, for any two triples of slopes, (s_1, s_2, s_3) and (s'_1, s'_2, s'_3), there exists an affine map of the plane turning any 3-slopes segment representation using slopes (s_1, s_2, s_3) into a 3-slopes segment representation using slopes (s'_1, s'_2, s'_3). We denote $\overrightarrow{a}$, $\overrightarrow{b}$, and $\overrightarrow{c}$ the vectors corresponding to slopes of the sets A, B, and C respectively. The magnitude of these vectors is chosen such that $\overrightarrow{a} + \overrightarrow{b} + \overrightarrow{c} = \overrightarrow{0}$.

3 TC-Representations and TC-Schemes

We begin with the definition of particular 3-slopes contact representations illustrated in Fig. 2.

Definition 1. *A* Triangular 3-slopes segment Contact representation *(*TC-representation *for short) is a 3-slopes contact segment representation using the same slopes as* $\overrightarrow{a}$, $\overrightarrow{b}$, *and* $\overrightarrow{c}$, *and where:*

- *Three segments* $\mathbf{a}_0$, $\mathbf{b}_0$, *and* $\mathbf{c}_0$, *form a triangle which contains all the other segments.*
- *Every inner region is a triangle, whose each side is contained in a segment of the representation.*

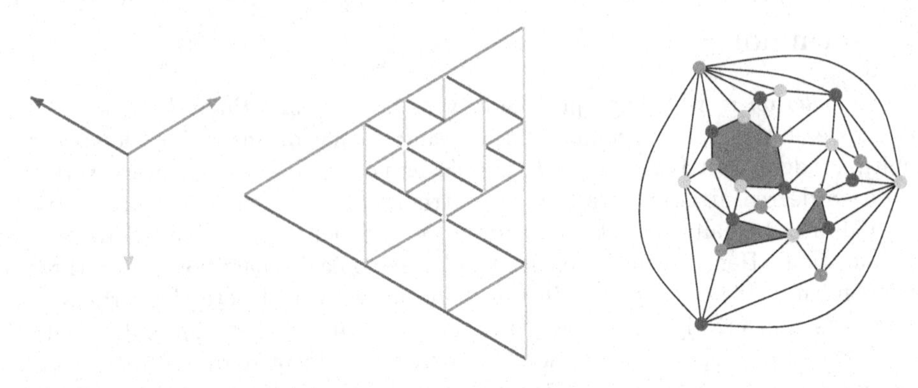

Fig. 2. (left) Vectors $\overrightarrow{a}$, $\overrightarrow{b}$, and $\overrightarrow{c}$. (middle) A TC-representation with various types of intersection points. (right) Its induced graph, where gray faces are particular degenerate faces. One has size six, and there are two faces of size three that correspond to the same intersection point.

– *Two parallel segments intersect on at most one point, their endpoint.*

Definition 2. *Let the plane graph $M(\mathcal{R})$* induced *by a TC-representation $\mathcal{R}$ be the map whose vertices correspond to the segments of the representation, and where two vertices are adjacent if and only if the corresponding segments form a corner of one of the inner triangles. The orders of the neighbors around a vertex v correspond to the order of the segments around* $\mathbf{v}$.

Note that the plane graph induced by a TC-representation has several properties. For example, two parallel segments correspond to non-adjacent vertices. The slopes hence define a 3-coloring of the graph. Note also that the dual graph of $M(\mathcal{R})$ is bipartite. Indeed such a map has two types of faces, one set contains the (triangular) faces corresponding to the inner regions of the TC-representation, and the other set contains the outerface and the faces corresponding to intersection points. Let us denote the latter faces *degenerate faces*, and note that those faces have size three or six. A size six face $(a_i, b_j, c_k, a_{i'}, b_{j'}, c_{k'})$ comes from the intersection point of six segments, and as those six segments go in distinct directions they do not intersect elsewhere, so this cycle has no chord in $M(\mathcal{R})$. Finally note that going clockwise in any inner region one successively follows $\alpha\overrightarrow{a}$, $\alpha\overrightarrow{b}$, and then $\alpha\overrightarrow{c}$, for some not necessarily positive value α.

Definition 3. *A TC-representation $\mathcal{R}$ is a* TC-scheme *of an Eulerian triangulation T if $M(\mathcal{R})$ is a subgraph of T with the same outer-face as T and such that the vertices and edges of $V(T) \setminus V(M(\mathcal{R}))$ lie inside degenerate faces of $M(\mathcal{R})$ (see Fig. 3).*

Actually as in $M(\mathcal{R})$, the inner faces around any vertex alternate among degenerate and non-degenerate. This implies that every edge of $M(\mathcal{R})$ bounds a non-degenerate face, and a face that is degenerate or that is the outerface. We thus have the following.

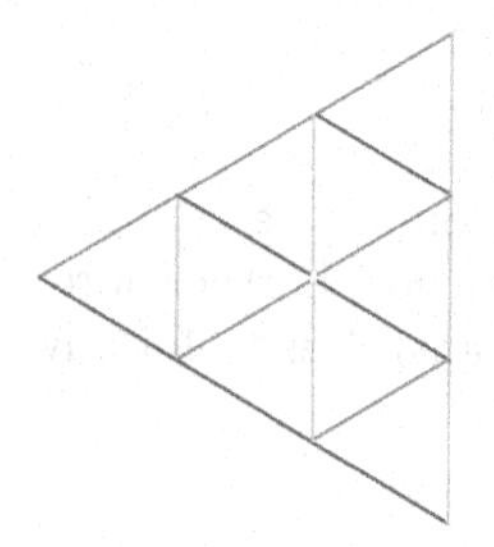
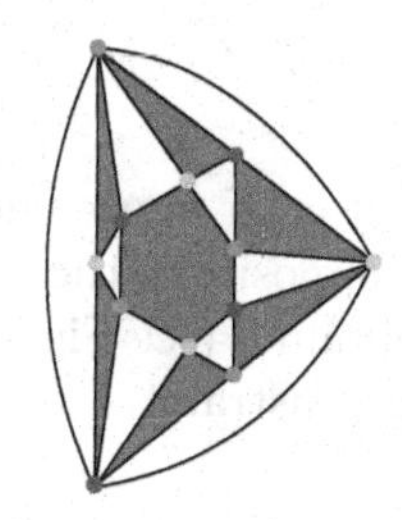
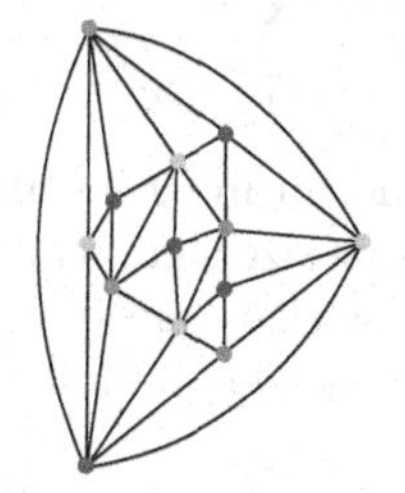
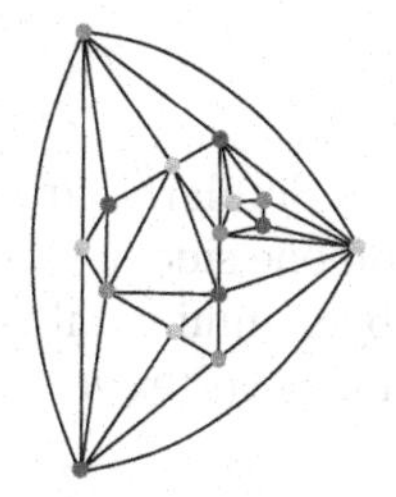

Fig. 3. From left to right. A TC-representation $\mathcal{R}$; its induced map $M(\mathcal{R})$, where gray faces are the degenerate faces; and two triangulations having $\mathcal{R}$ as TC-scheme.

Remark 1. A TC-representation $\mathcal{R}$ is a TC-scheme of T if and only if the non-degenerate faces of $M(\mathcal{R})$ and its outerface are faces of T.

The main ingredient in the proof of Theorem 1 is the following.

Theorem 2. *Every Eulerian triangulation T has a TC-scheme, and this scheme is unique.*

To prove this theorem we proceed by the following steps. We first model TC-schemes of T by means of a system of linear equations and we sketch out why this linear system always has a unique solution.

3.1 The Linear System Model

In a TC-representation all the triangles are homothetic. Let us define the *size* of a triangle as its relative size with respect to the outer-triangle. We may require that the outer-triangle has size 1, the triangles with a corner on the left have positive sizes, while the triangles with a corner on the right have negative sizes. The variables of our linear system correspond to the sizes of the triangular regions. So for each face $f \in F_1$ we have a variable x_f. Informally, the value of x_f will prescribe the size and shape of the corresponding triangle in a TC-representation. If $x_f < 0$, $x_f = 0$, or if $x_f > 0$ the corresponding triangle has a corner on the right, is missing, or has a corner on the left, respectively.

Let us denote by $F_1(v)$ the subset of faces of F_1 incident to v. As the outer triangle has size 1 and contains the other triangles, the faces in $F_1(a_0)$ should have non-negative sizes, and they should sum up to 1 (see Fig. 4 left). We hence consider the following constraint.

$$\sum_{f \in F_1(a_0)} x_f = 1 \qquad (a_0)$$

We add no constraint about the sign of these sizes. Note that similar constraints hold for b_0 and c_0.

$$\sum_{f \in F_1(b_0)} x_f = 1 \qquad (b_0)$$

$$\sum_{f \in F_1(c_0)} x_f = 1 \qquad (c_0)$$

Similarly, around an inner segment of a TC-representation all the triangles on one side have same size sign, which is opposite to the other side. Furthermore, by summing all these sizes one should obtain 0 (see Fig. 4 right). Hence, for any inner-vertex u we consider the following constraint.

$$\sum_{f \in F_1(u)} x_f = 0 \qquad (u)$$

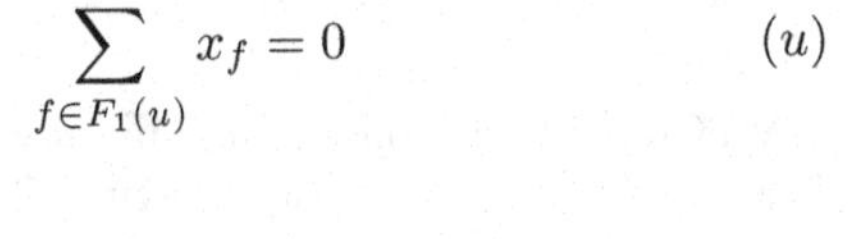

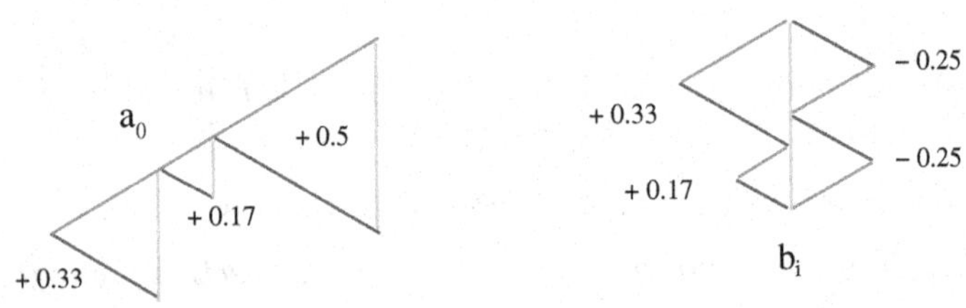

Fig. 4. (left) The size of the triangles around a_0. (right) The size of the triangles around some inner vertex b_i.

In the following, Equ. (a_j) will refer to Eq. (u) where vertex u is replaced by a_j. Note that every face $f \in F_1$ is incident to exactly one vertex of A, one vertex of B, and one vertex of C. Hence by summing Eqs. (a_0), (a_1),...,$(a_{|A|})$, one obtains that $\sum_{f \in F_1} x_f = 1$. The same holds with Eqs. (b_0), (b_1),...,$(b_{|B|})$, or with Eqs. (c_0), (c_1),...,$(c_{|C|})$. Equations (b_0) and (c_0) are hence implied by the others and thus we do not need to consider them anymore. Let us denote by $\mathcal{L}$ the obtained system of $n-2$ linear equations on $|F_1| = n-2$ variables.

Let us define the set $V' = V \setminus \{b_0, c_0\}$ of size $n-2$. Finding a solution to $\mathcal{L}$ is equivalent to finding a vector $S \in \mathbb{R}^{F_1}$ (that is a vector indexed by elements of F_1) such that $MS = I$, where $M \in \mathbb{R}^{V' \times F_1}$ (a square matrix indexed by elements of $V' \times F_1$) and $I \in \mathbb{R}^{V'}$ are defined by

$$M(x_i, f) = \begin{cases} 1 & \text{if } f \in F_1(x_i) \\ 0 & \text{otherwise.} \end{cases} \qquad I(x_i) = \begin{cases} 1 & \text{if } x_i = a_0 \\ 0 & \text{otherwise.} \end{cases}$$

Given some bijective mappings $g_{V'} : [1, \ldots, n-2] \longrightarrow V'$ and $g_{F_1} : [1, \ldots, n-2] \longrightarrow F_1$, one can index the elements of M by pairs $(i, j) \in [1, \ldots, n-2] \times [1, \ldots, n-2]$, and thus define the determinant of M. By the following lemma, $\mathcal{L}$ has a solution vector S, and this solution is unique.

Lemma 1. *The matrix M is non-degenerate, i.e.* $\det(M) \neq 0$.

The full proof of this lemma is inspired by the work of Felsner [4] on contact representations with homothetic triangles. The proof is not provided in this extended abstract but the main idea is to consider the bipartite graph T_M with independent sets V' and F_1 such that $v \in V'$ and $f \in F_1$ are adjacent if and only if v and f are incident in T. Note that M is the biadjacency matrix of T_M. From the embedding of T one can easily embed T_M in such a way that all the inner faces have size 6, and such that a_0 is on the outerboundary.

Note that every perfect matching of T_M (if any) corresponds to a permutation σ on $[1, \ldots, n-2]$ (we say σ belongs to the permutation group S_{n-2}) defined by $\sigma(g_{F_1}^{-1}(f)) = g_{V'}^{-1}(v)$, for any edge vf of the perfect matching. If the obtained permutation is even we call such perfect matching positive, otherwise it is negative. From the Leibniz formula for the determinant,

$$\det(M) = \sum_{\sigma \in S_{n-2}} sgn(\sigma) \prod_{i \in [1,\ldots,n-2]} M(g_{V'}(\sigma(i)), g_{F_1}(i)),$$

one can see that $\det(M)$ counts the number of positive perfect matchings of T_M minus its number of negative perfect matchings.

Claim. The graph T_M admits at least one perfect matching.

Given a graph G and a perfect matching M of G, an *alternating cycle* C is a cycle of G with edges alternating between M and $E(G) \setminus M$. Note that replacing in M the edges of $M \cap C$ by the edges of $C \setminus M$ yields another perfect matching. We call such operation a *cycle exchange*. It is folklore that the set of perfect matchings of a graph are linked by cycle exchanges. Actually, for T_M one can restrict itself to cycles of length six.

Claim. All the perfect matchings of T_M are linked by 6-cycle exchanges.

This implies that the perfect matchings of T_M are either all positive, or all negative. Thus $\det(M) \neq 0$. The following lemma (not proved in this extended abstract) allows us to conclude the proof of Theorem 2.

Lemma 2. *Every Eulerian triangulation T admits a TC-scheme $\mathcal{R}$ that corresponds to the solution of its linear system $\mathcal{L}$.*

4 3-Slopes Segment Representations

In this section we use Theorem 2 to prove the main theorem of the article, Theorem 1. As already mentioned it is sufficient to prove it for Eulerian triangulations. Given an Eulerian triangulation T, let us denote a_1, b_1 and c_1 the vertices forming a face with vertices b_0 and c_0, with a_0 and c_0, and with a_0 and b_0, respectively. Theorem 1 follows from the following technical proposition.

Proposition 1. *For every $\epsilon > 0$, every Eulerian triangulation T admits a 3-slopes segment representations $\mathcal{R}$ such that:*

- *The segments $\mathbf{a_0}$, $\mathbf{b_0}$, and $\mathbf{c_0}$ form a triangle Δ of size 1 (its sides are obtained by following $\overrightarrow{a}$, $\overrightarrow{b}$, and $\overrightarrow{c}$).*

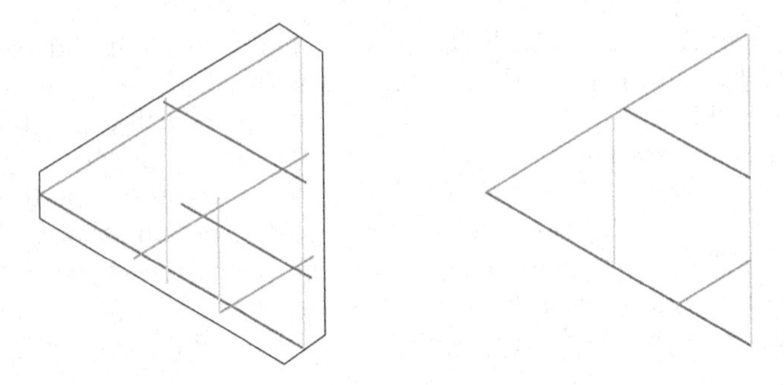

Fig. 5. (left) A 3-slopes segment representation inside an hexagon. (right) A scheme representing its shape.

- *Every segment is contained in the hexagon centered on Δ, obtained by successively following* $(1-\epsilon)\overrightarrow{a}$, $-2\epsilon\overrightarrow{c}$, $(1-\epsilon)\overrightarrow{b}$, $-2\epsilon\overrightarrow{a}$, $(1-\epsilon)\overrightarrow{c}$, *and* $-2\epsilon\overrightarrow{b}$ *(see Fig. 5).*

Given such representation $\mathcal{R}$, we define the *shape* of $\mathcal{R}$ as the triplet (s_a, s_b, s_c) of sizes in $\mathcal{R}$ (w.r.t. the outer-triangle) of the triangles corresponding to $a_1b_0c_0$, $a_0b_1c_0$, $a_0b_0c_1$, respectively. Note that if ϵ is chosen sufficiently small, that is for $\epsilon < 1$, as the vertices a_1, b_1, and c_1 have neighbors that are inner vertices, $\mathbf{a_1}$, $\mathbf{b_1}$, and $\mathbf{c_1}$ intersect Δ, and we have $s_a > 0$, $s_b > 0$, and $s_c > 0$.

Proof. We proceed by induction as we assume that the proposition holds for any Eulerian triangulation with less vertices. The initial case of this induction, when $|V(T)| = 3$ clearly holds.

Given an Eulerian triangulation T with at least four vertices, we consider a TC-scheme $\mathcal{R}$ of T (given by Theorem 2), and by successively *resolving* degenerate points (i.e. intersection points of at least three segments) from left to right, we eventually reach the sought representation. Here resolving means that the segments of a 3-degenerate point (resp. a 6-degenerate point) are moved to form a triangle (resp. a polygon) inside which we are going to draw a 3-slopes representation of the graph corresponding to this degenerate face of $M(\mathcal{R})$, this is possible by using the induction on this smaller graph. The degenerate points of $\mathcal{R}$ are resolved from left to right. This means that at a given stage of this process there is a vertical line (parallel with $\overrightarrow{b}$) $\mathcal{V}$ such that on its left there is no intersection point of three or more segments. This implies that on the left of $\mathcal{V}$ the representation handles some small perturbations: one can slightly move the segments without changing the intersections.

Let $\mathcal{V}$ be the leftmost vertical line containing degenerate points. We resolve those degenerate points by slightly moving segments on the left of or on $\mathcal{V}$, while maintaining the right side of the representation unchanged. We consider different cases according to the degenerate points on $\mathcal{V}$.

If $\mathcal{V}$ contains a 3-degenerate point $\mathbf{p}$ *in the interior of a (vertical) segment* $\mathbf{b_j}$ *and at the end of two segments* $\mathbf{a_i}$ *and* $\mathbf{c_k}$ *lying on the left of $\mathcal{V}$*, the situation is

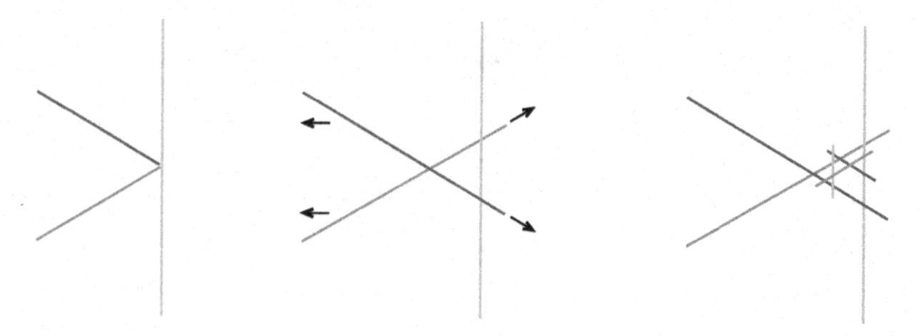

Fig. 6. (left) A 3-degenerate point on $\mathcal{V}$ (middle) Small perturbation of $\mathcal{R}$ (right) The addition of a representation inside the new triangle.

rather simple. Move these segments a little to the left and slightly prolong them to intersect $\mathbf{b_j}$ (see Fig. 6). As there is no degenerate point on the left of $\mathcal{V}$ these moves can be done while maintaining the existing intersections and avoiding new intersections. If $a_ib_jc_k$ is not a face of T, consider the triangulation T' induced by the vertices in the cycle $a_ib_jc_k$ of T. By induction T' has a representation that can be drawn inside the newly formed triangle bordered by the segments $\mathbf{a_i}$, $\mathbf{b_j}$ and $\mathbf{c_k}$.

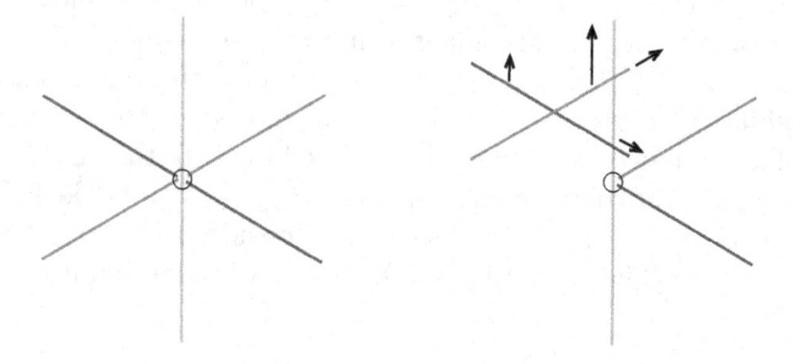

Fig. 7. (left) A double 3-degenerate point on $\mathcal{V}$ (right) Small perturbation of $\mathcal{R}$.

If $\mathcal{V}$ contains a double 3-degenerate point $\mathbf{p}$ *in the interior of a (vertical) segment* $\mathbf{b_j}$, the situation is similar to the previous one. Move the segments on the left of $\mathcal{V}$ as depicted in Fig. 7. If the new triangle is not a face of T, we add a representation inside. We are now left with a simple 3-degenerate point at $\mathbf{p}$. This corresponds to the following case.

If $\mathcal{V}$ contains a 3-degenerate point $\mathbf{p}$ *in the interior of a (vertical) segment* $\mathbf{b_j}$ *and at the end of two segments,* $\mathbf{a_i}$ *and* $\mathbf{c_k}$*, lying on the right of* $\mathcal{V}$, one can move $\mathbf{b_j}$ slightly to the right or slightly to the left and resolve these points without changing the right part of the representation. The choice of moving $\mathbf{b_j}$ to the right or to the left is explained in the next paragraph, but we can assume this move to be arbitrarily small. Whatever the direction $\mathbf{b_j}$ is moved one has to

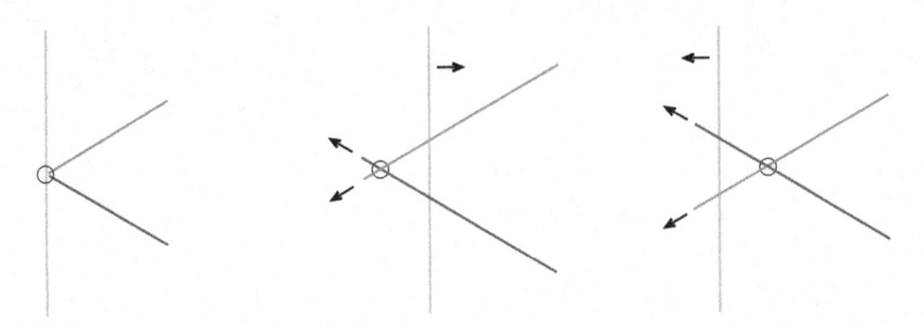

Fig. 8. (left) A 3-degenerate point on $\mathcal{V}$ (middle) Slightly moving $\mathbf{b_j}$ to the right (right) Slightly moving $\mathbf{b_j}$ to the left.

prolong $\mathbf{a_i}$ and $\mathbf{c_k}$ to have all the intersections, between these segments or with $\mathbf{b_j}$ (see Fig. 8). Note that in order to preserve the representation on the right of $\mathcal{V}$ the segments $\mathbf{a_i}$ and $\mathbf{c_k}$ are not moved, they are only prolonged around $\mathbf{p}$. Again, if $a_i b_j c_k$ is not a face of T, we draw a representation inside the newly formed triangle. Note that if $\mathbf{b_j}$ moves to the right, the triangle bordered by $\mathbf{a_i}$, $\mathbf{b_j}$ and $\mathbf{c_k}$ has negative size, but it suffices to apply a homothety with negative ratio to obtain a representation that can be drawn inside.

Consider now the degenerate points at the end of a (vertical) segment $\mathbf{b_j}$ *of* $\mathcal{V}$. Let $\mathbf{b_1}, \mathbf{b_2}, \ldots, \mathbf{b_t}$ be a maximal sequence of segments on $\mathcal{V}$ such that $\mathbf{b_j}$ and $\mathbf{b_{j+1}}$ intersect on a point. We are going to move these segments alternatively to the right and to the left, for example the segments with even index are moved to the left while the ones with odd index are moved to the right. The exact magnitude of these moves will be set later, but first note that the 3-degenerate points in the interior of the segments $\mathbf{b_j}$ with $1 \leq j \leq t$ can be dealt if the move of $\mathbf{b_j}$ is sufficiently small (see previous cases). Consider the intersection point $\mathbf{p}$ between $\mathbf{b_j}$ and $\mathbf{b_{j+1}}$. The case of $\mathbf{b_1}$ and $\mathbf{b_t}$'s end is similar and it is not detailed here.

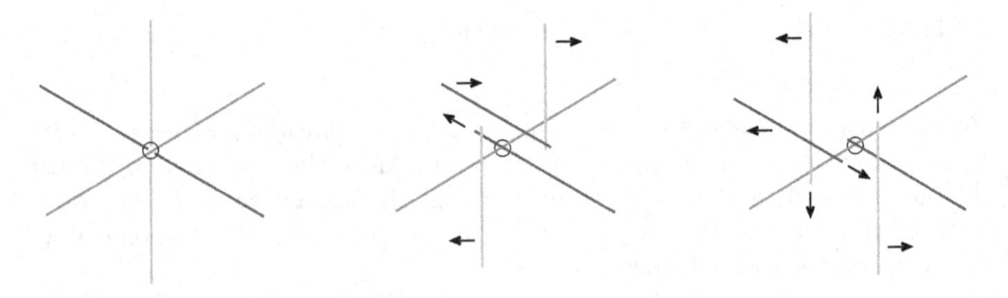

Fig. 9. (left) A double 3-degenerate point on $\mathcal{V}$ (middle) & (right) Small moves that resolve this point.

If there is a segment $\mathbf{a_i}$ going through $\mathbf{p}$. It is shown in Fig. 9 how to resolve these two overlapped 3-degenerate points, in order to create two triangles, where one can add a small representation if needed. The case where there is a segment $\mathbf{c_k}$ going through $\mathbf{p}$ is similar.

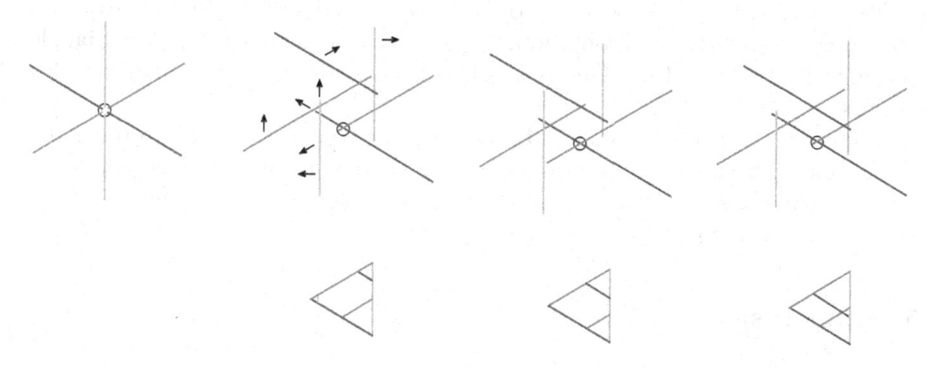

Fig. 10. From left to right: a 6-degenerate point on $\mathcal{V}$. Resolution if there is no chord in $b_j acb_{j+1}a'c'$ with the shape of $\mathcal{R}'$. Resolution if none of b_jc, ca', or $a'b_j$ is a chord, with the shape of $\mathcal{R}'$. Resolution if ac' and ca' are chords, with the shape of $\mathcal{R}_2$.

Assume now that six segments intersect at $\mathbf{p}$. Let $\mathbf{b_j}$ be the one below $\mathbf{p}$, and let $\mathbf{a}$, $\mathbf{c}$, $\mathbf{b_{j+1}}$, $\mathbf{a}'$, and $\mathbf{c}'$ be the other ones around $\mathbf{p}$ clockwisely. Let us assume wlog that $\mathbf{b_j}$ has to move to the left, while $\mathbf{b_{j+1}}$ has to move to the right. The degenerate face corresponding to $\mathbf{p}$ is bounded by these six vertices and there are several cases according to whether there are chords among them in T (see Fig. 10).

If there is no chord inside the cycle $b_jacb_{j+1}a'c'$ we consider the subgraph of T induced by the vertices on and inside this cycle, add we add the edges ab_{j+1}, $b_{j+1}c'$, and ac' outside the cycle, and we denote by T' the obtained simple Eulerian triangulation. By the induction we know that T' admits a 3-slope segment representation $\mathcal{R}'$, and let (s_a, s_b, s_c) be the shape of $\mathcal{R}'$. We resolve the point by moving the segments as depicted in Fig. 10, and the magnitude of each of these moves is prescribed by the shape (s_a, s_b, s_c) in order to allow us to copy $\mathcal{R}'$ inside the triangle formed by $\mathbf{a}$, $\mathbf{b_{j+1}}$, and $\mathbf{c}'$. Then we shorten $\mathbf{a}$, $\mathbf{b_{j+1}}$, and $\mathbf{c}'$ to avoid intersections among them. Actually, the case where none of ab_{j+1}, $b_{j+1}c'$, or ac' is a chord is similar.

If none of b_jc, ca', or $a'b_j$ is a chord of $b_jacb_{j+1}a'c'$ we proceed similarly. The only difference is that we add the edge b_jc, ca', or $a'b_j$ outside $b_jacb_{j+1}a'c'$ to obtain T', and that we have to perform a homothety with negative ratio to include $\mathcal{R}'$.

Finally, if there are two opposite chords on $b_jacb_{j+1}a'c'$, say ac' and ca', we consider two triangulations. Let T_1 be the one inside the cycle $c'b_ja$ and let T_2 be the one obtained from the interior of the 5-cycle $acb_{j+1}a'c'$ by adding the edges ab_{j+1} and $b_{j+1}c'$. By the induction we know that T_1 and T_2 admit 3-slopes segment representations $\mathcal{R}_1$, and $\mathcal{R}_2$, and let (s_a, s_b, s_c) be the shape of $\mathcal{R}_2$. We resolve the point by moving the segments as depicted in Fig. 10, and the magnitude of each of these moves, except for $\mathbf{b_j}$, is prescribed by the shape (s_a, s_b, s_c) in order to allow us to copy $\mathcal{R}_2$ inside the triangle formed by $\mathbf{a}$, $\mathbf{b_{j+1}}$, and $\mathbf{c}'$. Then we shorten $\mathbf{a}$, and $\mathbf{b_{j+1}}$ to avoid the intersections corresponding to ab_{j+1} and $b_{j+1}c'$. The segment $\mathbf{b_j}$ is moved sufficiently to the left to avoid the interior of the triangle containing $\mathcal{R}_2$. Then $\mathcal{R}_1$ is drawn inside the triangle bordered by $\mathbf{b_j}$, $\mathbf{a}$ and $\mathbf{c}'$. This is possible because $\mathcal{R}_2$ does not intersect this triangle.

Finally note that the moves of $\mathbf{b_j}$ and $\mathbf{b_{j+1}}$ are opposite but of proportional magnitudes (up to some constant depending on the shapes (s_a, s_b, s_c) of $\mathcal{R}'$ or $\mathcal{R}_2$). So it is clear that we can simultaneously move all the segments $\mathbf{b_j}$ on $\mathcal{V}$. This concludes the proof of the lemma.

5 Conclusion

West [15] and de Fraysseix and Ossona de Mendez [5] independently ask for a generalization of Scheinerman's Conjecture.

Conjecture 1. Planar graphs that are k-colorable admit a k-slopes segment representation.

The case $k = 1$ is trivial. We have seen that the case $k = 2$ holds with 2-slopes contact representations. We have seen that the case $k = 3$ also holds. For the final case $k = 4$ we would like to apply the same approach as here. This means that we would like to go through a TC-representation of the considered triangulation, and then resolve its degenerate points. There are at least two obstacles for this approach. The first one is to find an order on the degenerate points to resolve them. The left to right approach does not seem sufficient here. The second one is connected to degenerate points $\mathbf{p}$ such that going clockwisely around $\mathbf{p}$ we successively cross segments $\mathbf{a}$, $\mathbf{b}$, $\mathbf{c}$, $\mathbf{d}$, and $\mathbf{a}$ again. In that case its is impossible, restricting ourselves to small perturbations to create a region bordered by $\mathbf{a}$, $\mathbf{b}$, $\mathbf{c}$, and $\mathbf{d}$ in the clockwise or counterclockwise order. To avoid this issue we want 4-colorings (i.e. 4-slopes assignements) of planar graphs with particular properties.

Conjecture 2. Planar graphs admit a $\{1, 2, 3, 4\}$-coloring such that there is no induced C_4 colored with colors 1, 2, 3 and 4 in clockwise order.

Examples show that one cannot extend this condition to non-induced 4-cycles. A positive answer to this conjecture would imply that simple signed planar graphs have chromatic number at most 4, positively answering a conjecture of Máčajová, Raspaud, and Škoviera [11].

Acknowledgements. The author is thankful to Marc de Visme for fruitful discussions on this topic, and to Pascal Ochem for bringing [11] to his attention.

References

1. de Castro, N., Cobos, F., Dana, J.C., Márquez, A., Noy, M.: Triangle-free planar graphs as segment intersection graphs. J. Graph Algorithms Appl. **6**(1), 7–26 (2002)
2. Chalopin, J., Gonçalves, D.: Every planar graph is the intersection graph of segments in the plane: extended abstract. In: Proceedings of the Forty-first Annual ACM Symposium on Theory of Computing, pp. 631–638 (2009)
3. Czyzowicz, J., Kranakis, E., Urrutia, J.: A simple proof of the representation of bipartite planar graphs as the contact graphs of orthogonal straight line segments. Inform. Process. Lett. **66**(3), 125–126 (1998)
4. Felsner, S.: Triangle contact representations. In: Midsummer Combinatorial Workshop (2009)
5. de Fraysseix, H., Ossona de Mendez, P.: Representations by contact and intersection of segments. Algorithmica **47**, 453–463 (2007)
6. de Fraysseix, H., de Mendez, P.O., Pach, J.: Representation of planar graphs by segments. Colloq. Math. Soc. János Bolyai **63**, 109–117 (1994). Intuitive Geometry (Szeged, 1991)
7. Gonçalves, D., Isenmann, L., Pennarun, C.: Planar Graphs as L-intersection or L-contact graphs. In: Proceedings of SODA, pp. 172–184 (2018)
8. Hartman, I.B.-A., Newman, I., Ziv, R.: On grid intersection graphs. Discrete Math. **87**(1), 41–52 (1991)
9. Kratochvíl, J.: String graphs. II. Recognizing string graphs is NP-hard. J. Combin. Theory. Ser. B **52**, 67–78 (1991)
10. Kratochvíl, J., Matoušek, J.: Intersection graphs of segments. J. Combin. Theory. Ser. B **62**, 180–181 (1994)
11. Máčajová, E., Raspaud, A., Škoviera, M.: The chromatic number of a signed graph. Electr. J. Comb. **23**(1), P1, 1–14 (2016)
12. Pach, J., Solymosi, J.: Crossing patterns of segments. J. Combin. Theory. Ser. A **96**, 316–325 (2001)
13. Scheinerman, E.R.: Intersection classes and multiple intersection parameters of graphs. Ph.D., Thesis, Princeton University (1984)
14. Scheinerman, E.R.: Private communication to D. West (1993)
15. West, D.: Open problems. SIAM J. Discrete Math. Newslett. **2**(1), 10–12 (1991)

The Exponential-Time Complexity of Counting (Quantum) Graph Homomorphisms

Hubie Chen[1], Radu Curticapean[2,3](✉), and Holger Dell[3]

[1] Department of Computer Science and Information Systems, Birkbeck University of London, London, UK
[2] Basic Algorithms Research Copenhagen (BARC), Copenhagen, Denmark
[3] IT University of Copenhagen, Copenhagen, Denmark
racu@itu.dk

Abstract. Many graph parameters can be expressed as homomorphism counts to fixed target graphs; this includes the number of independent sets and the number of k-colorings for any fixed k. Dyer and Greenhill (RSA 2000) gave a sweeping complexity dichotomy for such problems, classifying which target graphs render the problem polynomial-time solvable or #P-hard. In this paper, we give a new and shorter proof of this theorem, with previously unknown tight lower bounds under the exponential-time hypothesis. We similarly strengthen complexity dichotomies by Focke, Goldberg, and Živný (SODA 2018) for counting *surjective* homomorphisms to fixed graphs. Both results crucially rely on our main contribution, a complexity dichotomy for evaluating linear combinations of homomorphism numbers to fixed graphs. In the terminology of Lovász (Colloquium Publications 2012), this amounts to counting homomorphisms to *quantum graphs*.

Keywords: Graph homomorphisms · Exponential-time hypothesis · Counting complexity · Complexity dichotomy · Surjective homomorphisms

1 Introduction

The classification program in counting complexity strives to identify comprehensive classes of counting problems that are well-behaved enough to allow for exhaustive complexity classifications [4–7,9,16]. Particularly good candidates for such classes are counting variants of the Constraint Satisfaction Problem (#CSP) [3,4]. In the general #CSP, a problem instance is defined by a set of *variables* $V = \{v_1, \ldots, v_n\}$, each taking values from a *domain* D. The computational task is to determine the number of assignments $a : V \to D$ from

Hubie Chen acknowledges the support of Spanish Project TIN2017-86727-C2-2-R. Radu Curticapean was partly supported by ERC grant SYSTEMATICGRAPH (No. 725978) and VILLUM Foundation grant 16582 while working on this project.

I. Sau and D. M. Thilikos (Eds.): WG 2019, LNCS 11789, pp. 364–378, 2019.
https://doi.org/10.1007/978-3-030-30786-8_28

variables to domain elements, subject to the requirement that a satisfies a set of *constraints* that are part of the input. Each constraint is applied to a tuple of variables and restricts the admissible assignments to that tuple.

In this full generality, the #CSP framework can easily express #P-hard problems such as counting satisfying assignments to Boolean formulas in CNF, or counting the proper k-colorings of graphs G, for fixed $k \in \mathbf{N}$. For instance, to count k-colorings, interpret the vertices of G as variables over the domain $\{1, \ldots, k\}$ and constrain variable pairs corresponding to adjacent vertices to have distinct assignments.

Among other properties, the complexity of #CSP depends on the types of constraints present in the instance. This motivates the study of #CSP($\mathcal{F}$) for fixed constraint sets $\mathcal{F}$, where only instances with constraints from $\mathcal{F}$ are allowed as input. After a wealth of research, a full dichotomy for these problems is known by now: For every finite set $\mathcal{F}$, the problem #CSP($\mathcal{F}$) has been shown to be either polynomial-time solvable or #P-hard, with an explicit decidable dichotomy criterion [5,17]. Dichotomies are known even in weighted settings [7] that arise in statistical physics in the context of *partition functions*.

1.1 Graph Homomorphisms

The full dichotomy for #CSP($\mathcal{F}$) was predated by numerous results for special cases, with a particular focus on *graph homomorphisms* [6,16,24]. Given graphs G and H, a homomorphism from G to H is a function $h : V(G) \to V(H)$ such that any edge $uv \in E(G)$ is mapped to an edge $h(u)h(v) \in E(H)$. Homomorphisms from G to H are sometimes also called H*-colorings* of G, since they generalize q-colorings for fixed $q \in \mathbf{N}$ by taking $H = K_q$.

We write $\mathrm{Hom}(G, H)$ for the number of homomorphisms from G to H. For a fixed graph H, the computational problem $\mathrm{Hom}(\star, H)$ asks to compute $\mathrm{Hom}(G, H)$ on input a graph G. This is indeed a particular #CSP($\mathcal{F}$) problem: Viewing $V(G)$ as variables and $V(H)$ as domain, a homomorphism h corresponds to an assignment from variables to domain elements that respects certain constraints on variable pairs: If u and v are connected by an edge in G, then its assignments $h(u)$ and $h(v)$ must be such that $h(u)h(v)$ is an edge of H. Following this interpretation, it can be seen that the class of problems $\mathrm{Hom}(\star, H)$ for fixed H correspond exactly to #CSP($\mathcal{F}$) problems where $\mathcal{F}$ contains only a single constraint, and this constraint depends only (symmetrically) on two variables.

Despite these restrictions, many interesting counting problems on graphs can be expressed as the $\mathrm{Hom}(\star, H)$ for suitable choices of H. This includes the number of independent sets in a graph, the number of k-colorings for fixed k, and certain partition functions from statistical physics. In a seminal result, Dyer and Greenhill proved a full classification for the complexity of $\mathrm{Hom}(\star, H)$ when H is an undirected graph that may contain self-loops. In the following, we say that a graph is *reflexive* if every vertex features a self-loop, and we say that it is *irreflexive* if no vertex does. Note that bipartite graphs are irreflexive.

Theorem 1 (Dyer and Greenhill [16]). *Let H be a fixed undirected graph. If each connected component of H is a bipartite complete graph or a reflexive complete graph, then* $\mathrm{Hom}(\star, H)$ *can be computed in polynomial time. Otherwise the problem is #P-hard, even on irreflexive input graphs.*

This exhaustive dichotomy was extended in numerous ways, including a setting where H has edge-weights and the weight of a homomorphism is the product of edge-weights in the image, counted with multiplicities [6,8,24]: Given an input graph G, the task is to determine the sum of weights of all homomorphisms from G to H, a quantity that occurs naturally in statistical physics. The case of directed graphs was also fully classified [7,15]. Furthermore, a variant was investigated that asks to determine the number of homomorphisms modulo a fixed prime [18,22,23], but a full dichotomy was not yet obtained for such problems.

Our Contribution: New proof of Theorem 1 *with tight lower bound under ETH.* Using techniques originally introduced by Lovász [27], we significantly shorten the proof of Theorem 1. Our new proof also gives tight conditional lower bounds on the running times needed to solve the #P-hard cases: For a k-vertex graph H and an n-vertex graph G, the quantity $\mathrm{Hom}(G, H)$ can be computed in time roughly $O(k^n)$ using exhaustive search. It was shown by Cygan et al. [13] that $\mathrm{Hom}(G, H)$ cannot be computed in time $\exp(o(n \log k))$ when both G and H are input, unless the widely-believed *exponential-time hypothesis* (ETH) by Impagliazzo and Paturi [25] fails. However, this result leaves open the possibility of $\exp(o(n))$-time algorithms for particular fixed graphs H for which $\mathrm{Hom}(\star, H)$ is #P-hard. We rule out such algorithms under ETH. In fact, we only require the counting exponential-time hypothesis #ETH, introduced in [14]. This makes the result slightly stronger, since ETH implies #ETH.

Theorem 2. *For every hard graph H in Theorem 1, the problem* $\mathrm{Hom}(\star, H)$ *cannot be computed in* $\exp(o(n))$ *time on n-vertex input graphs unless #ETH fails. This holds even for bipartite and irreflexive inputs with* $O(n)$ *edges.*

1.2 Surjective Homomorphisms

Focke, Goldberg, and Živný [20] used Theorem 1 as a starting point to classify the complexity of counting homomorphisms with surjectivity constraints. We call a homomorphism h from G to H *surjective* if its image contains every vertex and every edge of H. That is, for every vertex $v \in V(H)$, the preimage $h^{-1}(v)$ is non-empty, and for every edge $st \in E(H)$, there is at least one edge between the sets $h^{-1}(s)$ and $h^{-1}(t)$ in G. This notion can be relaxed by requiring surjectivity only on a subset of the vertices and edges of H. For instance, *vertex-surjective homomorphisms* only require every vertex to be hit. Likewise, a *compaction* is a vertex-surjective homomorphism from G to H that hits all non-loop edges of H.

The above authors proved a dichotomy theorem for counting vertex-surjective homomorphisms to fixed graphs H [20], discovering that the dichotomy criterion for these problems coincides with that for standard homomorphisms. They proved a similar dichotomy for counting compactions and showed that there are significantly fewer polynomial-time solvable cases.

Theorem 3 (Focke, Goldberg, and Živný [20]). *Let H be a fixed graph. The problem* $\mathrm{VertSurj}(\star, H)$ *is polynomial-time solvable if every connected component of H is a complete bipartite graph or a reflexive complete graph. The problem* $\mathrm{Comp}(\star, H)$ *is polynomial-time solvable if every component of H is an irreflexive star or a reflexive complete graph of size at most two. In all other cases, the problems are #P-hard, even on irreflexive inputs.*

Our Contribution: Simplified and strengthened version of Theorem 3. We define a problem that jointly generalizes the problems $\mathrm{VertSurj}(\star, H)$ and $\mathrm{Comp}(\star, H)$ in a natural way. To this end, we consider target graphs H in which some edges and vertices of H are *marked*. A *partially surjective homomorphism* then is a homomorphism h whose image includes all marked objects of H; we write $\mathrm{PartSurj}(G, H)$ for their number. With appropriate choices of markings, this can be seen to generalize various quantities, such as homomorphisms, surjective and vertex-surjective homomorphisms, and compactions. We obtain the following complexity dichotomy, from which Theorem 3 easily follows.

Theorem 4. *Let H be a graph in which some edges and/or vertices are marked, and let $\mathcal{D}(H)$ be the set of graphs obtainable from H by deleting marked objects.*

- *If every graph in $\mathcal{D}(H)$ is a disjoint union of bipartite complete graphs and reflexive complete graphs, then* $\mathrm{PartSurj}(\star, H)$ *is polynomial-time solvable.*
- *Otherwise,* $\mathrm{PartSurj}(\star, H)$ *is #P-hard and cannot be computed in* $\exp(o(n))$ *time on n-vertex input graphs unless #ETH fails. This holds even for bipartite and irreflexive inputs with $O(n)$ edges.*

1.3 Our Techniques: Homomorphisms to Quantum Graphs

While the class of homomorphism problems $\mathrm{Hom}(\star, H)$ to fixed H subsumes many interesting counting problems for graphs, there are also natural problems that cannot be expressed in this framework. This includes the number of perfect matchings in a graph [21,28]. To give another example that is more similar to homomorphism counts, recall that counting 3-colorings in a graph is expressible as $\mathrm{Hom}(\star, K_3)$. However, counting surjective 3-colorings (colorings that use all three colors) cannot be expressed as $\mathrm{Hom}(\star, H)$ for a fixed graph H. This is because, for any graph G, the number of surjective 3-colorings is

$$\mathrm{VertSurj}(G, K_3) = \mathrm{Hom}(G, K_3) - 3 \cdot \mathrm{Hom}(G, K_2) + 3 \cdot \mathrm{Hom}(G, K_1). \quad (1)$$

However, the expression of a graph parameter as a linear combination of homomorphism counts $\mathrm{Hom}(\star, H)$ is known to be unique, see [27, Exercise 5.51], ruling out the existence of a graph H with $\mathrm{VertSurj}(\star, K_3) = \mathrm{Hom}(\star, H)$.

More generally, the uniqueness of such expressions implies that closing the class of homomorphism counts under point-wise linear combinations gives a strictly richer class of graph parameters. Following Lovász's terminology [27,

Chapter 6], we call these graph parameters homomorphism counts to *quantum graphs*. Here, a quantum graph $\overline{H}$ is a formal linear combination

$$\overline{H} = \sum_{H \in \mathcal{C}} \alpha_H H$$

for a finite set of constituent graphs $\mathcal{C}$ where each $H \in \mathcal{C}$ has an associated coefficient $\alpha_H \in \mathbf{Q}$. The canonical linear extension of homomorphism counts to quantum graphs $\overline{H}$ then reads

$$\mathrm{Hom}(G, \overline{H}) = \sum_{H \in \mathcal{C}} \alpha_H \cdot \mathrm{Hom}(G, H).$$

In other words, every finite (point-wise) linear combination of homomorphism counts to fixed graphs can be expressed as a homomorphism count to a fixed quantum graph. The computational problem $\mathrm{Hom}(\star, \overline{H})$ for fixed $\overline{H}$ is to compute $\mathrm{Hom}(G, \overline{H})$ for a given input G.

As exemplified in (1), problems that do not immediately appear to be linear combinations of homomorphism counts may in fact be expressible in this format. For instance, all partially surjective homomorphism counts can be expressed as linear combinations of ordinary homomorphism counts.

Our Contribution: Dichotomy for homomorphisms to quantum graphs. We prove that the complexity of counting homomorphisms to fixed graphs enjoys a very favorable monotonicity property. (A similar phenomenon was already observed for linear combinations of homomorphism counts *from* fixed graphs [10,12].)

Let $\overline{H}$ be a fixed quantum graph that is properly normalized, that is, its constituents are pairwise non-isomorphic and all coefficients are non-zero. Then, for any constituent H of $\overline{H}$, the problem $\mathrm{Hom}(\star, H)$ reduces to $\mathrm{Hom}(\star, \overline{H})$ under polynomial-time Turing reductions. That is, given access to an oracle that delivers the quantity $\mathrm{Hom}(G', \overline{H})$ on any query G', we can compute $\mathrm{Hom}(G, H)$ for any input graph G and any constituent H of $\overline{H}$. In particular, if $\mathrm{Hom}(\star, H)$ is #P-hard, then any linear combination of homomorphism counts containing the summand $\mathrm{Hom}(\star, H)$ is #P-hard.

Moreover, to determine $\mathrm{Hom}(G, H)$ for an n-vertex graph G, our reduction only needs to query graphs G' with $n + c$ vertices, with c depending only on $\overline{H}$. This makes the reduction very suitable in the exponential-time setting: An algorithm with running time $O(b^n)$ for $\mathrm{Hom}(\star, \overline{H})$ would imply $O(b^n)$ time algorithms for any constituent problem $\mathrm{Hom}(\star, H)$. We use the complexity monotonicity of quantum graphs to obtain our final dichotomy theorem:

Theorem 5. *Let $\overline{H} = \sum_{i=1}^{k} \alpha_i H_i$ be a fixed quantum graph, where $H_1, \ldots, H_k$ are fixed pairwise non-isomorphic graphs and $\alpha_1, \ldots, \alpha_k \in \mathbf{Q} \setminus \{0\}$ are fixed.*

- *If the problem $\mathrm{Hom}(\star, H_i)$ can be solved in polynomial time for every $i \in [k]$, then so can $\mathrm{Hom}(\star, \overline{H})$.*

– *If there is some* $i \in [k]$ *such that* $\mathrm{Hom}(\star, H_i)$ *is* #P*-hard, then so is* $\mathrm{Hom}(\star, \overline{H})$*. In this case, unless #ETH fails,* $\mathrm{Hom}(\star, \overline{H})$ *cannot be solved in time* $\exp(o(n))$*, even for bipartite and irreflexive input graphs with* $O(n)$ *edges.*

The quantum graph $\overline{H}$ in this theorem may have negative coefficients; if $\overline{H}$ has only positive coefficients, the #P-hardness of $\mathrm{Hom}(\star, \overline{H})$ can already be derived from Theorem 1.

Organization of the Paper

After introducing notions related to homomorphisms and exponential-time complexity in Sect. 2, we prove the dichotomy theorem for homomorphisms to quantum graphs (Theorem 5) in Sect. 3. Using the complexity monotonicity of homomorphism numbers to quantum graphs, we sketch the proof of the exponential-time Dyer–Greenhill theorem (Theorem 2) in Sect. 4. Finally, we derive the dichotomy for partially surjective homomorphisms (Theorem 4) in Sect. 5. Due to lack of space, some proofs are deferred to the full version.

2 Preliminaries

Let $\mathcal{G}$ be the set of all unlabeled and undirected finite graphs. These graphs may have self-loops but no parallel edges. In the remainder of this section, let $G, H \in \mathcal{G}$. We denote the vertex set of G with $V(G)$ and the edge set with $E(G)$.

Homomorphisms and Graph Algebra: Let $\mathrm{Hom}(G, H)$ be the number of homomorphisms from G to H, that is, functions $h : V(G) \to V(H)$ such that any edge $uv \in E(G)$ is mapped to an edge $h(u)h(v) \in E(H)$. For fixed H, we write $\mathrm{Hom}(\star, H)$ for the graph parameter that maps input graphs G to $\mathrm{Hom}(G, H)$.

Our proofs rely on a result of Borgs et al. [2, Lemma 4.2], who show that the graph function Hom, when viewed as a matrix, has certain non-singular finite submatrices. We use the following extension, which we derive from the original result in the full version.

Lemma 6. *For any set of pairwise non-isomorphic graphs* $H_1, \ldots, H_k$*, there exist irreflexive graphs* $F_1, \ldots, F_k$ *such that the* $k \times k$ *matrix* M *with* $M[i, j] = \mathrm{Hom}(F_i, H_j)$ *is invertible.*

Even though $H_1, \ldots, H_k$ may feature self-loops, the lemma guarantees the existence of irreflexive graphs $F_1, \ldots, F_k$. In fact, these graphs can even be guaranteed to be 3-colorable.

Our proofs also rely upon two binary operations on graphs (which can be viewed as graph products) and their effects on homomorphism counts: The *disjoint union* of graphs, and its "dual", the *tensor product.*

Definition 7. *Let A, B be graphs on disjoint vertex sets. The disjoint union $A \cup B$ has vertex set $V(A) \cup V(B)$ and consists of a copy of A and one of B.*

The tensor product $A \otimes B$ is the graph on vertex set $V(A) \times V(B)$ where (u, v) and (u', v') are adjacent if and only if $(u, u') \in E(A)$ and $(v, v') \in E(B)$.

From a matrix perspective, the adjacency matrix of $A \cup B$ is a block matrix with blocks corresponding to A and B, and the adjacency matrix of $A \otimes B$ is the Kronecker product of the respective adjacency matrices. The following identities hold for all vertex-disjoint graphs G, F, A, B:

$$\mathrm{Hom}(G \cup F, A) = \mathrm{Hom}(G, A) \cdot \mathrm{Hom}(F, A)\,, \text{ and} \tag{2}$$

$$\mathrm{Hom}(G, A \otimes B) = \mathrm{Hom}(G, A) \cdot \mathrm{Hom}(G, B). \tag{3}$$

If additionally G is connected, then we also have

$$\mathrm{Hom}(G, A \cup B) = \mathrm{Hom}(G, A) + \mathrm{Hom}(G, B). \tag{4}$$

The proofs are elementary and can be found in [27, (5.28)–(5.30)].

Exponential-Time Complexity: The *counting exponential time hypothesis* (#ETH) of Dell et al. [14], adapted from the decision setting of Impagliazzo, Paturi, and Zane [25,26], asserts that there is no $\exp(o(m))$ time algorithm to count the satisfying assignments of a given 3-CNF formula with m clauses. We use the following stringent type of polynomial-time reduction:

Definition 8 (Linear Reduction). *Let $f, g : \mathcal{G} \to \mathbf{Q}$ be two graph parameters. We write $f \preceq g$ if there is a polynomial-time Turing reduction from f to g that, on input a graph with m edges, queries only graphs with at most $O(m)$ edges.*

Note that $\preceq$ is a reflexive and transitive relation; it is called *size-preserving* reducibility in [19, p. 422]. If $f \preceq g$, then an algorithm with running time $\exp(o(m))$ for g on m-edge graphs would imply one for f.

3 Counting Homomorphisms to Quantum Graphs

We are ready to prove Theorem 5, the dichotomy for counting homomorphisms to quantum graphs. We establish the theorem via the following proposition on the complexity monotonicity for counting homomorphisms to quantum graphs.

Proposition 9 (Complexity Monotonicity). *Fix any quantum graph*

$$\overline{H} = \sum_{j=1}^{k} \alpha_j H_j$$

with non-isomorphic graphs $H_1, \ldots, H_k$ and coefficients $\alpha_1, \ldots, \alpha_k \in \mathbf{Q} \setminus \{0\}$. For every fixed $j \in [k]$, we then have

$$\mathrm{Hom}(\star, H_j) \preceq \mathrm{Hom}(\star, \overline{H}).$$

Furthermore, if the input graph G for $\mathrm{Hom}(\star, H_j)$ is irreflexive, then all queries for $\mathrm{Hom}(\star, \overline{H})$ are irreflexive as well.

Proof. Without loss of generality, let $j = 1$. By Lemma 6, there exist irreflexive graphs $F_1, \ldots, F_k$ such that the matrix M with $M[i,j] = \mathrm{Hom}(F_i, H_j)$ is invertible. On input a graph G, we first construct the graphs $G \cup F_i$ for all $i \in [k]$. By (2), we obtain the following linear equation for every $i \in [k]$:

$$\mathrm{Hom}(G \cup F_i, \overline{H}) = \sum_{j=1}^{k} \alpha_j \,\mathrm{Hom}(G, H_j) \cdot M[i,j]. \tag{5}$$

The set of these equations for all $i \in [k]$ forms a linear equation system $b = Mx$, with $b_i = \mathrm{Hom}(G \cup F_i, \overline{H})$ for all $i \in [k]$ and $x_j = \alpha_j \,\mathrm{Hom}(G, H_j)$ for all $j \in [k]$. Thus if G is the input and we wish to compute $\mathrm{Hom}(G, H_1)$ using the oracle for $\mathrm{Hom}(\star, \overline{H})$, we use the following procedure:

1. Compute the vector $b \in \mathbf{Q}^k$ using k queries to $\mathrm{Hom}(\star, \overline{H})$.
2. Output the number $(M^{-1}b)_1/\alpha_1$.

This indeed yields $\mathrm{Hom}(G, H_1)$, because $\alpha_1 \,\mathrm{Hom}(G, H_1) = (M^{-1}b)_1$ and $\alpha_1 \neq 0$ hold. Since $H_1, \ldots, H_k$ is fixed, we can hard-code the constants α_j and graphs F_j, for $j \in [k]$, as well as the matrix M^{-1} into the reduction. The reduction itself runs in linear time to prepare the queries $G \cup F_i$. Given as input an m-edge graph G, it only issues queries on graphs with $m + C$ edges, where C is a fixed constant depending only on $\overline{H}$. If G is irreflexive, then so are all query graphs $G \cup F_i$ for $i \in [k]$, since all F_i are irreflexive. ∎

Theorem 5 follows easily from Proposition 9 and Theorem 2.

4 Revisiting the Dyer-Greenhill Dichotomy

We outline our new proof of Theorem 1 and classify the complexity of $\mathrm{Hom}(\star, H)$. Our proof also gives a tight lower bound under #ETH, resulting in Theorem 2.

Throughout this section, let us say that a graph H is *hom-easy* if every connected component of H is either a complete bipartite graph $K_{a,b}$ for $a, b \in \mathbf{N}$ or a reflexive complete graph K_q° for $q \in \mathbf{N}$. It is straightforward to check that $\mathrm{Hom}(\star, H)$ can be solved in linear time if H is a hom-easy graph. If H is not hom-easy, we call H *hom-hard.* In the remainder of the section, we show how to establish the #P-hardness of $\mathrm{Hom}(\star, H)$ for hom-hard graphs H in three steps.

1. Ensuring bipartiteness: Rather than working directly with H, we proceed to its *bipartite double cover* $H \otimes K_2$. Recall from (3) that

$$\mathrm{Hom}(G, H \otimes K_2) = \mathrm{Hom}(G, H) \cdot \mathrm{Hom}(G, K_2)$$

holds for all graphs G. Since K_2 is hom-easy, we can compute $\mathrm{Hom}(G, K_2)$ in linear time, and this readily implies $\mathrm{Hom}(\star, H \otimes K_2) \preceq \mathrm{Hom}(\star, H)$.

Hence, it suffices to establish hardness of $\mathrm{Hom}(\star, H \otimes K_2)$ for the bipartite graph $H \otimes K_2$. Note that $H \otimes K_2$ is hom-hard if H is hom-hard.

2. Isolating 2-neighborhoods: Similar to the original proof [16], we successively isolate induced subgraphs from $H \otimes K_2$ until reaching a hard base case.

Given a bipartite graph B and $v \in V(B)$, let B_v denote the subgraph induced by vertices of distance at most 2 from v. We show $\mathrm{Hom}(\star, B_v) \preceq \mathrm{Hom}(\star, B)$ for all $v \in V(B)$ by using the monotonicity for quantum graph homomorphisms. This reduction may happen to be useless for some vertices $v \in V(B)$, as B_v may be a complete bipartite graph $K_{a,b}$ or B itself. If this holds for all $v \in V(B)$, we call B an *impasse.*

Starting at $B = H \otimes K_2$, we repeatedly pick a vertex $v \in V(B)$ and set $B := B_v$ until reaching an impasse P. We show that the vertices in the above process can be chosen to ensure that P is not a $K_{a,b}$. Since $\mathrm{Hom}(\star, P) \preceq \mathrm{Hom}(\star, H \otimes K_2)$ follows, it remains to prove hardness for this impasse P.

3. Exploded four-vertex paths: A structural argument shows that any impasse P that is not a $K_{a,b}$ is in fact a 4-vertex path $P(a_1, a_2, a_3, a_4)$ in which the i-th vertex is replaced by a positive number a_i of clones. For example, $P(1, 3, 4, 2)$ is the following graph:

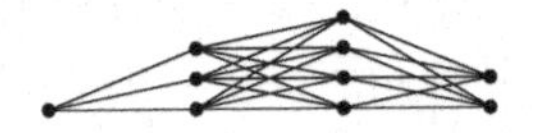

Due to space limitations, we defer the hardness proof for $\mathrm{Hom}(\star, P)$ with $P = P(a_1, a_2, a_3, a_4)$ for arbitrary fixed integers a_1, a_2, a_3, a_4 to the full version. The reduction proceeds from the #P-hard problem of counting independent sets, for which #ETH rules out $2^{o(m)}$ time algorithms [11].

In the special case $P = P(1, 1, 1, 1)$, a simple reduction is possible: Note that $P = \bullet\!\!-\!\!\circlearrowleft \otimes \bullet\!\!-\!\!\bullet$ holds, and hence

$$\mathrm{Hom}(G, P) = \mathrm{Hom}(G, \bullet\!\!-\!\!\circlearrowleft) \cdot \mathrm{Hom}(G, \bullet\!\!-\!\!\bullet).$$

Since $\mathrm{Hom}(G, \bullet\!\!-\!\!\circlearrowleft)$ counts precisely the independent sets of G, and $\mathrm{Hom}(G, \bullet\!\!-\!\!\bullet)$ can be computed in linear time, the reduction is immediate.

Overall, given a hom-hard graph H, the three steps outlined above identify a graph $P = P(a_1, a_2, a_3, a_4)$ for $a_1, a_2, a_3, a_4 \in \mathbf{N}$ such that

$$\mathrm{Hom}(\star, P) \preceq \ldots \preceq \mathrm{Hom}(\star, H \otimes K_2) \preceq \mathrm{Hom}(\star, H). \tag{6}$$

By establishing hardness of $\mathrm{Hom}(\star, P)$, we thus prove hardness of $\mathrm{Hom}(\star, H)$.

Details of Step 2: Successively Isolating 2-Neighborhoods

In the remainder of this section, we provide more details for the second step—details for the other steps are deferred to the full version.

After the first step, we may assume H to be bipartite, but not a complete bipartite graph $K_{a,b}$. We find a hom-hard impasse P with $\mathrm{Hom}(\star, P) \preceq \mathrm{Hom}(\star, H)$ by transitioning successively to proper induced subgraphs of H, in a manner similar to the bipartite case of [16, Theorem 1.1]. In the following, we describe one step of this process.

Let B be a bipartite graph; initially $B = H \otimes K_2$. For any vertex $v \in V(B)$, recall that B_v is the subgraph of B induced by vertices at distance at most 2 from v. We prove that $\mathrm{Hom}(\star, B_v) \preceq \mathrm{Hom}(\star, B)$ holds. To this end, we first show in Lemma 10 how to compute the sum $\sum_{v \in V(B)} \mathrm{Hom}(G, B_v)$ on input G with an oracle for $\mathrm{Hom}(\star, B)$. Combining this with Proposition 9, we will then extract $\mathrm{Hom}(G, B_v)$ for any fixed vertex $v \in V(B)$ from the sum in Proposition 11.

Lemma 10. *Let B be a bipartite graph and let G be a connected bipartite graph with bipartition $V(G) = L \cup R$. Let G^a_L be derived from G by adding an "apex" vertex a that is adjacent to all of R, and let G^a_R be derived by adding an apex vertex a adjacent to all of L. Then*

$$\mathrm{Hom}(G^a_L, B) + \mathrm{Hom}(G^a_R, B) = \sum_{v \in V(B)} \mathrm{Hom}(G, B_v) . \tag{7}$$

Proof. For any $v \in V(B)$, we write $\mathrm{Hom}(G^a_L, B \mid a \to v)$ for the number of homomorphisms from G^a_L to B that map a to v, with an analogous definition for G^a_R. We observe that

$$\mathrm{Hom}(G^a_L, B) + \mathrm{Hom}(G^a_R, B) \\ = \sum_{v \in V(B)} \mathrm{Hom}(G^a_L, B \mid a \to v) + \mathrm{Hom}(G^a_R, B \mid a \to v) , \tag{8}$$

because the set of homomorphisms h from G^a_L to B can be partitioned according to the image $h(a) = v$ and the same applies to homomorphisms from G^a_R. In the remainder of the proof, we establish that, for all $v \in V(B)$,

$$\mathrm{Hom}(G^a_L, B \mid a \to v) + \mathrm{Hom}(G^a_R, B \mid a \to v) = \mathrm{Hom}(G, B_v). \tag{9}$$

Together with (8), this implies (7). To prove (9), fix any vertex $v \in V(B)$. We say that a homomorphism h from G^a_L or G^a_R to B is an *extension* of a homomorphism g from G to B_v if h agrees with g on all of $V(G)$, and h also maps the additional vertex a in G^a_L or G^a_R to v.

We first claim that any homomorphism h from G^a_L or G^a_R to B with $h(a) = v$ is an extension of some homomorphism g from G to B_v. Secondly, we claim that for any homomorphism g from G to B_v, there is precisely one homomorphism h from either G^a_L or G^a_R to B that is an extension of g. Then (9) follows.

For the first claim, let h be a homomorphism from G^a_L to B with $h(a) = v$. (The argument for homomorphisms from G^a_R is analogous.) Then h maps R to the neighborhood of v in B: Since a has edges to all of R in G^a_L, there must be edges from $h(a) = v$ to all of $h(R)$ in B. Furthermore, since G is connected, $h(L)$ is contained in the neighborhood of $h(R)$. It follows that the entire image of h is contained in B_v, so the restriction g of h to $V(G)$ is a homomorphism from G to B_v. Hence h is an extension of g, proving the first claim.

For the second claim, let X be the bipartition side of B_v not containing v. Consider a homomorphism g from G to B_v. Since G is connected, either g maps R to X, or g maps L to X.

1. In the first case, we can extend g to a map h from G_L^a to B via $h(a) = v$, and we show that h is indeed a homomorphism: By definition, g preserves edges on G and the image of G is the subgraph B_v of B. Since g maps R to X, and X is the neighborhood of v in B_v by definition of B_v, we see that h maps the edges aw for $w \in R$ in G_L^a to edges of B_v. Thus h is an extension of g. Furthermore, the map h' from G_R^a to B obtained from g by setting $h'(a) = v$ is not a homomorphism, since v and R are all mapped to X, which is an independent set in B_v.
2. In the second case, we can extend g to a homomorphism h from G_R^a to B as above. By a symmetric argument, h maps the edges aw for $w \in L$ in G_R^a to edges of B_v. Thus h is an extension of g.

Hence, the homomorphisms h from G_L^a and G_R^a to B are extensions of homomorphisms g from G to B_v, and each g has precisely one extension from either G_L^a or G_R^a. This establishes (9), thus concluding the proof. ■

With Lemma 10 at hand, we can readily reduce $\mathrm{Hom}(\star, B_v)$ to $\mathrm{Hom}(\star, B)$.

Proposition 11. *For every bipartite graph B and every vertex $v \in V(B)$, we have* $\mathrm{Hom}(\star, B_v) \preceq \mathrm{Hom}(\star, B)$.

Proof. Let G be the input for $\mathrm{Hom}(\star, B_v)$. Without loss of generality, we can assume that G is connected and bipartite with $V(G) = L \cup R$.

Let G_L^a and G_R^a be the graphs derived from G in Lemma 10. Both have $O(n)$ vertices and $O(n + m)$ edges, with $n = |V(G)|$ and $m = |E(G)|$. By (7),

$$\mathrm{Hom}(G_L^a, B) + \mathrm{Hom}(G_R^a, B) = \sum_{v \in V(B)} \mathrm{Hom}(G, B_v)\,.$$

We can compute the left-hand side with an oracle for $\mathrm{Hom}(\star, B)$. On the right-hand side, no graphs cancel when collecting terms for isomorphic graphs B_v, as all coefficients in the sum are 1. Since B is fixed, all graphs and coefficients are fixed, and Proposition 9 gives $\mathrm{Hom}(\star, B_v) \preceq \mathrm{Hom}(\star, B)$. ■

To prove hardness of $\mathrm{Hom}(\star, H)$, we start with $B := H$ and establish hardness of $\mathrm{Hom}(\star, B)$ by reduction from $\mathrm{Hom}(\star, B_v)$ for some $v \in V(B)$ such that B_v is hom-hard and has less vertices than B. This is possible unless B is an impasse or hom-easy. We verify in the full version that any hom-hard impasse is actually an exploded 4-vertex path.

Lemma 12. *Let B be bipartite and connected, but not a $K_{a,b}$. Assume that for every $v \in V(B)$, the graph B_v is either a $K_{a,b}$ or equal to B. Then B is isomorphic to $P(a_1, a_2, a_3, a_4)$ for positive integers a_1, a_2, a_3, a_4.*

Thus, repeated applications of Proposition 11 give a reduction from $\mathrm{Hom}(\star, P)$ to $\mathrm{Hom}(\star, H)$ with $P = P(a_1, a_2, a_3, a_4)$ for some positive integers a_1, a_2, a_3, a_4. In the full version, we conclude the hardness proof for $\mathrm{Hom}(\star, H)$ by establishing hardness of counting homomorphisms to $P(a_1, a_2, a_3, a_4)$ for all fixed positive integers a_1, a_2, a_3, a_4.

5 Counting Partially Surjective Homomorphisms

Finally, we prove a dichotomy for $\mathrm{PartSurj}(\star, H)$, thus establishing Theorem 4. For a fixed graph H with marked vertices and edges, let $\mathcal{D}(H)$ denote the set of graphs obtainable from H by deleting marked objects. We first show in Lemma 13 that $\mathrm{PartSurj}(\star, H)$ can be expressed as a linear combination of functions $\mathrm{Hom}(\star, F)$ for $F \in \mathcal{D}(H)$. Then we apply Theorem 5 to classify the complexity of these linear combinations.

Lemma 13. *For every graph H with markings, there is a quantum graph $\overline{F} = \sum_{F \in \mathcal{D}(H)} \alpha_F F$ such that $\mathrm{PartSurj}(G, H) = \mathrm{Hom}(G, \overline{F})$ holds for all graphs G. After collecting for isomorphic graphs, we have $\alpha_H = 1$ and $\alpha_F < 0$ for every graph $F \in \mathcal{D}(H)$ obtained by deleting at most one marked edge from H.*

Lemma 13 is shown in the full version; it is a simple consequence of the inclusion-exclusion principle. Using it in combination with Theorem 5, we obtain the classification for partially surjective homomorphisms.

Proof of Theorem 4. By Lemma 13, there is a quantum graph $\overline{F}$ with constituents from $\mathcal{D}(H)$ such that $\mathrm{PartSurj}(G, H) = \mathrm{Hom}(G, \overline{F})$. It follows that $\mathrm{PartSurj}(\star, H)$ and $\mathrm{Hom}(\star, \overline{F})$ are the same problem.

Recall the notions of hom-easy and hom-hard graphs from Sect. 4. If every graph $F \in \mathcal{D}(H)$ is hom-easy, then $\mathrm{Hom}(\star, \overline{F})$ is polynomial-time solvable. Otherwise, there are hom-hard graphs $F \in \mathcal{D}(H)$, and it only remains to find one with $\alpha_F \neq 0$ in order for Theorem 5 to yield the hardness of $\mathrm{Hom}(\star, \overline{F})$.

If H itself is hom-hard, then we pick $F = H$ and obtain $\alpha_F \neq 0$ by Lemma 13. Otherwise, H is hom-easy, so every connected component of H is a K_q° or a $K_{a,b}$. We check that only one marked edge e^* needs to be deleted from H to obtain a hom-hard graph $F \in \mathcal{D}(H)$:

- If H contains a component C with marked edges and $C = K_q^\circ$ for $q \geq 3$ or $C = K_{a,b}$ for $a, b > 1$, we can choose e^* to be any marked edge in C.
- If H contains a component $C = K_2^\circ$ with at least one marked self-loop, we can choose e^* to be any marked self-loop in C.

If neither of these conditions applies to H, then it can be checked that $\mathcal{D}(H)$ contains only hom-easy graphs. Thus, if $\mathcal{D}(H)$ contains any hom-hard graphs at all, then there is an edge e^* such that $F = H - e^*$ is hom-hard. Lemma 13 then implies $\alpha_F \neq 0$, so Theorem 5 gives hardness of $\mathrm{Hom}(\star, \overline{F})$. ■

To conclude, we note that Theorem 3 can be easily derived from Theorem 4.

6 Conclusion

We consider Theorem 2 as an initial step towards a fine-grained understanding of general #CSP problems, and we believe that our shortened proof can be used to simplify and strengthen other dichotomy results for #CSP following in the wake of Dyer and Greenhill's seminal result [16]. Techniques based on quantum graphs might also advance the state of the art for open problems regarding approximate and modular homomorphism counting.

An interesting open problem is to improve Theorem 2 to more precise running time bounds under the strong exponential-time hypothesis. Doing so however is challenging, as non-trivial improvements upon the running time $O(k^n)$ are possible for some #P-hard patterns H. For example, Björklund et al. [1] prove that the number of proper k-colorings, which is equal to $\mathrm{Hom}(G, K_k)$, can be computed in time $2^n \cdot n^{O(1)}$ for any $k \in \mathbf{N}$.

References

1. Björklund, A., Husfeldt, T., Kaski, P., Koivisto, M.: Computing the Tutte polynomial in vertex-exponential time. In: 49th Annual IEEE Symposium on Foundations of Computer Science, FOCS 2008, 25–28 October 2008, Philadelphia, PA, USA, pp. 677–686. IEEE Computer Society (2008). https://doi.org/10.1109/FOCS.2008.40
2. Borgs, C., Chayes, J.T., Kahn, J., Lovász, L.: Left and right convergence of graphs with bounded degree. Random Struct. Algorithms **42**(1), 1–28 (2013). https://doi.org/10.1002/rsa.20414
3. Bulatov, A.A.: The complexity of the counting constraint satisfaction problem. In: Aceto, L., Damgård, I., Goldberg, L.A., Halldórsson, M.M., Ingólfsdóttir, A., Walukiewicz, I. (eds.) ICALP 2008. LNCS, vol. 5125, pp. 646–661. Springer, Heidelberg (2008). https://doi.org/10.1007/978-3-540-70575-8_53
4. Bulatov, A.A., Dalmau, V.: Towards a dichotomy theorem for the counting constraint satisfaction problem. In: Proceedings of 44th Symposium on Foundations of Computer Science (FOCS 2003), 11–14 October 2003, Cambridge, MA, USA, pp. 562–571 (2003). https://doi.org/10.1109/SFCS.2003.1238229
5. Bulatov, A.A., Dyer, M.E., Goldberg, L.A., Jalsenius, M., Jerrum, M., Richerby, D.: The complexity of weighted and unweighted #CSP. J. Comput. Syst. Sci. **78**(2), 681–688 (2012). https://doi.org/10.1016/j.jcss.2011.12.002
6. Bulatov, A.A., Grohe, M.: The complexity of partition functions. Theor. Comput. Sci. **348**(2–3), 148–186 (2005). https://doi.org/10.1016/j.tcs.2005.09.011
7. Cai, J.-Y., Chen, X.: Complexity of counting CSP with complex weights. J. ACM **64**(3), 19:1–19:39 (2017). https://doi.org/10.1145/2822891
8. Cai, J.-Y., Chen, X., Lu, P.: Graph homomorphisms with complex values: a dichotomy theorem. SIAM J. Comput. **42**(3), 924–1029 (2013). https://doi.org/10.1137/110840194
9. Cai, J.-Y., Lu, P., Xia, M.: The complexity of complex weighted boolean #CSP. J. Comput. Syst. Sci. **80**(1), 217–236 (2014). https://doi.org/10.1016/j.jcss.2013.07.003

10. Chen, H., Mengel, S.: Counting answers to existential positive queries: a complexity classification. In: Proceedings of the 35th ACM SIGMOD-SIGACT-SIGAI Symposium on Principles of Database Systems, PODS 2016, San Francisco, CA, USA, 26 June–01 July 2016, pp. 315–326 (2016). https://doi.org/10.1145/2902251.2902279
11. Curticapean, R.: Block interpolation: a framework for tight exponential-time counting complexity. Inf. Comput. **261**(Part), 265–280 (2018). https://doi.org/10.1016/j.ic.2018.02.008
12. Curticapean, R., Dell, H., Marx D.: Homomorphisms are a good basis for counting small subgraphs. In: Hatami, H., McKenzie, P., King, V., (eds.) Proceedings of the 49th Annual ACM SIGACT Symposium on Theory of Computing, STOC 2017, Montreal, QC, Canada, 19–23 June 2017, pp. 210–223. ACM (2017). https://doi.org/10.1145/3055399.3055502
13. Cygan, M., et al.: Tight bounds for graph homomorphism and subgraph isomorphism. In: SODA, pp. 1643–1649. SIAM (2016). https://doi.org/10.1137/1.9781611974331.ch112
14. Dell, H., Husfeldt, T., Marx, D., Taslaman, N., Wahlen, M.: Exponential time complexity of the permanent and the Tutte polynomial. ACM Trans. Algorithms **10**(4), 21:1–21:32 (2014). https://doi.org/10.1145/2635812
15. Dyer, M.E., Goldberg, L.A., Paterson, M.: On counting homomorphisms to directed acyclic graphs. J. ACM **54**(6), 27 (2007). https://doi.org/10.1145/1314690.1314691
16. Dyer, M.E., Greenhill, C.S.: The complexity of counting graph homomorphisms. Random Struct. Algorithms **17**(3–4), 260–289 (2000). https://doi.org/10.1002/1098-2418(200010/12)17:3/4⟨260::AID-RSA5⟩3.0.CO;2-W
17. Dyer, M.E., Richerby, D.: An effective dichotomy for the counting constraint satisfaction problem. SIAM J. Comput. **42**(3), 1245–1274 (2013). https://doi.org/10.1137/100811258
18. Faben, J., Jerrum, M.: The complexity of parity graph homomorphism: an initial investigation. Theory Comput. **11**, 35–57 (2015). https://doi.org/10.4086/toc.2015.v011a002
19. Flum, J., Grohe, M.: Parameterized Complexity Theory. TTCSAES. Springer, Heidelberg (2006). https://doi.org/10.1007/3-540-29953-X
20. Focke, J.,Goldberg, L.A., Živný, S.: The complexity of counting surjective homomorphisms and compactions. In: Czumaj, A. (ed.) Proceedings of the Twenty-Ninth Annual ACM-SIAM Symposium on Discrete Algorithms, SODA 2018, New Orleans, LA, USA, 7–10 January 2018, pp. 1772–1781. SIAM (2018). https://doi.org/10.1137/1.9781611975031.116
21. Freedman, M., Lovász, L., Schrijver, A.: Reflection positivity, rank connectivity, and homomorphism of graphs. J. Am. Math. Soc. **20**(1), 37–51 (2007). https://doi.org/10.1090/S0894-0347-06-00529-7
22. Göbel, A., Goldberg, L.A., Richerby, D.: The complexity of counting homomorphisms to cactus graphs modulo 2. TOCT **6**(4), 17:1–17:29 (2014). https://doi.org/10.1145/2635825
23. Göbel, A., Goldberg, L.A., Richerby, D.: Counting homomorphisms to square-free graphs, modulo 2. TOCT **8**(3), 12:1–12:29 (2016). https://doi.org/10.1145/2898441
24. Goldberg, L.A., Grohe, M., Jerrum, M., Thurley, M.: A complexity dichotomy for partition functions with mixed signs. SIAM J. Comput. **39**(7), 3336–3402 (2010). https://doi.org/10.1137/090757496
25. Impagliazzo, R., Paturi, R.: On the complexity of k-SAT. J. Comput. Syst. Sci. **62**(2), 367–375 (2001). https://doi.org/10.1006/jcss.2000.1727

26. Impagliazzo, R., Paturi, R., Zane, F.: Which problems have strongly exponential complexity? J. Comput. Syst. Sci. **63**(4), 512–530 (2001). https://doi.org/10.1006/jcss.2001.1774
27. Lovász, L.: Large networks and graph limits, volume 60 of colloquium publications. American Mathematical Society (2012). https://www.ams.org/bookstore-getitem/item=COLL-60
28. Schrijver, A.: Graph invariants in the edge model. In: Grötschel, M., Katona, O.H., Sági, G. (eds.) Building Bridges. BSMS, vol. 19, pp. 487–498. Springer, Berlin (2008). https://doi.org/10.1007/978-3-540-85221-6_16

Minimal Separators in Graph Classes Defined by Small Forbidden Induced Subgraphs

Martin Milanič[1,2] and Nevena Pivač[1(✉)]

[1] University of Primorska, IAM, Muzejski trg 2, 6000 Koper, Slovenia
martin.milanic@upr.si, nevena.pivac@iam.upr.si

[2] University of Primorska, FAMNIT, Glagoljaška 8, 6000 Koper, Slovenia

Abstract. Minimal separators in graphs are an important concept in algorithmic graph theory. In particular, many problems that are NP-hard for general graphs are known to become polynomial-time solvable for classes of graphs with a polynomially bounded number of minimal separators. Several well-known graph classes have this property, including chordal graphs, permutation graphs, circular-arc graphs, and circle graphs. We perform a systematic study of the question which classes of graphs defined by small forbidden induced subgraphs have a polynomially bounded number of minimal separators. We focus on sets of forbidden induced subgraphs with at most four vertices and obtain an almost complete dichotomy, leaving open only two cases.

Keywords: Minimal separator · Hereditary graph class · Forbidden induced subgraph

1 Introduction

The main concept studied in this paper is that of a minimal separator in a graph. Given a graph G, a *minimal separator* in G is a subset of vertices that separates some non-adjacent vertex pair a, b and is inclusion-minimal with respect to this property (separation of a and b). Minimal separators in graphs are important for reliability analysis of networks [42], for sparse matrix computations, via their connection with minimal triangulations (see [20] for a survey), and are related to other graph concepts such as potential maximal cliques [4]. Many graph algorithms are based on minimal separators, see, e.g., [1–4,9,12,23,26,35].

The work is supported in part by the Slovenian Research Agency (I0-0035, research program P1-0285 and research projects J1-9110, N1-0102, and a Young Researchers grant). Part of the work was done while M. M. was visiting Osaka Prefecture University in Japan, under the operation Mobility of Slovene higher education teachers 2018–2021, co-financed by the Republic of Slovenia and the European Union under the European Social Fund.

I. Sau and D. M. Thilikos (Eds.): WG 2019, LNCS 11789, pp. 379–391, 2019.
https://doi.org/10.1007/978-3-030-30786-8_29

In this work we focus on graphs with "few" minimal separators. Such graphs enjoy good algorithmic properties. Many problems that are NP-hard for general graphs become polynomial-time solvable for classes of graphs with a polynomially bounded number of minimal separators. This includes TREEWIDTH and MINIMUM FILL-IN [5], MAXIMUM INDEPENDENT SET and FEEDBACK VERTEX SET [14], DISTANCE-d INDEPENDENT SET for even d [32] and many other problems [13]. It is therefore important to identify classes of graphs with a polynomially bounded number of minimal separators. Many known graph classes have this property, including chordal graphs [38], chordal bipartite graphs [28], weakly chordal graphs [4], permutation graphs [1,24], circular-arc graphs [12,23,28], circle graphs [23,25,28], etc. Moreover, a class of graphs has a polynomially bounded number of minimal separators if and only if it has a polynomially bounded number of potential maximal cliques [5].

We perform a systematic study of the question which classes of graphs defined by small forbidden induced subgraphs have a polynomially bounded number of minimal separators. We focus on sets of forbidden induced subgraphs with at most four vertices and obtain an almost complete dichotomy, leaving open only two cases, the class of graphs of independence number at most three that are either C_4-free or $\{\text{claw}, C_4\}$-free. Our approach combines a variety of tools and techniques, including constructions of graph families with exponentially many minimal separators, applications of Ramsey's theorem, study of the behavior of minimal separators under various graph operations, and structural characterizations of graphs in hereditary classes.

Statement of the Main Result

Given two non-adjacent vertices a and b in a graph G, a set $S \subseteq V(G) \setminus \{a, b\}$ is an *a, b-separator* if a and b are contained in different components of $G - S$. If S contains no other a, b-separator as a proper subset, then S is a *minimal a, b-separator*. We denote by $\mathcal{S}_G(a, b)$ the set of all minimal a, b-separators. A *minimal separator* in G is a set $S \subseteq V(G)$ that is a minimal a, b-separator for some pair of non-adjacent vertices a and b. We denote by $\mathcal{S}_G$ the set of all minimal separators in G and by $s(G)$ the cardinality of $\mathcal{S}_G$. The main concept of study in this paper is the following property of graph classes.

Definition 1. *We say that a graph class $\mathcal{G}$ is* tame *if there exists a polynomial $p : \mathbb{R} \to \mathbb{R}$ such that for every graph $G \in \mathcal{G}$, we have $s(G) \leq p(|V(G)|)$.*

Given a set $\mathcal{F}$ of graphs, we say that a graph G is *$\mathcal{F}$-free* if no induced subgraph of G is isomorphic to a member of $\mathcal{F}$. Given two sets $\mathcal{F}$ and $\mathcal{F}'$ of graphs, we write $\mathcal{F} \trianglelefteq \mathcal{F}'$ if the class of $\mathcal{F}$-free graphs is contained in the class of $\mathcal{F}'$-free graphs, that is, if every $\mathcal{F}$-free graph is also $\mathcal{F}'$-free.

Observation 1. *Let $\mathcal{F}$ and $\mathcal{F}'$ be two sets of graphs such that $\mathcal{F} \trianglelefteq \mathcal{F}'$. If the class of $\mathcal{F}'$-free graphs is tame, then so is the class of $\mathcal{F}$-free graphs.*

It is well known and not difficult to see that relation $\mathcal{F} \trianglelefteq \mathcal{F}'$ can be checked by means of the following criterion, which becomes particularly simple for finite sets $\mathcal{F}$ and $\mathcal{F}'$.

Observation 2 (Folklore). *For every two sets of graphs $\mathcal{F}$ and $\mathcal{F}'$, we have $\mathcal{F} \trianglelefteq \mathcal{F}'$ if and only if every graph from $\mathcal{F}'$ contains an induced subgraph isomorphic to a member of $\mathcal{F}$.*

Our main result is Theorem 3. It deals with graph classes defined by sets of forbidden induced subgraphs having at most four vertices. The relevant graphs are named as in Fig. 1.

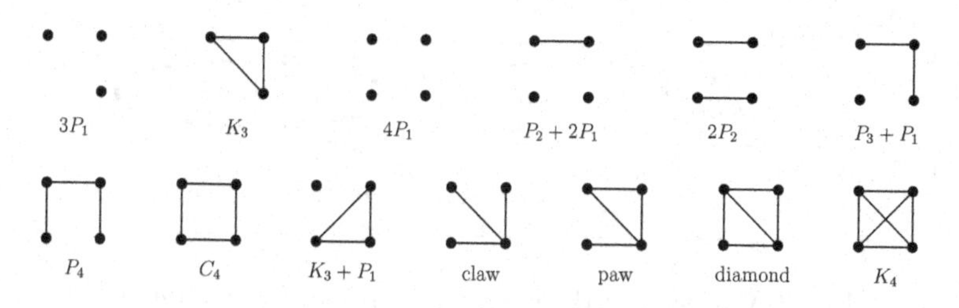

Fig. 1. Graphs on at most 4 vertices appearing in the statement of the main theorem.

Theorem 3. *Let $\mathcal{F}$ be a set of graphs with at most four vertices such that $\mathcal{F} \neq \{4P_1, C_4\}$ and $\mathcal{F} \neq \{4P_1, \text{claw}, C_4\}$. Then the class of $\mathcal{F}$-free graphs is tame if and only if $\mathcal{F} \trianglelefteq \mathcal{F}'$ for one of the following sets $\mathcal{F}'$:*

(i) $\mathcal{F}' = \{P_4\}$ *or* $\mathcal{F}' = \{2P_2\}$,
(ii) $\mathcal{F}' = \{F, \text{paw}\}$ *for some* $F \in \{4P_1,\ P_2+2P_1,\ P_3+P_1,\ \text{claw}\}$,
(iii) $\mathcal{F}' = \{F, K_3+P_1\}$ *for some* $F \in \{4P_1,\ P_2+2P_1,\ P_3+P_1,\ \text{claw}\}$,
(iv) $\mathcal{F}' = \{F, K_4\}$ *for some* $F \in \{4P_1,\ P_2+2P_1,\ P_3+P_1\}$,
(v) $\mathcal{F}' = \{F, C_4\}$ *for some* $F \in \{P_2+2P_1,\ P_3+P_1\}$,
(vi) $\mathcal{F}' = \{4P_1, C_4, \text{diamond}\}$.

Theorem 3 can be equivalently stated in a dual form, characterizing minimal classes of $\mathcal{F}$-graphs that are not tame.

Theorem 4. *Let $\mathcal{F}$ be a set of graphs with at most 4 vertices such that $\mathcal{F} \neq \{4P_1, C_4\}$ and $\mathcal{F} \neq \{4P_1, C_4, \text{claw}\}$. Then the class of $\mathcal{F}$-free graphs is not tame if and only if $\mathcal{F}' \trianglelefteq \mathcal{F}$ for one of the following sets $\mathcal{F}'$:*

(i) $\mathcal{F}' = \{3P_1, \text{diamond}\}$,
(ii) $\mathcal{F}' = \{\text{claw}, K_4, C_4, \text{diamond}\}$,
(iii) $\mathcal{F}' = \{K_3, C_4\}$.

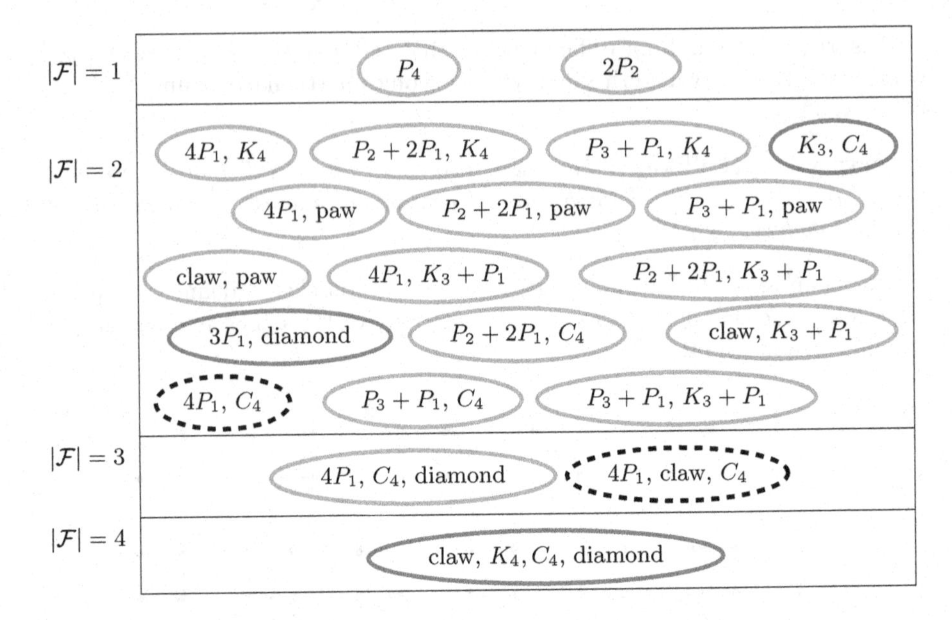

Fig. 2. Overview of the main result. (Color figure online)

In Fig. 2 we give an overview of maximal tame and minimal non-tame classes of $\mathcal{F}$-free graphs, where $\mathcal{F}$ contains graphs with at most four vertices. Maximal tame classes correspond to sets $\mathcal{F}$ of forbidden induced subgraphs depicted in green ellipses, while minimal non-tame classes correspond to sets depicted in red ellipses (in gray-scale printing, green and red appear in brighter, resp., darker ellipses). The two open cases are dashed. A similar figure with respect to boundedness of the clique-width can be found in [8].

Related Work. To the best of the authors' knowledge, this work represents the first systematic study of the problem of classifying hereditary graph classes with respect to the existence of a polynomial bound on the number of minimal separators of the graphs in the class. Dichotomy studies for many other problems in mathematics and computer science are available in the literature in general, as well as within the field of graph theory, for properties such as boundedness of the clique-width [6–8,11], price of connectivity [19], and polynomial-time solvability of various algorithmic problems such as Chromatic Number [16,27,29], Graph Homomorphism [21], Graph Isomorphism [39], and Dominating Set [30].

Structure of the Paper. We collect the main notations, definitions, and preliminary results in Sect. 2. In Sect. 3 we present several families of graphs with exponentially many minimal separators. In Sect. 4, we study the effect of various graph operations on the number of minimal separators. Our main result, given by Theorems 3 and 4, is proved in Sect. 5. Due to space limitations, most proofs are omitted in this extended abstract.

2 Preliminaries

All graphs in this paper will be finite, simple, undirected, and will have at least one vertex. A vertex v in a graph G is *universal* if it is adjacent to every other vertex in the graph, and *simplicial* if its neighborhood is a clique. Given a graph G and a set $S \subseteq V(G)$, we denote by $N_G(S)$ the set of all vertices in $V(G) \setminus S$ having a neighbor in S. For a vertex $v \in V(G)$, we write $N_G(v)$ for $N_G(\{v\})$ and $N_G[v]$ for $N_G(v) \cup \{v\}$. Two vertices u and v are said to be *true twins* if $N_G[u] = N_G[v]$. A graph is *co-connected* if its complement is connected. Given two graphs F and G, we write $F \subseteq_i G$ if F is an induced subgraph of G. A graph class $\mathcal{G}$ is *hereditary* if it is closed under vertex deletion, or, equivalently, if there exists a set $\mathcal{F}$ of graphs such that $\mathcal{G}$ is exactly the class of $\mathcal{F}$-free graphs. A graph G is the *join* of two vertex-disjoint graphs G_1 and G_2, written $G = G_1 * G_2$, if $V(G) = V(G_1) \cup V(G_2)$ and $E(G) = E(G_1) \cup E(G_2) \cup \{xy \mid x \in V(G_1) \text{ and } y \in V(G_2)\}$. As usual, we denote by P_n, C_n, K_n the path, the cycle, and the complete graph with n vertices, respectively. For positive integers m, n, we denote by $K_{m,n}$ the complete bipartite graph with m and n vertices in the two parts of the bipartition. For undefined terms related to graphs and graph classes, we refer the reader to [17,40].

An important ingredient for some of our proofs will be the following classical result [37].

Ramsey's Theorem. *For every two positive integers k and ℓ, there exists a least positive integer $R(k, \ell)$ such that every graph with at least $R(k, \ell)$ vertices contains either a clique of size k or an independent set of size ℓ.*

Given a graph G and a set $S \subseteq V(G)$, a component C of the graph $G - S$ is *S-full* if every vertex in S has a neighbor in C, or, equivalently, if $N_G(V(C)) = S$. The following well-known lemma characterizes minimal separators (see, e.g., [12,17,22]).

Lemma 1. *Given a graph $G = (V, E)$, a set $S \subseteq V$ is a minimal separator in G if and only if the graph $G - S$ contains at least two S-full components.*

Corollary 1. *Let S be a minimal separator in a graph G. Then for every $v \in S$ the set $S \setminus \{v\}$ is a minimal separator in $G - v$.*

The following result shows that the class of P_4-free graphs is tame.

Theorem 5 (Nikolopoulos and Palios [33]). *If G is a P_4-free graph, then $s(G) < 2/3|V(G)|$.*

We will also need the following result about the structure of paw-free graphs. A graph G is *complete multipartite* if there is some positive integer k such that the vertex set of G can be partitioned into k parts such that two vertices are adjacent if and only if they belong to different parts.

Theorem 6 (Olariu [34]). *A connected paw-free graph G is either K_3-free or complete multipartite.*

We conclude this section with a straightforward but useful simplification of the defining property of tame graph classes.

Lemma 2. *A graph class $\mathcal{G}$ is tame if and only if there exists a non-negative integer k such that $s(G) \leq |V(G)|^k$ for all $G \in \mathcal{G}$.*

An easy consequence of Lemma 2 is the fact that any union of finitely many tame graph classes is tame.

3 Graph Families with Exponentially Many Minimal Separators

In this section we identify some families of graphs with exponentially many minimal separators. We give two constructions with structurally different properties. The first construction, explained in Sect. 3.1, involves families of graphs of arbitrarily large maximum degree but without arbitrarily long induced paths. The second construction, explained in Sect. 3.2, involves two families of graphs with small maximum degree but with arbitrarily long induced paths. In both cases, we make use of line graphs.

3.1 Theta Graphs and Their Line Graphs

Given positive integers k and ℓ, the *k,ℓ-theta graph* is the graph $\theta_{k,\ell}$ obtained as the union of k internally disjoint paths of length ℓ with common endpoints a and b. For every positive integer ℓ, we define a family of graphs Θ_ℓ in the following way: $\Theta_\ell = \{\theta_{k,\ell} \mid k \geq 2\}$. Note that ℓ refers to the length of each of the a,b-paths and not to the number of paths, which is unrestricted.

Observation 7. *For every integer $\ell \geq 3$, the class Θ_ℓ is not tame.*

Corollary 2. *If $\mathcal{G}$ is a class of graphs such that $\Theta_\ell \subseteq \mathcal{G}$ for some $\ell \geq 3$, then $\mathcal{G}$ is not tame.*

Consider now the family of line graphs of theta graphs. More precisely, given positive integers k and ℓ, let $L_{k,\ell}$ denote the line graph of $\theta_{k,\ell}$ and let $\mathcal{L}_\ell = \{L_{k,\ell} \mid k \geq 2\}$.

Observation 8. *For every integer $\ell \geq 2$, the class $\mathcal{L}_\ell$ is not tame.*

Corollary 3. *If $\mathcal{G}$ is a class of graphs such that $\mathcal{L}_\ell \subseteq \mathcal{G}$ for some $\ell \geq 2$, then $\mathcal{G}$ is not tame.*

Corollary 4. *The class of $\{3P_1$, diamond$\}$-free graphs is not tame.*

3.2 Elementary Walls and Their Line Graphs

Let $r, s \geq 2$ be integers. An $r \times s$*-grid* is the graph with vertex set $\{0,\ldots,r-1\} \times \{0,\ldots,s-1\}$ in which two vertices (i,j) and (i',j') are adjacent if and only if $|i-i'|+|j-j'|=1$. Given an integer $h \geq 2$, an *elementary wall of height* h is the graph W_h obtained from the $(2h+2) \times (h+1)$-grid by deleting all edges with endpoints $(2i+1, 2j)$ and $(2i+1, 2j+1)$ for all $i \in \{0,1,\ldots,h\}$ and $j \in \{0,1,\ldots,\lfloor(h-1)/2\rfloor\}$, deleting all edges with endpoints $(2i, 2j-1)$ and $(2i, 2j)$ for all $i \in \{0,1,\ldots,h\}$ and $j \in \{1,\ldots,\lfloor h/2\rfloor\}$, and deleting the two resulting vertices of degree one. Note that an elementary wall of height h consists of h levels each containing h bricks, where a brick is a cycle of length six.

Grids contain exponentially many minimal separators [41]. A similar construction works for walls.

Proposition 1. *For every integer $h \geq 2$, an elementary wall of height h has at least 2^h minimal separators.*

Another useful family with exponentially many minimal separators is given by the line graphs of elementary walls.

Proposition 2. *For every even integer $h \geq 2$, the graph $L(W_h)$ has at least $2^{h/2}$ minimal separators.*

Corollary 5. *The class of {claw, K_4, C_4, diamond}-free graphs is not tame.*

4 Graph Operations

We now discuss the effect of various graph operations on the number of minimal separators. The set of minimal separators of a disconnected graph can be computed from the sets of minimal separators of its components, and a similar statement holds for graphs whose complements are disconnected. The correspondences are as follows, see [36, Theorem 3.1].

Theorem 9. *If G is a disconnected graph, with components $G_1,\ldots,G_k$, then $\mathcal{S}_G = \{\emptyset\} \cup \bigcup_{i=1}^{k} \mathcal{S}_{G_i}$. If G is the join of graphs $G_1,\ldots,G_k$, then $S \in \mathcal{S}_G$ if and only if there exists some $i \in \{1,\ldots,k\}$ and some $S_i \in \mathcal{S}_{G_i}$ such that $S = S_i \cup (V(G) \setminus V(G_i))$.*

Using this theorem we can derive the following corollaries.

Corollary 6. *Let $\mathcal{G}$ be a hereditary class of graphs and let $\mathcal{G}'$ be the class of connected graphs in $\mathcal{G}$. Then $\mathcal{G}$ is tame if and only if $\mathcal{G}'$ is tame.*

Corollary 7. *Let $\mathcal{G}$ be a hereditary class of graphs and let $\mathcal{G}'$ be the class of co-connected graphs in $\mathcal{G}$. Then $\mathcal{G}$ is tame if and only if $\mathcal{G}'$ is tame.*

McKee observed in [31] that if G_1 is an induced subgraph of G_2, then every minimal separator of G_1 is contained in a minimal separator in G_2. The proof actually shows that the following monotonicity property holds.

Proposition 3. *If G_1 is an induced subgraph of G_2, then $s(G_1) \leq s(G_2)$.*

In view of Proposition 3, it is natural to ask how large can the gap $s(G_2) - s(G_1)$ be if the graphs G_1 and G_2 are not "too different", for example, if G_1 is obtained from G_2 by deleting only one vertex. In the following three propositions we identify three properties of a vertex v in a graph G such that deleting v either leaves the minimal separators unchanged or decreases it by one.

Proposition 4. *Let G be a graph with at least two vertices and let v be a universal vertex in G. Then $s(G) = s(G - v)$.*

Proposition 5. *Let G be a graph having a pair of true twins v, w with $v \neq w$. Then $s(G) = s(G - v)$.*

Proposition 6. *Let G be a graph with at least two vertices and let v be a simplicial vertex in G. Then $s(G - v) \leq s(G) \leq s(G - v) + 1$.*

5 Proof of Theorem 3

In this section we sketch the proof of Theorem 3. We do this in several steps. We start with a proposition giving a necessary condition for a set $\mathcal{F}$ of graphs so that the class of $\mathcal{F}$-free graphs is tame.

Proposition 7. *Let $\mathcal{F}$ be a finite set of graphs such that for every $F \in \mathcal{F}$ we have $F \not\subseteq_i P_4$, $F \not\subseteq_i 2P_2$. If, in addition, all graphs in $\mathcal{F}$ contain cycles or all of them are of girth more than 5, then the class of $\mathcal{F}$-free graphs is not tame.*

The following sufficient condition is derived using Ramsey's theorem.

Proposition 8. *For every two positive integers k and ℓ, the class of $\{P_2 + kP_1, K_\ell + P_2\}$-free graphs is tame.*

Proof. By Observations 1 and 2, we may assume that $k \geq 2$ and $\ell \geq 2$. Then $R(\ell, k) \geq 2$. Let G be a $\{P_2 + kP_1, K_\ell + P_2\}$-free graph. We will prove that for every minimal separator S in G, there exists a set $X \subseteq V(G)$ such that $|X| \leq R(\ell, k) - 1$ and $S = N_G(X)$. Clearly, this will imply that G has at most $\binom{|V(G)|}{R(\ell,k)-1} = \mathcal{O}\left(|V(G)|^{R(\ell,k)-1}\right)$ minimal separators. Let S be a minimal separator in G and let C and D be two S-full components of $G - S$. Since $N_G(V(C)) = N_G(V(D)) = S$, it suffices to show that $|V(C)| \leq R(\ell, k) - 1$ or $|V(D)| \leq R(\ell, k) - 1$. Suppose that this is not the case. Then $|V(C)| \geq R(\ell, k)$ and $|V(D)| \geq R(\ell, k)$. By Ramsey's theorem, this implies that there exists a set $Z \subseteq V(C)$ such that Z is either a clique of size ℓ or an independent set of size k. But then Z together with a pair of adjacent vertices from D induces either a $K_\ell + P_2$ or $P_2 + kP_1$, respectively. Both cases lead to a contradiction. □

The next proposition simplifies the cases with $P_3 + P_1 \in \mathcal{F}$.

Proposition 9. *Let $\mathcal{F}$ be a set of graphs such that $P_3 + P_1 \in \mathcal{F}$ and let $\mathcal{F}' = (\mathcal{F} \setminus \{P_3 + P_1\}) \cup \{3P_1\}$. Then the class of $\mathcal{F}$-free graphs is tame if and only if the class of $\mathcal{F}'$-free graphs is tame.*

We now consider various families of forbidden induced subgraphs with at most four vertices. Propositions 8 and 9 can be used to prove the following.

Proposition 10. *For every $F \in \{4P_1,\ P_2 + 2P_1,\ P_3 + P_1,\ claw\}$, the class of $\{F, K_3 + P_1\}$-free graphs is tame.*

The next result follows from a structural property of $\{3P_1, C_4\}$-free graphs proved by Choudum and Shalu [10].

Proposition 11. *The class of $\{P_3 + P_1, C_4\}$-free graphs is tame.*

The next two propositions are proved using a structural analysis of graphs in the respective classes.

Proposition 12. *The class of $\{P_2 + 2P_1, C_4\}$-free graphs is tame.*

Proposition 13. *The class of $\{4P_1, C_4,$ diamond$\}$-free graphs is tame.*

Propositions 8 and 9 can be used to prove the following.

Proposition 14. *For every $F \in \{4P_1, P_2+2P_1, P_3+P_1\}$, the class of $\{F, K_4\}$-free graphs is tame.*

Proposition 15. *For every $F \in \{4P_1,\ P_2 + 2P_1,\ P_3 + P_1,\ claw\}$ the class of $\{F, paw\}$-free graphs is tame.*

Proof. Let G be an $\{F, \text{paw}\}$-free graph. By Corollary 6, we may assume that G is connected. Theorem 6 implies that G is either K_3-free, or complete multipartite. If G is K_3-free, then G is also K_4-free and Proposition 14 applies. If G is complete multipartite, then G is P_4-free, and Theorem 5 applies. □

The next proposition can be proved by showing that in a $2P_2$-free graph G, every minimal separator S is of the form $N_G(v)$ for some vertex $v \in V(G)$.

Proposition 16. *The class of $2P_2$-free graphs is tame.*

We now have all the ingredients ready to prove Theorem 3.

Proof (of Theorem 3). Let $\mathcal{F}$ be a set of graphs on at most 4 vertices such that $\mathcal{F} \neq \{C_4, 4P_1\}$ and $\mathcal{F} \neq \{4P_1,\ C_4,\ \text{claw}\}$. If $\mathcal{F}'$ is a set of graphs satisfying one of the conditions *(i)–(vi)* then the class of $\mathcal{F}'$-free graphs is tame by Theorem 5 and Propositions 10, 11, 12, 15, 13, 14, and 16. Thus, if $\mathcal{F} \trianglelefteq \mathcal{F}'$ for some set of graphs satisfying one of the conditions *(i)–(vi)*, then the class of $\mathcal{F}$-free graphs, being a subclass of the tame class of $\mathcal{F}'$-free graphs, is tame, too.

Suppose now that for all sets $\mathcal{F}'$ in *(i)–(vi)* we have $\mathcal{F} \ntrianglelefteq \mathcal{F}'$. We want to prove that the class of $\mathcal{F}$-free graphs is not tame. Since $\mathcal{F} \ntrianglelefteq \{2P_2\}$ and $\mathcal{F} \ntrianglelefteq \{P_4\}$, it follows that if $F \subseteq_i 2P_2$ or $F \subseteq_i P_4$, then $F \notin \mathcal{F}$. Let $A = \{K_3,\ C_4,\ K_3 + P_1,$ paw, diamond, $K_4\}$, $B = \{3P_1,\ 4P_1,\ P_2 + 2P_1,\ P_3 + P_1,$ claw$\}$. Since $\mathcal{F}$ does not contain any induced subgraph of either $2P_2$ or P_4, we infer that $\mathcal{F} \subseteq A \cup B$. Since Proposition 7 implies that the class of $\mathcal{F}$-free graphs is not tame if all graphs in $\mathcal{F}$ contain cycles or all of them are acyclic, we may assume that $\mathcal{F}$ contains two graphs F_1 and F_2 such that F_1 contains a cycle and F_2 is acyclic. Clearly, $F_1 \in A$ and $F_2 \in B$.

We claim that $\mathcal{F} \cap \{K_3,\ K_3 + P_1,$ paw$\} = \emptyset$. Indeed, if $F \in \{K_3,\ K_3 + P_1,$ paw$\}$, then $\{F, F_2\} \subseteq \mathcal{F}$, which implies that $\mathcal{F} \trianglelefteq \mathcal{F}'$ for $\mathcal{F}' = \{F',\ F''\}$ where $F' \in \{4P_1,\ P_2 + 2P_1,\ P_3 + P_1,$ claw$\}$ and $F'' \in \{$paw$,\ K_3 + P_1\}$, contrary to the assumptions on $\mathcal{F}$. It follows that $F_1 \in \mathcal{F} \cap A \subseteq \{K_4,\ C_4,$ diamond$\}$.

Suppose that $K_4 \in \mathcal{F}$. If there exists a graph $F \in \mathcal{F} \cap \{3P_1, 4P_1, P_2 + 2P_1, P_3 + P_1\}$, then $\mathcal{F} \trianglelefteq \mathcal{F}'$ where $\mathcal{F}'$ satisfies condition (iv), a contradiction. It follows that $F_2 \in \mathcal{F} \cap B \subseteq \{$claw$\}$, that is, F_2 is the claw. We also have $\mathcal{F} \setminus \{K_4,$ claw$\} \subseteq \{C_4,$ diamond$\}$. Consequently, $\{$claw, $K_4,\ C_4,$ diamond$\} \trianglelefteq \mathcal{F}$. By Corollary 5, the class of $\{$claw, $K_4,\ C_4,$ diamond$\}$ is not tame and hence by Observation 1, neither is the class of $\mathcal{F}$-free graphs.

From now on, we assume that $K_4 \notin \mathcal{F}$. Suppose that $C_4 \in \mathcal{F}$. If $\{3P_1, P_2 + 2P_1, P_3 + P_1\} \cap \mathcal{F} \neq \emptyset$, then $\mathcal{F} \trianglelefteq \mathcal{F}'$ where $\mathcal{F}'$ satisfies condition (v), a contradiction. It follows that $F_2 \in \mathcal{F} \cap B \subseteq \{4P_1,$ claw$\}$. Suppose first that $4P_1 \in \mathcal{F} \cap B$. If the diamond is not in $\mathcal{F}$, then $\mathcal{F} \neq \{4P_1, C_4\}$ or $\mathcal{F} \neq \{4P_1,$ claw, $C_4\}$, which is impossible. Thus, the diamond is in $\mathcal{F}$, which implies that $\{4P_1, C_4,$ diamond$\} \subseteq \mathcal{F}$, hence $\mathcal{F} \trianglelefteq \mathcal{F}'$ where $\mathcal{F}'$ satisfies condition (vi), a contradiction. We conclude that $4P_1 \notin \mathcal{F}$, which implies that $\mathcal{F} \cap B = \{$claw$\}$. Consequently, $\mathcal{F} \subseteq \{$claw, $C_4,$ diamond$\}$, which implies that $\{$claw, $K_4,\ C_4,$ diamond$\} \trianglelefteq \mathcal{F}$. By Corollary 5, the class of $\{$claw, $K_4,\ C_4,$ diamond$\}$ is not tame and hence by Observation 1, neither is the class of $\mathcal{F}$-free graphs.

From now on, we assume that $C_4 \notin \mathcal{F}$. It follows that $F_1 \in \mathcal{F} \cap A \subseteq \{$diamond$\}$, that is, $\mathcal{F} \cap A = \{$diamond$\}$. Clearly, $F_2 \in \mathcal{F} \cap B \subseteq \{3P_1,\ 4P_1,\ P_2 + 2P_1,\ P_3 + P_1,$ claw$\}$, which implies that every graph in $\mathcal{F} \cap B$ contains an induced $3P_1$. Consequently, $\{3P_1,$ diamond$\} \trianglelefteq \mathcal{F}$. From Corollary 4 it follows that the class of $\{3P_1,$ diamond$\}$-free graphs is not tame and by Observation 1, neither is the class of $\mathcal{F}$-free graphs.

This completes the proof. □

6 Conclusion

In this work we considered graphs with "few" minimal separators. Our main result was an almost complete dichotomy for the property of having a polynomially bounded number of minimal separators within the family of graph classes defined by forbidden induced subgraphs with at most four vertices. Two exceptional families for which the problem is still open are the class of $\{4P_1, C_4\}$-free graphs and the class of $\{4P_1, \text{claw}, C_4\}$-free graphs. Note that the class of $\{4P_1, C_4\}$-free graphs and their complements was already of interest to Erdős, who offered \$20 to determine whether the vertex set of every $\{4P_1, C_4\}$-free graph can be covered by 4 cliques (which was resolved in the affirmative by Nagy and Szentmiklóssy, see [18]). Moreover, the class of $\{4P_1, C_4\}$-free graphs is one of the only three graph classes defined by a set of four-vertex forbidden induced subgraphs for which the complexity of coloring is still open [15].

Our results have algorithmic consequences. In particular, the algorithmic metatheorem due to Fomin et al. [13] implies that for any tame graph class (in particular, for any graph class corresponding to one of the green ellipses in Fig. 2), the following optimization problem is solvable in polynomial time for graphs in the class. Let φ be a counting monadic second order logic formula and $t \geq 0$ be an integer. For a given graph G, the task is to find a set $|X| \subseteq V(G)$ maximizing $|X|$, subject to the following: there is a set $U \subseteq V(G)$ such that $X \subseteq U$, the subgraph of G induced by U is of treewidth at most t, and the structure $(G[U], X)$ models φ.

Some of the results given here (for example Proposition 8) are not restricted to forbidden induced subgraphs of at most four vertices, and they might prove useful for developing more general dichotomy studies related to minimal separators. In this respect, an interesting and challenging question would be to develop a dichotomy result for graph classes characterized by two forbidden induced subgraphs.

Note Added in Proof. Soon after this work was presented at WG 2019, the authors resolved the two open questions regarding the classes of $\{4P_1, C_4\}$-free and $\{4P_1, \text{claw}, C_4\}$-free graphs, by proving that both graph classes are tame. Furthermore, as kindly communicated to us by Daniel Lokshtanov, the same result was also obtained independently and at about the same time by Peter Gartland. He actually proved a more general result stating that any class of C_4-free graphs of bounded independence number is tame.

References

1. Bodlaender, H.L., Kloks, T., Kratsch, D.: Treewidth and pathwidth of permutation graphs. SIAM J. Discrete Math. **8**(4), 606–616 (1995)
2. Bodlaender, H.L., Koster, A.M.C.A.: Treewidth computations I. upper bounds. Inform. Comput. **208**(3), 259–275 (2010)
3. Bodlaender, H.L., Rotics, U.: Computing the treewidth and the minimum fill-in with the modular decomposition. Algorithmica **36**(4), 375–408 (2003)
4. Bouchitté, V., Todinca, I.: Treewidth and minimum fill-in: grouping the minimal separators. SIAM J. Comput. **31**(1), 212–232 (2001)
5. Bouchitté, V., Todinca, I.: Listing all potential maximal cliques of a graph. Theor. Comput. Sci. **276**(1–2), 17–32 (2002)
6. Brandstädt, A., Dabrowski, K.K., Huang, S., Paulusma, D.: Bounding the clique-width of H-free split graphs. Discrete Appl. Math. **211**, 30–39 (2016)
7. Brandstädt, A., Dabrowski, K.K., Huang, S., Paulusma, D.: Bounding the clique-width of H-free chordal graphs. J. Graph Theory **86**(1), 42–77 (2017)
8. Brandstädt, A., Engelfriet, J., Le, H.O., Lozin, V.V.: Clique-width for 4-vertex forbidden subgraphs. Theory Comput. Syst. **39**(4), 561–590 (2006)
9. Chiarelli, N., Milanič, M.: Linear separation of connected dominating sets in graphs. Ars Math. Contemp. **16**, 487–525 (2019)

10. Choudum, S.A., Shalu, M.A.: The class of $\{3K_1, C_4\}$-free graphs. Australas. J. Comb. **32**, 111–116 (2005)
11. Dabrowski, K.K., Paulusma, D.: Classifying the clique-width of H-free bipartite graphs. Discrete Appl. Math. **200**, 43–51 (2016)
12. Deogun, J.S., Kloks, T., Kratsch, D., Müller, H.: On the vertex ranking problem for trapezoid, circular-arc and other graphs. Discrete Appl. Math. **98**(1–2), 39–63 (1999)
13. Fomin, F.V., Todinca, I., Villanger, Y.: Large induced subgraphs via triangulations and CMSO. SIAM J. Comput. **44**(1), 54–87 (2015)
14. Fomin, F.V., Villanger, Y.: Finding induced subgraphs via minimal triangulations. In: STACS 2010: 27th International Symposium on Theoretical Aspects of Computer Science. Leibniz International Proceedings Informatics (LIPIcs), vol. 5, pp. 383–394. Schloss Dagstuhl. Leibniz-Zent. Inform., Wadern (2010)
15. Fraser, D.J., Hamel, A.M., Hoàng, C.T., Maffray, F.: A coloring algorithm for $4K_1$-free line graphs. Discrete Appl. Math. **234**, 76–85 (2018)
16. Golovach, P.A., Johnson, M., Paulusma, D., Song, J.: A survey on the computational complexity of coloring graphs with forbidden subgraphs. J. Graph Theory **84**(4), 331–363 (2017)
17. Golumbic, M.C.: Algorithmic Graph Theory and Perfect Graphs, Annals of Discrete Mathematics, vol. 57, 2nd edn. Elsevier, Amsterdam (2004)
18. Gyárfás, A.: Problems from the world surrounding perfect graphs. In: Proceedings of the International Conference on Combinatorial Analysis and its Applications, (Pokrzywna, 1985), vol. 19, pp. 413–441 (1988, 1987)
19. Hartinger, T.R., Johnson, M., Milanič, M., Paulusma, D.: The price of connectivity for cycle transversals. European J. Comb. **58**, 203–224 (2016)
20. Heggernes, P.: Minimal triangulations of graphs: a survey. Discrete Math. **306**(3), 297–317 (2006)
21. Hell, P., Nešetřil, J.: On the complexity of H-coloring. J. Comb. Theory Ser. B **48**(1), 92–110 (1990)
22. Kloks, T., Kratsch, D.: Finding all minimal separators of a graph. In: Enjalbert, P., Mayr, E.W., Wagner, K.W. (eds.) STACS 1994. LNCS, vol. 775, pp. 759–768. Springer, Heidelberg (1994). https://doi.org/10.1007/3-540-57785-8_188
23. Kloks, T., Kratsch, D., Wong, C.K.: Minimum fill-in on circle and circular-arc graphs. J. Algorithms **28**(2), 272–289 (1998)
24. Kloks, T. (ed.): Treewidth Computations and Approximations. LNCS, vol. 842. Springer, Heidelberg (1994). https://doi.org/10.1007/BFb0045375
25. Kloks, T.: Treewidth of circle graphs. Int. J. Found. Comput. Sci. **7**(02), 111–120 (1996)
26. Kloks, T., Kratsch, D., Spinrad, J.: On treewidth and minimum fill-in of asteroidal triple-free graphs. Theor. Comput. Sci. **175**(2), 309–335 (1997)
27. Král', D., Kratochvíl, J., Tuza, Z., Woeginger, G.J.: Complexity of coloring graphs without forbidden induced subgraphs. In: Brandstädt, A., Le, V.B. (eds.) WG 2001. LNCS, vol. 2204, pp. 254–262. Springer, Heidelberg (2001). https://doi.org/10.1007/3-540-45477-2_23
28. Kratsch, D.: The structure of graphs and the design of efficient algorithms. Habilitation thesis, Friedrich-Schiller-Universität, Jena (1996)
29. Lozin, V.V., Malyshev, D.S.: Vertex coloring of graphs with few obstructions. Discrete Appl. Math. **216**(part 1), 273–280 (2017)
30. Malyshev, D.S.: A complexity dichotomy and a new boundary class for the dominating set problem. J. Comb. Optim. **32**(1), 226–243 (2016). https://doi.org/10.1007/s10878-015-9872-z

31. McKee, T.A.: Requiring that minimal separators induce complete multipartite subgraphs. Discuss. Math. Graph Theory **38**(1), 263–273 (2018)
32. Montealegre, P., Todinca, I.: On Distance-d independent Set and other problems in graphs with "few" Minimal Separators. In: Heggernes, P. (ed.) WG 2016. LNCS, vol. 9941, pp. 183–194. Springer, Heidelberg (2016). https://doi.org/10.1007/978-3-662-53536-3_16
33. Nikolopoulos, S.D., Palios, L.: Minimal separators in P_4-sparse graphs. Discrete Math. **306**(3), 381–392 (2006)
34. Olariu, S.: Paw-free graphs. Inform. Process. Lett. **28**(1), 53–54 (1988)
35. Parra, A., Scheffler, P.: Characterizations and algorithmic applications of chordal graph embeddings. Discrete Appl. Math. **79**(1–3), 171–188 (1997)
36. Pedrotti, V., de Mello, C.P.: Minimal separators in extended P_4-laden graphs. Discrete Appl. Math. **160**(18), 2769–2777 (2012)
37. Ramsey, F.P.: On a problem of formal logic. Proc. London Math. Soc. **s2–30**(4), 264–286 (1929)
38. Rose, D.J., Tarjan, R.E., Lueker, G.S.: Algorithmic aspects of vertex elimination on graphs. SIAM J. Comput. **5**(2), 266–283 (1976)
39. Schweitzer, P.: Towards an isomorphism dichotomy for hereditary graph classes. Theory Comput. Syst. **61**(4), 1084–1127 (2017)
40. Spinrad, J.P.: Efficient Graph Representations, Fields Institute Monographs, vol. 19. American Mathematical Society, Providence (2003)
41. Suchan, K.: Minimal Separators in Intersection Graphs. Master's thesis, Akademia Górniczo-Hutnicza im. Stanisława Staszica w Krakowie (2003)
42. Zabłudowski, A.: A method for evaluating network reliability. Bull. Acad. Polon. Sci. Sér. Sci. Tech. **27**(7), 647–655 (1979)

Author Index

Printed in the United States
By Bookmasters